To Fausto

Luca Formaggia · Fausto Saleri · Alessandro Veneziani

Solving Numerical PDEs: Problems, Applications, Exercises

 Springer

Luca Formaggia
MOX – Dipartimento di Matematica
"F. Brioschi"
Politecnico di Milano, Italy

Fausto Saleri [†]
MOX – Dipartimento di Matematica
"F. Brioschi"
Politecnico di Milano, Italy

Alessandro Veneziani
Department of Mathematics
and Computer Science
Emory University, Atlanta, Georgia, USA

Additional material can be downloaded from http://extras.springer.com
Password: 978-88-470-2411-3

Translated by: Simon G. Chiossi, Dipartimento di Matematica, Politecnico di Torino, Italy,
Luca Formaggia and Alessandro Veneziani

The translation is based on the original Italian edition:
L. Formaggia, F. Saleri, A. Veneziani: Applicazioni ed esercizi di modellistica numerica per problemi
differenziali, © Springer-Verlag Italia, Milano 2005

UNITEXT – La Matematica per il 3+2
ISSN print edition: 2038-5722   ISSN electronic edition: 2038-5757

ISBN 978-88-470-2411-3   e-ISBN 978-88-470-2412-0
DOI 10.1007/978-88-470-2412-0

Library of Congress Control Number: 2011937784

Springer Milan Heidelberg New York Dordrecht London

Cover design: Beatrice Ꞵ, Milano
Cover image: Numerical simulation of the spreading of racoon rabies in the State of NY, USA. Simula-
tions carried out by J. Keller (Emory University) on a domain reconstructed from satellite map with a
SEI model. More details: see Remark 5.5.

Typesetting with LaTeX: PTP-Berlin, Protago TeX-Production GmbH, Germany
(www.ptp-berlin.eu)

Springer-Verlag Italia S.r.l., Via Decembrio 28, I-20137 Milano
Springer-Verlag fa parte di Springer Science+Business Media (www.springer.com)

Preface

This book results from the experience of the authors in the courses on Numerical Methods in Engineering, Partial Differential Equations and Numerical Partial Differential Equations given to undergraduate and graduate students at the Politecnico di Milano (Italy), EPFL Lausanne (Switzerland), the University of Bergamo (Italy) and Emory University (Atlanta, GA, USA). In these courses we introduce students to methods for the numerical approximation of Partial Differential Equations (PDEs). One of the difficulties in this field is to identify the right trade-off between theoretical concepts and their practical use. In this collection of examples and exercises we address this difficulty by providing exercises with an "academic" flavor, stemming directly from basic concepts of Numerical Analysis (and often inspired by exams of the courses mentioned above), as well as exercises, marked by the symbol *, which address a practical problem that the student is encouraged to formalize in terms of PDE, analyze and solve numerically. These problems are derived from the knowledge gained by the authors in projects developed in collaboration with scientists of different fields (geology, mechanical engineering, biology, etc.) and industries.

The book consists of four parts. The *first part* recalls basic concepts and introduces the notation which will be used throughout the text, which is however quite standard. Therefore, it may be skipped by readers already acquainted with the general subject and more interested in applications and exercises. More precisely, in Chapter 1 we give a brief overview of elementary functional analysis and fundamentals of linear algebra, while Chapter 2 addresses fundamentals of composite interpolation, which is the basis of the finite element method, and of numerical differentiation, basis for the finite difference method. The *second part* covers steady elliptic problems, with a dominance of diffusion effects in Chapter 3, the treatment of advection and reaction terms being given in Chapter 4. Finite elements is the discretization method of reference for those chapters, but some exercises are developed with finite differences. The *third part* covers time-dependent problems, parabolic equations Chapter 5, linear hyperbolic equations Chapter 6, as well as basic problems in incompressible fluid dynamics (Navier-Stokes equations) in Chapter 7. Time discretization is carried out mainly with finite differences, but we address space-time finite elements as well.

The *fourth part* contains complementary material collected into two appendices. Appendix A recalls some implementation details of sparse matrices, which is of interest for readers more interested in programming numerical methods for Partial DIfferential Equations. Finally, Appendix B gives a short biography of mathematicians and scientists often cited in the text. It is not intended to be an exhaustive collection, but rather to give an historical perspective and an insight on people who gave fundamental contributions to numerical analysis and applied mathematics.

The approach we have followed is to allow different ways of using this book, by splitting the solution of each exercise into three parts, *Mathematical Analysis* of the problem, *Numerical Approximation*, *Analysis of the results*. Exercises marked by * have a preliminary section dedicated to the *Model formulation*. The rationale behind this subdivision is to make easier for the reader to select the specific aspects on which (s)he is interested.

Many exercises are taken from the companion book *Numerical Modeling for Partial Differential Equations*, by A. Quarteroni (Springer, 2010), where they are proposed without solution. Therefore, the cited monograph provides the main background of this book. However, the summary of the main theoretical results provided in the introductory chapter should make this text self-contained.

We have assumed different levels for the reader also for the software tools adopted in the book. For a basic reader interested in a quick check of the theoretical results, we provide MATLAB programs, most of which are described in the text and downloadable from the Springer website: http://extras.springer.com. In particular, the code `fem1D`, written by the authors, is equipped with a simple graphical user interface for 1D (elliptic, parabolic, hyperbolic) problems, so that the user does not need any specific programming knowledge for obtaining the numerical results.

A reader interested in more advanced problems but still not on implementation details would appreciate the exercises developed with the code `Freefem++`, authored by O.Pironneau, F.Hecht and A.Le Hyaric. The main advantages of using this code is the proximity of the syntax to the weak formulation of the problems and the relative easiness of installation of the code with different operating systems.

Finally, for the sake of readers with an interest in more advanced numerical programming, a few (simple) exercises have been completed with a 3D extension of the problem proposed. In this case, the code we use is the Finite Element library `LifeV` (see www.lifev.org). This library is a joint project of different research groups (EPFL, Politecnico di Milano, Emory and INRIA Rocquencourt, France) working in different fields of Scientific Computing. The purpose of these examples is to encourage the reader to explore the fascinating (yet difficult) world of professional implementation of numerical

methods for PDEs. As a matter of fact, we truly believe that a deep compre-
hension of the Numerical Modeling of Differential Problems requires a good
knowledge of all the aspects, ranging from the most theoretical to the most
practical ones, including implementation techniques.

Notice that for the sake of space, we give in the text only the synopsis of
the programs, or some snapshots of the codes as well as programming details
whenever necessary for the understanding of the algorithm. The reader is
encouraged to have a look at the complete code at the companion web sites
mentioned above.

We thank all the colleagues and students who contributed to improve this
text. In particular, we thank Alfio Quarteroni for many fruitful suggestions,
the team of developers of the code LifeV, in particular Alessio Fumagalli at
MOX, Politecnico di Milano, Tiziano Passerini at the Department of Mathematics and Computer Science at Emory University. Nicola Parolini of MOX is
gratefully acknowledged for providing an interesting exercise for the chapter
on Navier-Stokes, and Joshua Keller of Emory University for the simulations
and the images of the cover. We thank Francesca Bonadei at Springer for the
editorial work and the patience during the preparation of the text. Alessandro Veneziani wishes to thank the Emory's team, Lucia Mirabella, Marina
Piccinelli, Tiziano Passerini, Luca Gerardo Giorda, Mauro Perego, Alexis
Aposporidis, Marta D'Elia, Umberto Villa and Luca Bertagna for reading
and supporting the improvement of the manuscript. Luca Bertagna has also
kindly provided a 3D Navier-Stokes computation carried out with LifeV and
shown in one of the exercises. The help of Michele Benzi (and his impressive
knowledge of the history of Numerical Analysis) is greatly acknowledged. Errors or imprecision are obviously under our own responsibility and we thank
in advance all the colleagues and students who will give us their feedback.

*When we wrote the Italian edition of this book in 2005, Fausto was a
fundamental component of the Authors' team in conveying to students our
passion for Mathematics and its applications in any field of science (and
life). We wrote this English edition, which is a revision and an extension
of the Italian one, hoping that this book contributes in keeping his and our
enthusiasm alive. To his memory, friendship and passion we dedicate this
work.*

Milan, Atlanta, August 2011   *Luca Formaggia*
                                                          Alessandro Veneziani

Contents

Part I
Basic Material

1

Some fundamental tools

Mathematical modeling of real-life problems in engineering, physics or life sciences often gives rise to partial differential problems that cannot be solved analytically but need a numerical scheme to obtain a suitable approximation. Dealing with numerical modeling requires first of all an understanding of the underlying differential problem. The type of differential problem, as well as issues of well-posedness and regularity of the solution may indeed drive the selection of the appropriate simulation tool. A second requirement is the analysis of numerical schemes, in particular their stability and convergence characteristics. Last, but not least, numerical schemes must be implemented in a computer language, and often aspects which look easy "on paper" arise complex implementation issues, particularly when computational efficiency is at stake.

These considerations have driven the selection of the exercises in this book and their solution, with the precise intent of giving to the reader practical examples of mathematical and numerical modeling, as well as computer implementation.

We will consider differential problems of the following general form: find $u : \Omega \to \mathbb{R}$ such that

$$L(u) = f \text{ on } \Omega, \tag{1.1}$$

where L is a differential operator (often linear in u) and Ω is an open subset of $\mathbb{R}^d$, with d either 1 or 2. We will also give some examples of problems in three dimensions.

In most applications of numerical methods with practical interest the computational domain is bounded. Therefore, when not otherwise stated we consider Ω bounded. Problem (1.1) will be supplemented by suitable boundary conditions and – for time-dependent problems – by initial conditions.

According to the type of formulation chosen for (1.1) it will be necessary to introduce certain function spaces, besides discussing the regularity required

Formaggia L., Saleri F., Veneziani A.: Solving Numerical PDEs: Problems, Applications, Exercises. DOI 10.1007/978-88-470-2412-0_1, © Springer-Verlag Italia 2012

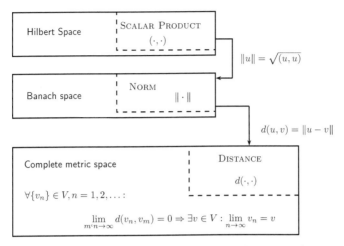

Fig. 1.1 The relationship among a complete metric space (bottom box), a Banach space (middle) and a Hilbert space (top)

for the problem data and the domain Ω itself, and give an appropriate meaning to the operations of differentiation and integration.

In this introductory chapter we recall just some basic results that are instrumental to the understanding of the solution of the proposed exercises. The interested reader may find more complete treatment of mathematical modeling and analysis of partial differential equations in [RR04, Sal08, Eva10], while advanced text in functional analysis are [Bre11, AF03, Leo09]. For the numerical aspects the main reference is [Qua09]. For the linear algebra part we refer to [GV96, Str03].

1.1 Hilbert spaces

Let V be a real vector space equipped with a metric defined by a *distance* $d : V \times V \to \mathbb{R}$ between elements $u, v \in V$. A sequence of elements v_n of V with $n = 1, 2, \ldots$, is called *Cauchy sequence* if for any $\varepsilon > 0$ there is a q such that for $m, n \geq q$ we have $d(v_m, v_n) \leq \varepsilon$. The space V is called *complete* if any Cauchy sequence converges to an element $v \in V$.

A real vector space V where a *norm* $\|\cdot\|$ can be defined is equipped by a distance $d(u, v) \equiv \|u - v\|$. And, if the space is complete is called a *Banach space*. Finally, a Banach space V equipped with a scalar product, denoted by $(\cdot, \cdot)_V$ has a norm, namely $\|v\| = \sqrt{(v, v)}$ for any $v \in V$. Hence it features a metric for which V is complete. In this case, V is called *Hilbert space*, see Fig. 1.1.

Given two Hilbert spaces V_1 and V_2 one may consider an operator

$$\mathcal{L} : V_1 \to V_2,$$

mapping $v \in V_1$ to $w = \mathcal{L}v \in V_2$. The operator $\mathcal{L}$ is *linear* if $\mathcal{L}(\alpha u + \beta v) = \alpha \mathcal{L}(u) + \beta \mathcal{L}(v)$ for any $\alpha, \beta \in \mathbb{R}$, $u, v \in V_1$. It is called *bounded* if there exists a constant $C > 0$ such that $\|\mathcal{L}(v)\|_{V_2} \leq C\|v\|_{V_1} \; \forall v \in V_1$. If $V_2 \equiv \mathbb{R}$, the operator is called *functional*. One can prove that if $\mathcal{L}$ is a bounded linear operator, then it is continuous. More precisely, there exists a constant $C_1 > 0$ such that $\|\mathcal{L}v - \mathcal{L}w\|_{V_2} \leq C_1\varepsilon$ for any $v, w \in V_1$ with $\|v - w\|_{V_1} \leq \varepsilon$.

The space of linear and continuous functionals on a Hilbert space V is the *dual space* of V, indicated by V'. It is a Banach space when equipped with the norm

$$\|\mathcal{L}\|_{V'} = \sup_{v \in V} \frac{\mathcal{L}(v)}{\|v\|_V}.$$

Riesz' representation theorem guarantees that V' is isometric[1] to V and that for any $v \in V$ the linear functional $\mathcal{L}_v(w)$ given by $(v, w)_V$ is bounded and $\|\mathcal{L}_v\|_{V'} = \|v\|_V$. Moreover, for any $\mathcal{L} \in V'$ there is a unique element $v \in V$ such that $\mathcal{L}(w) = (v, w)_V$ for all $w \in V$ and $\|v\|_V = \|\mathcal{L}\|_{V'}$. As a consequence, if we identify V' with V we deduce that the dual of a Hilbert space is still a Hilbert space.

The application of a functional $\mathcal{L}$ to an element $v \in V$ is often indicated by $\langle \mathcal{L}, v \rangle$, instead of $\mathcal{L}(v)$, to stress the duality between V and its dual space.

1.2 Distributions

Let Ω be an open set in $\mathbb{R}^d$ and let $\mathcal{D}(\Omega)$ denote the space of C^∞ functions with compact support[2] in Ω. A *distribution* is a linear and continuous functional T on $\mathcal{D}(\Omega)$. We will typically denote the application of T to an element $\phi \in \mathcal{D}(\Omega)$ by $\langle T, \phi \rangle$. The space of distributions is thus the dual space to $\mathcal{D}(\Omega)$, hence it is indicated by $\mathcal{D}'(\Omega)$. Let f be a function such that $\int_K |f| \, d\Omega < \infty$ for any compact K contained in Ω (i.e. f is locally integrable). We may associate to it the distribution T_f given by

$$\langle T_f, \phi \rangle \equiv \int_\Omega f\phi \, d\Omega. \tag{1.2}$$

Distributions which are defined in this way are called *regular*. Yet the set of distribution is much larger. An important example is the so-called *Dirac distribution* (also known as *Dirac delta*): if $\mathbf{a} \in \Omega$, the Dirac delta centered at $\mathbf{a}$, written $\delta_{\mathbf{a}}$, is the distribution defined by

$$\langle \delta_{\mathbf{a}}, \phi \rangle = \phi(\mathbf{a}) \; \forall \phi \in \mathcal{D}(\Omega).$$

This distribution is not regular since it cannot be represented by an integrable function.

---

[1] An isometry is a map between metric spaces that preserves distances.

[2] The support of a function is the closure of the set of points where the function is not equal to zero.

Therefore, a distribution can be regarded as a generalization of the concept of a function (that's why distributions are sometimes called *generalized functions*). The derivative in the variable x_k of a distribution T is the distribution $\partial T/\partial x_k$ defined by

$$\left\langle \frac{\partial T}{\partial x_k}, \phi \right\rangle \equiv - \left\langle T, \frac{\partial \phi}{\partial x_k} \right\rangle \qquad \forall \phi \in \mathcal{D}(\Omega). \tag{1.3}$$

A distribution is always differentiable, in the sense given above.

A sequence of distributions $T_n \in \mathcal{D}'(\Omega)$ is said to converge to T in $\mathcal{D}'(\Omega)$ if $\langle T_n, \phi \rangle \to \langle T, \phi \rangle$ for all $\phi \in \mathcal{D}(\Omega)$.

The *derivative in the sense of distributions* or *weak derivative* of a locally integrable function f is the distribution $v = f'$ such that

$$\int_\Omega v\phi \; dx = -\int_\Omega f\phi' \; dx \quad \forall \phi \in \mathcal{D}(\Omega). \tag{1.4}$$

1.3 The spaces L^p and H^s

A function $f : \Omega \to \mathbb{R}$, with Ω open in $\mathbb{R}^n$, belongs to the space $L^p(\Omega)$ with $1 \le p \le \infty$ if

$$\|f\|_{L^p(\Omega)} \equiv \left(\int_\Omega |f(\mathbf{x})|^p \; d\Omega \right)^{1/p} < \infty \quad \text{when } p \in [1, \infty),$$

$$\|f\|_{L^\infty(\Omega)} \equiv \operatorname*{ess\,sup}_{\mathbf{x} \in \Omega} |f(\mathbf{x})| < \infty \qquad \text{when } p = \infty.$$

The functions belonging to these spaces are, to be precise, representatives of equivalence classes (two functions being equivalent if they differ on a zero-measure set only). The symbol ess sup (essential supremum) indicates that the supremum of f on Ω is computed up to sets of zero measure.

The spaces $L^p(\Omega)$ are all Banach spaces, and among them only $L^2(\Omega)$ is a Hilbert space, with scalar product $(f, g) \equiv \int_\Omega f \, g \; d\Omega$. The dual space to $L^2(\Omega)$ is $L^2(\Omega)$ itself. Furthermore, one can prove that $\mathcal{D}(\Omega)$ is dense in $L^p(\Omega)$ for any finite p.

An important role for Lebesgue spaces is played by *Hölder's inequality*, according to which for any $f \in L^p(\Omega)$, $g \in L^q(\Omega)$ and integers p, q such that $1/p + 1/q = 1$ we have

$$\int_\Omega fg \; d\Omega \le \left(\int_\Omega |f|^p d\Omega \right)^{1/p} \left(\int_\Omega |g|^q d\Omega \right)^{1/q} = \|f\|_{L^p(\Omega)} \|g\|_{L^q(\Omega)}. \tag{1.5}$$

The special case $p = q = 1/2$ yields the Cauchy-Schwarz inequality

$$\left| \int_\Omega uv \; d\Omega \right| \le \|u\|_{L^2(\Omega)} \|v\|_{L^2(\Omega)}, \tag{1.6}$$

valid for any $u, v \in L^2(\Omega)$. Using Hölder's inequality we can identify the dual space to $L^p(\Omega)$, with p finite, as the space $L^q(\Omega)$ with $1/p + 1/q = 1$: in fact, given a functional $f \in (L^p(\Omega))'$ one can always find a function $u_f \in L^q(\Omega)$ such that $\langle f, v \rangle = \int_\Omega u_f v \, d\Omega$ and $\|f\|_{(L^p(\Omega))'} = \|u_f\|_{L^q(\Omega)}$.

Another relationship with a paramount role in the study of the stability of PDEs is Young's inequality. It may be derived from the simple observation that $(A - B)^2 \geq 0$ implies $2AB \leq A^2 + B^2$ for given real numbers A, B. This result can be generalized: for any $\epsilon > 0$, setting $a = A\sqrt{\epsilon}$ and $b = B/\sqrt{\epsilon}$ transforms the previous inequality into $ab \leq a^2/(2\epsilon) + b\epsilon/2$. Now assuming $a \equiv \|u\|_{L^2(\Omega)}$ and $b \equiv \|v\|_{L^2(\Omega)}$, from (1.6) it follows, for any $u, v \in L^2(\Omega)$ and $\epsilon > 0$,

$$\left| \int_\Omega uv \, d\Omega \right| \leq \frac{1}{2\epsilon} \|u\|^2_{L^2(\Omega)} + \frac{\epsilon}{2} \|v\|^2_{L^2(\Omega)}. \tag{1.7}$$

Sobolev spaces. Suppose $\Omega \subset \mathbb{R}^d$ is an open set with sufficiently regular boundary. More precisely, denote by V a normed space of functions on $\mathbb{R}^{d-1}$ and by $B(\mathbf{y}, R) = \{\mathbf{x} : \|\mathbf{x} - \mathbf{y}\| < R\}$ the ball centered at $\mathbf{y} \in \mathbb{R}^d$ with radius R. We say $\partial\Omega$ is of class V if, for any $\mathbf{x} \in \partial\Omega$, there exists $R > 0$ and a function $g \in V$ such that the set $\Omega \cap B(\mathbf{x}, R)$ can be represented by the inequality $x_d > g(x_1, x_2, \ldots, x_{d-1})$. An open set will thus be, for instance, C^1 if its boundary can be defined by C^1 functions. In the context of PDEs treated in this book we will normally assume Ω to be Lipschitz, in this case the boundary is defined by Lipschitz-continuous functions.

Given a positive integer k, a function v belongs to the Sobolev space $H^k(\Omega)$ if f together with all its weak derivatives up to order k belong to $L^2(\Omega)$. The spaces $H^k(\Omega)$ are Hilbert spaces, equipped with scalar product

$$(u, v)_{H^k(\Omega)} \equiv \sum_{|\alpha| \leq k} \int_\Omega D^\alpha u \, D^\alpha v \, d\Omega, \tag{1.8}$$

where $\alpha = (\alpha_1, \ldots, \alpha_d)$ is a *multi-index*, $\alpha_i \in \mathbb{N}$, $|\alpha| = \sum_{i=1}^d \alpha_i$ and

$$D^\alpha u(\mathbf{x}) \equiv \frac{\partial^{|\alpha|} u(\mathbf{x})}{\partial x_1^{\alpha_1} \ldots \partial x_d^{\alpha_d}}.$$

For example, for $d = 3$ if $\alpha = (1, 0, 1)$, then $|\alpha| = 2$ and $D^\alpha u = \partial^2 u/(\partial x_1 \partial x_3)$.

The norm $\|v\|_{H^k(\Omega)}$ of $v \in H^k(\Omega)$ is induced by the scalar product,

$$\|v\|_{H^k(\Omega)} = (v, v)_{H^k(\Omega)}^{\frac{1}{2}} = \sqrt{\sum_{|\alpha| \leq k} \int_\Omega (D^\alpha v)^2 \, d\Omega}. \tag{1.9}$$

It is useful to introduce also the following *seminorm*:

$$|v|_{H^k(\Omega)} = \sqrt{\sum_{|\alpha|=k} \int_\Omega (D^\alpha v)^2 \, d\Omega}, \qquad (1.10)$$

in particular, for $\Omega \in \mathbb{R}^2$ we have

$$|v|_{H^2(\Omega)} = \sqrt{\int_\Omega \left(\frac{\partial^2 v}{\partial x^2}\right)^2 + 2\frac{\partial^2 v}{\partial x \partial y} + \left(\frac{\partial^2 v}{\partial y^2}\right)^2 \, d\Omega}.$$

In the context of Sobolev space the classical concept of C^k regularity is replaced by the, somewhat weaker, concept of H^k regularity. The regularity of functions in $H^k(\Omega)$ is determined by a collection of Functional Analysis results known as *Sobolev's embedding theorems*.

Given two Banach spaces V_1, V_2 with $V_1 \subseteq V_2$, one says V_1 is *embedded* in V_2, written $V_1 \hookrightarrow V_2$, if $v \in V_1 \Rightarrow v \in V_2$ and there is a positive constant C such that $\|v\|_{V_2} \leq C\|v\|_{V_1}$ for any $v \in V_1$. One embedding theorem in particular states that

---

Theorem 1.1 (Sobolev embedding theorem). *Given a bounded open set $\Omega \subset \mathbb{R}^d$ with non-empty and Lipschitz-continuous boundary:*

1. *if $k < d/2$, $H^k(\Omega) \hookrightarrow L^q(\Omega)$ for any $q \leq p^* = 2d/(d - 2k)$;*
2. *if $k = d/2$, $H^k(\Omega) \hookrightarrow L^q(\Omega)$ for any $q \in [2, \infty)$;*
3. *if $k > d/2$, $H^k(\Omega) \hookrightarrow C^0(\bar{\Omega})$.*

In dimension one, therefore, functions in H^1 are continuous or, more precisely, every element of H^1 (which is in fact an equivalence class of functions) admits a continuous representative.

In many differential problems it is relevant to restrict a function to a measurable portion $\Gamma \subset \partial\Omega$ of the boundary of the domain Ω in order to assign boundary conditions. It becomes crucial to give a meaning to "value on the boundary", as functions in $H^k(\Omega)$, are not necessarily continuous in $\bar{\Omega}$. Indeed we may define an operator that coincides, in the case of continuous functions, with the function's restriction to the boundary. Precisely, there is a linear operator γ from $H^1(\Omega)$ to $L^2(\Gamma)$, called *trace operator*, such that $\gamma u = u_{|\Gamma}$ if $u \in H^1(\Omega) \cap C(\bar{\Omega})$ and that extends with continuity to the entire $H^1(\Omega)$, i.e. $\exists \gamma_T > 0$ such that for any $u \in H^1(\Omega)$

$$\|\gamma u\|_{L^2(\Gamma)} \leq \gamma_T \|u\|_{H^1(\Omega)}. \qquad (1.11)$$

This is said *trace inequality*.

One proves that the image of $H^1(\Omega)$ under γ is not the whole $L^2(\Gamma)$, rather a subspace, which we denote $H^{1/2}(\Gamma)$ and whose norm can be defined as

$$||g||_{H^{1/2}(\Gamma)} = \inf_{\substack{v \in H^1(\Omega) \\ \gamma v = g}} ||v||_{H^1(\Omega)}.$$

Conversely, given a measurable portion $\Gamma \subset \partial\Omega$ there exists a $c_\gamma > 0$ such that for every $g \in H^{1/2}(\Gamma)$ one can find a function $G \in H^1(\Omega)$ with $\gamma G = g$ and

$$||G||_{H^1(\Omega)} \le c_\gamma ||g||_{H^{1/2}(\Gamma)}. \tag{1.12}$$

The map G is called *extension* or *lifting* of g.

The results listed above hold if the domain is sufficiently regular, like when Ω is Lipschitz. Abusing the notation, the trace γu is often written simply as $u|_\Gamma$.

Sobolev vector spaces. For time-dependent problems one often uses special vector spaces, which we recall below. If V is a Sobolev space of functions on $\Omega \subset \mathbb{R}^d$, and $I = (0, T)$ is a time interval, we set

$$L^2(I; V) \equiv \{v : I \to V| \ v \text{ is measurable and } \int_0^T ||v(t)||_V dt < \infty\},$$

and

$$L^\infty(I; V) \equiv \{v : I \to V| \ v \text{ is measurable and ess } \sup_{t \in I} ||v(t)||_V < \infty\}.$$

The former is the space of functions in V whose V-norm is square integrable. For the latter, the V-norm is bounded. They come equipped with the following norms

$$||u||_{L^2(I;V)} = \sqrt{\int_0^T ||u(t)||_V^2 dt}, \ ||u||_{L^\infty(I;V)} = \text{ess } \sup_{t \in I} ||u(t)||_V. \tag{1.13}$$

If $||u(t)||_V$ is continuous on $[0, T]$, then $||u||_{L^\infty(I;V)} = \max_{x \in [0,T]} ||u(t)||_V$.

For time dependent problems it is useful to define spaces of function with values in Sobolev spaces. Their definition is rather technical we give here only an "operative" definition, the reader may find a more formal and precise treatment in [Eva10]. Let I a time interval $I \subset \mathbb{R}$ and V a Sobolev space on Ω. For a function

$$f : \Omega \times I \to \mathbb{R}, \quad (x, t) \to f(x, t)$$

we define the norm

$$||f||_{L^p(I;V)} = \left(\int_I ||f(t)||_V^p \right)^{1/p}, \quad \text{for } 1 \le p < \infty$$

and

$$||f||_{L^\infty(I;V)} = \operatorname*{ess\,sup}_{I} ||f(t)||_V.$$

Function f belongs to the space $L^p(I;V)$, the space of L^p functions with values in V, if those norms are well defined and finite. This space is a Banach space, equipped with the given norm. Here $f(t)$ indicates the function

$$f(t) : \Omega \to \mathbb{R}, \quad x \to f(t)(x) = f(x,t).$$

In practice, $f(t)$ may be though as a "snapshot" of $f(x,t)$ taken at time t. This snapshot is still a function of x. So, for instance $f \in L^2(I;H^1(\Omega))$ if $f(t)$ belongs to $H^1(\Omega)$ and if $||f(t)||_{H^1(\Omega)}$, taken as a function of time t, is in $L^2(I)$.

1.4 Sequences in l^p

In the analysis of numerical schemes for some PDE problems it may be useful to consider functions defined only on points of a lattice. In a way analogous to that used in the continuum setting we may introduce the space l^p. Given a partition of the real axis into intervals $[x_j, x_{j+1}]$, $j = 0, \pm1, \pm2\ldots$, of length h, we say that a sequence $\mathbf{y} = \{y_i \in \mathbb{R}, \quad -\infty < i < \infty\}$ belongs to l^p for some $1 \leq p \leq \infty$ if $||\mathbf{y}||_{\triangle,p} < \infty$, where

$$||\mathbf{y}||_{\triangle,p} \equiv \left(h \sum_{j=-\infty}^{\infty} |y_j|^p \right)^{\frac{1}{p}}, \quad \text{when } 1 \leq p < \infty, \tag{1.14}$$

$$||\mathbf{y}||_{\triangle,\infty} \equiv \max_{-\infty < j < \infty} (|y_j|).$$

For $u \in L^p(\mathbb{R}) \cap C^0(\mathbb{R})$, $1 \leq p \leq \infty$, the vector $\mathbf{u} = \{u(x_i), -\infty < i < \infty\} \in l^p$. Moreover, if $\mathbf{y} \in l^p$, $1 \leq p < \infty$, then $\lim_{j \to \infty} |y_j| = 0$ and the following property, called *telescopic sum*, holds:

$$\sum_{j=-\infty}^{\infty} (y_{j+1} - y_j) = 0.$$

This property can be proved by noticing that $S_N = \sum_{j=-N}^{N} (y_{j+1} - y_j) = y_{N+1} - y_{-N}$ and that $|\lim_{N\to\infty} S_N| \leq \lim_{N\to\infty} |S_N| = \leq \lim_{N\to\infty} (|y_{N+1}| + |y_{-N}|) = 0$. Similarly one can prove the relationships

$$\sum_{j=-\infty}^{\infty} (y_{j+1} - y_{j-1}) = 0, \qquad \sum_{j=-\infty}^{\infty} (y_{j+1} - y_{j-1})y_j = 0,$$

which will be useful in the analysis of schemes hyperbolic equations in Chapter 6.

The same type of norms can be defined when one considers only a finite number of values y_i, for $i = 0, \ldots, N$ on a grid of spacing h. In this case we will set

$$\|\mathbf{y}\|_{\triangle,p} \equiv \left(h \sum_{j=0}^{N} |y_j|^p \right)^{\frac{1}{p}}, \quad \text{when } 1 \le p < \infty,$$

$$\|\mathbf{y}\|_{\triangle,\infty} \equiv \max_{0 \le j \le N} (|y_j|).$$

The $\|\cdot\|_{\triangle,\infty}$ norm is often called *discrete infinity norm*.

1.5 Important inequalities

In the analysis of PDEs we will often make use of an important inequality due to Poincaré.

Theorem 1.2 (Poincaré inequality). *Assume that Ω is a bounded connected open set of $\mathbb{R}^d$ and that Σ is a non-empty sufficiently regular (Lipschitz continuous is enough) subset of $\partial\Omega$. Let us set*

$$H_\Sigma(\Omega) = \{v \in H^1(\Omega) : v|_\Sigma = 0\},$$

where $v|_\Sigma$ indicates the trace of v on Σ. Then, there exists a positive constant C_Ω, which depends only on Ω and Σ, such that

$$\|v\|_{L^2(\Omega)} \le C_\Omega \|\nabla v\|_{L^2(\Omega)}, \quad \forall v \in H_\Sigma(\Omega). \tag{1.15}$$

Clearly, if $d = 1$ the gradient is in fact the derivative of v.

An extension of Poincaré inequality, sometimes called *Poincaré-Wittinger inequality*, will become handy when treating Navier-Stokes problems. It reads

---

Theorem 1.3 (Poincaré-Wittinger inequality). *Assume that Ω is a bounded connected open set of $\mathbb{R}^d$. Let us set*

$$H^1(\Omega) \setminus \mathbb{R} = \{v \in H^1(\Omega) : \int_\Omega v| = 0\}.$$

Then, there exists a positive constant C such that

$$\|v\|_{L^2(\Omega)} \le C \|\nabla v\|_{L^2(\Omega)}, \quad \forall v \in H^1(\Omega) \setminus \mathbb{R}. \tag{1.16}$$

An important consequence of Poincaré inequalities is the equivalence of H^1 norm and seminorm for functions in $H_\Sigma(\Omega)$ (or $H^1(\Omega) \setminus \mathbb{R}$). Indeed, for a $v \in H_\Sigma(\Omega)$ we have

$$\|v\|_{H^1(\Omega)}^2 = \|v\|_{L^2(\Omega)}^2 + \|\nabla v\|_{L^2(\Omega)}^2 \le (1+C_\Omega^2)\|\nabla v\|_{L^2(\Omega)}^2 = (1+C_\Omega^2)|v|_{H^1(\Omega)}^2,$$

while obviously $|v|_{H^1(\Omega)} \le \|v\|_{H^1(\Omega)}$. Thus, we have that $|v|_{H^1(\Omega)} \le \|v\|_{H^1(\Omega)}$ $\le \sqrt{1 + C_\Omega^2}|v|_{H^1(\Omega)}$ for any $v \in H_\Sigma(\Omega)$. The same applies to $v \in H^1(\Omega) \setminus \mathbb{R}$.

Another result, very useful for the study of time-dependent equations, is Grönwall's Lemma (see for example Lemma 6.1 in [Qua09], or Lemma 1.4.1 in [QV94]). We recall below one of the versions present in the literature.

Lemma 1.1 (Grönwall's Lemma). *Let* $f \in L^1(t_0, T]$ *be non-negative,* g *and* φ *continuous on* $[t_0, T]$. *If* φ *satisfies the inequality*

$$\varphi(t) \le g(t) + \int_{t_0}^t f(\tau)\varphi(\tau)d\tau$$

for all $t \in [t_0, T]$, *then*

$$\varphi(t) \le g(t) + \int_{t_0}^t f(s)g(s)e^{\int_s^t f(\tau)d\tau}ds \qquad (1.17)$$

for any $t \in [t_0, T]$. *Moreover if* g *is non-decreasing, the inequality*

$$\varphi(t) \le g(t)e^{\int_{t_0}^t f(\tau)d\tau}, \quad \forall t \in [t_0, T]$$

holds.

From a quantitative point of view this inequality turns out to be not so significant for large t, because of the exponential growth of the term on the right.

However, under more restrictive hypotheses we may obtain a different estimate.

Proposition 1.1. *Let* $f : I \to \mathbb{R}^+$ *be a positive summable function in the interval* $I = (0, T)$, g *a non negative constant and* y *a continuous non negative function in* I *satisfying for all* $t \in I$ *the inequality*

$$y^2(t) \le g + \int_0^t f(\tau)y(\tau)\,d\tau.$$

Then,

$$y(t) \le \frac{1}{2}\int_0^t f(\tau)\,d\tau + \sqrt{g}, \quad \forall t \in I.$$

We mention that discrete versions of this Lemma for time discrete functions are available, see e.g. [GR86].

1.6 Brief overview of matrix algebra

Unless otherwise stated we will denote by $D \in \mathbb{R}^{m \times n}$ a generic matrix with $m \geq 1$ rows and $n \geq 1$ columns and by $A \in \mathbb{R}^{n \times n}$ a non-singular square matrix of size $n \geq 1$ over the field $\mathbb{R}$, i.e. with real entries.

The symbol $||\mathbf{v}||_p = (\sum_{i=1}^{n} |v_i|^p)^{1/p}$ is used for the p-norm of a vector $\mathbf{v} \in \mathbb{R}^n$, for some $p \in [1, \infty)$, while $||\mathbf{v}||_\infty = \max_{1 \leq i \leq n} |v_i|$. The 2-norm, also known as *Euclidean norm*, is denoted simply by $||\mathbf{v}||$. Given two vectors $\mathbf{u}$ and $\mathbf{v}$ in $\mathbb{R}^n$ the symbol $(\mathbf{u}, \mathbf{v})$, or alternatively $\mathbf{u}^T \mathbf{v}$, will be their Euclidean dot product. More precisely, $(\mathbf{u}, \mathbf{v}) = \mathbf{u}^T \mathbf{v} = \sum_{i=1}^{n} u_i v_i$.

The p-norm of A is, by definition,

$$||A||_p = \max_{\substack{\mathbf{v} \in \mathbb{R}^n \\ \mathbf{v} \neq 0}} \frac{||A\mathbf{v}||_p}{||\mathbf{v}||_p} \tag{1.18}$$

for an integer $p \geq 1$. In this case as well the 2-norm is usually written $||A||$. For all $p \in [1, \infty]$ the inequality $||A\mathbf{v}||_p \leq ||A||_p ||\mathbf{v}||_p$ holds.

The *range* of a rectangular matrix D is the subspace of vectors $\mathbf{y} \in \mathbb{R}^m$ s.t. $\mathbf{y} = D\mathbf{x}$ for some $\mathbf{x} \in \mathbb{R}^n$. The dimension of the range of D is called *rank* of D. It corresponds to the number of linearly independent columns of D. It can be shown that

$$\text{rank}(D) = \text{rank}(D^T).$$

Consequently, $\text{rank}(D) \leq \min(m, n)$. If this inequality is strict, we say that D is *rank-deficient*. If the equality holds, we say that D is a *full rank* matrix. For a square matrix, to be full rank is equivalent to non-singularity.

The null-space or *kernel* of a matrix D is the sub-space of $\mathbb{R}^n$ s.t.

$$\text{Ker}(D) = \{\mathbf{x} \in \mathbb{R}^n : D\mathbf{x} = \mathbf{0}\}. \tag{1.19}$$

If the kernel reduces to the zero vector solely, we say that the kernel is *trivial*.

For a generic matrix D it is possible to prove that

$$\text{rank}(D) + \dim \text{Ker}(D) = n.$$

A square matrix is therefore non-singular if and only if its kernel is trivial (so the dimension of the kernel is 0).

If $\pi(x) = \sum_{i=0}^{r} a_i x^i$ is a polynomial of degree r in the variable x, then $\pi(B)$ indicates the corresponding *matrix polynomial* applied to the square matrix B, defined as $\pi(B) = \sum_{i=0}^{r} a_i B^i$, where $B^0 = I$ by convention.

A complex number $\lambda \in \mathbb{C}$ and a non-null complex vector $\mathbf{r} \in \mathbb{C}^n$ are respectively called an eigenvalue and a right eigenvector (or simply eigenvector) of A if $A\mathbf{v} = \lambda \mathbf{r}$. An eigenvalue λ satisfies $\pi_A(\lambda) = 0$, where π_A is the *characteristic polynomial* of A, defined by $\pi_A(z) = |A - zI|$, I being the identity matrix. To any pair $(\lambda, \mathbf{r})$ a pair $(\lambda, \mathbf{l})$ is associated, where $\mathbf{l}$ is the *left eigenvector* of A and satisfies $\mathbf{l}^T A = \lambda \mathbf{l}^T$.

A matrix of size n admits n distinct eigenvectors at most. If an eigenvalue is a root of the characteristic polynomial of multiplicity m_a, one says that it has *algebraic multiplicity* $m_a(\lambda)$. The *geometric multiplicity* $m_g(\lambda)$ of an eigenvalue λ is the dimension of $\mathrm{Ker}(A - \lambda I)$, in other terms the maximum number of *linearly independent* right eigenvectors of the matrix. Therefore $m_g \le m_a$. The set of eigenvalues $\sigma(A)$ of matrix A is called the *spectrum* of A. We recall that eigenvectors are known up to a multiplicative constant, so if $\mathbf{r}$ is an eigenvector, so is $c\mathbf{r}$ for any real $c \ne 0$.

A matrix A is said to be *similar* to a square matrix of the same size B (possibly with complex entries) if there exists a non-singular (possibly complex) matrix R so that $R^{-1}AR = B$. If R is a *unitary matrix*, i.e. $RR^T = R^T R = I$, then A and B are said *unitarily similar*; they enjoy the property $||B||_2 = ||A||_2$. If $B = \Lambda$, diagonal matrix, then A is said to be *diagonalizable*. The diagonal entries of Λ are the eigenvalues of A, while the columns of R and the rows of R^{-1} (sometimes R^{-1} is indicated by L) are right and left eigenvectors of A, respectively. Therefore, by appropriate scaling the left and right eigenvectors of a diagonalizable matrix satisfy

$$\mathbf{l}_j^T \mathbf{r}_i = \begin{cases} 0 & \text{if } i \ne j, \\ 1 & \text{if } i = j. \end{cases}$$

A matrix with nd distinct eigenvalues λ_i is diagonalizable if and only if $\sum_{i=1}^{nd} m_g(\lambda_i) = n$, which implies that the eigenvectors of A form a basis of $\mathbb{R}^d$. If the eigenvalues of A are all real numbers also the eigenvectors have only real components, and thus R is real. We recall that if $nd = n$ then the matrix is diagonalizable (but a diagonalizable matrix can have conversely not all distinct eigenvalues).

The spectral radius $\rho(A)$ of a matrix A is $\rho(A) = \max\limits_{\lambda \in \sigma(A)} |\lambda|$. The following result holds: $||A||_2 = \sqrt{\rho(A^T A)} = \sqrt{\rho(AA^T)}$.

The eigenvalues (and eigenvectors) of a symmetric matrix A (i.e. such that $A = A^T$) are all real, and A is called *symmetric positive-definite* (we will often use the shorthand notation s.p.d.) if the eigenvalues are all positive. This is equivalent to say that $\mathbf{v}^T A\mathbf{v} > 0$ for any vector $\mathbf{v} \ne \mathbf{0}$.

If A is s.p.d. it is possible to define the *A-norm* of a vector $\mathbf{v}$ by $||\mathbf{v}||_A = \sqrt{\mathbf{v}^T A\mathbf{v}}$, and similarly the associated matrix norm. The space $\mathbb{R}^n$ with norm $||\cdot||_A$ can be equipped with the induced inner product $(\mathbf{w}, \mathbf{v})_A \equiv (\mathbf{w}, A\mathbf{v}) = (A\mathbf{w}, \mathbf{v})$. Two orthogonal vectors with respect to this product, i.e. such that $(\mathbf{w}, \mathbf{v})_A = 0$, are called A-conjugate.

In general, the rank of the product of two matrices is bounded by the minimum of the rank of the two factors. In many applications (see Chapter 7), we have to solve a $m \times m$ matrix in the form DAD^T. The rank of this matrix is then bounded by the rank of D and A. This implies that

$$\mathrm{rank}(DAD^T) \le \min(m, n).$$

We promptly conclude that DAD^T is rank deficient if $n < m$. If we assume that $m \leq n$ and A is s.p.d., then we conclude that DAD^T is non-singular if and only if D is a full-rank matrix. In this case, in fact, $\text{rank}(D^T) = m$ implies that the kernel of D^T is trivial. For any non-zero vector $\mathbf{y} \in \mathbb{R}^m$, $\mathbf{z} = D^T\mathbf{y}$ is therefore non null. Since DAD^T is symmetric and $\mathbf{y}^T DAD^T \mathbf{y} = \mathbf{z}^T A\mathbf{z} > 0$, we conclude that it is also p.d., so it is non-singular.

It is useful to recall the following *Neumann expansion*[3], valid for any matrix A with $\rho(A) < 1$

$$(I - A)^{-1} = \sum_{k=0}^{\infty} A^k, \qquad (1.20)$$

where I is the identity matrix. Consequently

$$(I - A)^{-1} \approx I + A.$$

We recall at last that the condition number $K_p(A)$ of a non-singular matrix A is given, for any integer $p \geq 1$, by $K_p(A) = ||A||_p ||A^{-1}||_p$. The condition number $K_2(A)$ is particularly relevant, so we will simply write it as $K(A)$. For *symmetric positive-definite matrices* we have $K_2(A) = \lambda_{\max}/\lambda_{\min}$ where $\lambda_{\max}$ and $\lambda_{\min}$ are the maximum and minimum eigenvalues of A, respectively. The condition number actually plays a double role. On one hand it describes how the solution of the associated linear system is sensitive on the perturbation of the coefficients and of the right-hand side. On the other one can be seen as a 'cost indicator' in many iterative methods: the more ill-conditioned a matrix is, the more iterations will be necessary, in general, to solve the system with a given accuracy. It is possible to show that the inverse of the condition number $K_2(A)^{-1}$ is in fact a measure of the "distance" of A to the set of singular matrices. So we can say that as the condition number gets larger the matrix approaches singularity.

---

[3] The expansion has a well-known scalar analogue: if $|q| < 1$, the identity $\sum_{j=0}^n q^j = (1 - q^{n+1})/(1 - q)$ implies $\sum_{j=0}^{\infty} q^j = (1 - q)^{-1}$.

2

Fundamentals of finite elements and finite differences

In this introductory chapter we summarize some fundamental topics of approximation theory by finite elements and finite differences. The aim is to introduce the reader to notations and conventions used in the book and to gather in an organic way some basic results that are scattered in the vast literature concerning finite elements and finite differences.

In the finite element method an approximate solution of a partial differential problem is selected from a suitable finite dimensional space of functions V_h. For most of the problems addressed in this book, functions in V_h are globally continuous piecewise polynomials. The space V_h is often called the *discrete space* since any function in V_h can be identified by a discrete number of parameters, as we will make precise later on.

The convergence properties of the finite element method depend directly on the approximation properties of the discrete space V_h, which we are going to describe in the following sections.

2.1 The one dimensional case: approximation by piecewise polynomials

Let us consider a given bounded interval $[a, b] \subset \mathbb{R}$ and $n + 1$ points v_i, $i = 0, \ldots, n$, with $a = v_0 < v_1, \ldots, v_{n-1} < v_n = b$. We may consider the interval as partitioned into n sub-intervals $K_j = [v_{j-1}, v_j]$ of length $h_j = v_j - v_{j-1}$, $j = 1, \ldots, n$, which we will call *elements*. The partition will be called *mesh* and indicated by $\mathcal{T}_h(a, b)$, or simply by $\mathcal{T}_h$, the points v_i being the mesh *vertexes*.

To avoid ambiguity we use the letter v to indicate the mesh vertexes, while the letter x will be employed to indicate the interpolation *nodes*, which are in general a super-set of the vertexes. The letter K will normally indicate a generic element of the mesh.

Given a continuous function $f : [a, b] \rightarrow \mathbb{R}$ we can use the given mesh to build an approximation by piecewise polynomials. If $r \geq 1$ is the degree of

Formaggia L., Saleri F., Veneziani A.: Solving Numerical PDEs: Problems, Applications, Exercises. DOI 10.1007/978-88-470-2412-0_2, © Springer-Verlag Italia 2012

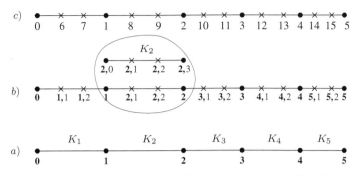

Fig. 2.1 The nodes of a piecewise polynomial interpolation. In the bottom line a) we show the mesh with the vertexes of the elements K_j, indicated by •. In b) we have the nodes for the case of a piecewise cubic interpolation, with a detail of the local numbering of the nodes belonging to element K_2. In c) we show the global numbering obtained using the second of the proposed schemes

the polynomial, we need $r-1$ distinct nodes in the interior of each element, which added to the two end nodes will give the $r+1$ nodes $x_{j,s} = v_j + sh_j/r$, with $s = 0, \ldots, r$, required to generate on each element a polynomial of degree r.

The numbering $s = 0, \ldots, r$ of the nodes on each element is called *local numbering*. It is however convenient to provide the nodes with a unique identifier, called *global numbering*. There is not a single way of doing it. A possibility is to number them consecutively by setting the global numbering k so that $x_k = x_{j,s}$ with $k = r(j-1) + s$, for $s = 0, \ldots r-1$ and $j = 1, \ldots, n$, while $x_{rn+1} = x_{n,r}$. We have then a total of $N = nr + 1$ interpolation nodes satisfying $x_0 < x_1 < \ldots < x_N$.

Another scheme for the global numbering which may seem odd, yet is more akin to what is done in the context of finite elements, is illustrated in line c) of Fig. 2.1. The association between the local numbering (j, s), for $j = 1, \ldots, n$ and $s = 1, \ldots, r-1$, and the global numbering is given by

$$(j, s) \Rightarrow \begin{cases} j, & \text{for } j = 0, \ldots n, \text{ and } s = 0, \\ n + (j-1)(r-1) + s, & \text{for } j = 1, \ldots, n-1, \text{ and } s = 1, \ldots, r-1, \\ N = nr + 1, & \text{if } j = n \text{ and } s = r-1. \end{cases}$$

The advantage of this scheme is that the mesh vertexes coincide with the first $n+1$ nodes, that is $x_i = v_i$, for $i = 0, \ldots, n$. Clearly, the two numbering schemes coincide in the case of piecewise linear interpolation.

Given a grid $\mathcal{T}_h$ of nodes $x_0, \ldots x_N$, we define the *piecewise (or composite) polynomial interpolant of degree r* of a function $f \in C^0([a, b])$, the function $\Pi_h^r f$ such that

$$\Pi_h^r f|_{K_j} \in \mathbb{P}^r(K_j), \quad j = 1, \ldots, n, \quad \Pi_h^r f(x_i) = f(x_i), \quad i = 0, \ldots N.$$

Here, $\mathbb{P}^r(K_j)$ indicates the space of polynomials of at most degree r in the interval K_j[1], while $h = \max_{1 \le j \le n} h_j$ is the *grid spacing*.. The interpolant $\Pi_h^r f$ may be expressed uniquely as a linear combination of the Lagrange characteristic polynomials of degree r, ϕ_i, $i = 0, 1, \ldots, N$. The Lagrange characteristic polynomials are piecewise composite polynomials of degree (at most) r on the given grid that satisfy the fundamental property

$$\phi_i(x_j) = \delta_{ij}, \quad i, j = 0, \ldots, N, \tag{2.1}$$

where δ_{ij} is the *Kronecker symbol*,

$$\delta_{ij} = \begin{cases} 1 & \text{if } i = j \\ 0 & \text{otherwise.} \end{cases}$$

Then we have that, for all $x \in [a, b]$

$$\Pi_h^r f(x) = \sum_{i=0}^{N} f(x_i) \phi_i(x). \tag{2.2}$$

This procedure is called *(Lagrange) piecewise polynomial interpolation*.

It is immediate to verify that $\phi_i(x) = 0$ whenever x does not belong to an element containing node x_i. Therefore, the Lagrange characteristic polynomials have a *small support*, more precisely the support coincides with the union of the elements containing node x_i. Another interesting property is

$$\sum_{i=0}^{N} \phi_i(x) = 1, \quad \forall x \in [a, b],$$

which can derived by considering the (exact) interpolation of the constant function $f = 1$ and using equation (2.2).

The piecewise polynomial interpolation can also be interpreted as the application of the operator $\Pi_h^r : C^0([a, b]) \to X_h^r(a, b)$, defined by equation (2.2), $X_h^r(a, b)$ being the linear space of all piecewise polynomials of degree (at most) r that can be defined on the grid $\mathcal{T}_h$, that is

$$X_h^r(a, b) \equiv \{v_h \in C^0([a, b]) : v_h|_{K_j} \in \mathbb{P}^r(K_j), j = 1, \ldots, n\}. \tag{2.3}$$

The characteristic Lagrange polynomials are a basis for $X_h^r(a, b)$, whose dimension is then $nr + 1$. The operator Π_h^r is linear and bounded uniformly with respect to h in the norm

$$\|f\|_{C^0([a,b])} \equiv \max_{a \le x \le b} |f(x)|.$$

---

[1] Obviously a polynomial function is defined on the whole real axis. Here, hoverer, we are considering only its restriction in the given interval.

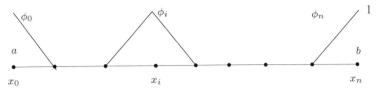

Fig. 2.2 Some examples of Lagrange characteristic polynomials in the case of piecewise linear interpolation

Indeed, in the case of linear interpolation we have that

$$\|\Pi_h^1 v\|_{C^0([a,b])} \leq \|v\|_{C^0([a,b])}, \quad \forall v \in C^0([a,b]).$$

Thank to the Sobolev immersion theorem (see Chapter 1) the operator can be extended to functions in $H^1(a,b)$. In fact, in the one dimensional case it is a linear and bounded operator also with respect to the H^1 norm defined in (1.9). More precisely, there exists a constant $C_r > 0$ (which depends on r) such that $\forall v \in H^1(a,b)$

$$\|\Pi_h^r v\|_{H^1(a,b)} \leq C_r \|v\|_{H^1(a,b)}. \tag{2.4}$$

We warn that this is not anymore true in more than one dimension, however also in that case it is possible to find other interpolation operators which satisfy the stated inequality (see e.g. [EG04, BS02, QV94]).

We can now recall some fundamental results for the interpolation error, valid for any $v \in H^{p+1}(a,b)$, integers $p > 0$, $r > 0$, and $s = \min(r,p)$.

$$|v - \Pi_h^r v|_{H^1(a,b)} \leq C_{s,r,1} \sqrt{\sum_{j=1}^{n} h_j^{2s} |v|^2_{H^{s+1}(K_j)}} \leq C_{s,r,1} h^s |v|_{H^{s+1}(a,b)}, \tag{2.5}$$

$$\tag{2.6}$$
$$\|v - \Pi_h^r v\|_{L^2(a,b)} \leq C_{s,r,0} \sqrt{\sum_{j=1}^{n} h_j^{2(s+1)} |v|^2_{H^{s+1}(K_j)}} \leq C_{s,r,0} h^{s+1} |v|_{H^{s+1}(a,b)}.$$

Here, the $C_{r,s,t}$ are positive constants independent from h.

In the case of piecewise linear interpolation ($r = 1$) and $v \in H^2(a,b)$ we have $C_{1,1,0} = \sqrt{5/24}$ and $C_{1,1,1} = \sqrt{2}/2$.

If $v \in H^1(a,b)$, and $v \notin H^2(a,b)$, we can give a convergence result for the L^2 norm of the interpolation error, namely

$$\|v - \Pi_h^r v\|_{L^2(a,b)} \leq C_r \sqrt{\sum_{j=1}^{n} h_j^2 |v|^2_{H^1(K_j)}} \leq C_r h |v|_{H^1(a,b)},$$

and, in particular, $C_1 = \sqrt{2}$. If $v \in H^1(a,b)$ but $v \notin H^2(a,b)$ we can still say that

$$\lim_{h \to 0} |v - \Pi_h^r v|_{H^1(a,b)} = 0, \tag{2.7}$$

i.e. convergence is assured. From this results we may conclude that for a piecewise interpolation of degree r the maximal order of convergence attainable in norm H^1 is r (and is called *optimal order of convergence* in norm H^1) and is obtained if the function belongs to $H^{r+1}(a,b)$. On a less regular functions we may obtain only a sub-optimal order of convergence.

We mention that other type of interpolation are possible, like splines or Hermite interpolation [QSS00, BS02, EG04].

Exercise 2.1.1. Verify inequality (2.5) through numerical experiments applied to the following functions,

$$f_1(x) = \sin^2(3x) \quad \text{and} \quad f_2(x) = |\sin(3x)|\sin(3x),$$

with $x \in [1,3]$ and $r = 1,2,3,4$. Justify the results by taking into account that $f_1 \in H^s(1,3)$ for all $s \geq 0$, while $f_2 \in H^2(1,3)$, but $f_2 \notin H^3(1,3)$. You can use the MATLAB programs of the suite comppolyXX (which is provided in the book web site), grids with uniform spacing and number of elements equal to 4, 8, 16, 32 e 64.

Solution 2.1.1. The MATLAB functions comppolyXX, where XX is either fit, val or err, extend to composite polynomial interpolation of degree $r > 1$ the already existing MATLAB functionality for linear piecewise interpolation provided by commands polyfit and polyval. More precisely, comppolyfit computes the coefficients of the piecewise polynomial interpolation Π_h^r, comppolyval computes $\Pi_h^r(x)$ at given points[2] $x \in \mathbb{R}$, while comppolyerr calculates the L^2 or H^s norm of the interpolation error in the given interval $[a,b]$ and for a given integer[3] $s > 0$.

The MATLAB script contained in the file exercise_comppoly.m gives a possible solution of the exercise. We provide here the main instructions

```
N=4;
f1='sin(3*x)*sin(3*x)'; f2='abs(sin(3*x))*sin(3*x)';
fd2='3*cos(3*x)*(abs(sin(3*x))+sin(3*x)*sign(sin(3*x)))';
for k=1:ntimes
 mesh=linspace(1,3,N+1); h(k)=2/N;
 coeff1=comppolyfit(mesh,f1,degree);
 coeff2=comppolyfit(mesh,f2,degree);
 err1(k)=comppolyerr(coeff1,[1,3],f1,mesh,norm);
 err2(k)=comppolyerr(coeff2,[1,3],{f2,fd2},mesh,norm); N=2*N
end
```

---

[2] If $x \notin [a,b]$ it provides an extrapolation.

[3] To obtain results in accordance with the theory comppolyerr computes the error using numerical quadrature formulas of a sufficient accuracy to guarantee that the quadrature error is at most of the same order of the optimal interpolation error. Therefore, the computational cost of comppolyerr may be high for large values of s.

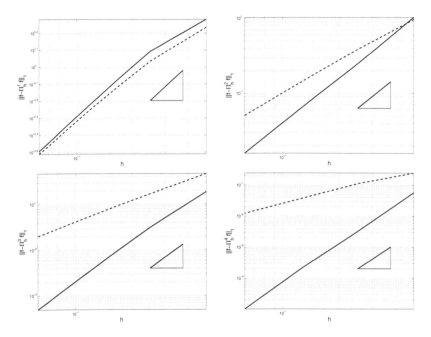

Fig. 2.3 $H^1(1,3)$ norm of the interpolation error n function of the spacing h. A uniform mesh in the interval $[1,3]$ has been considered. The full line refers to function f_1, the dashed line to the less regular function f_2. From top to bottom and from left to right we show the results corresponding to $r = 1, 2, 3$ and 4. The pictures show also a reference line with slope corresponding to the optimal convergence rate

The plots of the errors versus h computed by the script are shown in Fig. 2.3. We can note that in the case of linear interpolation the errors are similar for the two functions. The situation changes completely when we consider polynomials of higher degree. Here, the smaller regularity of function f_2 limits the order of convergence w.r.t. h to 1, while the interpolation of f_1 shows an optimal convergence rate for all r.   ◊

2.2 Interpolation in higher dimension using finite elements

2.2.1 Geometric preliminary definitions

A (geometric) *finite element* K is a compact set of $\mathbb{R}^d$ obtained from a given *reference finite element* $\widehat{K} \subset \mathbb{R}^d$ through the application of a one-to-one and regular map $T_K : \mathbb{R}^d \to \mathbb{R}^d$ (refer to Fig. 2.4), such that

$$K = T_K(\widehat{K}). \tag{2.8}$$

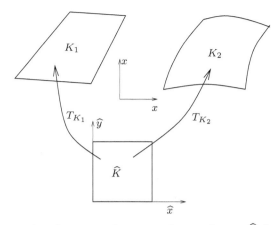

Fig. 2.4 Two examples of maps on the same reference element $\widehat{K}$ (here $\widehat{K}$ is the unit square). T_{K_1} defines the quadrilateral element $K_1 = T_{K_1}(\widehat{K})$ with straight sides, while $K_2 = T_{K_2}(\widehat{K})$ has curved sides

The reference element $\widehat{K}$ is typically a *polygon* (polyhedron in the three dimensional case). We can think T_K as a change from a reference coordinate system $\widehat{\mathbf{x}} = (\widehat{x}_1, \ldots, \widehat{x}_d)$ to the current reference system $\mathbf{x} = (x_1, \ldots, x_d)$. Consequently, in the following we will use the symbol $\widehat{}$ to indicate quantities associated to the reference coordinates. Since in practice $d = 2$ or $d = 3$, whenever convenient we will use the simpler notation x, y and z instead of x_1, x_2 and x_3, respectively.

Two very common choices for $\widehat{K}$ are the *unit simplex*[4] or the *unit square (cube in 3D)*, i.e. a square (cube) with unit side length. The regularity requested on the map T_K depends on the regularity of approximation space we wish to construct. A minimal request is that $T_K \in C^1(\widehat{K})$ with inverse $T_K^{-1} \in C^1(K)$, and that the determinant of the Jacobian

$$J(T_K) = \det\left(\begin{bmatrix} \frac{\partial x_1}{\partial \widehat{x}_1} & \cdots & \frac{\partial x_1}{\partial \widehat{x}_d} \\ & \cdots & \\ \frac{\partial x_d}{\partial \widehat{x}_1} & \cdots & \frac{\partial x_d}{\partial \widehat{x}_d} \end{bmatrix} \right) = \det\left(\begin{bmatrix} \frac{\partial x_i}{\partial \widehat{x}_j} \end{bmatrix}_{i,j=1,\ldots,d} \right)$$

be strictly positive for all $\widehat{\mathbf{x}} \in \widehat{K}$. Here, $\mathbf{x} = T_K(\widehat{\mathbf{x}})$. The request on the Jacobian matrix is equivalent to say that the mapping is orientation-preserving. Most often $T_K \in C^\infty(\widehat{K})$.

A particular, yet rather frequent, case is that of an *affine* map, where T_K takes the form

$$T_K(\widehat{\mathbf{x}}) = \mathbf{a}_K + \mathrm{F}_K \widehat{\mathbf{x}}, \tag{2.9}$$

---

[4] A simplex of $\mathbb{R}^d$ is a polygon with $d + 1$ vertexes. In a unit simplex a vertex is conventionally placed at the origin of the reference coordinate system, while the other d vertexes are located on the Cartesian axes at a unit distance from the origin.

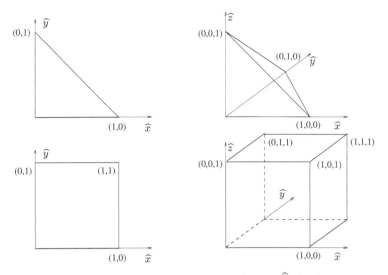

Fig. 2.5 The most common shapes for the reference element $\widehat{K}$. At the top we have the unit simplex in $\mathbb{R}^2$ and $\mathbb{R}^3$, respectively. At the bottom we show the unit square and the unit cube

with $\mathbf{a}_K \in \mathbb{R}^d$ and $F_K \in \mathbb{R}^{d \times d}$ a matrix with $|F_K| > 0$. In this case, if $\widehat{K}$ is a polygon, so it will be K.

We will indicate with h_K the *diameter* of K, defined by $h_K = \max_{\mathbf{x}_1, \mathbf{x}_2 \in K} \|\mathbf{x}_1 - \mathbf{x}_2\|$, and with ρ_K the radius of the maximal circle inscribed in K. Analogous definitions may be given for the three-dimensional case. We will denote by $\|\mathbf{v}\|$ the Euclidean norm of a vector $\mathbf{v} \in \mathbb{R}^d$.

A *triangulation* (also called *grid* or *mesh*) $\mathcal{T}_h(\Omega)$ of Ω is a set of N_e elements, $\mathcal{T}_h(\Omega) = \{K_i : K_i = T_{K_i}(\widehat{K}), i = 1, \ldots N_e\}$, such that the *discretized domain* (or triangulated domain) Ω_h, defined as

$$\Omega_h = \text{int}\left(\bigcup_{K \in \mathcal{T}_h(\Omega)} K\right),$$

is an approximation of Ω. Here $\text{int}(A)$ indicates the interior of set A. More precisely, it is possible to build a family of triangulations, parametrized with $h = \max_{K \in \mathcal{T}_h(\Omega)} h_K$, such that $\lim_{h \to 0} \text{d}(\partial\Omega, \partial\Omega_h) = 0$, $\text{d}(A, B)$ indicating the distance between two subsets A and B of $\mathbb{R}^d$.

In general, $\Omega_h \neq \Omega$, yet they coincide in some cases. For instance, if Ω is polygonal the use of an affine map for T_K is enough to allow an exact discretization of the domain. For more general domains however, we may have to accept that the discretized domain is only an approximation of Ω, even if the use of more complex mapping than the affine map may help to better approximate curved boundaries.

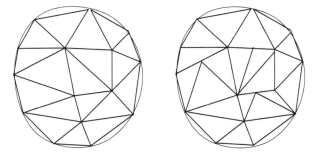

Fig. 2.6 An example of conforming (left) and non-conforming (right) grid

The use of more complex mappings like non uniform rational B-splines (NURBS) may allow to represent exactly most geometries produced by CAD (Computer Aided Design) systems. The description of this technique, however, goes beyond the objectives of this book, the interested reader may refer to [HCB05]. For the sake of simplicity, in this book we will often assume that $\Omega_h = \Omega$. We will also write $\mathcal{T}_h$ instead of $\mathcal{T}_h(\Omega)$ whenever convenient.

If the map T_K is affine the grid is said to be a (geometrically) *affine grid*. The affine map transform a polynomial in the $\widehat{\mathbf{x}}$ coordinates in a polynomial of the same order in the $\mathbf{x}$ coordinates.

In $\mathcal{T}_h$ we can identify the set $\mathcal{V}_h$ of *vertexes*, the set of sides $\mathcal{S}_h$ (also called *edges*) and (in 3D) the set of *faces* $\mathcal{F}_h$. A grid is said to be *conforming* if for any couple of elements $K_1, K_2 \in \mathcal{T}_h$ with $K_1 \neq K_2$ we have that (see also Fig. 2.6)

$$\overset{\circ}{K}_1 \cap \overset{\circ}{K}_2 = \emptyset, \quad K_1 \cap K_2 \neq \emptyset \Rightarrow K_1 \cap K_2 \in \mathcal{V}_h \cup \mathcal{S}_h \cup \mathcal{F}_h$$

where $\overset{\circ}{K}_i$ for $i = 1, 2$ denotes the internal part of K_i. In other words, the intersection of two distinct elements of a conforming mesh is either null or a common vertex, or a common side or a common face.

In this book we will consider only conforming grids. Non conforming grids lead to more complex approximating spaces and more complex algorithms. They are of certain interest, however, because they allow to implement fast grid adaptation techniques (see e.g. [EG04]).

We will indicate with N_e, N_v, N_s and N_f the number of elements, vertexes, sides (edges) and faces of a mesh, respectively. Clearly N_f is only relevant in the three dimensional case. We will add the suffix b to indicate boundary items, for instance $N_{v,b}$ and $N_{s,b}$ are the number of vertexes and edges laying on $\partial\Omega_h$.

Furthermore, let n_v, n_s and n_f be the number of vertexes, sides and faces of an element, respectively. If m indicates the number of "holes" in the triangulated domain (in particular $m = 0$ if the domain is simply connected), and, for the 3D case, c_b is the number of connected components of the boundary, we have the following relations [FG00] for or a 2D grid

$$
\begin{array}{ll}
(a)\ N_e - N_s + N_v = 1 - m, & (b)\ 2N_s - N_{s,b} = n_v N_e, \\
(c)\ N_{v,b} = N_{s,b},
\end{array}
\tag{2.10}
$$

while for a three-dimensional grid,

$$
\begin{array}{ll}
(a)\ N_e - N_f + N_s - N_v = m - c_b - 1, & (b)\ 2N_f - N_{f,b} = n_v N_e, \\
(c)\ N_{v,b} + N_{f,b} = N_{s,b} + 2(c_b - m).
\end{array}
\tag{2.11}
$$

In both cases we have assumed that Ω is connected. However, the relations, which are derived from the formula for the Euler characteristic, can be easily extended to domains composed by several disjoint connected sets, by considering each set separately.

Finally, we recall that a family of grids $\{\mathcal{T}_h(\Omega)\}_h$ parametrized with h is said to be *regular* if there exists a $\gamma > 0$ such that

$$
\gamma_K = \frac{h_K}{\rho_K} \geq \gamma, \quad \forall K \in \mathcal{T}_h(\Omega),
\tag{2.12}
$$

for all h. The quantity γ is the *regularity constant* (or also *sphericity*). We will say that a grid is *regular* if it belongs to a family of regular grids.

A minimal data structure required to store a grid is composed by the list of vertexes, with their coordinates, and a table that gives the numbering of the vertexes of each element. This table is usually called *connectivity matrix* or *incidence matrix*. Normally, a similar structure is provided also for the boundary faces (boundary connectivity matrix).

Exercise 2.2.1. A three dimensional mesh generator provides the number of elements N_e, vertexes N_v and boundary faces $N_{f,b}$ of a grid made of tetrahedrons (most grid generators provide these data). Using just this information and making some assumption on the topology of the domain (when necessary), calculate the total number of edges and faces of the grid (this information is of practical use, for instance to allocate arrays which store information related to those quantities).

Solution 2.2.1. In a tetrahedron $n_v = 4$, then from (2.11-b) we deduce that $N_f = 2N_e + \dfrac{N_{f,b}}{2}$. The boundary of a tetrahedral grid is a closed surface made of triangles. We can thus exploit (2.10-b), by reformulating it appropriately. Indeed, for a triangulation which is the boundary of a three-dimensional grid the parameter N_s in (2.10-b) is in fact $N_{s,b}$, while $N_{s,b}$ in (2.10-b) is here zero since the boundary surface is closed. Finally N_e in (2.10-b) is here equal to $N_{f,b}$ since the elements of the triangulated boundary surface are indeed the boundary faces of $\mathcal{T}_h$. We obtain that $2N_{s,b} = 3N_{f,b}$ and thus $N_{s,b} = \frac{3}{2}N_{f,b}$.

The exact number of other geometrical entities cannot be computed without having some more information about the topology of the domain. Let assume that the boundary is formed by just one connected component, $c_b = 1$, and that the domain is simply connected, $m = 0$. Then, $N_{v,b} = 2 + N_{s,b} - N_{f,b} = 2 + \frac{1}{2}N_{f,b}$ and $N_s = N_v + N_f - N_e - 2$.   ◇

If K is a simplex it is common to adopt a particular system of coordinates, called *barycentric coordinates*. $\boldsymbol{\xi} = (\xi_i, \ldots, \xi_{d+1})$. Let $\{\mathbf{v}_i, i = 1, \ldots, d+1\}$ be the vertexes of K of coordinates $v_{i,j}$, $j = 1, \ldots, d$, and $\mathbf{x}$ a point of $\mathbb{R}^d$. We set

$$\langle K \rangle \equiv \begin{vmatrix} 1 & \cdots & 1 & 1 \\ v_{1,1} & \cdots & v_{d,1} & v_{d+1,1} \\ \vdots & & & \vdots \\ v_{1,d} & \cdots & v_{d,d} & v_{d+1,d} \end{vmatrix}. \tag{2.13}$$

Let us note that $\langle K \rangle = \pm d! |K|$ where the operator $| \cdot |$ indicates here the area (or volume) of a measurable set of R^n, the sign being determined by the orientation of the simplex K. In particular, the convention normally adopted is that a a triangle is positively oriented if its vertexes are numbered anti-clock-wise. For a tetrahedron, we follow the so called "right-hand rule".

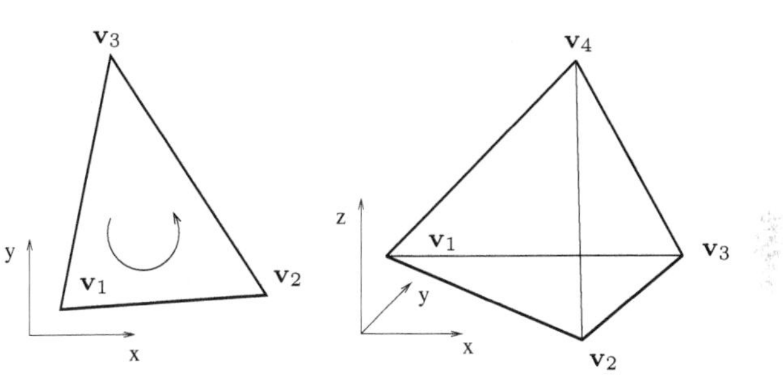

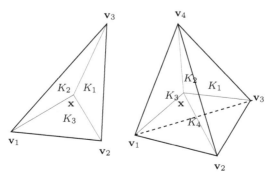

Fig. 2.7 Conventional numbering of vertexes for a positively oriented triangle and tetrahedron

Fig. 2.8 The construction of the barycentric coordinates of a point $\mathbf{x}$, in the case of a triangle (left) and of a tetrahedron (right)

If $\langle K \rangle$ is positive, we say that K is *positively oriented* (see Fig. 2.7). Now, for $j = 1, \ldots, d+1$ we indicate with $K_j(\mathbf{x})$ the simplex obtained from K by replacing vertex $\mathbf{v}_j$ with $\mathbf{x}$, and we define

$$\xi_j(\mathbf{x}) \equiv \frac{\langle K_j(\mathbf{x}) \rangle}{\langle K \rangle} \tag{2.14}$$

as the j-th *barycentric coordinate* of point $\mathbf{x}$ associated to simplex K (refer to Fig. 2.8)[5].

Program 1 - barcoord : Computation of the barycentric coordinates of a point $\mathbf{x}$, relative to a simplex K

```
function [xi]=barcoord(K,x)
% K is a matrix of dimension (d,d+1) containing on each column
% the coordinates of the corresponding vertex of a simplex
% x is a column vector of dimension (d,1) with the coordinates
% of x
% The column vector xi returns the d+1 barycentric coordinates
dd=size(K,2);
A=[ones(1,dd);K];
b=[1;x];
xi=A\b;
```

Program 2 - coorbar : Computation of the Cartesian coordinates of point $\mathbf{x}$, given its barycentric coordinates relative to a simplex K

```
function [x]=coorbar(K,xi)
% K is a matrix of dimension (d,d+1) containing on each column
% the coordinates of the corresponding vertex of the simplex
% xi ia a column vector with the barycentric coordinates
% x ia a column vector returning the corresponding Cartesian coordinates
x=K*xi;
```

We have that $\sum_{j=1}^{d+1} \xi_j = 1$. Moreover the barycentric coordinates are all positive if the point is interior to K. The barycentric coordinates are invariant by affine transformations. If T_K is the affine map that transform the unitary simplex $\widehat{K}$ into K, we have the following relations between the barycentric coordinates of a point $\mathbf{x}$ and the coordinates $\widehat{\mathbf{x}} = T_K^{-1}(\mathbf{x})$ in the reference coordinate system,

$$\xi_1 = 1 - \sum_{s=1}^{d} \widehat{x}_s, \quad \xi_{j+1} = \widehat{x}_j, \ j = 1, \ldots, d. \tag{2.15}$$

---

[5] The barycentric coordinates of a point $\mathbf{x}$ depend on the the simplex K under consideration. For the sake of simplicity, however, we have avoided to indicate the dependence on K explicitly.

Finally, if f is a continuous function and we set $f_j = f(\mathbf{v}_j)$, for $j = 1, \ldots, d+1$, then the quantity $\sum_{j=1}^{d+1} x_j f_j$ is the linear interpolation of f at point $\mathbf{x}$.

Programs 1 and 2 show how it is possible to implement in MATLAB the computation of the barycentric coordinates of a point (and of the Cartesian coordinates given the barycentric ones).

2.2.2 The finite element

The finite element method is based on the construction of a finite dimensional space of functions of the type

$$X_h(\Omega_h) \equiv \{v_h : v_h \in V(\Omega_h), \quad v_h|_K \in P(K), \ \forall K \in \mathcal{T}_h\}, \tag{2.16}$$

where $V(\Omega_h)$ is a given functional space, typically a Hilbert space which in turn depends on the type of differential problem we wish to solve. $P(K)$ is a space of dimension n_l made of regular functions (typically C^∞) defined on the (geometrical) element K. Therefore, $X_h(\Omega_h)$, which in the following we will often indicate simply by X_h, is built by "assembling" functions defined on each element K of a given triangulation of Ω. The construction is similar, yet somehow more complex, to what we have already illustrated for the one-dimensional piecewise polynomial interpolation. Indeed, the formulation we present in this section is more general, and in particular it includes piecewise polynomials as a special (yet quite frequent) case.

We have requested that X_h be a subspace of $V(\Omega_h)$, in this case we say that the finite elements are *conforming* to $V(\Omega_h)$, or, more simply, V-conforming.

An important case is the C^0-conformity, which means that the functions in X_h are continuous. The latter implies H^1-conformity as long as $P(K) \subset H^1(K)$ for all K [QV94]. The dimension of space $X_h(\Omega_h)$ will be indicated by N_h.

We say that X_h is a space *approximating* V if $X_h \to V$ when $h \to 0$. More precisely, if for all $v \in V$ and $\epsilon > 0$ there is a $h_0 > 0$ such that

$$\inf_{v_h \in X_h} \|v - v_h\|_V \leq \epsilon \quad \text{if } h \leq h_0.$$

The degrees of freedom

The *degrees of freedom*

$$\Sigma = \{\sigma_i : X_h \to \mathbb{R}, i = 1, \ldots, N_h\}$$

are a set of continuous and linear functionals on X_h such that the application

$$v_h \in X_h \ \to \ (\sigma_1(v_h), \ldots, \sigma_{N_h}(v_h)) \in \mathbb{R}^{N_h}$$

is an isomorphism. Consequently, a function $v_h \in X_h$ is uniquely identified by the value of the degrees of freedom applied to itself, i.e. $\sigma_i(v_h)$, $i = 1, \ldots, N_h$.

The definition of a finite element space we are giving here is rather general and includes, as a particular case, the *Lagrangian finite elements*, which are probably the most common type of finite elements. In a Lagrangian finite elements the degrees of freedom are the values of v_h at particular points $\mathbf{N}_i$ called *nodes*. That is, $\sigma_i(v_h) = v_h(\mathbf{N}_i)$, for $i = 1, \ldots, N_h$.

Other choices are possible. For instance in the Raviart-Thomas finite elements used to approximate vector fields, the degrees of freedom are the fluxes across the edges (faces in 3D) of the grid. Or, in the C^1-conforming Hermite finite elements, the degrees of freedom include also the first derivatives of v_h. More details may be found, for instance, in [EG04, BS02].

Having defined the degrees of freedom, we can find a basis for X_h by selecting the functions $\{\phi_i, \quad i = 1, \ldots, N_h\}$ with $\phi_i \in X_h$ and that satisfy

$$\sigma_i(\phi_j) = \delta_{ij}, \quad i, j = 1, \ldots, N_h. \tag{2.17}$$

This choice is unique.

The basis functions ϕ_i are often called *shape functions*. Relation (2.17) establishes a duality between the shape functions and the degrees of freedom: once one of the two is given (2.17) allows to determine the other. In particular, in Lagrangian finite elements we have

$$\phi_i(\mathbf{N}_j) = \delta_{ij}, \quad i, j = 1, \ldots, N_h, \tag{2.18}$$

similarly to what we have already seen for the one-dimensional case.

Every $v_h \in X_h$ can then be written in the form

$$v_h(\mathbf{x}) = \sum_{i=1}^{N_h} \sigma_i(v_h)\phi_i(\mathbf{x}). \tag{2.19}$$

The *finite element* is the building block to construct the space X_h. It consists of the geometric element K, a space of functions $P(K)$ of dimension n_l and the *local degrees of freedom* $\Sigma_K = \{\sigma_{K,i} : P(K) \to \mathbb{R}, i = 1, \ldots, n_l\}$, such that the application

$$p \in P(K) \longrightarrow (\sigma_{K,1}(p), \sigma_{K,2}(p), \ldots, \sigma_{K,n_l}(p)) \in \mathbb{R}^{n_l}$$

is an isomorphism. Again, we can identify a particular basis $\{\phi_{K,i}, i = 1, \ldots n_l\}$ of $P(K)$, associated to Σ_K, whose components satisfy $\sigma_{K,i}(\phi_{K,j}) = \delta_{ij}$. The function $\phi_{K,i}$ are called *local shape functions* or *local basis functions*, associated to element K.

Consequently, any $p \in P(K)$ can be expressed as

$$p(\mathbf{x}) = \sum_{i=1}^{N_l} \sigma_{K,i}(p)\phi_{K,i}(\mathbf{x}) \quad \text{for } \mathbf{x} \in K.$$

Local and global numbering

We have to link the local quantities at element level to the global ones defined previously. To this aim, we need to relate local and global numbering, as we did in the 1D case. Yet, here things are more involved since we cannot find a simple "natural" ordering of the nodes.

Let $\nu_{K,i}$, for $i = 1, \ldots, n_l$, be an entry of the *connectivity matrix*, which returns for a given element K and local numbering i the *global numbering* of the local degree of freedom $\sigma_{K,i}$. The union of all entries in the connectivity matrix is the set $\{1, \ldots, N_h\}$ and for any $1 \le j \le N_h$ there is at least one K and one index i such that $j = \nu_{K,i}$.

Then, for all K and $i = 1, \ldots, n_l$

$$\sigma_{\nu_{K,i}}(v) = \sigma_{K,i}(v|_K), \quad \forall v \in X_h,$$

while local and global shape functions are linked by relation

$$\begin{cases} \phi_j|_K = \phi_{K,i}, & \text{if } \exists K, i \quad \text{such that } j = \nu_{K,i}, \\ 0 & \text{otherwise.} \end{cases} \tag{2.20}$$

Using (2.20) we infer that the support of shape function ϕ_j is limited to the union ("patch") of the elements K for which $\nu_{K,i} = j$ for a $1 \le i \le n_l$. The finite element method exploits this *local support property* to be able to work on each element separately, as it will be explained later on.

To have a V-conforming space is necessary that $P(K) \subset V(K)$, for instance we need that the local shape functions be continuous if we want X_h to C^0-conforming. Yet, this is clearly not enough. In general we will need to impose some further conditions, as it will be detailed in the exercises.

The reference element

Another important characteristics of the finite element method, particularly relevant in the multidimensional case, is that the construction of the local space $P(K)$ is usually done by considering the reference element $\widehat{K}$ and a polynomial space $\widehat{P}(\widehat{K}) \subset \mathbb{P}^r(\widehat{K})$ for a integer $r \ge 0$, defined on $\widehat{K}$. There are practical reasons for this, for instance it is easier to define $\widehat{P}(\widehat{K})$ than $P(K)$ directly, since $\widehat{K}$ has a simple (and fixed) shape. We will indicate the degrees of freedom in the reference element as $\widehat{\Sigma} = \{\widehat{\sigma}_1, \widehat{\sigma}_2, \ldots, \widehat{\sigma}_{n_l}\}$ and the corresponding shape functions as $\{\widehat{\phi}_1, \ldots, \widehat{\phi}_{n_l}\}$.

We have the fundamental relation $\widehat{\sigma}_i(\widehat{\phi}_j) = \delta_{ij}$ and thus

$$\widehat{p}(\widehat{\mathbf{x}}) = \sum_{i=1}^{n_l} \widehat{\sigma}_i(\widehat{p}) \widehat{\phi}_i(\widehat{\mathbf{x}}).$$

Lagrangian finite elements

In the case of Lagrangian finite elements the link between a function $\widehat{p} \in \widehat{P}(\widehat{K})$ and the corresponding one in the current element K is simply given by the change of coordinates provided by the map T_K, that is

$$p(\mathbf{x}) = \widehat{p}(T_K^{-1}(\mathbf{x})). \tag{2.21}$$

The shape functions may transform similarly, $\phi_{K,i} = \widehat{\phi}_i \circ T_K^{-1}$, yet we wish to point out that this choice is not always the optimal one for non-Lagrangian finite elements, the interested reader may consult [EG04].

We will now analyze more in detail how to build C^0-conforming finite element spaces using Lagrangian finite elements. First of all, we note that to have non trivial C^0-conforming spaces we need that $r \geq 1$. In a Lagrangian finite element of degree $r \geq 1$ we have that $\widehat{P}(\widehat{K}) \subset \mathbb{P}^r(\widehat{K})$ and we can identify the local degrees of freedom through a set of points $\widehat{\mathbf{N}}_i \in \widehat{K}$, $i = 1, \ldots, n_l$, called *nodes*, by setting $\widehat{\sigma}_i(\widehat{p}) = \widehat{p}(\widehat{\mathbf{N}}_i)$. The corresponding shape functions and degrees of freedom on the current element K are given by $\phi_{K,i} = \widehat{\phi}_i \circ T_K^{-1}$ and $\sigma_{K,i}(p) = p(\mathbf{N}_{K,i})$, where $\mathbf{N}_{K,i} = T_K(\widehat{\mathbf{N}}_i)$ is the ith node of element K.

Lagrangian finite elements of most common usage are the so called $\mathbb{P}^r$ finite elements, where $\widehat{K}$ is the unit simplex, $\widehat{P}(\widehat{K}) = \mathbb{P}^r(\widehat{K})$ and T_K is an affine map. The corresponding finite element space will be indicated by $X_h^r(\Omega_h)$, or simply X_h^r. Since the affine map transforms a polynomial into a polynomial of equal degree, X_h^r can equivalently be defined as

$$X_h^r = \{v_h \in C^0(\Omega_h) : \quad v_h|_K \in \mathbb{P}^r(K), \quad \forall K \in \mathcal{T}_h\}. \tag{2.22}$$

Fig. 2.9 illustrates the most common $\mathbb{P}^r$ finite elements, of which we give the corresponding local shape functions expressed as function of the *barycentric coordinates*. We have adopted the following scheme for the local numbering of the nodes. The node index, here indicated by n, advances at each increment of the indexes i, j e k present in the formula. When more than one index is present, index k is the fastest running index, then j and finally i. Beware of the constraints imposed on the indexes and that $n_l = \frac{1}{d!} \prod_{j=1}^{d}(r+j)$. In particular, we have the correct correspondence between the expressions for the shape functions given in the following and the node numbering illustrated in Fig. 2.9. With this scheme we have numbered the nodes in correspondence of the vertexes first, than the node interior to the sides, then those interior to the faces and eventually those interior to the element. The choice adopted here is convenient because it allows to easily identify on which geometry entity (vertex, side, face or element) a node belongs to (the local numbering of the nodes is in fact arbitrary). The expressions for the shape functions are valid in both 2 $(d = 2)$ and 3 $(d = 3)$ dimensions.

For $r = 1$ ($\mathbb{P}^1$ or *linear* elements)

$$\widehat{\phi}_n = \xi_i, \quad 1 \leq i \leq d+1 \quad \text{and} \quad n = 1, \ldots, d+1.$$

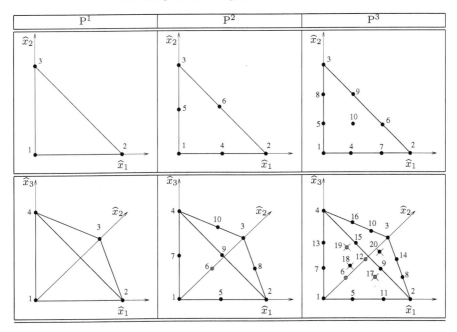

Fig. 2.9 The main P^r reference finite elements with the local numbering of the nodes in 2D (top) and 3D (bottom). To ease the reading of the image corresponding to the tetrahedron P^3, the nodes located at the center of a face have been marked with a cross

For $r = 2$ (P^2 or *quadratic* elements)

$$
\begin{aligned}
\widehat{\phi}_n &= \xi_i(2\xi_i - 1), \, 1 \leq i \leq d+1 \quad \text{and } n = 1, \ldots, d+1, \\
\widehat{\phi}_n &= 4\xi_i\xi_j, \qquad\qquad 1 \leq i < j \leq d+1 \quad \text{and } n = d+2, \ldots, n_l.
\end{aligned}
\tag{2.23}
$$

For $r = 3$ (P^3 or *cubic* elements)

$$
\begin{aligned}
\widehat{\phi}_n &= \tfrac{1}{2}\xi_i(3\xi_i - 1)(3\xi_i - 2), \, 1 \leq i \leq d+1 \quad \text{and } n = 1, \ldots, d+1, \\
\widehat{\phi}_n &= \tfrac{9}{2}\xi_i(3\xi_i - 1)\xi_j, \qquad\qquad 1 \leq i < j \leq d+1 \quad \text{and } n = d+2, \ldots, 3d+1, \\
\widehat{\phi}_n &= \tfrac{9}{2}\xi_j(3\xi_j - 1)\xi_i, \qquad\qquad 1 \leq i < j \leq d+1 \quad \text{and } n = 3d+2, \ldots, 5d+1, \\
\widehat{\phi}_n &= 27\xi_i\xi_j\xi_k, \qquad\qquad\qquad 1 \leq i < j < k \leq d+1 \quad \text{and } n = 5d+2, \ldots, n_l.
\end{aligned}
$$

Notice that a C^0 global approximating space requires certain symmetries in the position of the nodes. For instance, nodes interior to a side should be symmetric with respect to the midpoint of the side. Nodes interior to a face should be symmetric with respect to the median lines of the face. Indeed, if the symmetry is not satisfied it would be in general impossible to build a continuous function by assembling the local shape functions defined on each element of an arbitrary grid.

> **Exercise 2.2.2.** Verify that the functions defined in (2.23) are indeed the shape functions for P² Lagrangian finite elements in two and three dimensions by noting that $\widehat{\phi}_i(\mathbf{N}_j) = \delta_{ij}$, where the $\widehat{\mathbf{N}}_j$ are the nodes illustrated in Fig. 2.9.

Solution 2.2.2. Let us consider the two-dimensional case first. The number of local degrees of freedom is in this case $n_l = 6$ and the nodes are either on the vertexes or in the midpoint of a side. Recalling relations (2.15), we can thus determine the barycentric coordinates of the nodes as listed below:

i	ξ_1	ξ_2	ξ_3	i	ξ_1	ξ_2	ξ_3
1	1	0	0	4	1/2	1/2	0
2	0	1	0	5	1/2	0	1/2
3	0	0	1	6	0	1/2	1/2

Let us consider the first three shape functions associated to the vertexes of the triangle, namely $\widehat{\phi}_i = \xi_i(2\xi_i - 1)$, $i = 1, 2, 3$. Condition $\widehat{\phi}_i = 0$ is satisfied if $\xi_i = 0$ or $\xi_i = 1/2$. If we look at the table we realize that this is true at all nodes beside $\widehat{\mathbf{N}}_i$. Furthermore, in this node $\xi_i = 1$, so $\phi_i(\widehat{\mathbf{N}}_i) = 1(2 - 1) = 1$. The other shape functions are $\phi_4 = 4\xi_1\xi_2$, $\phi_5 = 4\xi_1\xi_3$ e $\phi_6 = 4\xi_2\xi_3$. They are clearly zero in the first three nodes, and by the help of the given table it can be verified that they are zero in the other nodes as well, a part the node with the same numbering, where they take the value 1.

The three-dimensional case is analogous, the barycentric coordinates of the nodes are, referring again to Fig. 2.9,

i	ξ_1	ξ_2	ξ_3	ξ_4	i	ξ_1	ξ_2	ξ_3	ξ_4
1	1	0	0	0	6	1/2	0	1/2	0
2	0	1	0	0	7	1/2	0	0	1/2
3	0	0	1	0	8	0	1/2	1/2	0
4	0	0	0	1	9	0	1/2	0	1/2
5	1/2	1/2	0	0	10	0	0	1/2	1/2

We can now repeat the same considerations made for the two-dimensional case. ◇

If $\widehat{K}$ is the unit square (cube in 3D), a Lagrangian finite element can be built using as shape functions the tensorial product of the corresponding one dimensional shape functions. Then, $\widehat{P}(\widehat{K})$ is equal to $\mathbb{Q}^r(\widehat{K})$, the space of polynomials where the maximal exponent for each variables $\widehat{x}_s$, $s = 1, \ldots, d$, is r. For instance, $x_1^2 x_2^2 \in \mathbb{Q}^2(\widehat{K})$, while $x_1^3 \notin \mathbb{Q}^2(\widehat{K})$. We can note that $\mathbb{P}^r(\widehat{K}) \subset \mathbb{Q}^r(\widehat{K}) \subset \mathbb{P}^{dr}(\widehat{K})$.

In the case of quadrilaterals, the use of an affine map to generate the current element K is very limiting. Indeed K will necessarily have parallel sides (faces). Therefore, we need more general maps.

2.2.3 Parametric Finite Elements

If $\mathbf{N}_i$ e $\widehat{\phi}_i$ are the nodes and the corresponding finite element shape functions on the reference element, the function

$$T_K(\widehat{\mathbf{x}}) = \sum_{s=1}^{n_l} \mathbf{N}_i \widehat{\phi}_i(\widehat{\mathbf{x}}), \qquad (2.24)$$

when it is invertible defines a one-to-one map $\widehat{K} \to K$. Indeed if we consider $\mathbb{P}^1$ shape functions T_K is the affine map.

This technique is very practical and it allows to use the same shape functions adopted to discretize the differential problem for the discretization of the domain. In general, a finite element where the transformation from the reference element is given by (2.24) is called *parametric* finite element. In particular, it is of geometric degree m if $\widehat{\phi}_i \in \mathbb{P}^m(\widehat{K})$ in the case of simplex elements (then $T_K \in [\mathbb{P}^m(\widehat{K})]^d$), or if $\widehat{\phi}_i \in \mathbb{Q}^m(\widehat{K})$ in the case of quadrilateral (cubic) elements, (in which case $T_K \in [\mathbb{Q}^m(\widehat{K})]^d$).

We can relate the physical coordinates to the reference coordinates as follows,

$$\mathbf{x} = \sum_{i=1}^{n_l} \mathbf{N}_i \widehat{\phi}_i(\widehat{\mathbf{x}}), \quad \widehat{\mathbf{x}} \in \widehat{K}.$$

The invertibility of the map may require some conditions on the shape of K, as illustrated in Exercise 2.2.6. The map $T_K \in [\mathbb{Q}^1(\widehat{K})]^d$ is called *bilinear* if $d = 2$, and *trilinear* if $d = 3$. We have an *isoparametric finite element* if we choose $m = r$, that is if we use the same shape functions for the map and for the finite element itself. If instead $m < r$ we have a *subparametric finite element*. The case $m > r$ (*superparametric finite elements*) is less common. Indeed, the most common type of finite elements are either affine (or bilinear/trilinear in the case of quadrilateral/cubic elements) or isoparametric. The use of parametric elements with $m > 1$ allows a better approximation of domains with curved boundaries.

The $\mathbb{Q}^r$ finite element (with $r > 0$) is the quadrilateral (or cubic) Lagrangian finite element whose shape functions in the reference element belong to $\mathbb{Q}^r$, i.e. $\widehat{P}(\widehat{K}) = \mathbb{Q}^r(\widehat{K})$, while the map to the current element is bilinear (trilinear), i.e. $T_K \in [\mathbb{Q}^1(\widehat{K})]^d$. The corresponding finite element space is then

$$h^r = \{v_h \in C^0(\overline{\Omega}_h) : \quad v_h|_K = \widehat{v}_h \circ T_K^{-1}; \widehat{v}_h \in \mathbb{Q}^r(\widehat{K}), \quad \forall K \in \mathcal{T}_h\}.$$

The shape functions on $\widehat{K}$ are the tensor product of the one dimensional Lagrange characteristic piecewise polynomials $\widehat{\phi}_i^1$, for $i = 0, \ldots, r$. If ξ_i, $i = 0, \ldots, r$ indicate the coordinates of the nodes, which are evenly-distributed in the interval $[0, 1]$, a generic shape function in $\mathbb{Q}^r(\widehat{K})$ takes the form $\widehat{\phi}_i^1(x)\widehat{\phi}_j^1(y)$ for $d = 2$ and $\widehat{\phi}_i^1(x)\widehat{\phi}_j^1(y)\widehat{\phi}_k^1(z)$ for $d = 3$, with $0 \leq i, j, k \leq r$.

We note that in this case $v_h|_K$ is *not* in general a polynomial, since T_K is not an affine map.

Exercise 2.2.3. The number of local degrees of freedom of a $\mathbb{P}^r$ element is given by $n_l = \frac{1}{d!} \prod_{j=1}^{d}(r+j) = \binom{r+d}{d} = \binom{r+d}{r}$, while for elements $\mathbb{Q}^r$ we have $n_l = (r+1)^d$. Verify this statement for $d = 2$ and $d = 3$.

Recall that $\sum_{i=1}^{n} i = n(n+1)/2$ and that $\sum_{i=1}^{n} i^2 = n(n+1)(2n+1)/6$.

Solution 2.2.3. For a $\mathbb{P}^r$ element we have $P(K) = \mathbb{P}^r(K)$. Then the number of degrees of freedom equals the dimension of $\mathbb{P}^r(K)$, that is the number of independent monomials in a polynomial of degree r. Let us consider the case of $d = 2$. A polynomial of degree r is formed by monomials of degree at most r of the form $x_1^i x_2^j$, with $i \geq 0$, $j \geq 0$, $i + j = s$ and $s = 0, \ldots, r$. The number of monomials of degree s is $s + 1$. Indeed for each i between 0 and s we have only a possible value of j, that is $j = s - i$. Consequently the total number of monomials of degree at most r is given by

$$\sum_{s=0}^{r}(s+1) = \sum_{s=1}^{r+1} s = (r+1)(r+2)/2.$$

In the three-dimensional case the monomials of degree s take the form $x_1^i x_2^j x_3^k$, with $i + j + k = s$. Therefore, in correspondence to the index i varying between 0 and s, the indexes j e k have to satisfy the constraint $j + k = s - i$. We can now use the result found for the two-dimensional case to deduce that for each i we have $s - i + 1$ possible combinations of j and k. Consequently the number of monomials of degree s is

$$\sum_{i=0}^{s}(s+1-i) = \sum_{i=0}^{s}(s+1) - \sum_{i=0}^{s}i = (s+1)^2 - s(s+1)/2 = \frac{1}{2}(s^2 + 3s + 2).$$

Finally, the total number of monomials of degree at most r is given by $\frac{1}{2}\sum_{s=0}^{r} s^2 + 3s + 2$. We obtain the desired result by applying the formulas suggested in the text.

In the case of $\mathbb{Q}^r$ finite elements the dimension of the local finite element space derives directly from the definition of $\mathbb{Q}^r(\widehat{K})$. Indeed the dimension of the corresponding one-dimensional finite element space is $r + 1$, and the function of $\mathbb{Q}^r$ are computed from their tensor product, Thus $n_l = (r+1)^d$.   ◇

Exercise 2.2.4. Show that the dimension of the finite element space X_h^r defined in (2.22) is given by the following table

d	$r=1$	$r=2$	$r=3$
2	N_v	$N_v + N_s$	$N_v + 2N_s + N_e$
3	N_v	$N_v + N_s$	$N_v + 2N_s + N_f$

Solution 2.2.4. We have seen that a function $p \in X_h^r$ satisfies the condition $p|_K = \widehat{p} \circ T_K^{-1}$, with $\widehat{p} \in \mathbb{P}^r$. Being T_K continuous and invertible with a continuous inverse, the continuity of $\widehat{p}$ implies the continuity of p in the interior of each element K.

Since a function in X_h^r is globally continuous we need now to ensure the continuity at the interface between elements. Let us consider the two-dimensional case first and two elements K_1 and K_2 which intersect each other. Since the grid is geometrically conforming, their intersection is either a vertex or a whole side. Let $\mathbf{V}$ then be a common vertex. A generic function $p \in X_h^r$ is continuous only if $p|_{K1}(\mathbf{V}) = p|_{K2}(\mathbf{V})$. Since this must be true for all functions in X_h^r, $\mathbf{V}$ must be a node, and one of the global degrees of freedom will correspond to the value of p at $\mathbf{V}$. As this consideration can be repeated for any couple of elements sharing a vertex, we conclude that the mesh vertexes must be nodes.

Let us now assume that K_1 and K_2 share a whole side Γ_{12}. Continuity of p implies $p|_{K_1}(\mathbf{x}) = p|_{K_2}(\mathbf{x})$ for $\mathbf{x} \in \Gamma_{12}$. This is possible for a generic function p in X_h^r only if the restriction of p on Γ_{12} is uniquely determined by the nodes on Γ_{12}. The restriction is a one-dimensional polynomial still of degree at most r, thus we need to have $r+1$ nodes on the side. Since we have already found that vertexes are nodes, we need additional $r-1$ nodes at the interior of the side.

Therefore, on the reference element a total of $3 + 3(r-1) = 3r$ are located on the vertexes and on the interior of the sides. From Exercise 2.2.3 we know that the space $\mathbb{P}^r(\widehat{K})$ has dimension $(r+1)(r+2)/2$, therefore the possible $(r+1)(r+2)/2 - 3r$ remaining nodes should be placed at the interior of the element. Then, for $r=1$ the 3 nodes will be placed at the vertexes, consequently the dimension of X_h^1 is equal to the total number of vertexes in the grid N_v. For $r=2$, we have that $n_l = 6$ and we need an additional node in the midpoint of each side. Consequently, $N_h = N_v + N_s$. For $r=3$, we need two nodes at the interior of each side, and the additional node required to reach $n_l = 10$ is placed in the middle of the element, consequently $N_h = N_v + 2N_s + N_e$.

The result for the three-dimensional case is obtained following the same considerations. In this case, we need to consider also the continuity across the element faces.

As explained in Exercise 2.2.5, the conditions here described are only necessary to reach C^0-conformity. We have indeed some additional constraints on the position of the nodes on $\partial \widehat{K}$ to ensure the proper "gluing" of the nodes of adjacent grid elements. ◇

Exercise 2.2.5. Let K_1 and K_2 be two triangles of a grid $\mathcal{T}_h$ which share a common side $e = K_1 \cap K_2$. They are obtained by applying to the reference element $\widehat{K}$ the affine maps T_{K_1} and T_{K_2}, respectively. Show that there exists a numbering of the nodes such that $e = T_{K_1}(\widehat{e}) = T_{K_2}(\widehat{e})$, $\widehat{e}$ being a side of $\widehat{K}$.
Furthermore, if $\widehat{\mathbf{x}}_1, \widehat{\mathbf{x}}_2 \in \widehat{e}$ are placed symmetrically with respect to the midpoint of $\widehat{e}$, then $T_{K_1}(\widehat{\mathbf{x}}_1) = T_{K_2}(\widehat{\mathbf{x}}_2) = \mathbf{x} \in e$. Deduce that to have C^0-conforming P^r finite elements we need the nodes on the sides of $\widehat{K}$ to be symmetric with respect to the midpoint of the side.

Solution 2.2.5. Let us refer to Fig. 2.10. Since the maps T_{K_1} and T_{K_2} are continuous we have that $T_{K_1}(\partial \widehat{K}) = \partial K_1$ and $T_{K_2}(\partial \widehat{K}) = \partial K_2$. Furthermore, the vertexes of K_1 and K_2 are the images of the vertexes of $\widehat{K}$, by the corresponding maps. We will identify with $\widehat{\mathbf{V}}_j$, $j = 1, 2, 3$, the vertexes of the reference element $\widehat{K}$, and with $\widehat{\mathbf{V}}_j^{K_i}$, $j = 1, 2, 3$ the vertexes of K_i, for $i = 1, 2$. Let $\mathbf{A}$ and $\mathbf{B}$ be the vertexes of the common side e. Obviously, there exists a numbering of the vertexes of K_1 such that $T_{K_1}(\widehat{\mathbf{V}}_1) = \mathbf{V}_1^{K_1} = \mathbf{A}$. Since the map preserves the orientation of the element we have necessarily that $T_{K_1}(\widehat{\mathbf{V}}_2) = \mathbf{V}_2^{K_1} = \mathbf{B}$. Indeed, the vertexes $\mathbf{V}_j^{K_1}$ have an anticlockwise orientation when j runs from 1 to 3, and the same must be true for their counter-images on $\widehat{K}$. For what concerns K_2, if we set $T_{K_2}(\widehat{\mathbf{V}}_2) = \mathbf{V}_2^{K_2} = \mathbf{A}$, then $T_{K_2}(\widehat{\mathbf{V}}_1) = \mathbf{V}_1^{K_2} = \mathbf{B}$. In fact, setting $T_{K_2}(\widehat{\mathbf{V}}_1) = \mathbf{A}$ and $T_{K_2}(\widehat{\mathbf{V}}_2) = \mathbf{B}$ would be incompatible with the requirement of having an anticlockwise orientation of the vertexes of K_2.

If $\widehat{e}$ indicates the side of $\widehat{K}$ of vertexes $\widehat{\mathbf{V}}_1$ and $\widehat{\mathbf{V}}_2$, we can conclude that $T_{K_1}(\widehat{e})$ and $T_{K_2}(\widehat{e})$ coincide with side e, but with opposite orientation.

Using the definition of affine map (2.9), and exploiting the fact that the vertex $\widehat{\mathbf{V}}_1$ has coordinates[6] $(0,0)$ and that $T_{K_1}(\widehat{\mathbf{V}}_1) = \mathbf{A}$ and $T_{K_2}(\widehat{\mathbf{V}}_1) = \mathbf{B}$ we deduce that

$$T_{K_1}(\widehat{\mathbf{x}}) = \mathbf{A} + \mathrm{F}_{K_1}\widehat{\mathbf{x}}, \quad T_{K_2}(\widehat{\mathbf{x}}) = \mathbf{B} + \mathrm{F}_{K_2}\widehat{\mathbf{x}}.$$

Imposing that $T_{K_1}(\widehat{\mathbf{V}}_2) = \mathbf{B}$ and $T_{K_2}(\widehat{\mathbf{V}}_2) = \mathbf{A}$ we finally obtain that

$$\mathrm{F}_{K_1}\widehat{\mathbf{V}}_2 = \mathbf{B} - \mathbf{A}, \quad \mathrm{F}_{K_2}\widehat{\mathbf{V}}_2 = \mathbf{A} - \mathbf{B}.$$

---

[6] Similar considerations may however been made for any choice for the vertex $\widehat{\mathbf{V}}_1$.

In particular, $F_{K_1}\widehat{\mathbf{V}}_2 = -F_{K_2}\widehat{\mathbf{V}}_2$. Let us consider two points of side $\widehat{e}$, which we will indicate by $\widehat{\mathbf{x}}_1$ and $\widehat{\mathbf{x}}_2$, respectively, placed symmetrically with respect to the mid point of the side. Consequently, it exists $0 \le \alpha \le 1$ such that $\widehat{\mathbf{x}}_1 = \alpha(\widehat{\mathbf{V}}_2 - \widehat{\mathbf{V}}_1)$ and $\widehat{\mathbf{x}}_2 = (1 - \alpha)(\widehat{\mathbf{V}}_2 - \widehat{\mathbf{V}}_1)$. Thanks to the relations found previously and exploiting again the fact that $\widehat{\mathbf{V}}_1$ has coordinates $(0,0)$) we have

$$T_{K_1}(\widehat{\mathbf{x}}_1) = \mathbf{A} + \alpha F_{K_1}\widehat{\mathbf{V}}_2 = \mathbf{A} - \alpha F_{K_2}\widehat{\mathbf{V}}_2 = \mathbf{B} + (\mathbf{A} - \mathbf{B}) - \alpha F_{K_2}\widehat{\mathbf{V}}_2 =$$
$$\mathbf{B} + (1 - \alpha)F_{K_2}\widehat{\mathbf{V}}_2 = T_{K_2}[(1 - \alpha)\widehat{\mathbf{V}}_2] = T_{K_2}(\widehat{\mathbf{x}}_2).$$

As X_h is C^0-conforming we must have $v_h|_{K_1}(\mathbf{x}) = v_h|_{K_2}(\mathbf{x})$, for all $\mathbf{x} \in e$. This is clearly possible only if the images of the local nodes $\widehat{\mathbf{N}}_i \in \widehat{e}$ obtained using the maps T_{K_1} and T_{K_2}, respectively, contain the same points (which are, in fact, the nodes $\mathbf{N}_i$). The relations just found lead us to the conclusion that this is possible only if the local nodes are placed symmetrically with respect to the mid point of the side $\widehat{e}$. If the number of nodes on the side is odd, one of the nodes is necessarily given by the mid-point. Fig. 2.11 illustrates this fact graphically.   ◇

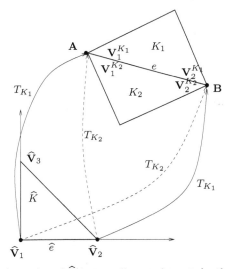

Fig. 2.10 The transformation of $\widehat{K}$ in two adjacent elements by the affine maps T_{K_1} and T_{K_2}

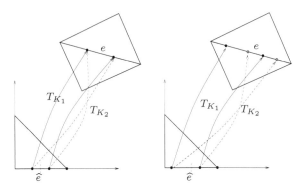

Fig. 2.11 The images of two points $\widehat{e}$ placed symmetrically with respect to the mid-point and obtained by the affine maps associated to the two elements sharing a side (the image of $\widehat{e}$) are formed by the same set of points (left picture). If the two points are not symmetric, they are instead mapped on two distinct couples of points (right picture)

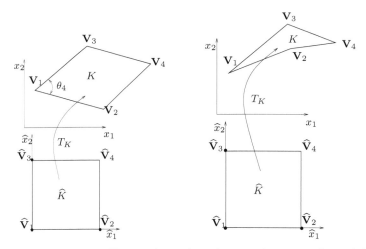

Fig. 2.12 The bilinear map T_K transforms the unit square into a generic quadrilateral. The right picture illustrates the case of a degenerate map

Exercise 2.2.6. In Fig. 2.12 we show the action of the bilinear map $T_K : \widehat{K} \to K$ associated to a Q^1 quadrilateral element. Show that K has necessarily straight sides and discuss the conditions that should be satisfied to have a one-to-one map.

Trace the isolines at constant $\widehat{x}_1$ and $\widehat{x}_2$ on the Q_1 quadrilaterals K_1 e K_2 defined by the following vertexes: for K_1: $(0,0)$, $(1.5,0)$, $(0.1,1.4)$ and $(1.6,1.8)$, for K_2: $(1.3,1.1)$, $(1.5,0)$, $(0.1,1.4)$, $(1.6,1.8)$. The spacing between isolines should be of 0.02. To this purpose, use program 3 `Q1Trasf.m`. Comment the result.

Solution 2.2.6. In the two-dimensional case, the shape functions of a Q^1 element are given by

$$\widehat{\phi}_1(\widehat{x}_1, \widehat{x}_2) = (1 - \widehat{x}_1)(1 - \widehat{x}_2), \quad \widehat{\phi}_2(\widehat{x}_1, \widehat{x}_2) = \widehat{x}_1(1 - \widehat{x}_2),$$
$$\widehat{\phi}_3(\widehat{x}_1, \widehat{x}_2) = (1 - \widehat{x}_1)\widehat{x}_2, \quad \widehat{\phi}_4(\widehat{x}_1, \widehat{x}_2) = \widehat{x}_1\widehat{x}_2.$$

Setting $\mathbf{l}_{ij} = \mathbf{V}_j - \mathbf{V}_i$, using the definition of bilinear map and the fact that $\sum_{i=1}^{n_l} \widehat{\phi}_i = 1$, we may write

$$T_K(\widehat{x}_1, \widehat{x}_2) = \sum_{i=1}^{4} \widehat{\phi}_i(\widehat{x}_1, \widehat{x}_2)\mathbf{V}_i = [1 - \sum_{i=2}^{4} \widehat{\phi}_i(\widehat{x}_1, \widehat{x}_2)]\mathbf{V}_1 + \sum_{i=2}^{4} \widehat{\phi}_i(\widehat{x}_1, \widehat{x}_2)\mathbf{V}_i$$
$$= \mathbf{V}_1 + \widehat{\phi}_2(\widehat{x}_1, \widehat{x}_2)\mathbf{l}_{12} + \widehat{\phi}_3(\widehat{x}_1, \widehat{x}_2)\mathbf{l}_{13} + \widehat{\phi}_4(\widehat{x}_1, \widehat{x}_2)\mathbf{l}_{14}.$$

Let us consider the restriction of T_K on a side of $\widehat{K}$. For simplicity, and without loss of generality, we consider here just side $\widehat{l}_{12} = \{\widehat{x}_2 = 0, \ 0 \leq \widehat{x}_1 \leq 1\}$ and we set $T_K^{12} = T_K|_{\widehat{l}_{12}}$.

We have that

$$T_K^{12}(\widehat{x}_1) = T_K(\widehat{x}_1, 0) = \mathbf{V}_1 + \widehat{\phi}_2(\widehat{x}_1, 0)\mathbf{l}_{12} + \widehat{\phi}_3(\widehat{x}_1, 0)\mathbf{l}_{13} + \widehat{\phi}_4(\widehat{x}_1, 0)\mathbf{l}_{14}.$$

The definition of the shape functions $\widehat{\phi}_i$ gives that $\widehat{\phi}_3(\widehat{x}_1, 0) = \widehat{\phi}_4(\widehat{x}_1, 0) = 0$. Consequently, $T_K^{12}(\widehat{x}_1) = T_K(\widehat{x}_1, 0) = \mathbf{V}_1 + \widehat{\phi}_2(\widehat{x}_1, 0)\mathbf{l}_{12} = \mathbf{V}_1 + \widehat{x}_1\mathbf{l}_{12}$, which is indeed the equation of the line passing through $\mathbf{V}_1$ and oriented along $\mathbf{l}_{12}$. Moreover, $T_K^{12}(0) = \mathbf{V}_1$ and $T_K^{12}(1) = \mathbf{V}_2$. Consequently the image $\mathbf{x} = T_K^{12}(\widehat{x}_1)$ of any point $(\widehat{x}_1, 0) \in \widehat{l}_{12}$ is on the straight segment defined by $\mathbf{V}_1$ and $\mathbf{V}_2$.

However, T_K is not an affine map in the interior of $\widehat{K}$. In the expression of T_K we have the presence of monomials of the form $\widehat{x}_1\widehat{x}_2$ (or $\widehat{x}_1\widehat{x}_2\widehat{x}_3$ in 3D), originating from the tensor products of one-dimensional linear polynomials. Then, the Jacobian matrix $J(T_K)$ is not constant in the element and we must verify the condition $|J(T_K)| > 0$ at all points interior to $\widehat{K}$ (we remind that $|A|$ here indicates the determinant of matrix A).

If $T_{K,1}$ and $T_{K,2}$ indicate the two components of the vector T_K, we have (by definition) that

$$|J(T_K)| = \begin{vmatrix} \dfrac{\partial T_{K,1}}{\partial \widehat{x}_1} & \dfrac{\partial T_{K,1}}{\partial \widehat{x}_2} \\ \dfrac{\partial T_{K,2}}{\partial \widehat{x}_1} & \dfrac{\partial T_{K,2}}{\partial \widehat{x}_2} \end{vmatrix}$$

or, equivalently, $|J(T_K)| = \dfrac{\partial T_K}{\partial \widehat{x}_1} \times \dfrac{\partial T_K}{\partial \widehat{x}_2}$, where $\times$ indicates the vector product in two dimensions. Furthermore, we can verify that the previous expressions lead to

$$\frac{\partial T_K}{\partial x_1} = (1 - \widehat{x}_2)\mathbf{l}_{12} + \widehat{x}_2\mathbf{l}_{34}, \quad \frac{\partial T_K}{\partial x_2} = (1 - \widehat{x}_1)\mathbf{l}_{13} + \widehat{x}_1\mathbf{l}_{24}.$$

Setting $l_{ij}^{kl} = \mathbf{l}_{ij} \times \mathbf{l}_{kl}$, we may write $|J(T_K)| = \widehat{\phi}_1 l_{12}^{13} + \widehat{\phi}_2 l_{12}^{24} + \widehat{\phi}_3 l_{34}^{13} + \widehat{\phi}_4 l_{34}^{24}$.

The shape functions $\widehat{\phi}_j$, $j = 1, \ldots, 4$ are positive in the interior of $\widehat{K}$, since they are product of functions that are positive in the interval $(0, 1)$ (note that this is true only for bilinear maps). Moreover they sum to 1. Consequently $|J(T_K)|$ satisfy

$$\min(l_{ij}^{kl}) \leq |J(T_K)| \leq \max(l_{ij}^{kl}), \tag{2.25}$$

on $\widehat{K}$. Moreover, the inequalities are sharp, i.e. for each of the two inequalities there is a point in K where the equality holds.

Therefore, a sufficient condition to have $|J(T_K)| > 0$ is that all products l_{ij}^{kl} are positive. As the angle θ between two vectors $\mathbf{v}$ e $\mathbf{w}$ of $\mathbb{R}^2$ satisfies the equation $\mathbf{v} \times \mathbf{w} = |\mathbf{v}||\mathbf{w}| \sin(\theta)$, we infer that the positivity condition is satisfied when the quadrilateral K has all interior angles *smaller than* π. Conversely, if an interior angle greater that π the map is not invertible (we normally say that it is degenerate). Indeed in this case from (2.25) and the continuity of $J(T_K)$ we may deduce $|J(T_K)| < 0$ in an open set contained in K and that there are points in the interior of K where $|J(T_K)| = 0$. The case of an interior angle equal to π is borderline, the map is continuous in the interior of the element but not up to the border, and this is usually considered unacceptable.

In the picture on the right of Fig. 2.12 we show an example of degenerate case where the map Q_1 is not one-to-one.

Program 3 - Q1trasf : It applies the transformation $Q1$ associated to a given quadrilateral to a set of points of $\mathbb{R}^2$

```
function [x1,x2]=Q1trasf(K,xh1,xh2)
%[x1,x2]=Q1trasf(K,xh1,xh2)
% It applies the map Q1 associated to a quadrilateral K to the set of points
% (xh1,xh2) . K is a matrix of dimension $4 x 2$ containing
% in each row the coordinates of the vertexes of the quadrilateral,
% with the following convention for the numbering of the vertexes:
%        3              4
%        .              .
%        1              2
%
% xh1 and xh2 are column vectors containing the coordinates of the
% points in the reference plane. The vectors x1 e x2 contains the
% corresponding coordinates in the physical plane.
```

The MATLAB script `exercise_quad.m` (a copy of which may be found on the book site) contains the solution of the last part of the exercise. We comment here the main parts. First of all we define the elements according to the format required by program `Q1trasf`

```
K1=[0 0;1.5 0;0.1 1.4;1.6 1.8];K2=[1.3 1.1;1.5 0;0.1 1.4;1.6 1.8]
```

Then, using the MATLAB command meshgrid (see the MATLAB help to have more information) we define the matrices xh and yh containing the coordinates in the reference space of the isolines we want to visualize.

```
[xh,yh]=meshgrid(0:0.02:1,0:0.02:1);
```

The command [x1,x2]=Q1trasf(K1,xh,yh) computes the corresponding points in the physical plane. The MATLAB commands surf and view plot the desired isolines.

```
surf(x1,x2,ones(size(x1)));view([0,0,1]);
```

For the second quadrilateral,

```
[x1,x2]=Q1trasf(K2,xh,yh);
surf(x1,x2,ones(size(x1)));view([0,0,1]);
```

We obtain the pictures given in Fig. 2.13 (in the figure the plots have been slightly modified for editing purposes).

We can note that in the right picture, corresponding to element K_2, the isolines at constant $\widehat{x}_2$ intersect. This indicates that the map $T_{K_2} : \widehat{K} \to K_2$ is not one-to-one. Indeed each intersection is the image of two distinct points of the $(\widehat{x}_1, \widehat{x}_2)$ plane. $\diamond$

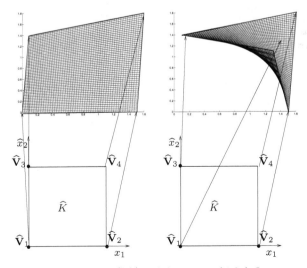

Fig. 2.13 The cases of a one-to-one (left) and degenerate (right) Q_1 map

2.2.4 Function approximation by finite elements

Having defined the finite element space $X_h \subset V(\Omega_h)$ conforming to the space $V(\Omega_h)$, it is now natural to introduce an *interpolation operator* Π_h,

$$\Pi_h : V(\Omega_h) \to X_h, \quad \Pi_h v(\mathbf{x}) = \sum_{i=0}^{N_h} \sigma_i(v)\phi_i(\mathbf{x}).$$

We can associate to Π_h its local counterpart $\Pi_K : V(K) \to P_K$ operating on a generic element $K \in \mathcal{T}_h$ as

$$\Pi_K v(\mathbf{x}) = \sum_{i=0}^{\widehat{N}} \sigma_i(v)\phi_i(\mathbf{x}), \quad \mathbf{x} \in K.$$

This operator is normally defined for functions in Ω_h. Its extension to functions in Ω is immediate whenever $\Omega_h \subseteq \Omega$. In the more general cases the extension is more critical and the interested reader may consult [Cia78].

The definition given here is very general. We now make it more precise by considering the case of Lagrangian finite elements of degree $r > 0$ and affine map T_K. This case includes the P^r elements and elements of type Q^r with parallel sides (faces). In this context, we may use as Π_h^r the interpolation operator that acts on functions $v \in C^0(\overline{\Omega_h})$ in the following way,

$$\Pi_h^r v(\mathbf{x}) = \sum_{i=1}^{N_h} v(\mathbf{N}_i)\phi_i(\mathbf{x}), \quad \mathbf{x} \in \overline{\Omega_h},$$

ϕ_i being the i-th shape function and $\mathbf{N}_i$ the corresponding node.

We may now state a fundamental result on the *interpolation error* on a single element: let K be a Lagrangian finite element of degree $r \geq 1$ whose geometry is defined by an affine map T_K. We consider a function $v \in H^{q+1}(K)$ for a $q > 0$, and set $l = \min(r, q)$.

Then, there exists a $C > 0$, independent from h_K, such that for $0 \leq m \leq l+1$

$$|v - \Pi_h^r(v)|_{H^m(K)} \leq C h_K^{l+1-m} \gamma_K^m |v|_{H^{l+1}(K)}. \tag{2.26}$$

Let us note that for $m > 0$ the error depends also on the sphericity γ_K of the given element. The result in fact holds in three dimensions.

To obtain an estimate of the global interpolation error we observe that X_h^r is H^1-conforming but it does not conform to H^m for $m > 1$. Consequently, it does not make sense to seek the $H^m(\Omega_h)$ norm of the interpolation error for $m > 1$. In the following we will then assume that m is either 0 or 1.

Under the same hypotheses stated for the local error, for a $v \in H^{q+1}(\Omega_h)$, with $q > 0$, we have

$$|v - \Pi_h^r(v)|_{H^m(\Omega_h)} \leq C \left[\sum_{K \in \mathcal{T}_h} h_K^{2(l+1-m)} \gamma_K^{2m} |v|_{H^{l+1}(K)}^2 \right]^{1/2} \leq$$

$$C\gamma_h^m h^{l+1-m} |v|_{H^{l+1}(\Omega_h)}, \quad (2.27)$$

where either $m = 0$ or $m = 1$, $l = \min(r, q)$ and $\gamma_h = \max_{K \in \mathcal{T}_h} \gamma_K$. If the mesh is regular $\gamma_K \leq \gamma$ for all h and then the inequality takes a form very similar to the one-dimensional case, namely

$$|v - \Pi_h^r(v)|_{H^m(\Omega_h)} \leq C h^{l+1-m} |v|_{H^{l+1}(\Omega_h)}. \quad (2.28)$$

These results can be extended to the more general case of isoparametric elements and one may also account for the errors linked to the discretization of the domain, see [QV94, EG04].

For the sake of completeness we recall the error estimates in the case of Q^r Lagrangian finite elements with $r \geq 1$ and for the two-dimensional case only. Clearly the grid should be such that the bilinear map T_K is well defined on each element (see Exercise 2.2.6). Let K indicate a generic quadrilateral element of $\mathcal{T}_h$ with vertexes $\mathbf{V}_j$, with $j = 1, \ldots, 4$. The triangle that is obtained by joining the three vertexes different from $\mathbf{V}_j$ is indicated by K_j. Clearly we have 4 of such triangles as $j = 1, \ldots, 4$. We set

$$\tilde{\rho}_K = \min_{1 \leq j \leq 4} \rho_{K_j} \quad \text{and} \quad \gamma_K = \frac{h_K}{\tilde{\rho}_K}.$$

The coefficient of regularity γ_K has a role similar to that of the sphericity for a simplex element. A grid of quadrilateral elements is *regular*[7] if $\gamma_k > \gamma$ for a $\gamma > 0$.

Let $\Pi_K^{Q^r}$ be the interpolation operator on element K. It exists a $C > 0$ such that, for $0 \leq m \leq r + 1$ and $\forall v \in H^{r+1}(K)$

$$|v - \Pi_K^{Q^r}|_{H^m(K)} \leq C\gamma_K^{\max(4m-1,1)} h^{k+1-1} |v|_{H^{r+1}(K)}. \quad (2.29)$$

It is important to notice that in this case the coefficient of regularity γ_K appears with an exponent higher than the analogous coefficient in the estimates for a simplex elements and affine map. Furthermore, here the coefficient of regularity influences also the L^2 estimate (where $m = 0$).

We may note that *on regular grids the order of convergence of Q^r finite elements is equal to that found for P^r elements, despite the fact that the polynomial space of Q^r elements is "richer" than that of its P^r counterpart.*

---

[7] As in the case of simplex elements, it is more appropriate to speak of a family of regular meshes.

2.3 The method of finite differences

The finite-difference method is one of the most popular discretization techniques and is particularly appealing for problems on simple geometries. The first step is to create a grid, whose vertexes make up the set of *nodes* of the discretization, on the computational domain where the differential equation must be solved. In correspondence of the nodes, the differential equation to be approximated is evaluated (*collocated*), and eventually each derivative in differential operators is approximated by suitable difference quotients. Here we propose exercises on the formulation and accuracy of the approximation of derivatives using difference quotients. In Chapters 3 and 4 we propose some boundary value problems to be solved with finite differences for elliptic problems. Finite difference time discretization of unsteady problems is considered in Chapters 5, 6 and 7.

2.3.1 Difference quotients in dimension one

We intend to discuss the problem of finding approximations for the derivatives of a function $u : \mathbb{R} \rightarrow \mathbb{R}$ with various accuracy. We will denote by u_i the value $u(x_i)$ where $x_i = ih$ ($i \in \mathbb{Z}$) is the ith discretization node, and $h > 0$ is the mesh size, also called grid spacing or discretization step. We consider here approximate derivatives over the entire $\mathbb{R}$, and in the next chapters the case of bounded intervals. To keep the notation simple we will set $D^p u_i \equiv d^p u / dx^p(x_i)$. When no confusion can arise we will write $D_i^p \equiv D^p u_i$. Furthermore, $\|\mathbf{u}\|_{\Delta,\infty} \equiv \max_i |u_i|$ will be the discrete maximum norm (see Chapter 1).

The numerical approximation of the derivative of a function f can be defined following two alternatives:

1. using suitable truncations of the Taylor expansion of f;
2. substituting f with a suitable interpolating polynomial Πf and approximating f' by the exact derivative of Πf.

The first strategy requires minimizing the truncation error by the *indeterminate coefficient* method (see Exercises 2.3.1–2.3.4). The foremost instance of the second strategy is furnished by the so-called *pseudo-spectral derivative* (cf. Exercise 2.3.5), where the derivative of a function is approximated by the derivative of Lagrange's interpolating polynomial on a distribution of Gauss-type interpolating nodes.

We use the notation $o(h^p)$ to indicate an infinitesimal with respect to h of order greater than p, i.e. if $f(h) = o(h^p)$ then $\lim_{h \to 0} \dfrac{f(h)}{h^p} = 0$. While, $O(h^p)$ indicates an infinitesimal of order at least p: if $f(h) = O(h^p)$ then there exist two constants $C \geq 0$ and $\alpha > 0$ such that $|f(h)| \leq C|h^p|$ whenever $|h| \leq \alpha$. If $C > 0$ then $f(h)$ is an infinitesimal with respect to h of the same order of h^p.

We recall that if $D(u_i, u_{i\pm 1}, u_{i\pm 2}, \ldots)$ is a formula providing an approximation of the pth derivative of a function u at point x_i, then the local truncation error is the functional $\tau_i(u; h) = d^p u(x_i)/dx^p - D(u(x_i), u(x_{i\pm 1}), u(x_{i\pm 2}), \ldots)$, and the truncation error is defined as $\tau(u; h) = \max_i |\tau_i(u; h)|$. A difference formula is *convergent* with order p if $\tau(u; h)$ is an infinitesimal of order p with respect to h for all u sufficiently smooth. The maximal p attainable under conditions on the regularity of u is the *order* of the scheme.

We recall as well the basic formulas for approximating first and second derivatives of a function u at a point $x_i = ih$, $h > 0$ being the constant spacing,

$$\frac{du}{dx}(x_i) \simeq \begin{cases} \dfrac{u_i - u_{i-1}}{h} & \text{(backward difference)}, \\[2mm] \dfrac{u_{i+1} - u_i}{h} & \text{(forward difference)}, \\[2mm] \dfrac{u_{i+1} - u_{i-1}}{2h} & \text{(centered difference)}, \end{cases}$$

and

$$\frac{d^2 u}{dx^2}(x_i) \simeq \frac{u_{i-1} - 2u_i + u_{i+1}}{h^2} \quad \text{(centered difference)}.$$

The first two formulas are of first order (provided u is continuously differentiable), while the centered formulas are both of second order, (provided $u \in C^2(\mathbb{R})$ and $u \in C^4(\mathbb{R})$, respectively).

For the theoretical backgrounds the reader should consult [QSS00, Ch. 9], and [Str04].

Pseudo-spectral differentiation

The pseudo-spectral derivative (see [Qua09, Ch. 10]) $\mathcal{D}_n f$ of a function f in the interval $[a, b]$ is defined as the derivative of the polynomial of degree n interpolating f at the Gauss points, either Gauss-Legendre-Lobatto (GLL) or Gauss-Chebyshev-Lobatto (GCL) points, $x_0, \ldots, x_n$. That is $\mathcal{D}_n f = D \Pi_n f$, being $\Pi_n f$ the interpolating polynomial. Polynomial $\Pi_n f$ may be expressed in terms of the Chebyshev-Lobatto (or Lagrange-Lobatto) characteristic polynomials ψ_i, $\Pi_n f(x) = \sum_{i=0}^n f_i \psi_i(x)$. The characteristic polynomials are polynomials of degree n which satisfy $\psi_i(x_j) = \delta_{ij}$.

The Gauss-Chebyshev-Lobatto nodes in the interval $[-1, 1]$ are $\hat{x}_j = -\cos(\pi j/n)$, $j = 0, \ldots, n$. The corresponding values in the interval $[a, b]$ are given by the affine transformation $x_j = (a+b)/2 + \hat{x}_j(b-a)/2$. We may wish to approximate $\mathcal{D}_n f$ itself by a polynomial interpolating at the same points: $\mathcal{D}_n f \simeq \sum_{i=0}^n d_i \psi_i$.

The vector $\mathbf{d}$ containing the value of the pseudo-spectral derivative at the Gauss points is then given by

$$\mathbf{d} = D_n \mathbf{f},$$

being $f_j = f(x_j)$ for $j = 0, \ldots, n$, and D_n the so called pseudo-spectral matrix of order n of components $D_{ij} = \psi_i'(x_j)$.

For the Gauss-Chebyshev-Lobatto points in the interval $[-1, 1]$ we have

$$
D_{lj} = \begin{cases}
\dfrac{c_l}{c_j} \dfrac{(-1)^{l+j}}{\hat{x}_l - \hat{x}_j}, & l \neq j, \\[3mm]
\dfrac{-\hat{x}_j}{2(1 - \hat{x}_j^2)}, & 1 \leq l = j \leq n-1, \\[3mm]
-\dfrac{2n^2 + 1}{6}, & l = j = 0, \\[3mm]
\dfrac{2n^2 + 1}{6}, & l = j = n,
\end{cases}
\tag{2.30}
$$

where $c_j = 1$ for $j = 1, \ldots, n-1$ and $c_0 = c_n = 2$.

Exercise 2.3.1. Justify the following approximations of the first derivative of u, and check they are accurate to order 2 and 4 respectively, with respect to the grid size h

$$
Du_i \simeq \frac{1}{2h}(-u_{i+2} + 4u_{i+1} - 3u_i),
$$

$$
Du_i \simeq \frac{1}{12h}(-u_{i+2} + 8u_{i+1} - 8u_{i-1} + u_{i-2}).
\tag{2.31}
$$

Check the results, as h varies, taking $u(x) = \sin(2\pi x)$ on the interval $(0, 1)$.

Solution 2.3.1.

Numerical approximation

Let us consider the following Taylor expansions, valid if $u \in C^5(\mathbb{R})$:

$$
u_{i\pm1} = u_i \pm hD_i + \frac{h^2}{2}D_i^2 \pm \frac{h^3}{6}D_i^3 + \frac{h^4}{24}D^4 u_i \pm \frac{h^5}{120}D_i^5 + o(h^5),
$$

$$
u_{i\pm2} = u_i \pm 2hD_i + 2h^2 D_i^2 \pm \frac{4h^3}{3}D_i^3 + \frac{2h^4}{3}D_i^4 \pm \frac{4h^5}{15}D_i^5 + o(h^5),
\tag{2.32}
$$

Formula $(2.31)_1$ is a consequence of combining the two of (2.32) considering only the terms up to the second order, for, if $u \in C^3$,

$$
-u_{i+2} + 4u_{i+1} - 3u_i = 2hDu_i - \frac{2}{3}h^3 \frac{d^3u}{dx^3}(\xi),
$$

where $\xi \in (x_i, x_i + 2h)$. Similarly, using again the Taylor expansion,

$$u_{i+2} - u_{i-2} = 4hDu_i + \frac{8}{3}h^3 D^3 u_i + o(h^5),$$

$$u_{i+1} - u_{i-1} = 2hDu_i + \frac{1}{3}h^3 D^3 u_i + o(h^5),$$

(2.33)

we recover $(2.31)_2$ because

$$u_{i-2} - u_{i+2} + 8(u_{i+1} - u_{i-1}) = 12hDu_i - \frac{2}{5}h^5 \frac{d^5 u}{dx^5}(\bar{\xi}),$$

where $\bar{\xi} \in (x_i - 2h, x_i + 2h)$. If u is sufficiently regular (it should be at least $C^3(\mathbb{R})$ and $C^5(\mathbb{R})$, respectively) the two methods are accurate to order 2 and 4 with respect to h, since the truncation errors equal $(h^2/3)\, d^3\, u(\xi)/dx^3$ and $(h^4/30)\, d^5\, u(\bar{\xi})/dx^5$ respectively.

As we have mentioned above, there is a different way for obtaining (2.31), based on the Lagrange interpolation The idea is to compute an interpolation polynomial of the available data of the function and then we differentiate it in the point where we want to approximate the derivative. The error analysis descends from the interpolation theory.

Let us consider the first of (2.31). Set $\xi = x - x_i$. We clearly we have

$$\frac{df}{dx}\Big|_{x=x_i} = \frac{df}{d\xi}\Big|_{\xi=0} \frac{d\xi}{dx}\Big|_{x=x_i} = \frac{df}{d\xi}\Big|_{\xi=0}.$$

We work therefore with the ξ coordinate using the nodes located in $0, h, 2h$, corresponding to x_i, x_{i+1}, x_{i+2} respectively. Let $p(\xi)$ be the polynomial such that

$$p(0) = u_i, \quad p(h) = u_{i+1}, \quad p(2h) = u_{i+2}.$$

(2.34)

If we write this polynomial in the usual form

$$p(\xi) = c_2 \xi^2 + c_1 \xi + c_0,$$

(2.35)

then

$$\frac{dp}{d\xi} = 2c_2 \xi + c_1 \Rightarrow \frac{dp}{d\xi}\Big|_{\xi=0} = c_1.$$

In this simple case, we can compute the polynomial directly following the definition, by forcing (2.34) to (2.35). We obtain the Vandermonde system [QSS00]

$$\begin{bmatrix} 0 & 0 & 1 \\ h^2 & h & 1 \\ 4h^2 & 2h & 1 \end{bmatrix} \begin{bmatrix} c_2 \\ c_1 \\ c_0 \end{bmatrix} = \begin{bmatrix} u_i \\ u_{i+1} \\ u_{i+2} \end{bmatrix}.$$

By solving this system, we get $c_1 = \frac{1}{2h}(-u_{i+2} + 4u_{i+1} - 3u_i)$, i.e. the first of (2.31).

For the second of (2.31), the Vandermonde approach is not convenient since we have to solve a symbolic 5×5 system. We build therefore the interpolation by means of Lagrange polynomials. Let $\xi = -2h, -h, 0, h, 2h$ be the nodes corresponding to the indexes $i-2, i-1, i, i+1, i+2$ respectively. We have

$$p(x) = \sum_{k=i-2}^{i+2} u_k \varphi_k(\xi)$$

where the Lagrange polynomial reads

$$\varphi_k(\xi) = \frac{\prod\limits_{\substack{j=i-2 \\ j\neq k}}^{i+2} (\xi - \xi_j)}{\prod\limits_{\substack{j=i-2 \\ j\neq k}}^{i+2} (\xi_k - \xi_j)} = \sum_{l=0}^{4} c_{4-l,k} \xi^{4-l}.$$

Here we have also introduced the coefficients $c_{j,k}$ of the usual power expansion of φ_k. Notice that $\dfrac{d\varphi_k}{d\xi}\big|_{\xi=0} = c_{1,k}$, so that

$$\frac{dp}{d\xi}\big|_{\xi=0} = \sum_{k=i-2}^{i+2} u_k \frac{d\varphi_k}{d\xi}\bigg|_{\xi=0} = \sum_{k=i-2}^{i+2} u_k c_{1,k}.$$

By direct inspection, we have that

$$c_{1,i-2} = \frac{1}{12h}, \quad c_{1,i-1} = -\frac{2}{3h}, \quad c_{1,i} = 0, \quad c_{1,i+1} = \frac{2}{3h}, \quad c_{1,i+2} = -\frac{1}{12h},$$

that leads to the second of (2.31).

Numerical results

As $u(x) = \sin(2\pi x)$ is periodic of period 1, the first derivative may be approximated using (2.31) on the whole interval $[0, 1]$, exploiting the periodicity.

Program 4 returns in the vector dfdxp the approximate values of the pth derivative of a periodic function, given in a string or the *inline* function fun, at nh+1 equidistant nodes on the interval (xspan(1),xspan(2)). The discretization nodes are given in the output vector x, assuming x(1) = xspan(1) and x(nh+1)=xspan(2). The approximation method of Program 4 is of the general form

$$Du_i \simeq \frac{1}{h^p} \sum_{k=i-N}^{i+N} c_k u_k,$$

where the values of the coefficients $\{c_k\}$ must be given in input in the vector coeff.

Program 4 - fddudx : Approximation by finite differences of the pth derivative of a periodic function

```
function [x,dfdxp]=fddudx(xspan,nh,coeff,p,fun,varargin)
%FDDUDX evaluates numerically the pth derivative of a periodic function
%   [X,DFDXP]=FDDUDX(XSPAN,NH,COEFF,P,FUN) approximates the Pth derivative
%    of the function FUN at NH+1 equidistant nodes contained in
%   the interval [XSPAN(1),XSPAN(2)] using a method of the form
%
%                   DFDX(X(I)) = (1/H^P)*SUM(COEFF.*FUNXI)
%
%   where FUNXI is the vector containing the values of  FUN at the nodes
%    X(I-(NC-1)/2:I+(NC+1)/2),  NC being an odd number equal to the length of
%    COEFF.
%
%   [X,DFDXP]=FDDUDX(XSPAN,NH,COEFF,P,FUN,P1,P2,...) passes the
%    additional parameters P1,P2, ... to the function FUN(X,P1,P2,...).
```

We use this program to check the properties of the methods discussed in the exercise. It will suffice to write the following instructions[8]

```
>> nh=10; fun=inline('sin(2*pi*x)');
for n=1:6
  [x,dfdx1]=fddudx([0,1],nh,coeff,1,fun);
  error(n) = norm(dfdx1-2*pi*cos(2*pi*x),'inf');
  nh=2*nh;
end
```

where the vector `coeff` equals $[0\ 0\ -3/2\ 2\ -1/2]$ by formula $(2.31)_1$, and $[1/12\ -2/3\ 0\ 2/3\ -1/12]$ by $(2.31)_2$. With the instruction[9]

```
>> q=log(error(1:end-1)./error(2:end))/log(2);
```

we can also give an estimate of the order of convergence q of the formula.

Table 2.1 shows the results obtained: as we halve h, the error is divided by 4 in the first case, by 16 in the second, which confirms the theoretical properties of the two formulas considered.

Table 2.1 Behavior of the error in discrete sup norm with respect to the grid's spacing h and estimate of the order of accuracy q for $(2.31)_1$ (rows E_1 and $q(E_1)$) and $(2.31)_2$ (rows E_2 and $q(E_2)$)

h	1/10	1/20	1/40	1/80	1/160	1/320
E_1	7.9e-01	2.1e-01	5.2e-02	1.3e-02	3.2e-03	8.1e-04
$q(E_1)$	–	1.95698	1.98930	1.99733	1.99933	1.99983
E_2	3.1e-02	2.0e-03	1.3e-04	8.0e-06	5.0e-07	3.1e-08
$q(E_2)$	–	3.94912	3.98729	3.99682	3.99921	3.99980

$\diamondsuit$

---

[8] The `inline(expr)` command declares the string `expr` as an inline function that can be evaluated in MATLAB.

[9] This expression is an approximation obtained by assuming that the error is Ch^p for a given (in general unknown) constant C.

Exercise 2.3.2. Consider the following approximations of the second derivative of a function u at $x_i = ih$, $i \in \mathbb{Z}$ ($h > 0$)

$$D^2 u_i \simeq d_0 u_{i+1} + d_1 u_i + d_2 u_{i-1},$$

$$D^2 u_i \simeq a_0 u_i + a_1 u_{i-1} + a_2 u_{i-2}, \tag{2.36}$$

$$D^2 u_i \simeq b_0 u_i + b_1 u_{i-1} + b_2 u_{i-2} + b_3 u_{i-3}.$$

Using the indeterminate coefficient method find the coefficient values that guarantee the greatest accuracy, under suitable regularity assumptions on u. Check their accuracy numerically by taking $u(x) = \sin(2\pi x)$, $x \in (0,1)$.

Solution 2.3.2.

Numerical approximation

If we assume $u \in C^3(\mathbb{R})$ the Taylor series expansion gives

$$u_{i\pm k} = u_i \pm khDu_i + \frac{(kh)^2}{2} D^2 u_i \pm \frac{(kh)^3}{6} D^3 u_i + o(h^3). \tag{2.37}$$

If we plug the previous expression in $(2.36)_1$ and rearrange the terms we obtain

$$D^2 u_i \simeq d_1 u_i + d_0 \left(u_i + hDu_i + \frac{h^2}{2} D^2 u_i + \frac{h^3}{6} D^3 u_i + o(h^3) \right)$$

$$+ d_2 \left(u_i - hDu_i + \frac{h^2}{2} D^2 u_i + \frac{h^3}{6} D^3 u_i + o(h^3) \right).$$

Let us gather the terms with the same order, we obtain

$$D^2 u_i \simeq (d_1 + d_1 + d_2) u_i + h (d_1 - d_2) Du_i + \frac{h^2}{2} (d_0 + d_2) D^2 u_i$$

$$\tag{2.38}$$

$$+ \frac{h^3}{6} (d_0 - d_2) D^3 u_i + (d_0 + d_2) o(h^3).$$

To approximate the second derivative we require the coefficients of u_i and Du_i to vanish, while the coefficient of $D^2 u_i$ in the left hand side should equal 1. In this way we come up to the linear system in the unknown d_i (non-singular, as one easily checks)

$$\begin{cases} d_0 + d_1 + d_2 = 0, \\ d_0 - d_2 = 0, \\ d_0 + d_2 = \dfrac{2}{h^2}, \end{cases}$$

by which we recover the well-known centered formula

$$D^2 u_i \simeq \frac{u_{i-1} - 2u_i + u_{i+1}}{h^2}. \tag{2.39}$$

As for the order, we may note that also the coefficient in front of the third derivative in (2.38) vanishes. Since the reminder is pre-multiplied by $d_0 + d_2 = \frac{2}{h^2}$, if u is just $C^3(\mathbb{R})$ we can only say that the convergence is more than linear. Yet, if $u \in C^4(\mathbb{R})$ then $o(h^3)$ is effectively replaced by $O(h^4)$ and the formula[10] has order 2.

We now deal with $(2.36)_2$. By a similar procedure we get

$$D^2 u_i \simeq a_0 u_i + a_1 \left(u_i - h D u_i + \frac{h^2}{2} D^2 u_i - \frac{h^3}{6} D^3 u_i + o(h^3) \right)$$

$$+ a_2 \left(u_i - 2h D u_i + 2h^2 D^2 u_i - \frac{4h^3}{3} D^3 u_i + o(h^3) \right)$$

and thus

$$D^2 u_i \simeq (a_0 + a_1 + a_2) u_i - h(a_1 + 2a_2) D u_i + \frac{h^2}{2} (a_1 + 4a_2) D^2 u_i$$

$$\tag{2.40}$$

$$- \frac{h^3}{6} (a_1 + 8a_2) D^3 u_i + o(h^3).$$

Again, to approximate the second derivative we require the coefficients of u_i and $D u_i$ to vanish and the coefficient of $D^2 u_i$ to equal 1, obtaining

$$\begin{cases} a_0 + a_1 + a_2 = 0, \\ a_1 + 2a_2 = 0, \\ a_1 + 4a_2 = \dfrac{2}{h^2}, \end{cases}$$

whose solution is $a_0 = a_2 = 1/h^2$, $a_1 = -2/h^2$. Neglecting higher order terms we obtain the following backward approximation of the second derivative:

$$D^2 u_i \simeq \frac{1}{h^2} (u_i - 2u_{i-1} + u_{i-2}). \tag{2.41}$$

The order of accuracy is 1 (provided $u \in C^3(\mathbb{R})$), since the coefficient of h^3 in (2.40), $a_1 + 8a_2$, is not zero for the values a_0, a_1, a_2 found above.

---

[10] If $f(h) = O(h^p)$ then $h^{-q} f(h) = O(h^{p-q})$ for all $q < p$.

Now to $(2.36)_3$. Similar computations yield

$$D^2 u_i \simeq b_0 u_i + b_1 \left(u_i - hDu_i + \frac{h^2}{2} D^2 u_i - \frac{h^3}{6} D^3 u_i + o(h^3) \right)$$

$$+ b_2 \left(u_i - 2hDu_i + 2h^2 D^2 u_i - \frac{4h^3}{3} D^3 u_i + o(h^3) \right) \qquad (2.42)$$

$$+ b_3 \left(u_i - 3hDu_i + \frac{9}{4} h^2 D^2 u_i - \frac{27h^3}{6} D^3 u_i + o(h^3) \right).$$

Since we have an extra coefficient to determine, we request that also the coefficient of $D^3 u_i$ in (2.42) vanishes. We obtain the following system

$$\begin{cases} b_0 + b_1 + b_2 + b_3 = 0, \\ b_1 + 2b_2 + 3b_3 = 0, \\ b_1 + 4b_2 + \dfrac{9}{2} b_3 = \dfrac{2}{h^2}, \\ b_1 + 8b_2 + 27b_3 = 0, \end{cases}$$

which furnishes the following values

$$b_0 = \frac{2}{h^2}, \, b_1 = -\frac{5}{h^2}, \, b_2 = \frac{4}{h^2}, \, b_3 = -\frac{1}{h^2}.$$

The corresponding approximation

$$D^2 u_i \simeq \frac{1}{h^2} (2u_i - 5u_{i-1} + 4u_{i-2} - u_{i-3}) \qquad (2.43)$$

is accurate to order 2 if $u \in C^4(\mathbb{R})$. The higher level of regularity requested on u in this case is due to the fact that to guarantee a second order scheme we need to have a remainder $O(h^4)$, and not just $o(h^3)$, in the Taylor's expansion.

The last two formulas derived in this exercise are not centered, indeed the derivative at node x_i depends only on backward values. This type of formulas is useful to evaluate derivatives at the ends of a bounded interval, since in this case we do not have all the data required to use a centered scheme.

Numerical results

To check the accuracy of the formulas with respect to h we use again Program 4, setting p=2 and assigning the scheme's coefficients in `coeff`. For the first scheme we set `coeff=[0 1 -2 1 0 ]`, for the second `coeff=[1 -2 1 0 0]` and `coeff=[-1 4 -5 2 0 0 0]` for the third. With the instructions

```
>> nh=10; u=inline('sin(2*pi*x)');
for n=1:6, [x,dfdx]=fddudx([0,1],nh,coeff,2,u);
  error(n) = norm(dfdx+4*pi^2*sin(2*pi*x),inf); nh=2*nh;
end, q=log(error(1:end-1)./error(2:end))/log(2);
```

we produce the results of Table 2.2, in agreement with the theoretical predictions of the schemes. ◇

Table 2.2 Behavior of the error in discrete infinity norm with respect to h, and order of accuracy q for formulas (2.39) (rows E_1, $q(E_1)$) (2.41) (rows E_2, $q(E_2)$) and (2.43) (rows E_3, $q(E_3)$))

h	1/10	1/20	1/40	1/80	1/160	1/320
E_1	1.2e+00	3.2e-01	8.1e-02	2.0e-02	5.1e-03	1.3e-03
$q(E_1)$	–	1.91337	1.99644	1.99911	1.99978	1.99994
E_2	2.3e+01	1.2e+01	6.2e+00	3.1e+00	1.6e+00	7.8e-01
$q(E_2)$	–	0.92760	0.98213	0.99555	0.99889	0.99972
E_3	1.3e+01	3.6e+00	8.9e-01	2.2e-01	5.6e-02	1.4e-02
$q(E_3)$	–	1.89634	1.98575	1.99644	1.99911	1.99978

Exercise 2.3.3. To approximate the fourth derivative of a function u, determine the coefficients in

$$D^4 u_i \simeq a_0 u_{i-2} + a_1 u_{i-1} + a_2 u_i + a_3 u_{i+1} + a_4 u_{i+2}, \qquad (2.44)$$

so to achieve the maximum order of accuracy (in norm $\| \cdot \|_{\Delta,\infty}$). Check the result using $u(x) = \sin(2\pi x)$ on the interval $(0, 1)$.

Solution 2.3.3.

Numerical approximation

The solution proceeds along the lines of the previous exercise, using a Taylor expansion up to the fourth order term. We finally obtain

$$\begin{cases} a_0 + a_1 + a_2 + a_3 + a_4 = 0, \\ 2a_0 + a_1 - a_3 - 2a_4 = 0, \\ 4a_0 + a_1 + a_3 + 4a_4 = 0, \\ 8a_0 + a_1 - a_3 - 8a_4 = 0, \\ 16a_0 + a_1 + a_3 + 16a_4 = \dfrac{24}{h^4}, \end{cases}$$

solved by $a_0 = a_4 = h^{-4}$, $a_1 = a_3 = -4h^{-4}$, $a_2 = 6h^{-4}$. The corresponding scheme

$$D^4 u_i \simeq \frac{1}{h^4}(u_{i-2} - 4u_{i-1} + 6u_i - 4u_{i+1} + u_{i+2}), \qquad (2.45)$$

has order 2 if $u \in C^6(\mathbb{R})$.

The order can be checked using Program 4 p=4, coeff = [1 -4 6 -4 1].
The outcome, in Table 2.3, essentially validates the expectation (order 2 for
(2.45)).

Table 2.3 Behavior of the error in discrete infinity norm for various grid sizes h and order
of accuracy q for formula (2.45) (rows E_4, $q(E_4)$)

h	1/10	1/20	1/40	1/80	1/160	1/320
E_4	9.5e+01	2.5e+01	6.4e+00	1.6e+00	4.0e-01	1.0e-01
$q(E_4)$	–	1.89568	1.99200	1.99800	1.99950	1.99993

We point out that a further decrease of h makes round-off errors appear that
degrade the accuracy of scheme (2.45), as shown in Fig. 2.14. Indeed, we
recall that all convergence results given here (a well as in the rest of the
book) assume "exact arithmetic". The actual error obtained on a computer,
however, depends also on the round-off error due to the finite representation
of the real numbers (in particular, MATLAB uses 8 bytes to represent real
numbers). This error may become important when h is very small. For a
thorough discussion on computer representation of real numbers and the
effects of round-off errors in numerical computations, the reader may consult
[Hig02].   ◇

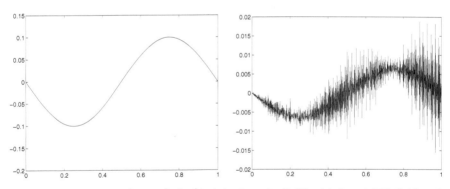

Fig. 2.14 The error $D^4 u_i - d^4 u/dx^4(x_i)$ for formula (2.45) with $h = 1/320$ (left) and
$h = 1/1280$ (right). Note the round-off errors in the second picture

Exercise 2.3.4. Let d_i indicate the approximation of the first derivative of u at $x_i = ih$ with $i \in \mathbb{Z}$ and $h > 0$. Compute these values using the formula

$$a_1 d_{i-1} + a_0 d_i + a_1 d_{i+1} = b_1 u_{i+1} + b_0 u_i - b_1 u_{i-1}, \qquad (2.46)$$

where the coefficients a_0, a_1, b_0, b_1 are chosen so that we obtain the maximal order of accuracy. Check the result by approximating the derivative of $f(x) = \sin(2\pi x)$ for $x \in (0,1)$.

Solution 2.3.4.

Numerical approximation

To find the coefficients that maximize the accuracy we compute the *truncation error* $\tau_i(u)$. This is the error made if we force the exact solution to obey the numerical scheme, i.e. if we set $d_i = Du(x_i) = Du_i$. In our case

$$\tau_i(u) = a_1(Du_{i-1} + Du_{i+1}) + a_0 Du_i - b_1(u_{i+1} - u_{i-1}) - b_0 u_i.$$

Recalling the expansions $(2.32)_1$ and

$$Du_{i\pm1} = Du_i \pm hD^2 u_i + \frac{h^2}{2} D^3 u_i \pm \frac{h^3}{6} D^4 u_i + \frac{h^4}{24} D^5 u_i + \frac{h^5}{120} D^6 u_i + o(h^5),$$

one finds

$$\tau_i(u) = -b_0 u_i + (2a_1 + a_0 - 2hb_1)Du_i + \left(a_1 h^2 - \frac{h^3}{3} b_1\right) D^3 u_i +$$
$$\left(a_1 \frac{h^4}{12} - b_1 \frac{h^5}{60}\right) D^5 u_i + o(h^5).$$

In order for the scheme to be consistent τ_i should tend to 0 as h goes to 0. So, we must impose $b_0 = 0$. To increase the accuracy we have to eliminate as many terms as possible in the error expansion. It is easy to realize that the linear system of unknown coefficients coming from annihilating all terms up to the fifth derivative is inconsistent (one of the equations would force $b_1 = 3a_1$, the other $b_1 = 5a_1$). So the best we can do is annihilating terms up to the third derivative, which gives the (under-determined) system

$$\begin{cases} 2a_1 + a_0 - 2hb_1 = 0, \\ a_1 h^2 - \dfrac{h^3}{3} b_1 = 0. \end{cases}$$

Solving in terms of a_0 we find $a_1 = a_0/4$, $b_1 = 3/4ha_0$; substituting in (2.46), and choosing $a_0 = 4$, produces the required finite-difference scheme

$$d_{i-1} + 4d_i + d_{i+1} = \frac{3}{h}(u_{i+1} - u_{i-1}). \tag{2.47}$$

Now recalling the expression of the truncation error, we see that the scheme (2.47) is accurate to order 4 in h.

Equation (2.46) is an example of *compact* or *Hermitian finite differences*. The adjective 'compact' owes to the fact that the required stencil, i.e. the number of nodes required for a given order of accuracy, is much smaller than the one required by a traditional finite-difference method (think for instance at $(2.31)_2$). The price to pay is that computing the derivative needs now solving a linear system, because (2.47) is equivalent to finding the vector $\mathbf{d}$, of components d_i, solution to

$$\mathbf{Cd} = \frac{1}{h}\mathbf{M}_1\mathbf{u}, \tag{2.48}$$

where $\mathbf{u} = [u_i]$. The matrices C and $\mathbf{M}_1$ are tridiagonal, both have on the subdiagonal 1, on the superdiagonal ∓ 3, on the main diagonal 4 and 0 respectively.

We have found (2.48) for a generic index $i \in \mathbb{Z}$. It is obvious that if i varied in a subset of $\mathbb{Z}$ (e.g., between 0 and n) we would have to ask about constructing approximations of order 4 for the boundary nodes x_0 and x_n, too, to which the scheme clearly does not apply. In such a case one looks for compact decentred approximations. For instance at x_0 we will construct a scheme like $a_0 D_0 + a_1 D_1 = b_0 u_1 + b_1 u_2 + b_2 u_3 + b_3 u_4$, accurate to order 4 if the coefficients (check this fact) satisfy

$$a_0 = 1, \, b_0 = -\frac{3 + a_1 + 2b_3}{2}, \, b_1 = 2 + 3b_3, \, b_2 = -\frac{1 - a_1 + 6b_3}{2}.$$

In a similar way one proceeds for the node x_n. Matrix C is still tridiagonal. We recall that tridiagonal linear systems can be solved efficiently by the Thomas' algorithm (see for instance [QSS00]).

Periodic functions are an exception: for this class, in fact, we may force the periodicity of the function and its derivatives by imposing $u_i = u_{i+n}$ and $D_i = D_{i+n}$ for any $i \in \mathbb{Z}$, thus completing system (2.48). This is precisely what Program 5 does: it computes the approximation d_i^p of the pth derivative of a periodic function u using a compact-finite-difference method of the general form

$$\sum_{k=-r}^{r} a_k d_{i+k}^p = \frac{1}{h^p} \sum_{k=-s}^{s} b_k u_{i+k}$$

with r, s, p positive integers. The coefficients $\{a_k\}_{k=-r}^{r}$ and $\{b_k\}_{k=-s}^{s}$ must be assigned in this order in the input vectors coeffC and coeffM, respectively.

The other parameters are the vector **xspan**, the scalar **nh** and the function **fun** as in Program 4.

Program 5 - fdcompact : Approximation by compact finite differences of the derivative of a periodic function

```
%FDCOMPACT approximates the pth derivative of a periodic function.
%  [X,DU]=FDCOMPACT(XSPAN,NH,COEFFC,COEFFM,FUN,P) approximates the
%  Pth derivative of the function FUN at NH+1 equidistant nodes in the
%   interval [XSPAN(1),XSPAN(2)] using compact finite differences of the form
%   SUM(COEFFC.*DU) = (1/H^P)*SUM(COEFFM.*FUNXI),
%  where FUNXI is the vector with the values of FUN at the nodes
%   X(I-(NC-1)/2:I+(NC+1)/2),  NC being an odd number equal to the length of
%   COEFFM. [X,DU]=FDCOMPACT(XSPAN,NH,COEFFC,COEFFM,FUN,P,P1,P2,...) passes
%  the additional parameters P1,P2,.. to the function FUN(X,P1,P2,..).
```

Formula (2.46) can be obtained also with another approach. Let us change the independent variable in a convenient way, as we have done in Exercise 2.3.1, $\xi = x - x_i$, so that the derivative we want to compute corresponds to the derivative of f with respect to ξ in $\xi = 0$. We will obtain this derivative as the differentiation of an interpolating polynomial of Hermitian type (see [QSS00]). Let $p(\xi) = \sum_{j=0}^{4} c_{4-j} \xi^{4-j}$ a polynomial such that (we temporarily assume to know $d_{i\pm1}$)

$$p(-h) = u_{i-1}, p(0) = u_i, \quad p(h) = u_{i+1},$$

$$\frac{dp}{d\xi}\Big|_{\xi=-h} = d_{i-1}, \quad \frac{dp}{d\xi}\Big|_{\xi=h} = d_{i+1}.$$

Coherently with this notation we call $d_i = \frac{dp}{d\xi}\Big|_{\xi=0}$ the derivative we want to compute. By a direct inspection, it is possible to check that the coefficients of the interpolating polynomial solve the system

$$c_0 = u_1, \quad \begin{bmatrix} 2h^4 & 2h^2 & 0 & 0 \\ 8h^3 & 4h & 0 & 0 \\ 0 & 0 & 2h^3 & 2h \\ 0 & 0 & 6h^2 & 2 \end{bmatrix} \begin{bmatrix} c_4 \\ c_2 \\ c_3 \\ c_1 \end{bmatrix} \begin{bmatrix} u_{i+1} + u_{i-1} \\ d_{i+1} - d_{i-1} \\ u_{i+1} - u_{i-1} \\ d_{i+1} + d_{i-1} \end{bmatrix}.$$

From here, it is promptly computed that

$$d_i = c_1 = \frac{3}{4h}(u_{i+1} - u_{i-1}) - \frac{1}{4}(d_{i+1} + d_{i-1}).$$

If we now move the second term on the right hand side (which is actually unknown) to the left hand side, we obtain (2.47).

Table 2.4 Behavior of the error E in discrete infinity norm with respect to h for the compact formula (2.47), and the corresponding estimate of the convergence order q with respect to h

h	1/10	1/20	1/40	1/80	1/160	1/320
E	5.7e-03	3.4e-04	2.1e-05	1.3e-06	8.3e-08	5.2e-09
q	–	4.0508	4.0127	4.0032	4.0008	4.0002

Computing compact finite differences for $f(x) = \sin(2\pi x)$ and checking the order can be done by

```
>> fun=inline('sin(2*pi*x)');
>> nh = 10;
>> for n=1:6
     [x,dfdx]=fdcompact([0,1],nh,[1 4 1],[-3 0 3],fun,1);
     error(n) = norm(dfdx'-2*pi*cos(2*pi*x),inf);
     nh=2*nh;
   end
>> log(error(1:end-1)./error(2:end))/log(2)
```

The results of Table 2.4 confirm the scheme's order of accuracy 4. Note how the compact-finite-difference scheme is more accurate than method $(2.31)_2$ introduced in Exercise 2.3.1, although it has a smaller *stencil*; the two methods have the same order of accuracy, though.   ◇

Exercise 2.3.5. Compute the spectral pseudo-derivative (using GCL points) of

$$f(x) = e^x \text{ and } g(x) = \begin{cases} 0 & x \in [-1, 1/3], \\ -27x^3 + 27x^2 - 9x + 1 & x \in (1/3, 1] \end{cases}$$

in the interval $(-1, 1)$. For both functions determine the error in the discrete infinity norm as the number of points n grows and discuss the results.

Solution 2.3.5.

With Programs 6 and 7 we construct the $N(= n + 1)$ GCL nodes and the corresponding matrix of pseudo-spectral differentiation D_n (given in (2.30)). It is enough to give the instructions x = xwglc(N); D = derglc(x,N); and just set `df=D*exp(x)` to compute the spectral pseudo-derivative of f. As for for g, we may define it using

```
g=inline('(x>1/3 && x<= 1).*(-27*x^3+27*x^2-9*x+1)')
```

the derivative being then approximated by `dg=D*g(x)`. We recall that a logical expression in MATLAB returns 0 if false and 1 if true.

Program 6 - xwgcl : Gauss-Chebyshev-Lobatto nodes

```
%XWGCL Nodes of the  Gauss-Chebyshev-Lobatto formula
%    [X]=XWGCL(NP,A,B) computes the nodes on the generic interval [A,B].
```

Program 7 - dergcl : Matrix of pseudo-spectral differentiation at Gauss-Chebyshev-Lobatto nodes

```
%DERGCL Matrix of pseudo-spectral differentiation at
%    Gauss-Chebyshev-Lobatto nodes.
%
%    [D]=DERGCL(X,NP) calculates the matrix of pseudo-spectral
%    differentiation D at the Gauss-Chebyshev-Lobatto nodes X.
%    on the interval [-1,1]. X is of dimension NP
%
%    NP-1 is the degree of the polynomial employed.
%
```

Numerical results

Fig. 2.15 shows the behavior of the error in function of n for f and g.

In the first case, since the function is analytic, the error decreases exponentially, thus faster than any power of n, as n grows, until "machine zero" is reached and the error stagnates because of round-off error.

As for g, it belongs to $C^2([-1,1])$ (but not to $C^3(-1,1)$), so the error behaves as n^{-2}, in accordance with the theory (see for instance [Qua09, Chapter 10]). $\diamond$

Remark 2.1 As a synopsis of numerical differentiation, notice that all the formulas we have introduced can in fact be regarded as the differentiation of a polynomial obtained by an appropriate interpolation method:

1. classical formulas use Lagrange polynomials;
2. compact finite differences use Hermite polynomials;
3. pseudo-spectral differentiation adopts interpolation at Gauss points.

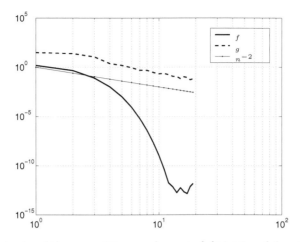

Fig. 2.15 Behavior of the error of the pseudo-spectral derivative of f and g. The graph of n^{-2} (solid curve with dots) is shown for comparison

Part II
Stationary Problems

3

Galerkin-finite element method for elliptic problems

In this chapter we discuss the approximation of elliptic problems whose *weak formulation* reads

$$\text{find } u \in V : a(u,v) = F(v) \ \forall v \in V, \tag{3.1}$$

where V is an Hilbert space.

A *sufficient* condition for the well posedness of this problem is given by the following *Lax-Milgram Lemma* [Qua09].

Lemma 3.1 (Lax-Milgram). *Let V be a Hilbert space, $a(\cdot,\cdot) : V \times V \to \mathbb{R}$ a bilinear form that is assumed to be continuous and coercive, with coercivity constant α, $F : V \to \mathbb{R}$ a linear and continuous functional. Then problem (3.1) admits one unique solution. Moreover*

$$||u||_V \leq \frac{1}{\alpha}||F||_{V'},$$

where

$$||F||_{V'} = \sup_{v \in V, ||v||_V \neq 0} \frac{|F(v)|}{||v||_V}.$$

In this chapter we consider in particular second order elliptic problems formulated in the bounded domain $\Omega \subset \mathbb{R}^d$ (mostly with $d = 1, 2$). In this case $V \subset H^1(\Omega)$.

Different kind of boundary conditions can be considered for this problem. When we assume homogeneous Dirichlet conditions on the portion of the boundary Γ_D, i.e. $u_{\Gamma_D} = 0$,, the space V is given by

$$V = H^1_{\Gamma_D} \equiv \{v \in H^1(\Omega) : \ v_{|\Gamma_D} = 0\}. \tag{3.2}$$

Formaggia L., Saleri F., Veneziani A.: Solving Numerical PDEs: Problems, Applications, Exercises. DOI 10.1007/978-88-470-2412-0_3, © Springer-Verlag Italia 2012

We remind that if $\Gamma_D \neq \emptyset$ the Poincaré inequality (1.15) holds. On the contrary, when meas(Γ_D) $= 0$, $V = H^1(\Omega)$ and the Poincaré inequality in general does not hold.

For non homogeneous Dirichlet conditions, i.e. $u = g$ on Γ_D, with $g \neq 0$. If Ω and g are smooth enough, we can reformulate the original problem into an equivalent one with homogeneous data. If Ω is an interval of $\mathbb{R}$ (and g is a real number) regularity assumptions are always verified. For higher dimensional problems, it is sufficient to require a C^1 or a polygonal boundary for Ω and $g \in H^{1/2}(\Gamma_D)$. By exploiting inequality (1.12), we introduce a *lifting* $G \in H^1(\Omega)$, such that $G_{|\Gamma_D} = g$ (in the sense of traces, see [Qua09]). With the notation

$$\overset{\circ}{u} \equiv u - G,$$

we obtain the following problem: find $\overset{\circ}{u} \in V \equiv H^1_{\Gamma_D}(\Omega)$ such that

$$a(\overset{\circ}{u}, v) = F_g(v) \equiv F(v) - a(G, v) \qquad \forall v \in V.$$

To this problem problem we still apply the Lax-Milgram Lemma, since F_g is linear and continuous provided that $F(\cdot)$ and $a(\cdot, \cdot)$ are linear and continuous and the Poincaré inequality holds.

Notice that the set

$$W_g \equiv \{v \in H^1(\Omega) : \ v_{\Gamma_D} = g\} \tag{3.3}$$

is not a space, since the null function and the sum of two elements of the set do not belong to it. This is usually called an *affine variety*. For the sake of brevity, for non homogeneous Dirichlet problems we will refer to the following corollary of the Lax-Milgram Lemma.

Corollary 3.1. *Consider the problem*

$$\text{find } u \in W_g : \ a(u, v) = F(v) \quad \forall v \in V \equiv H^1_{\Gamma_D}, \tag{3.4}$$

where $W_g \equiv \{v \in H^1(\Omega) : \ v_{\Gamma_D} = g\}$, $g \in H^{1/2}(\Gamma_D)$ *and* $\partial\Omega$ *regular enough. If* $a(\cdot, \cdot)$ *is a bilinear continuous coercive form in* $V \times V$ *and* $F(\cdot)$ *is a linear and continuous functional in* V, *then (3.3) has a unique solution, continuously depending on the data. More precisely*

$$\|u\|_V \leq C(\|F\|_{V'} + \|g\|_{H^{1/2}(\Gamma_D)}),$$

where C *depends on* α *and* c_γ *in (1.12).*

Given the extension G an alternative (and effective) way for denoting the variety W_g, is simply $G + V$. In general, the extension G can be selected according to different criteria. In the framework of a finite element discretiza-

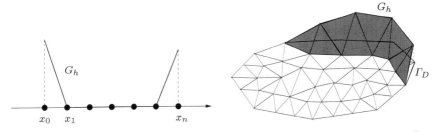

Fig. 3.1 Two possible piecewise linear finite element extensions of boundary data: 1D on the left, 2D on the right

tion, a natural extension is $G_h = \sum_{i \in B_{\Gamma_D}} g(\mathbf{x}_i)\phi_i|_{\Gamma_D}$, where B_{Γ_D} is the set of nodes $\mathbf{x}_i$ on Γ_D and ϕ_i are the associated basis functions. This extension clearly belongs to the finite element space V_h and the support is limited to the elements having at least one node on Γ_D (see Fig. 3.1). This lifting is associated to some algebraic techniques for prescribing Dirichlet boundary conditions discussed in Appendix A.

When referring to the discretization of (3.4) with the Galerkin finite element method, we will assume therefore to have the extension $G_h \in X_h^r$ (this space is introduced in (2.22)) and to rewrite the finite dimensional problem in the form: find $u_h \in G_h + V_h \equiv W_{g,h}$ with

$$G_h + V_h \equiv \{v_h \in X_h^r : v_h|_{\Gamma_D} = G_h\} \qquad (3.5)$$

such that $a(u_h, v_h) = F(v_h)$ for any $v_h \in V_h$, where

$$V_h = \{v_h \in X_h^r : v_h|_{\Gamma_D} = 0\}.$$

Remark 3.1 The Lax-Milgram Lemma is a sufficient condition. It is actually a consequence of what has been called the Banach-Necas-Babuska (BNB) Theorem in [EG04]. Most of the exercises in this chapter (actually in the book) relies upon the Lax-Milgram Lemma. However, it is worth remarking that there are many problems that require different tools. The most remarkable in this book is given by the incompressible Navier-Stokes equations (Chapter 7).

3.1 Approximation of 1D elliptic problems

In this section we solve problems in the form: find u such that

$$-\frac{d}{dx}\left(\nu \frac{du}{dx}\right) + \beta \frac{du}{dx} + \sigma u = f, \qquad x \in (a, b), \qquad (3.6)$$

under suitable boundary conditions, where coefficients ν (viscosity), β (advection), σ (reaction) and the forcing term f are in general functions of x. We solve the problems with the finite element method, implemented in the MATLAB code `fem1d` illustrated in the first exercise.

The same problem will be considered in Chapter 4 where we specifically analyze the critical case (from the viewpoint of numerical solution) of either an advection or reaction dominated case.

In some example we will exploit the following results: if $f \in H^m(a,b)$ for $m \geq 0$, then solution to (3.6) with constant coefficients belongs to $H^{m+2}(a,b)$ (see [Sal08]).

Exercise 3.1.1. Let us consider the following boundary problem

$$\begin{cases} -u'' + u = 0 \text{ for } x \in (0,1), \\ u(0) = 1, \quad u(1) = e. \end{cases} \tag{3.7}$$

1. Write the weak formulation of the problem and perform a well posedness analysis. Find the exact solution.
2. Approximate the problem with the Galerkin finite element method and solve it with **fem1d** by using piecewise linear, quadratic and cubic polynomials on a grid with constant size h. Verify the dependence of the error in the norms $H^1(0,1)$ and $L^2(0,1)$ on h in the interval $[1/320, 1/10]$.

Solution 3.1.1.

Mathematical analysis

For the weak formulation of the problem, notice that we have non homogeneous Dirichlet conditions. We multiply the equation by a test function $v \in V \equiv H_0^1(0,1)$; integrate over $(0,1)$ to obtain

$$-\int_0^1 u'' v \, dx + \int_0^1 uv \, dx = 0 \qquad \forall v \in V.$$

We integrate by parts the first term on the left. Since $v(0) = v(1) = 0$ we have

$$\int_0^1 u'v' \, dx + \int_0^1 uv \, dx = 0 \qquad \forall v \in V. \tag{3.8}$$

For $\Gamma_D = \{0,1\}$ the weak form reads: find $u \in W_g$ (introduced in (3.3)) such that $a(u,v) = 0 \ \forall v \in V$, where $a : V \times V \to \mathbb{R}$ is the bilinear symmetric form

$$a(u,v) \equiv \int_0^1 u'v' \, dx + \int_0^1 uv \, dx. \tag{3.9}$$

Well posedness of the problem (3.9) can be proved thanks to Corollary 3.1. Let us verify that the bilinear form $a(\cdot, \cdot)$ is continuous and coercive on $V \times V$. Observe that in fact:

1. a is *continuous* since

$$|a(u, v)| \leq \left| \int_0^1 u'v' \, dx \right| + \left| \int_0^1 uv \, dx \right|$$
$$\leq \|u'\|_{L^2(0,1)} \|v'\|_{L^2(0,1)} + \|u\|_{L^2(0,1)} \|v\|_{L^2(0,1)} \leq 2\|u\|_V \|v\|_V;$$

2. a is *coercive* since

$$a(u, u) = \|u'\|_{L^2(0,1)}^2 + \|u\|_{L^2(0,1)}^2 = \|u\|_V^2.$$

Thanks to Corollary 3.1 we conclude that problem (3.9) features a unique stable solution.

In this case, we can explicitly compute the solution. Equation (3.7) is actually a homogeneous ordinary differential equation with constant coefficients. To find the general solution, we find the root of the associated characteristic equation $-\lambda^2 + \lambda = 0$. The equation has two distinct roots, $\lambda_{1,2} = \pm 1$. The general solution to the equation reads therefore [BD70]

$$u = C_1 e^x + C_2 e^{-x} \tag{3.10}$$

where C_1 and C_2 are two arbitrary constants to be determined by the boundary conditions. In this case it is promptly verified that the solution of the boundary value problem features $C_1 = 1$ and $C_2 = 0$, so that $u = e^x$.

Numerical approximation

Let V_h be a finite dimensional subspace of V, such that, when $\dim(V_h) \to \infty$, then $V_h \to V$. Galerkin approximation of (3.9) reads: find $u_h \in G_h + V_h$ such that $a(u_h, v_h) = 0 \; \forall v_h \in V_h$. With finite elements P^r with $r = 1, 2, 3$, V_h reads

$$V_h \equiv \{v_h \in X_h^r : v_h = 0 \text{ in } \Gamma_D\}, \tag{3.11}$$

where X_h^r is defined in (2.3) of Chapter 2.
As usual, $\mathcal{T}_h$ denotes a partition of $(0,1)$ in N_h sub-intervals $K_j \equiv (x_{j-1}, x_j)$ with size h.

The graphical user interface of `fem1d` is organized as indicated in the diagram of Fig. 3.2.

Let us run the program in MATLAB and select `Elliptic` in the window (Fig. 3.3, left). Then we select the button "constant coefficient" and select either the conservative or non conservative formulations[1] (in this case they are equivalent). We write the coefficients in the appropriate editable windows

---

[1] The conservative formulation refers to a first order term in the form $(\beta u)'$, the non conservative one to a term reading as $\beta u'$, where β is a coefficient function of x.

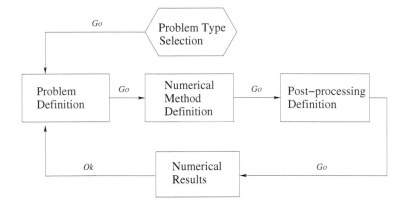

Fig. 3.2 Flow chart for the GUI of `fem1d`

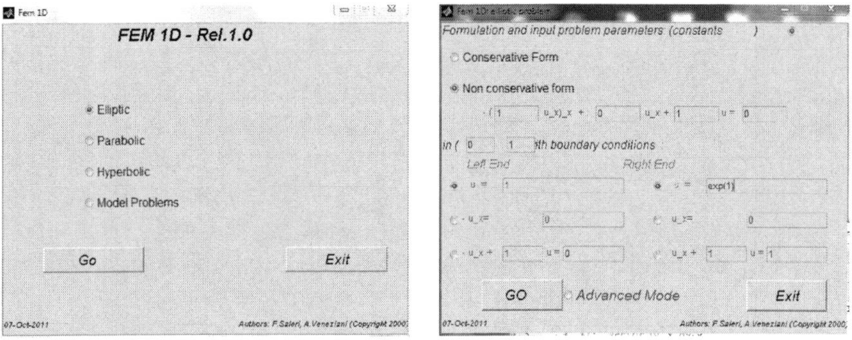

Fig. 3.3 First two windows of `fem1d`. In the second one we select the appropriate values for the problem (3.6)

and the boundary conditions, in this case Dirichlet in both the end points (Fig. 3.3, right). Push the button `Go` to go to the next window.

In the new window we have to select the degree for finite element (among $r = 1, 2, 3$) and the (constant) mesh size h. Default values are `P1` and `0.1`. Push `Go` (Fig. 3.4, left).

In the last window we are invited to select some *post-processing* options (e.g. the condition number of the matrix of the discrete problem). If we ask for the error computation, we have two editable windows where we can write the exact solution and its first derivative (for the computation of the error in the H^1 norm). The other two windows will host the computed errors. In this case the exact solution and the first derivative are both `exp(x)`. Push `Go`.

Running `fem1d` we obtain the results in Fig. 3.5. Errors are 0.0015393 in the L^2 norm and 0.051594 in the H^1 norm. Push `OK` in the window with the graph of the solution (Fig. 3.5, right). We come back to the second window and we can restart the computations. To quit, push `Exit` here and in the initial window.

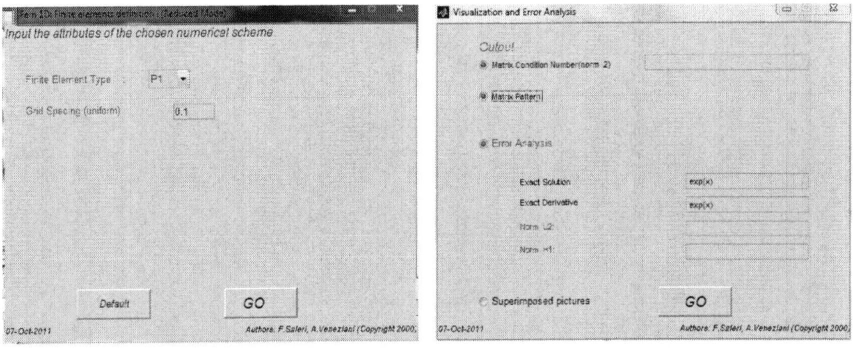

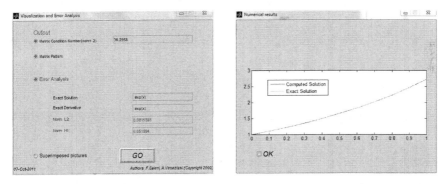

Fig. 3.4 Third and fourth windows of `fem1d`. Left, values for finite element and mesh size. Right, *post-processing* window with error computation

Fig. 3.5 The two final windows of `fem1d`. Left, computed errors, right, graph of the approximate solution and of the exact solution (evaluated in the discretization nodes)

Table 3.1 Error for different degree of Finite Elements as function of h both with respect to the $L^2(0,1)$ and $H^1(0,1)$ norms

h	P1		P2		P3	
	$\|u - u_h\|_{L^2}$	$\|u - u_h\|_{H^1}$	$\|u - u_h\|_{L^2}$	$\|u - u_h\|_{H^1}$	$\|u - u_h\|_{L^2}$	$\|u - u_h\|_{H^1}$
0.1	0.0015393	0.051594	1.0271e-05	0.00066576	5.9293e-08	5.6261e-06
0.05	0.00038484	0.025798	1.2845e-06	0.0001665	3.708e-09	7.0358e-07
0.025	9.6211e-05	0.012899	1.6059e-07	4.163e-05	2.3179e-10	8.7958e-08
0.0125	2.4053e-05	0.0064494	2.0074e-08	1.0408e-05	1.4489e-11	1.0995e-08
0.00625	6.0132e-06	0.0032247	2.5093e-09	2.6019e-06	2.2452e-12	1.3743e-09
0.003125	1.5033e-06	0.0016124	3.1367e-10	6.5048e-07	1.714e-11	1.9308e-10

In order to verify the expected convergence rate, let us half the mesh size progressively, taking note of the errors obtained in this way. We report the errors in the second and third columns of Table 3.1. Similarly, if we select elements P^2 or P^3 we obtain the other columns in Table 3.1. The agreement with the theory is evident being the solution $C^\infty(0,1)$ (order r for H^1 norm, $r + 1$ for L^2 norm when using finite elements of degree r). $\diamondsuit$

Exercise 3.1.2. Let us consider equation $(3.7)_1$ with boundary conditions

$$2u(0) - 5u'(0) = -3, \qquad u'(1) = e. \tag{3.12}$$

1. Write the weak formulation and analyze the well posedness. Find the exact solution.
2. Approximate the problem with **femld** and estimate the convergence order with the mesh size h .

Solution 3.1.2.

Mathematical analysis

The prescribed boundary conditions are of Robin type on the left end and Neumann on the right end. We look for $u \in V \equiv H^1(0,1)$ such that

$$\int_0^1 u'v' \, dx - u'(1)v(1) + u'(0)v(0) + \int_0^1 uv \, dx = 0 \qquad \forall v \in V.$$

If we prescribe boundary conditions (3.12) we obtain the weak formulation: find $u \in V$ such that

$$\int_0^1 u'v' \, dx + \frac{2}{5}u(0)v(0) + \int_0^1 uv \, dx = ev(1) - \frac{3}{5}v(0) \quad \forall v \in V, \tag{3.13}$$

to which we associate the bilinear form

$$a(u,v) \equiv \int_0^1 u'v' \, dx + \int_0^1 uv \, dx + \frac{2}{5}u(0)v(0)$$

and the functional $F(v) \equiv ev(1) - 3/5v(0)$. The bilinear form is coercive since

$$a(u,u) = \|u'\|_{L^2(0,1)}^2 + \|u\|_{L^2(0,1)}^2 + \frac{2}{5}u^2(0) \geq \|u\|_V^2.$$

Moreover, it is continuous, since

$$|a(u,v)| \leq \|u'\|_{L^2(0,1)}\|v'\|_{L^2(0,1)} + \|u\|_{L^2(0,1)}\|v\|_{L^2(0,1)} + \frac{2}{5}|u(0)| \, |v(0)|$$
$$\leq (2 + 2C/5)\|u\|_{H^1(0,1)}\|v\|_{H^1(0,1)},$$

thanks to the Cauchy-Schwarz and trace inequalities (see Section 1.3). With the same arguments, F is readily proved to be continuous.

The exact solution is found by prescribing the boundary conditions to the general solution (3.10). We obtain the system

$$\begin{cases} 3C_1 - 7C_2 = 3 \\ C_1 + e^{-2}C_1 = 1 \end{cases}$$

with solution $C_1 = 1, C_2 = 0$.

Numerical approximation

The Galerkin piecewise linear finite element formulation reads: find $u_h \in V_h$ such that $a(u_h, v_h) = F(v_h)$ for any $v_h \in V_h$, with $V_h = X_h^1$. Let A be the *stiffness matrix* with entries a_{ij} given by $a_{ij} \equiv a(\varphi_j, \varphi_i)$ and **f** the vector with entries $f_i \equiv F(\varphi_i)$. If **u** denotes the vector with components u_i, the discretized problem corresponds to solving the linear system (see [Qua09])

$$\mathbf{Au} = \mathbf{f}. \tag{3.14}$$

Boundary conditions are automatically included in the Galerkin formulation and the size of the system corresponds to the number of nodes of the mesh.

Let us solve the problem with `fem1d`. In the window of the parameters of the problem we select the Robin condition in the left end and the Neumann condition in the right end (see Fig. 3.6). Notice that in `fem1d` these conditions are in the form $-u'(0) + \alpha u(0) = \beta$ and $u'(1) = \gamma$. In our case, $\alpha = 2/5$, $\beta = -3/5$ and $\gamma = e$ (`=exp(1)` in MATLAB).

Let us select piecewise linear finite elements and compute the solution with a progressively small mesh size h starting from $h = 0.1$. We obtain the errors reported in Fig. 3.7 and indicated in Table 3.2.

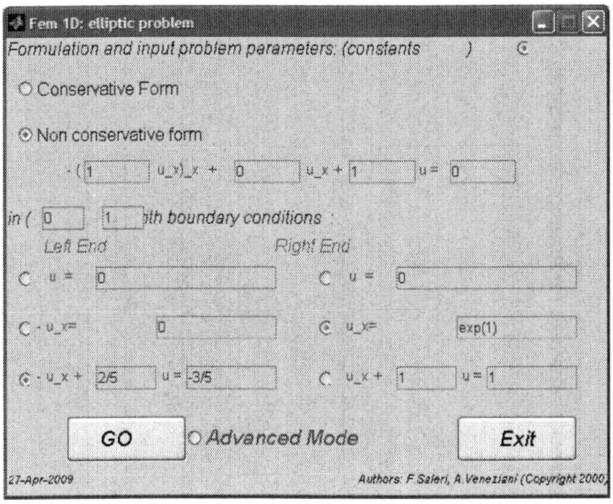

Fig. 3.6 Window of `fem1d` with the prescription of the boundary conditions for problem $(3.7)_1$-(3.12)

Table 3.2 Errors for Exercise 3.1.2

h	0.1	0.05	0.025	0.0125	0.00625	0.003125
$\|u - u_h\|_{L^2}$	0.00082538	0.00020652	5.164e-05	1.2911e-05	3.2277e-06	8.0692e-07
$\|u - u_h\|_{H^1}$	0.051577	0.025796	0.012899	0.0064494	0.0032247	0.0016124

Table 3.3 Estimated order of convergence for Exercise 3.1.2

h_1, h_2	0.05,0.01	0.025,0.05	0.0125,0.025	0.00625,0.0125	0.003125,0.00625
q_{L^2}	1.9988	1.9997	1.9999	2.0000	2.0000
q_{H^1}	0.9996	0.9999	1.0000	1.0000	1.0000

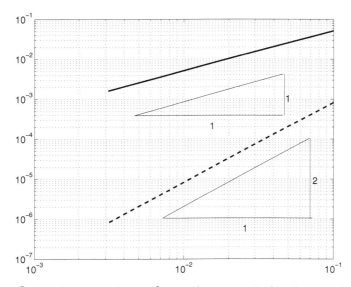

Fig. 3.7 L^2 errors (dashed line) and H^1 errors (continuous line) as function of h. We use a log-log scale (`loglog` in MATLAB)

Numerical results

Notice that if a method features a convergence order q, the error is such that $e(h) \simeq Ch^q$ (at least for h small enough), where C is independent of h. If we have two values for the error corresponding to $h = h_1$ and $h = h_2$ respectively, we can approximate q as follows

$$\frac{e(h_1)}{e(h_2)} \simeq \left(\frac{h_1}{h_2}\right)^q \qquad \Rightarrow \qquad q \simeq \log(e(h_1)/e(h_2))/\log(h_1/h_2).$$

In our case, we get Table 3.3, in agreement with the theory.   ◇

Exercise 3.1.3. The Temperature T of a thin rod with length L and constant section A for $x \in (0, L)$ is given by the solution of the differential problem

$$\begin{cases} -kAT'' + \sigma pT = 0, \ x \in (0, L), \\ T(0) = T_0, \qquad\qquad T'(L) = 0, \end{cases} \tag{3.15}$$

where p is the perimeter of A, k the thermal conductivity coefficient, σ is the convective transfer coefficient and T_0 is a given value.

1. Find the analytical solution.
2. Write the weak formulation of (3.15), and its Galerkin finite element approximation. Identify the dependence of the approximation error in norm $H_0^1(0, L)$ on k, σ, p, T_0 and the length L of the domain.
3. Assuming that the rod has $L = 1$m, with circular section with radius 1cm, solve the problem with piecewise linear finite element on uniform grids and compute the approximation error ($T_0 = 10$, $\sigma = 2$ e $k = 200$).

Solution 3.1.3.

Mathematical analysis

General solution of (3.15) has the form

$$T(x) = C_0 e^{\lambda_0 x} + C_1 e^{\lambda_1 x}, \tag{3.16}$$

where C_0 and C_1 are arbitrary constants determined by the boundary conditions, and λ_0 and λ_1 are the roots of the equation

$$-kA\lambda^2 + \sigma p = 0.$$

For $m = \sqrt{\sigma p/(kA)}$, the two roots are $\lambda_0 = -m$ and $\lambda_1 = m$ so that solution reads $T(x) = C_0 e^{-mx} + C_1 e^{mx}$.

Coefficients C_0 and C_1 are obtained by prescribing the boundary conditions, i.e. $C_0 + C_1 = T_0$ and $-C_0 e^{-mL} + C_1 e^{mL} = 0$. After some computations we get

$$C_0 = \frac{T_0 e^{mL}}{e^{-mL} + e^{mL}} = \frac{T_0 e^{mL}}{2\cosh(mL)}, \quad C_1 = \frac{T_0 e^{-mL}}{e^{-mL} + e^{mL}} = \frac{T_0 e^{-mL}}{2\cosh(mL)},$$

so that

$$T(x) = \frac{T_0}{2\cosh(mL)}\left(e^{m(L-x)} + e^{-m(L-x)}\right) = T_0 \frac{\cosh(m(x-L))}{\cosh(mL)}.$$

For the weak formulation of (3.15), we set $\Gamma_D = \{0\}$ and multiply (3.15) by a test function $v \in V \equiv H^1_{\Gamma_D}$ so that

$$-kA \int_0^L T''(x)v(x) \, dx + \sigma p \int_0^L T(x)v(x) \, dx = 0 \; \forall v \in V.$$

We integrate by parts the first term

$$kA \int_0^L T'(x)v'(x) \, dx + \sigma p \int_0^L T(x)v(x) \, dx - kA[T'v]_0^L = 0, \qquad (3.17)$$

where for a generic f, $[f]_0^L$ stands for $f(L) - f(0)$. The boundary terms here vanish since $T'(L) = 0$ and $v(0) = 0$. With the usual notation, the weak form reads: find $T \in W_g$ such that

$$a(T, v) = 0 \qquad \forall v \in V, \qquad (3.18)$$

where $a(\cdot, \cdot) : V \times V \to \mathbb{R}$ is the bilinear form

$$a(u, v) \equiv kA \int_0^L u'(x)v'(x) \, dx + \sigma p \int_0^L u(x)v(x) \, dx.$$

This bilinear form is *continuous* since $|a(u, v)| \leq M\|u\|_V\|v\|_V$ with $M = \max(kA, |\sigma|p)$ and it is *coercive* since

$$a(u, u) = kA\|u'\|^2_{L^2(0,L)} + \sigma p\|u\|^2_{L^2(0,L)} \geq kA\|u'\|^2_{L^2(0,L)} \geq \alpha\|u\|^2_V, \quad (3.19)$$

with $\alpha = kA/(1 + L^2/2)$. More precisely, since $v(0) = 0$

$$|v(x)|^2 = \left| \int_0^x v'(x)dx \right|^2 \leq \int_0^x 1^2 dx \int_0^x (v'(x))^2 dx \leq x\|v'\|^2_{L^2(0,L)}$$

so that by integrating over $[0, L]$ we have

$$\|v\|_{L^2} \leq \frac{L}{\sqrt{2}}\|v'\|^2_{L^2(0,L)},$$

that is the Poincaré inequality. Therefore,

$$\|v\|^2_{H^1} = \|v\|^2_{L^2(0,L)} + \|v'\|^2_{L^2(0,L)} \leq \left(1 + \frac{L^2}{2}\right)\|v'\|^2_{L^2(0,L)}$$

which is the inequality used in the last step of (3.19). Notice that a different way for proving inequality is to observe that

$$a(u, u) \geq \min(kA, \sigma p)\|u\|^2_V.$$

In this way, coercivity does not rely upon the Dirichlet condition.

On the other hand, coercivity based on the Poincaré inequality does not depend on σp (even if it depends on the length of the domain) and it holds true also for $\sigma \to 0$.

Numerical approximation

Let V_h be a finite dimensional subspace of V with dimension N_h. If G_h is an extension of the boundary data, Galerkin discretization of (3.18) reads: find $T_h \in G_h + V_h$ such that

$$a(T_h, v_h) = 0 \qquad \forall v_h \in V_h.$$

For Galerkin finite elements, once we have a partition $\mathcal{T}_h$ of $(0, L)$ in N_h sub-intervals $\{K_j\}$ (here we assume they all have the same length h) V_h is given by

$$V_h \equiv \{v_h \in X_h^r(0, L) : \ v_h(0) = 0\},$$

where X_h^r has been introduced in (2.3). The error behavior is stated in (2.28), where C is proportional to the quotient M/α. The error depends on parameters k, σ, p and on the size of the domain by means of M/α. In particular, notice that if σp gets larger, M/α can be a large number as well. In this case, we have a *reaction dominated* problem. The error can be large in this case if the mesh size is not small enough. We will come back to this kind of problems in Chapter 4.

Numerical results

We solve the problem with `fem1d` setting the diffusion coefficient to $kA = 0.02\pi$ and the advection one to $\sigma p = 0.04\pi$. We compute the error for h ranging from $1/10$ to $1/320$. Results are reported in Fig. 3.8 as function of

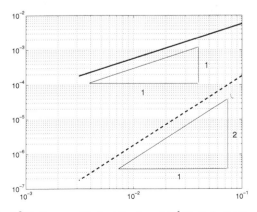

Fig. 3.8 Errors in $L^2(0,1)$ norm (dashed line) and $H^1(0,1)$ (solid line) as function of h

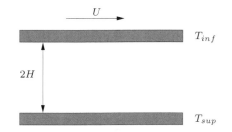

Fig. 3.9 Domain for the problem in Exercise 3.1.4

h, with a logarithmic scale. Slope of the curves (outlined by the triangles) confirms the expected orders 1 and 2, in norms H^1 and L^2 respectively.   ◇

Exercise 3.1.4. A viscous fluid is flowing between two horizontal parallel planes at a distance $2H$ (see Fig. 3.9). Assume that the upper plane, at temperature T_{sup} moves with a velocity U respect to the bottom one. A possible model for the temperature $T(y) : (0, 2H) \rightarrow \mathbb{R}$ in the fluid reads

$$\begin{cases} -\dfrac{d^2T}{dy^2} = \alpha(H-y)^2 \text{ in } (0, 2H), \\ T(0) = T_{inf}, \qquad T(2H) = T_{sup}, \end{cases} \qquad (3.20)$$

where $\alpha = 4U^2\mu/(H^4\kappa)$, $\kappa = 0.60$ is the thermal conductivity coefficient and $\mu = 0.14$kg s/m^2 is the fluid viscosity.

1. Find an analytical solution; write the weak formulation of the problem and prove that the solution obtained is unique.
2. Write the Galerkin finite-element formulation of the problem.
3. Verify the convergence properties of finite elements with degree 3 when $H = 1$m, $T_{inf} = 273K$, $T_{sup} = 293K$, $U = 10$m/s. Which polynomial degree for finite elements guarantees that the computed solution is exact?

Solution 3.1.4.

Mathematical analysis

Since the right hand side is a second order polynomial, the general solution is obtained after two integrations

$$T = -\frac{\alpha}{12}(H-y)^4 + Cy + D,$$

where C and D are two arbitrary constants. By forcing the boundary conditions we obtain

$$T(y) = -\frac{\alpha}{12}(H - y)^4 + \frac{T_{sup} - T_{inf}}{2H}y + T_{inf} + \frac{\alpha H^4}{12}.$$

The weak formulation is obtained by multiplying equation (3.20) by a test function $v \in V = H^1_{\Gamma_D} \equiv H^1_0(0, 2H)$ with $\Gamma_D = \{0, 2H\}$. Then we integrate between 0 and $2H$ and apply integration by parts to the second order differential term. Let us indicate with T' the derivative dT/dy. We obtain

$$\int_0^{2H} T'v' \, dy - [T'v]_0^{2H} = \alpha \int_0^{2H} (H - y)^2 v \, dy.$$

The weak formulation reads therefore: find $T \in W_g$ s.t. $a(T, v) = F(v)$ for any $v \in V$, where

$$a(T, v) \equiv \int_0^{2H} T'v' \, dy, \quad F(v) \equiv \alpha \int_0^{2H} (H - y)^2 v \, dy.$$

The bilinear form $a : V \times V \to \mathbb{R}$ is continuous and coercive (see Exercise 3.1.3) with continuity constant $= 1$ and coercivity constant $= 1/(1 + C_P^2)$, where C_P is the Poincaré constant. Functional $F : V \to \mathbb{R}$ is continuous as well, so that solution exists and is unique as a consequence of Corollary 3.1. The solution analytically computed above does coincide with the weak one.

Numerical approximation

Galerkin finite elements approximation with degree 3 is similar to the one introduced in the previous exercise. We take $V_h \equiv \{v_h \in X_h^3 : v_h(0) = v_h(2H) = 0\}$ and look for $T_h \in W_{g,h}$ s.t. $a(T_h, v_h) = F(v_h)$ for any $v_h \in V_h$.

Numerical results

In order to solve the problem with `fem1d` we follow the same guidelines of the previous exercises. We take cubic finite elements for a Dirichlet problem on the domain $(0, 2)$. In Fig. 3.10 we report the errors in logarithmic scale in norm $L^2(0, 2)$ and $H^1(0, 2)$ as a function of the grid size h. Convergence rates are h^4 and h^3 respectively, as expected.

Finally, notice that, since the exact solution is a fourth order polynomial, if we use finite elements of order 4 we get the exact solution (up to the rounding errors). By the way, observe that in the picture on the right of Fig. 3.10 the error in norm $L^2(0, 2H)$ slightly gets larger when the mesh size is small and

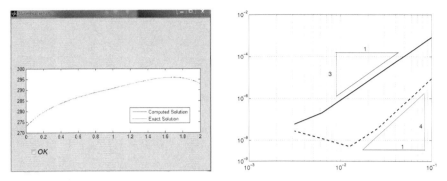

Fig. 3.10 Left: computed solution with cubic finite elements on a uniform grid with step $h = 0.1$. Right: dynamics of the errors w.r.t. $L^2(0,2)$ (dashed line) and $H^1(0,2)$ (solid line) norms as a function of h (logarithmic scale)

decreasing. This is a consequence of rounding errors (and the larger condition number of the linear system when $h \to 0$). ◇

Exercise 3.1.5. Find the temperature in a rod with length L and constant thermal conductivity $\mu > 0$. Assume that for chemical or nuclear reactions the rod produces thermal energy with a rate per volume Q given by

$$Q(x) = \begin{cases} 0 & x \in (0, L/2), \\ 6(x - L/2) & x \in (L/2, L). \end{cases}$$

Temperature T obeys the following equation

$$-\mu T'' = Q, \quad x \in (0, L). \tag{3.21}$$

Assume $\mu = 1$.

1. Verify that

$$T(x) = -1/2 \left[|x - L/2|^3 + (x - L/2)^3 \right] \tag{3.22}$$

is the only weak solution of (3.21) with boundary conditions given by

$$T(0) = 0, \ T'(L) = -3L^2/4. \tag{3.23}$$

2. For $L = 1$, solve the problem numerically with `fem1d` using piecewise finite elements of degree 1,2 and 3. Compare the convergence rate of the errors in norm $L^2(0,1)$ and $H^1(0,1)$ for different discretization steps h.
3. Comment on the results.

Solution 3.1.5.

Mathematical analysis

The weak formulation of (3.21) is obtained by multiplying the differential equation by a test function $v \in V$, where $V = H^1_{\Gamma_D}$ with $\Gamma_D = \{0\}$, integrating over $(0, L)$ and integrating by parts the differential term. We get

$$\mu \int_0^L T'v' \, dx - \mu \, [T'v]_0^L = \int_0^L Q \, v \, dx. \tag{3.24}$$

The first integral in (3.24) is finite provided that T and v belong to $H^1(0, L)$. The integral on the right hand side is finite if $v \in L^2(0, L)$ since $Q \in L^2(0, L)$ as can be verified by direct inspection. Including the boundary conditions, we obtain the weak formulation: find $u \in V$ such that for any $v \in V$

$$\mu \int_0^L T'v' \, dx = \mu[T'v]_0^L + \int_0^L Q \, v \, dx = 3\mu(L^2/4)v(L) + \int_0^L Q \, v \, dx.$$

Well posedness analysis follows from the Lax-Milgram Lemma. Bilinear form $a(T, v) \equiv \mu \int_0^L T'v' \, dx$ is indeed:

1. *Continuous*: by application of the Cauchy-Schwarz inequality

    $$|a(T, v)| \leq \mu \|T'\|_{L^2(0,L)} \|v'\|_{L^2(0,L)} \leq \mu \|T\|_{H^1(0,L)} \|v\|_{H^1(0,L)}.$$

2. *Coercive* thanks to the Poincaré inequality

    $$a(T, T) = \mu \int_0^L T'T' \, dx = \mu \|T'\|_{L^2(0,L)}^2 \geq \frac{\mu}{\sqrt{1 + C_P^2}} \|T\|_{H^1(0,L)}^2$$

    for any $T \in H^1(0, L)$.

Functional $F(v) \equiv 3\mu(L^2/4)v(L) + \int_0^L Q \, v \, dx$ is continuous being $Q \in L^2(0, L)$, as a consequence of the trace inequality (1.11) and the Cauchy-Schwarz one

$$|F(v)| \leq \left(3\gamma\mu(L^2/4) + \|Q\|_{L^2(0,L)}\right) \|v\|_{H^1(0,L)}.$$

Set $\mu = 1$. Analytical solution (3.22) is twice continuously differentiable since

$$T'(x) = \begin{cases} 0 & x \in (0, L/2), \\ -3(x - L/2)^2 & x \in (L/2, L), \end{cases}$$

$$T''(x) = \begin{cases} 0 & x \in (0, L/2), \\ -6(x - L/2) & x \in (L/2, L), \end{cases} \quad T'''(x) = \begin{cases} 0 & x \in (0, L/2), \\ -6 & x \in (L/2, L). \end{cases}$$

It is readily verified that $-T'' = Q$. Since T belongs to $C^2(0,L)$, it is regular enough to verify the strong formulation of the equation. Being a strong solution, it is also a weak one. Moreover, notice that $T''' \in L^2(0,L)$, while the fourth derivative does not. Therefore, we have $T \in H^3(0,L)$ (and $T \notin H^4(0,L)$).

Numerical approximation

Galerkin approximation is obtained by introducing the finite-dimensional subspace V_h of V and solving the problem: find $T_h \in V_h$ such that for any $v_h \in V_h$

$$\mu \int_0^1 T_h' v_h' \, dx = 3\mu(L^2/4)v_h(1) + \int_0^1 Q\, v_h \, dx. \qquad (3.25)$$

In particular, if we select piecewise polynomial spaces on a mesh of the domain $(0,1)$ we have a finite element approximation. Let us resort to `fem1d`, by specifying the coefficients of the problem in the appropriate editable windows and selecting linear, quadratic and cubic elements. In Fig. 3.11 we report the errors in $L^2(0,1)$ and $H^1(0,1)$ norms for linear and quadratic finite elements.

If we use cubic finite elements with a node of the reticulation placed in $x = L/2$, we get errors ranging from `5.1003e-16` in L^2 norm for $h = 0.1$ to `1.0129e-12` in the same norm for $h = 0.003125$ and similarly for the H^1 norm from `1.1601e-15` to `1.9477e-12`.

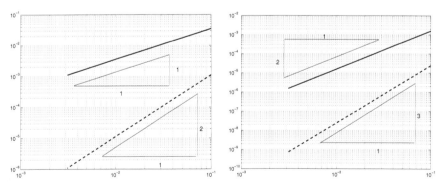

Fig. 3.11 Errors as a function of the mesh size plotted in logarithmic scale for piecewise linear (left) and quadratic (right) elements in $L^2(0,1)$ (dashed line) and $H^1(0,1)$ (continuous line) norms

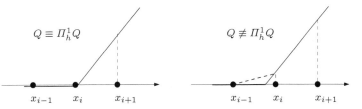

Fig. 3.12 The two possible approximations of Q for linear finite elements

---

Numerical results

Error behavior associated with linear and quadratic finite elements confirms the expected convergence rate for a solution in $H^s(0,1)$, as can be seen in Fig. 3.11. For cubic finite elements, notice that the exact solution is piecewise cubic, so that, for a node placed in $x = L/2$, it belongs to the finite elements space. The discretization error is therefore null and the only error component is associated with rounding errors and the consequent errors in solving linear systems. This explains the increasing errors when h decreases, being the condition number of the finite element matrix proportional to h^{-2}.

Remark 3.2 If the step h is reduced without the constraint of having a node in $x = L/2$ we can observe some anomalous error dynamics (see Fig. 3.13). Oscillations are explained by observing that in `fem1d` the right hand side is integrated by the following quadrature formula

$$\int_0^L Q(x)\varphi_i(x) \, dx \simeq \int_0^L \Pi_h^r f(x)\varphi_i(x) \, dx,$$

where r is the degree of the finite element used and $\Pi_h^r f$ is the piecewise interpolating polynomial with degree r on the mesh. In our case $Q(x)$ is a piecewise linear polynomial. If a node of the mesh is placed in $x = 0.5$, the integral on the right hand side is exact (see Fig. 3.12). If $x = 0.5$ is not a node of the mesh, a quadrature error is introduced with the same order of the discretization error (in norm L^2). For quadratic finite elements, the quadrature error is of lower order than the discretization one. This explains the oscillations for quadratic finite elements. If $x = 0.5$ is a node the L^2 norm error is of order 3, in the other case it features order 2. In practice, if a discontinuity point (for a function or its derivatives) is known *a-priori*, it is worth placing a mesh point on it. $\diamondsuit$

Exercise 3.1.6. Find the temperature T in the same rod of the Exercise 3.1.5, when the forcing term is given by

$$Q(x) = \begin{cases} 0 & x \in (0, L/2), \\ -\pi^2 \cos(\pi(x - L/2)) & x \in (L/2, L). \end{cases}$$

Set again $\mu = 1$.

1. Find the Sobolev space $H^s(0,L)$ the possible solution $T(x)$ belongs to.
2. Check that

$$T = \begin{cases} 0 & x \in (0, L/2], \\ 1 - \cos\left(\pi\left(x - L/2\right)\right) & x \in (L/2, L), \end{cases} \qquad (3.26)$$

   is the only weak solution of (3.21) with boundary conditions given by
   $T(0) = 0$ and $T'(L) = \pi$.
3. For $L = 1$, solve numerically the problem with linear, quadratic and
   cubic finite elements. Compare the error behavior in $H^1(0,1)$ norm for
   different values of the mesh size and comment.

Solution 3.1.6.

Mathematical analysis

The weak formulation of the problem is obtained as in Exercise 3.1.5.
Using the same notations, we look for a solution $u \in V$ such that for any
$v \in V$

$$\mu \int_0^L T'v' \, dx = \mu\pi v(L) + \int_0^L Q \, v \, dx. \qquad (3.27)$$

Using the same arguments used in the previous exercise, it is proved that
the bilinear form defined by the left hand side of (3.27) is continuous and
coercive. Since $Q \in L^2(0,L)$, the functional on the right hand side is linear
and continuous. We conclude that the problem at hand is well posed thanks
to the Lax-Milgram Lemma. In this exercise, however, $Q \in L^2(0,L)$ and
$Q \notin H^1(0,L)$, since Q is discontinuous. Therefore, the exact solution belongs
to $H^2(0,L)$ and not to $H^3(0,L)$.

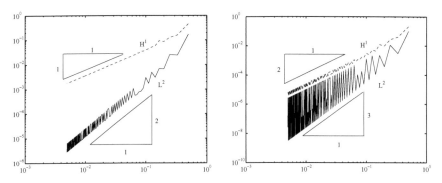

Fig. 3.13 Anomalous error behavior for linear (left) and quadratic (right) finite elements
with respect to $L^2(0,1)$ and $H^1(0,1)$ norms (see Remark 3.2)

Let us check that (3.26) is weak solution of the problem (it cannot be a strong solution, since the second derivative is discontinuous). Function T in (3.26) belongs to V and, in particular, it fulfills the boundary conditions. First derivative (in the classical sense) exists and is 0 for $x \in (0, L/2]$ and $\pi \sin[\pi(x - L/2)]$ for $x \in (L/2, L)$. Plugging this function into the left hand side of (3.27) (with $\mu = 1$), we get for a generic $v \in V$,

$$\int_0^L T'v'\, dx = \pi \int_{L/2}^L \sin\left(\pi\left(x - \frac{L}{2}\right)\right) v'\, dx.$$

Since on the integration domain the function is regular, we can integrate by parts,

$$\pi \left[\sin\left(\pi\left(x - \frac{L}{2}\right)\right) v\right]_{L/2}^L - \pi^2 \int_{L/2}^L \cos\left(\pi\left(x - \frac{L}{2}\right)\right) v\, dx$$

$$= \pi v(L) - \pi^2 \int_{L/2}^L \cos\left(\pi\left(x - \frac{L}{2}\right)\right) v\, dx.$$

This is exactly the right hand side of (3.27) and the check is completed.

Numerical approximation

Galerkin finite element approximation is obtained like in the previous exercise. From Remark 3.2 it is more convenient to place a node of the grid in $x = L/2 = 1/2$, so to reduce quadrature errors (which in this case are however unavoidable, since we have a non polynomial forcing term). Solution with `fem1d` is quite straightforward and yields results like in Fig. 3.14, left, for linear finite elements on a uniform grid with step $h = 0.1$.

In Fig. 3.14, right, we report the diagrams of the error in $H^1(0, 1)$ norm, for linear (solid line), quadratic (circles) and cubic (diamonds) finite elements.

Numerical results

Looking at Fig. 3.14 we notice that we have first order accuracy in all the cases. This confirms the theory, stating that the convergence order $q = \min(r, p - 1)$ being r the polynomial degree of the finite element and p the index of the Sobolev space analytical solution belongs to. In our case, $T \in H^2(0, 1)$, and $T \notin H^3(0, 1)$, so the order is 1 independently of the degree (1, 2 or 3) of the finite element.   ◇

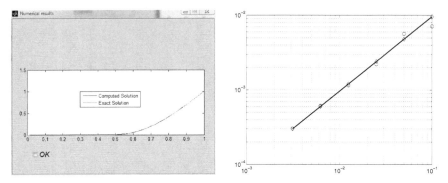

Fig. 3.14 Left, computed solution with linear finite elements with step $h = 0.1$. Right, errors in $H^1(0,1)$ norm for linear (solid line), quadratic (circles) and cubic (diamonds) finite elements for solving Exercise 3.1.6

Exercise 3.1.7. The wall of a furnace (Fig. 3.15) is composed by two materials with two different thermal conductivities. Temperature u along axis x obeys equation

$$(\chi u')' = 0 \quad \text{in } (0, L), \tag{3.28}$$

where

$$\chi = \begin{cases} \chi_1 & \text{in } (0, M], \\ \chi_2 & \text{in } (M, L). \end{cases}, \chi_1, \chi_2 > 0.$$

Assume that in $u(0) = u_0$ where u_0 is given and in $x = L$ furnace wall diffuse heat to the surrounding air that features temperature u_a as stated by the equation

$$-\chi_2 u'(L) = s(u(L) - u_a),$$

where s is the coefficient of convective heat transmission.

1. Write the weak formulation of the problem.
2. Compute the analytical weak solution and find conditions that guarantee that it is also a strong solution.
3. Solve the problem with **fem1d** with a proper selection of finite element degree. Numerical data: $L = 1$, $M = 0.2$, $\chi_1 = 1$, $\chi_2 = 10$, $u_0 = 100$, $u_a = 10$, $s = 0.1$.

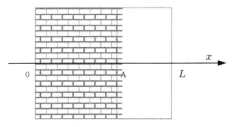

Fig. 3.15 Part of a wall of a furnace made of two different materials

Solution 3.1.7.

Mathematical analysis

The weak formulation is obtained in the usual way. Multiply (3.28) by a test function $v \in V \equiv H^1_{\Gamma_D}$ with $\Gamma_D = \{0\}$ and integrate over $(0, L)$. We find

$$\int_0^L \chi u'v' \, dx - \chi(L)u'(L)v(L) = 0.$$

If we prescribe the Robin condition in $x = L$ we get the weak formulation (3.4) with

$$a(u,v) \equiv \int_0^L \chi u'v'dx + su(L)v(L) \text{ and } F(v) \equiv su_a v(L).$$

Since χ is bounded and positive well posedness is proved with the same arguments of the previous exercises. The solution belongs to $H^1(0, L)$ and therefore it is continuous thanks to embedding Sobolev Theorem (see Chapter 1).

In $\Omega_1 = (0, M)$ and in $\Omega_2 = (M, L)$ separately χ is constant. Therefore (3.28) implies $u'' = 0$ in each subdomain, so that u is locally linear. Weak solution of (3.28) will be piecewise linear with continuous matching in $x = M$. Moreover, in $x = M$ thermal fluxes are continuous, therefore $\chi_1 u' = \chi_2 u'$ for $x = M$. This can be verified by considering the weak form associated with (3.28), and taking a test function $v \in C^\infty(0, L)$ with a support strictly in $(0, L)$. We obtain

$$0 = \int_0^L \chi u'v' \, dx = \int_0^M \chi u'v' \, dx + \int_M^L \chi u'v' \, dx =$$

$$[\chi u'v]_0^M + [\chi u'v]_M^L - \int_0^M (\chi u')' v \, dx - \int_M^L (\chi u')' v \, dx.$$

$$(3.29)$$

The last two terms are null, since on any sub-interval we have $(\chi u')' = 0$. Set $[(\chi u)']_M = (\chi u)'(M^+) - (\chi u)'(M^-)$ with $(\chi u')(M^\pm) = \lim_{\varepsilon \to 0}(\chi u)'(M \pm \varepsilon)$. Since $v(0) = v(L) = 0$, we get from (3.29) that $0 = -[(\chi u)']_M \, v(M)$. Therefore $[\chi u']_M = 0$, being v a generic function. From

$$u(x) = \begin{cases} u_1(x) = C_{0,1} + xC_{1,1} \text{ for } x \in (0, M), \\ \\ u_2(x) = C_{0,2} + xC_{1,2} \text{ for } x \in (M, L), \end{cases} \tag{3.30}$$

the conditions found yield

$$\begin{cases} u_1(0) = u_0 & \Rightarrow C_{0,1} = u_0, \\ \\ u_1(M) = u_2(M) & \Rightarrow C_{0,1} + MC_{1,1} = C_{0,2} + MC_{1,2}, \\ \\ \chi_1 u_1'(M) = \chi_2 u_2'(M) & \Rightarrow \chi_1 C_{1,1} = \chi_2 C_{1,2}, \\ \\ -\chi_2 u_2'(L) = s(u_2(L) - u_a) & \Rightarrow -\chi_2 C_{1,2} = s(C_{0,2} + LC_{1,2} - u_a). \end{cases}$$

After some algebra we find

$$C_{0,1} = u_0,$$

$$\begin{aligned} C_{1,1} &= \frac{s\chi_2(u_a - u_0)}{\chi_2\chi_1 + L\chi_1 - Ms(\chi_1 - \chi_2)}, \\ \\ C_{0,2} &= \frac{u_0(\chi_2 + L)\chi_1 - Msu_a(\chi_1 - \chi_2)}{(\chi_2 + L)\chi_1 - Ms(\chi_1 - \chi_2)}, \\ \\ C_{1,2} &= \frac{s\chi_1(u_a - u_0)}{\chi_2\chi_1 + L\chi_1 - Ms(\chi_1 - \chi_2)}. \end{aligned} \tag{3.31}$$

Notice that for $\chi_1 = \chi_2$ we have $C_{0,1} = C_{0,2}$ e $C_{1,1} = C_{1,2}$, as expected.

Even though function u in (3.30) is a weak solution, in general it is not a strong solution, since it is only a continuous function in $(0, L)$. A strong solution is supposed to be $C^2(0, L)$, that is true only for $\chi_1 = \chi_2$.

Numerical approximation

For the numerical solution of this problem, low regularity of the solution suggests using linear finite elements. We obtain

$$u(x) = \begin{cases} 100 - \dfrac{4500}{559}x & \text{if } x \in (0, 0.2), \\ \\ \dfrac{55090}{559} - \dfrac{450}{559}x & \text{if } x \in (0.2, 1). \end{cases}$$

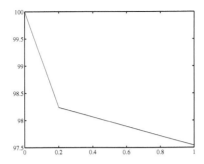

Fig. 3.16 Numerical solution with P^1 finite elements with $h = 1/10$ for Exercise 3.1.7

First derivative of the solution is clearly discontinuous, so that it does not belong to $H^1(0,1)$. This means that $u \in H^1(0,1)$ and $u \notin H^2(0,1)$. In this case, finite element theory ensures only sublinear convergence. However, for the present problem, if the point of discontinuity corresponds to a node of the mesh, analytical solution belongs to the finite element space. The numerical solution coincides with the exact solution (up to the errors due to rounding and linear system solution).

Numerical results

Using `fem1d` with linear finite elements on a uniform grid with step $h = 1/10$, where $x = 0.2$ is a node of the mesh, we find the solution in Fig. 3.16. This solution coincides with the exact one. ◇

Remark 3.3 The previous problem can be reformulated as a sequence of constant coefficient problems by subdomain splitting. This approach is discussed in Section 3.3.

3.1.1 Finite differences in 1D

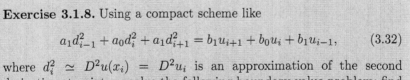

Exercise 3.1.8. Using a compact scheme like

$$a_1 d_{i-1}^2 + a_0 d_i^2 + a_1 d_{i+1}^2 = b_1 u_{i+1} + b_0 u_i + b_1 u_{i-1}, \qquad (3.32)$$

where $d_i^2 \simeq D^2 u(x_i) = D^2 u_i$ is an approximation of the second derivative at point x_i, solve the following boundary-value problem: find $u : (0,1) \to \mathbb{R}$ such that

$$\begin{cases} -\dfrac{d^2 u}{dx^2} = 4\pi^2 \sin(2\pi x) \text{ with } x \in (0,1), \\ u(0) = u(1) = 0. \end{cases}$$

Find the exact solution and check the order of accuracy.

Solution 3.1.8.

Mathematical analysis

By twice integration of the right hand side and computing the arbitrary constants by forcing the boundary conditions, one readily obtains the exact solution, $u(x) = \sin(2\pi x)$.

Numerical results

We seek an approximation for the second derivative of the form (3.32) by minimizing the truncation error

$$\tau_i(u) = a_1(D^2 u_{i-1} + D^2 u_{i+1}) + a_0 D^2 u_i - b_1(u_{i+1} + u_{i-1}) - b_0 u_i.$$

Using the series expansions $(2.32)_1$ and

$$D^2 u_{i\pm1} = D^2 u_i \pm h D^3 u_i + \frac{h^2}{2} D^4 u_i \pm \frac{h^3}{6} D^5 u_i + \frac{h^4}{24} D^6 u_i + o(h^4),$$

we find

$$\tau_i(u) = -(b_0 + 2b_1)u_i + (2a_1 + a_0 - h^2 b_1)D^2 u_i + \left(a_1 h^2 - \frac{h^4}{12} b_1\right) D^4 u_i + \dots$$

Imposing $\tau_i(u)$ be zero up to the term relative to $D^4 u_i$ we determine the value of the coefficients (in function of a_0)

$$a_1 = \frac{a_0}{10}, \ b_0 = -\frac{12}{5h^2} a_0, \ b_1 = \frac{6}{5h^2} a_0,$$

and consequently, we obtain the following compact-finite-difference scheme (accurate to order 4 in h)

$$\frac{1}{12}(d_{i-1}^2 + 10d_i^2 + d_{i+1}^2) = \frac{u_{i+1} - 2u_i + u_{i-1}}{h^2}.$$

Let $n_h + 1$ be the total number of nodes. We indicate with **d** and **u** the vectors of length $n_h - 1$ containing the approximated second derivatives d_i^2 and the function values u_i at the internal nodes $(i = 1, \dots, n_h - 1)$, respectively. Let D be the tridiagonal matrix with diagonal entries equal $10/12$ and off-diagonals entries $1/12$, and T the tridiagonal matrix with $-2/h^2$ on the main diagonal and h^{-2} elsewhere.

Since the boundary conditions are of homogeneous Dirichlet type, the compact-finite-difference approximation satisfies the linear system $\mathbf{Dd} = \mathbf{Tu}$. The full solution is then obtained by adding the known values at the end points, $u_0 = u_N = 0$.

The approximation of the boundary-value problem at the internal nodes will thus read

$$-\mathbf{d} = \mathbf{f},$$

where $\mathbf{f} = (f(x_i))$, thus

$$-D^{-1}T\mathbf{u} = \mathbf{f} \Rightarrow -T\mathbf{u} = D\mathbf{f}. \qquad (3.33)$$

We may note that matrix T in fact represents the second order finite difference operator δ^2 defined by

$$\delta^2 u(x_i) = \frac{u_{i-1} - 2u_i + u_{i+1}}{h^2}$$

while, by simple inspection,

$$D = I + \frac{h^2}{12} T.$$

As a consequence, we may formally write equation (3.33) as

$$-\delta^2 u = \left(1 + \frac{h^2}{12}\delta^2\right) f.$$

Equivalently,

$$-\left(1 + \frac{h^2}{12}\delta^2\right)^{-1} \delta^2 u = f. \qquad (3.34)$$

Equation (3.34) will become handy when we look to the extension of this scheme to the multidimensional case.

As for the MATLAB implementation, we tip off the instructions for building the matrices of concern:

```
h = (xspan(2) - xspan(1))/nh;ni = nh - 1;e = ones(ni,1);
T = - spdiags([e -2*e e],-1:1,ni,ni)/h^2;
D = spdiags([e 10*e e],-1:1,ni,ni)/12;
```

having in our case set **xspan**=[0 1] and denoted by **nh** the number of intervals decomposing $(0, 1)$ evenly.

Numerical results

Computing the error in norm $\|\cdot\|_{\Delta,\infty}$ (**error**) and checking the order (**q**), for h ranging from $1/10$ to $1/160$, gives the following results, in accordance with the expected order 4

```
>> error =
   6.1191e-04 3.9740e-05 2.4765e-06 1.5466e-07 9.6648e-09
>> q =
   3.9447   4.0042   4.0011   4.0003
```
◇

Exercise 3.1.9. Making use of (2.45) solve the following fourth-order boundary-value problem: find $u : [0, 1] \to \mathbb{R}$ such that

$$\begin{cases} \dfrac{d^4u}{dx^4} - \dfrac{d^2u}{dx^2} + u = (4\pi^2(4\pi^2 + 1) + 1)\sin(2\pi x) & \text{on } (0,1), \\ u(0) = u(1) = 0, & \\ \dfrac{du}{dx}(0) = \dfrac{du}{dx}(1) = 2\pi. & \end{cases} \qquad (3.35)$$

Find the exact solution. Introduce suitable finite difference formulas to approximate derivatives at the boundary. Give an estimate of the convergence order, in terms of h, of the method used.

Solution 3.1.9.

Mathematical analysis

The differential equation is linear with constant coefficients and the left hand side is of the form $C \sin(2\pi x)$. Furthermore, the coefficient of x in the sinusoidal function, 2π, is clearly not a solution of the characteristic equation $\lambda^4 - \lambda^2 + 1 = 0$. Consequently [BD70], a particular solution of the differential equation is of the form $u_p(x) = A\sin(2\pi x) + B\cos(2\pi x)$. As no odd derivatives are present, we may discard the cosine term directly and by substitution obtain $u_p(x) = \sin(2\pi x)$. Since this function satisfies also the boundary conditions we have in fact found the exact solution of our problem, $u(x) = \sin(2\pi x)$.

Numerical approximation

Given a discretization of $(0,1)$ with uniform spacing h and nodes x_i, $i = 0, \ldots, n$, the centered-finite-difference discretization of (3.35) is, for $i = 2, \ldots, n-2$,

$$\frac{u_{i-2} - 4u_{i-1} + 6u_i - 4u_{i+1} + u_{i+2}}{h^4} - \frac{u_{i-1} - 2u_i + u_{i+1}}{h^2} + u_i \qquad (3.36)$$
$$= 4\pi^2(4\pi^2 + 1)\sin(2\pi x_i) + \sin(2\pi x_i).$$

The point is to complete it by forcing the boundary conditions on u and its first derivative. For the latter two, in particular, we use the decentred difference quotient $(2.31)_1$, which is accurate to order 2 precisely like (2.45). We therefore add the equations

$$-\frac{-3u_n + 4u_{n-1} - u_{n-2}}{2h} = 2\pi, \quad \frac{-3u_0 + 4u_1 - u_2}{2h} = 2\pi \qquad (3.37)$$

to (3.36), while $u_0 = u_n = 0$.

Numerical results

To check the order of convergence of the scheme we can build the corresponding linear system directly. As an alternative, we may use Program 8, which solves with the given difference quotients, a general problem of the form

$$
\begin{cases}
\alpha \dfrac{d^4 u}{dx^4} + \beta \dfrac{d^2 u}{dx^2} + \sigma u = f \text{ with } x \in (a, b), \\[2mm]
u(a) = g_D(a), \qquad\qquad u(b) = g_D(b), \\[2mm]
\dfrac{du}{dx}(a) = g_N(a), \qquad\qquad \dfrac{du}{dx}(b) = g_N(b),
\end{cases}
\tag{3.38}
$$

where $\alpha, \beta, \sigma \in \mathbb{R}$ and f, g_D, g_N are functions in x. The input parameters are

xspan: vector with two entries, xspan(1)=a, xspan(2)=b;
nh: an integer equal to the number of equidistant sub-intervals of (a, b);
alpha, beta, sigma: the values of the constants α, β, σ in (3.38);
fun, gD, gN: the names of the *inline functions* yielding f, g_D, g_N.

In output we obtain the array x of the discretization nodes and the corresponding array uh of values u_i.

Program 8 - bvp4fd : Approximation by finite differences of a fourth-order problem

```
function [x,uh]=bvp4fd(xspan,nh,alpha,beta,sigma,fun,gD,gN,varargin)
%BVP4FD solves a boundary-value problem of order  4
%    [X,UH]=BVP4FD(XSPAN,NH,ALPHA,BETA,SIGMA,FUN,GD,GN) solves
%    with the centred-finite-difference method of order 2
%    the problem
%       ALPHA D^4U + BETA D^2U + SIGMA U = FUN   in (XSPAN(1),XSPAN(2))
%       U(XSPAN(:))  = GD(XSPAN(:))
%       DU(XSPAN(:)) = GN(XSPAN(:))
%
%    ALPHA, BETA, SIGMA are reals, FUN, GD, GN inline functions.
%================
%    [X,UH]=BVP4FD(XSPAN,NH,ALPHA,BETA,SIGMA,FUN,GD,GN,P1,P2,...)
%    passes the optional parameters P1,P2,.. to the functions FUN,GD,GN.
%
```

We then check the convergence order, in discrete max norm, using the instructions

```
>> fun=inline('(4*pi^2+1)*4*pi^2*sin(2*pi*x)+sin(2*pi*x)');
>> gD=inline('sin(2*pi*x)');
>> gN=inline('2*pi*cos(2*pi*x)');
>> xspan=[0 1]; alpha = 1; beta = -1; sigma = 1;
```

```
>> nh=10;
>> for n=1:6
      [x,uh]=bvp4fd(xspan,nh,alpha,beta,sigma,fun,gD,gN);
      error(n) = norm(uh-sin(2*pi*x)',inf);
      nh=2*nh;
   end
   q=log(error(1:end-1)./error(2:end))/log(2)
   q =
      1.9359   1.9890   1.9956   1.9996   1.9994
```

The step h varies between $1/10$ and $1/320$. Order 2 is confirmed (had we taken a smaller h we would start seeing errors due to round-offs, as we have already noted in Exercise 2.3.3).   ◇

Remark 3.4 Assume that the boundary conditions $(3.35)_3$ for the problem in the previous exercise are replaced by conditions on the second derivative like

$$\frac{d^2u}{dx^2}|_{x=0} = \frac{d^2u}{dx^2}|_{x=1} = 0.$$

Set $v = -\dfrac{d^2u}{dx^2}$ and $\mathbf{w} = [u, v]^T$. Equation $(3.35)_1$ reads

$$-\frac{d^2}{dx^2}\mathbf{w} + \begin{bmatrix} 0 & 1 \\ 1 & -1 \end{bmatrix}\mathbf{w} = \begin{bmatrix} 0 \\ -4\pi^2(4\pi^2 + 1)\sin(2\pi x) \end{bmatrix},$$

with $\mathbf{w}(0) = \mathbf{w}(1) = \mathbf{0}$. In this way, the problem can be promptly solved with methods for linear second order problems. We will see a similar situation in Exercise 5.1.9 and Remark 5.3.

Exercise 3.1.10. The temperature distribution u on a cylindrical cross-section of fuel of radius R, placed in a nuclear reactor, where the cylinder's lateral surface is kept at temperature U is governed by the following differential problem

$$\begin{cases} -\dfrac{d^2u}{dx^2} - \dfrac{1}{x}\dfrac{du}{dx} = 1 + \dfrac{1}{2}\left(\dfrac{x}{R}\right)^2, & x \in (0, R), \\[2mm] \dfrac{du}{dx}(0) = 0, & u(R) = U, \end{cases} \tag{3.39}$$

where the condition at $x = 0$ comes from the symmetry of the cylinder. Define an approximation by finite differences using a *stencil* with 3 nodes for this problem. For the discretization nodes use $x_i = R\sin(\pi i/(2n))$, $i = 0, \ldots, n$, and set $R = 1$, $U = 1$, so that the grid is not uniform. Compare the result with the exact solution.

Solution 3.1.10.

Mathematical analysis

Since the right hand side is a quadratic polynomial, we compute the analytic solution by looking for a polynomial solution of the form

$$u(x) = a_0 x^4 + a_1 x^3 + a_2 x^2 + a_3 x + a_4. \tag{3.40}$$

We observe first that $du/dx(0) = 0$ holds if $a_3 = 0$. We demand that (3.40) satisfies the differential equation, so

$$-16a_0 x^2 - 9a_1 x - 4a_2 = 1 + \frac{1}{2R^2} x^2,$$

and hence $a_0 = -\frac{1}{32R^2}$, $a_1 = 0$, $a_2 = -1/4$. Imposing the Dirichlet condition at $x = R$ we eventually get $a_4 = U + 9R^2/32$. Therefore the analytic solution is

$$u(x) = -\frac{1}{32} \frac{x^4}{R^2} - \frac{1}{4} x^2 + U + \frac{9}{32} R^2.$$

Numerical approximation

The exercise asks to implement the finite-difference method on a non-uniform grid. Let us then put

$$x_{i+1} = x_i + h_i, \qquad i = 0, \ldots, n-1,$$

where the $h_i > 0$ are given in such a way that x_i fulfill the law indicated in the exercise.

To find a scheme that approximates D_i^2 with 3 nodes we use the method of indeterminate coefficients. Namely, we seek an approximation of the first derivative of the form

$$\frac{du}{dx}(x_i) \simeq a_0 u_{i-1} + a_1 u_i + a_2 u_{i+1}. \tag{3.41}$$

Using the Taylor series expansions

$$u_{i+1} = u_i + h_i D u_i + \frac{h_i^2}{2} D^2 u_i + \frac{h_i^3}{6} \frac{d^3 u}{dx^3}(\xi_+),$$

$$u_{i-1} = u_i - h_{i-1} D u_i + \frac{h_{i-1}^2}{2} D^2 u_i - \frac{h_{i-1}^3}{6} \frac{d^3 u}{dx^3}(\xi_-)$$

we find

$$a_0 u_{i-1} + a_1 u_i + a_2 u_{i+1} = u_i(a_0 + a_1 + a_2)$$

$$+ D u_i(-a_0 h_{i-1} + a_2 h_i) + \frac{1}{2} D^2 u_i(a_0 h_{i-1}^2 + a_2 h_i^2) + o(h_{i-1}^2 + h_i^2).$$

Now, we impose that the required scheme approximates the first derivative of u at x_i. We will have to impose the constraints

$$
\begin{cases}
a_0 + a_1 + a_2 = 0, \\
-a_0 h_{i-1} + a_2 h_i = 1, \\
a_0 h_{i-1}^2 + a_2 h_i^2 = 0.
\end{cases}
$$

The solution to this system reads

$$
a_0 = -\frac{h_i}{h_{i-1}(h_i + h_{i-1})}, \quad a_1 = \frac{h_i^2 - h_{i-1}^2}{h_{i-1}h_i(h_i + h_{i-1})}, \quad a_2 = \frac{h_{i-1}}{h_i(h_i + h_{i-1})},
$$

so

$$
Du_i \simeq \frac{(h_{i-1}/h_i)(u_{i+1} - u_i) - (h_i/h_{i-1})(u_{i-1} - u_i)}{h_i + h_{i-1}}, \tag{3.42}
$$

which is accurate to order 2, as the error goes (up to higher-order infinitesimals) as $\frac{h_{i-1}h_i}{6} d^3 u/dx^3 u(\xi)$ with $\xi \in (x_{i-1}, x_{i+1})$. We have assumed that $u \in C^3(0, R)$, a condition satisfied by the exact solution.

Similar calculations will give us the approximation of the second derivative

$$
D^2 u_i \simeq \frac{2u_{i-1}}{h_{i-1}(h_i + h_{i-1})} - \frac{2u_i}{h_{i-1}h_i} + \frac{2u_{i+1}}{h_i(h_i + h_{i-1})} \tag{3.43}
$$

which is, though, accurate to order 1 only, if $h_i \neq h_{i-1}$, because the error behaves like $(h_i - h_{i-1})/3 d^3 u(\xi)/dx^3$ with $\xi \in (x_{i-1}, x_{i+1})$ (always assuming $u \in C^3$).

Thus, at the internal nodes, the finite-difference scheme will lead to the linear equations

$$
\frac{u_{i-1}}{h_{i-1}(h_i + h_{i-1})}\left(-2 + \frac{h_i}{x_i}\right) + \frac{u_i}{h_i h_{i-1}}\left(2 - \frac{h_i - h_{i-1}}{x_i}\right)
$$
$$
+ \frac{u_{i+1}}{h_i(h_i + h_{i-1})}\left(-2 - \frac{h_{i-1}}{x_i}\right) = 1 + \frac{1}{2}\left(\frac{x_i}{R}\right)^2
$$

for $i = 1, \ldots, n-1$. To these we add the equations at the nodes $x_0 = 0$ and $x_n = R$ derived by applying the boundary conditions. Whereas the Dirichlet condition at $x = R$ is trivially obtained by setting $u_n = U$, the homogeneous Neumann condition at $x = 0$ is more delicate. We introduce a decentred difference quotient at x_0, still with a three-node *stencil* on the non-uniform grid. Analogous considerations to those done before produce an approximation of the form

$$
\frac{du}{dx}(x_0) \simeq -\frac{2h_0 + h_1}{h_0(h_1 + h_0)}u_0 + \frac{h_1 + h_0}{h_1 h_0}u_1 - \frac{h_0}{h_1(h_1 + h_0)}u_2 \tag{3.44}
$$

(notice that it coincides with (2.31) when $h_0 = h_1 = h$). The boundary condition is imposed by setting the right-hand side of (3.44) to zero.

Creating a code for the non-uniform case is slightly more involved than in the uniform case, since now the the matrix coefficients changes on each row. We construct the coefficients with a suitable for-loop, as is done in Program 9. As input we must provide: the distribution of nodes in x, the interval's length in the number R and the temperature at $x = R$ in the scalar U. The output is the temperature distribution, stored by the array uh.

Program 9 - FDnonunif : Finite differences on a non-uniform grid

```
function uh=FDnonunif(x,R,U)
%FDNONUNIF solves exercise (4.2.4) on a non-uniform grid.
%    UH=FDNONUNIF(X,R,U) calcolates the solution UH with a
%    finite-difference method with three nodes on a non-uniform grid given by
%    the array X. R is the radius of the cylinder of radioactive substance,
%    , U the given temperature at x=R.
```

Let us compute the solution on a sparse grid with 21 nodes distributed as required. It suffices to give the instructions

```
>> R=1; U=1; n=20; i=[0:n]; xnu=R*(sin(pi*i/(2*n)));
>> uhnu=FDnonunif(xnu,R,1);
```

The graph, in Fig. 3.17, shows how the solution found computationally is practically indistinguishable from the exact one.   ◇

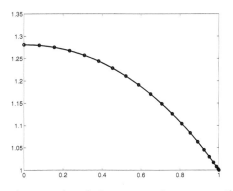

Fig. 3.17 Comparison between the solution computed on a non-uniform grid (dots) using 21 nodes and the analytic solution

Exercise 3.1.11. Find an approximation by the spectral collocation method on the Gauss-Chebyshev-Lobatto grid for the differential problem of Exercise 3.1.8. Analyze the error's behavior in discrete infinity norm as the number of nodes increases.

Solution 3.1.11.

Numerical approximation

Let D_n be the pseudo-spectral differentiation matrix introduced in (2.30) and computable by Program 7. Note first of all that the matrix generated by such program refers to the interval $(-1, 1)$, whereas our problem is defined on $(0, 1)$: therefore we need to change variables to $x = (1 + t)/2$, for $t \in [-1, 1]$, and then transform derivatives by the rule[2] $du/dx = (du/dt)(dt/dx) = 2du/dt$, applied twice. The matrix $4D_n^2 = 4D_n \cdot D_n$ will then correspond to the second derivative's approximation in $(0, 1)$. Denote by $\mathbf{u}$ the vector of unknown values of the approximate solution at the Gauss-Chebyshev-Lobatto nodes: owing to the boundary conditions we know $u_0 = u_n = 0$, and consequently we can reduce $\mathbf{u}$ to $\tilde{\mathbf{u}}$ at internal nodes only and eliminate the first and last columns of D_n^2. If D is the MATLAB matrix containing D^2 this can be done with the command

```
D=D(2:end-1,2:end-1);
```

With this new definition of the matrix the problem discretized by spectral collocation consists in finding $\tilde{\mathbf{u}} \in \mathbb{R}^{n-1}$ such that

$$-4D_n^2\tilde{\mathbf{u}} = 4\pi^2 \sin(2\pi\tilde{\mathbf{x}}),$$

where $\tilde{\mathbf{x}}$ is the vector of the coordinates of the Gauss-Chebyshev-Lobatto nodes interior to $(0, 1)$. The final vector $\mathbf{u}$ is obtained from $\tilde{\mathbf{u}}$ by adding the 2 boundary values.

For C^∞ solutions, the approximation error has an exponential behavior in n, so as n tends to infinity it goes to zero faster than any power of $1/n$ (see [Qua09, Ch. 10]).

Numerical results

Program 10 realizes precisely the proposed scheme, and requires as input parameters the end points of the problem's interval (given by **xspan**), the number of nodes n, the boundary values and a *function* defining the forcing term. The program is however more general since it allows also for non-homogeneous Dirichlet boundary conditions.

---

[2] For a generic interval (a, b) the transformation will be $dt/dx = 2/(b - a)$ with $x = t(b - a)/2 + (b + a)/2$.

Program 10 - diff1Dsp : Spectral collocation method for $-u'' = f$

```
function [x,us]=diff1Dsp(xspan,n,ulimits,f,varargin)
% DIFF1DSP solves -u"=f by spectral collocation.
%  [US,XGLC]=DIFF1DSP(XSPAN,N,ULIMTS,F) approximates -u"(x)=F(x)
%  for x in (XSPAN(1),XSPAN(2)) by the spectral-collocation method
%  on the grid on  N Gauss-Chebyshev-Lobatto nodes.
%  US is the  approximated solution.
%
%  The boundary conditions are u(XSPAN(1))=ULIMITS(1) and
%  u(XSPAN(2))=ULIMITS(2). F can be an inline function.
%
%  [US,XGLC]=DIFF1DSP(XSPAN,N,ULIMITS,F,P1,P2,..) passes P1,P2,..
%  as optional parameters to the inline function F.
```

In the case of concern the possible instructions are

```
>> xspan = [0 1]; ulimits = [0 0]; f = inline('4*pi^2*sin(2*pi*x)');
>> [x,us] = diff1Dspect(xspan,n,ulimits,f);
```

with varying n. Fig. 3.18 shows the error in discrete infinity norm as n grows: as predicted by the regularity of the analytic solution, we have spectral (exponential) behavior.   ◇

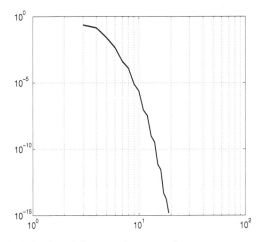

Fig. 3.18 Log-scale behavior of the error in terms of n

3.2 Elliptic problems in 2D

Exercise 3.2.1. On the unit square $\Omega \equiv (0,1) \times (0,1)$ we consider the following problem

$$\begin{cases} -\Delta u + 2u = 0 \ \text{ in } \Omega, \\ \\ u = g \equiv e^{x+y} \quad \text{ on } \partial\Omega. \end{cases} \tag{3.45}$$

1. Give the weak formulation and prove the well posedness of the problem. Find the analytical solution.
2. Give the linear finite element approximation using `Freefem++` on uniform grids.
3. Compute the $H^1(\Omega)$ and $L^2(\Omega)$ errors as function of the mesh grid h.

Solution 3.2.1.

Mathematical analysis

Problem (3.45) features non homogeneous Dirichlet conditions. The weak formulation is obtained multiplying equation (3.45)$_1$ by a generic function $v \in V \equiv H^1_{\Gamma_D}$ (with $\Gamma_D \equiv \partial\Omega$) and integrating by parts the Laplacian term. We get

$$\int_\Omega \nabla u \cdot \nabla v \, d\omega + 2\int_\Omega uv \, d\omega = 0.$$

As in the 1D case, we look for $u \in W_g = \{w \in H^1(\Omega) : w_{|\partial\Omega} = g\}$ s.t.

$$a(u,v) \equiv \int_\Omega \nabla u \cdot \nabla v \, d\omega + 2\int_\Omega uv \, d\omega = 0 \quad \forall v \in V. \tag{3.46}$$

The bilinear form is symmetric, continuous, coercive in V. We have indeed

$$|a(u,v)| \leq \|\nabla u\|_{L^2(\Omega)}\|\nabla v\|_{L^2(\Omega)} + 2\|u\|_{L^2(\Omega)}\|v\|_{L^2(\Omega)}$$
$$\leq 2\|u\|_V\|v\|_V.$$

Moreover for any $u \in V$ we have

$$a(u,u) \geq \|\nabla u\|^2_{L^2(\Omega)} + 2\|u\|^2_{L^2(\Omega)} \geq \|u\|^2_V.$$

From Corollary 3.1 solution to (3.46) exists and is unique.

We assume that the solution has the form $u = X(x)Y(y)$. Notice that in particular $g = B_x(x)B_y(y)$, where the boundary functions are $B_x = e^x$ and

$B_y = e^y$. With these positions, we write

$$-X''Y - XY'' + 2XY = 0.$$

Assuming $X \neq 0$ and $Y \neq 0$, we reformulate the equation in the form

$$\frac{X''}{X} = 2 - \frac{Y''}{Y}.$$

The two sides of the equation above are functions of two different variables, so they are constants. For a real constant K, we have

$$\frac{X''}{X} = K.$$

Assume that $K > 0$, so to obtain

$$X = C_{1,K}e^{\sqrt{K}x} + C_{2,K}e^{-\sqrt{K}x}.$$

By prescribing the boundary function B_x we conclude that

$$C_{1,K} = \begin{cases} 1 \text{ for } K = 1 \\ 0 \text{ for } K \neq 1 \end{cases}, C_{2,K} = 0, \quad \forall K > 0.$$

Proceeding similarly for Y (for $K = 1$) we obtain that $Y = e^y$ and conclude that $u_{ex} = e^{x+y}$. Our theoretical analysis concludes that this is the only solution to the problem.

Numerical approximation

Let V_h be the subspace of V of linear finite elements. The Galerkin finite element formulation of (3.46) reads: find $u_h \in W_{g,h}$ s.t.

$$a(u_h, v_h) = 0 \quad \forall v_h \in V_h. \tag{3.47}$$

We solve the problem with the code **Freefem++**. We write a file with extension **edp**. In this case, we call this file **ellextut1.edp**.

First of all, let us define the mesh on the square domain at hand. To this aim, it is enough to write in **ellextut1.edp** the following instruction

```
mesh Th=square(10,10);
```

We are building an object called **Th** of type **mesh** corresponding to a decomposition of a square unit in $10 \times 10 \times 2$ rectangular triangles. Once **Th** has been computed, we define the finite element space P^1, with the instruction

```
fespace Xh(Th,P1);
```

and declare in **Freefem++** the unknown u_h and the generic test function v_h as element of **Vh**

```
Xh uh,vh;
```

Instruction **problem** defines the variational problem at hand. We add the following code

```
func g=exp(x+y);
problem Problem1(uh,vh) =
    int2d(Th)(dx(uh)*dx(vh) + dy(uh)*dy(vh))
  + int2d(Th)(2*uh*vh)
  + on(1,2,3,4,uh=g);
```

The first two terms after the instruction **problem** are the two integrals of the Galerkin formulation (3.47). The last line prescribe Dirichlet conditions[3]. Function **g** (instantiated with the command **func**) is the boundary data. Finally, error computation in **ellextut1.edp** is given by the instructions

```
func ue=exp(x+y);
func dxue=exp(x+y);
func dyue=exp(x+y);
real errL2 = sqrt(int2d(Th)((uh-ue)^2));
real errH1 = sqrt(int2d(Th)((uh-ue)^2)+
    int2d(Th)((dx(uh)-dxue)^2)+int2d(Th)((dy(uh)-dyue)^2));
```

Command **plot(uh)** generates a plot with the isolines of the solution.

Numerical results

In Fig. 3.19 we report $H^1(\Omega)$ and $L^2(\Omega)$ norm errors as a function of the discretization step h. This is in perfect agreement with the theory.   ◇

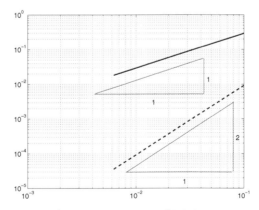

Fig. 3.19 Behavior of the $H^1(\Omega)$ (solid line) and $L^2(\Omega)$ (dashed line) errors as a function of h for the Exercise 3.2.1

---

[3] **Freefem++** prescribes these conditions as a *penalization* of the matrix obtained by the discretization. See Appendix A.

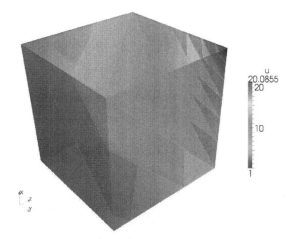

Fig. 3.20 Isosurfaces of the solution for the 3D extension of Exercise 3.2.1

Remark 3.5 In 3D we could consider the following similar problem[4]

$$-\triangle u + 3u = 0, \quad [0,1] \times [0,1] \times [0,1]$$

with boundary data $u = e^{x+y+z}$ on $\partial \Omega$. Function $u = e^{x+y+z}$ is also the analytical solution.

In Fig. 3.20 we report the solution computed with the library LifeV (see [lif10]).

We report a snapshot of the main code generating these results in Program 11. The main code (in C++), after reading the mesh, selecting some quantities for the computation (the degree of finite elements, the quadrature rules for the volume and the boundary) and managing the boundary data (*pre-processing*), performs the assembling of the matrix of the problem (*assembling*). Successively the assembled system is solved by a linear solver (*solution*) and the solution exported for the visualization (*post-processing*). LifeV resorts to some external library for some tasks, like GetPot for the definitions and the input of some coefficients and Trilinos/AztecOO[HBH+05] for the solution of the linear system.

The reader interested in more implementation details is recommended to check out the web site of LifeV.

Program 11 - main3-2-1 : 3D version of Exercise 3.2.1

```
int main( int argc, char** argv ) {

  typedef RegionMesh3D<LinearTetra> mesh_Type;
  // =====================================================
  // Load user preferences from file
  // =====================================================
  GetPot command_line( argc, argv );
  const char* data_file_name =
    command_line.follow( "data", 2, "-f", "--file" );
  GetPot data_file( data_file_name );
  Real reactionCoeff(data_file( "problem/reactionCoefficient",3.));
```

---

[4] We thank Tiziano Passerini of Emory University for this remark.

```
Real diffusionCoeff(data_file( "problem/diffusionCoefficient",1.));
  (...)
// ====================================================
// Finite elements definitions
// ====================================================
// the reference finite element
RefFE referenceFE =
strToRefFE( data_file("space_discretization/referenceFE",
                      "feTetraP1") );
RefFE referenceBoundarydFE = referenceFE.boundaryFE();
// choice of the quadrature rule (volume-surface integration)
const QuadRule& quadRule        = quadRuleTetra15pt;
const QuadRule& quadRuleBoundary = quadRuleTria3pt;
// ====================================================
// Mesh construction
// ====================================================
mesh_Type mesh;
readMeshFile( mesh, data_file, "mesh" );
// ====================================================
// Current FE classes for the problem under study with
// mapping and quadrature rules
// ====================================================
  CurrentFE   fe  ( referenceFE, getGeoMap(mesh),quadRule);
  CurrentBdFE feBd( referenceBoundarydFE,
              getGeoMap(mesh).boundaryMap(), quadRuleBoundary);
// ====================================================
// Update of the DOF local to global maps
// and the list of DOFs for the boundary conditions
// ====================================================
 Dof dof( referenceFE );
 dof.update( mesh );
 BCh.bdUpdate( mesh, feBd, dof );
// ====================================================
// Initialization of the vectors of unknowns and rhs
// ====================================================
UInt dim = dof.numTotalDof();
std::cout << "Total number of DOFs = " << dim << std::endl << std::endl;
ScalUnknown<Vector> unknown( dim ), rhs( dim ), exactSolution( dim );
unknown       = ZeroVector( dim );
rhs           = ZeroVector( dim );
exactSolution = ZeroVector( dim );
// ====================================================
// Exporter for the results
// ====================================================
 Ensight< mesh_Type > exporter ( data_file, &mesh, "elliptic3D" );
exporter.addVariable(ExporterData::ScalarData, "u", &unknown,0, dim );
// ====================================================
// Pattern construction and matrix assembling
// ====================================================
// pattern for the linear operator
 MSRPatt pattA( dof );
// A: stiff matrix + mass matrix
 MSRMatr<double> A( pattA );
 A.zeros();
```

```
  ElemMat elmat( fe.nbNode, 1, 1 );
  for( UInt i = 1; i <= mesh.numElements(); i++ ){
     fe.updateFirstDerivQuadPt( mesh.element(i) );
     elmat.zero();
     mass ( reactionCoeff,  elmat, fe );
     stiff( diffusionCoeff, elmat, fe );
     assemb_mat( A, elmat, fe, dof, 0, 0 );}
  // ====================================
  // Treatment of the Boundary conditions
  // ====================================
  std::cout << "*** BC Management\t\t: "<<std::endl;
  bcManage( A, rhs, mesh, dof, BCh, feBd, 1., 0. );
  // =============================
  // Resolution of the linear system
  // =============================
  // build the linear solver
  SolverAztec linearSolver;
  linearSolver.setMatrix( A );
  linearSolver.setOptionsFromGetPot( data_file, "aztec" );
  linearSolver.solve( unknown.giveVec(), rhs.giveVec() );
  // =========================================
  // Export of the results and comparison with the analytical unknown
  // =========================================
     exporter.postProcess( 0. );
     (...)
}
```

Exercise 3.2.2. Consider the problem (see [SV09])

$$\begin{cases} -\triangle u = -y & \text{in } \Omega, \\ u = 1 & \text{on } \partial\Omega, \end{cases} \tag{3.48}$$

formulated on a circular domain centered in the origin, with radius 1. Write the weak formulation of the problem and prove that it is well posed. Find the analytical solution. Write a code in **Freefem++** for solving the problem with finite elements of order 1 and compute the errors.

Solution 3.2.2.

Mathematical analysis

Let us write $v = u - 1$. The problem in v reads

$$\begin{cases} -\triangle v = -y & \text{in } \Omega, \\ v = 0 & \text{on } \partial\Omega. \end{cases} \tag{3.49}$$

The weak formulation for this problem is obtained in the usual way. Let V be the $H_0^1(\Omega)$ space. We have: find $v \in V$ s.t. for any $\varphi \in V$

$$\int_\Omega \nabla v \cdot \nabla \varphi = -\int_\Omega y\varphi.$$

The well posedness of the problem is an immediate consequence of the Lax-Milgram Lemma.

In this case, we can find the analytical solution (see [SV09]) by writing the problem in polar coordinates r, ϑ such that

$$x = r\cos\vartheta, \quad y = r\sin\vartheta.$$

The solution $V(r, \vartheta) = v(r\cos\vartheta, r\sin\vartheta)$ is 2π periodic with respect to the angle ϑ. After writing the Laplace operator in polar coordinates, the problem at hand reads

$$\Delta v = \frac{\partial^2 V}{\partial r^2} + \frac{1}{r}\frac{\partial V}{\partial r} + \frac{1}{r^2}\frac{\partial^2 V}{\partial \vartheta^2} = r\sin\vartheta$$

with the boundary conditions $V(1, \vartheta) = 0$. The functional form of the right hand side is written in terms of a finite Fourier expansion and this suggests to look for solutions in the form

$$V(r, \vartheta) = b(r)\sin\vartheta$$

with $b(1) = 0$. In fact, when we plug this expression into the equation, we obtain (the prime denotes here differentiation with respect to r)

$$\left(b'' + \frac{1}{r}b' - \frac{1}{r^2}b\right)\sin\vartheta = r\sin\vartheta.$$

This leads to the Cauchy-Euler equation

$$r^2 b'' + rb' - b = r^3.$$

The homogeneous equation associated (of Cauchy-Euler type) can be solved (see e.g. [BD70]) admits solutions in the form

$$b_H = r^k$$

for some k real. In particular, we have

$$r^2 b_H'' + rb_H' - b_H = k(k-1)r^k + kr^k - r^k = (k^2 - 1)r^k = 0.$$

Nontrivial solutions are obtained therefore for $k = \pm 1$, to have

$$b_H = C_1 r + C_2 \frac{1}{r}.$$

A particular solution in consideration of the cubic right hand side can be written in the form

$$b_P = Ar^3,$$

leading to

$$(6A + 3A - A)r^3 = r^3 \Rightarrow A = \frac{1}{8}.$$

The general solution reads therefore

$$b(r) = \frac{1}{8}r^3 + C_1 r + C_2 \frac{1}{r}.$$

Since we expect the solution to be bounded for $r = 0$, we take $C_2 = 0$ and the boundary condition in $r = 1$ imposes that $C_1 = -\frac{1}{8}$. We obtain therefore the analytical solution

$$V(r, \vartheta) = \frac{1}{8}r(r^2 - 1)\sin\vartheta \Rightarrow v(x, y) = \frac{1}{8}(x^2 + y^2 - 1)y.$$

The solution for u reads therefore

$$u = 1 + \frac{1}{8}(x^2 + y^2 - 1)y.$$

Numerical approximation

In **Freefem++** we define the boundary of the circular domain Ω with the following instruction

```
border gamma(t=0,2*pi)x=cos(t);y=sin(t);label=1;
```

This describes the circular domain in the usual parametric form. Variable **label** associates a code (1 in this case) to the boundary. Different labels will be assigned for prescribing different boundary conditions on different parts of the boundary.

The mesh is generated by the instruction

```
mesh Th=buildmesh(gamma(n));
```

where **n** is the integer number specifying the number of uniform intervals in the range $[0, 2\pi]$ of parameter **t**. We report the code for the sake of completeness.

```
border gamma(t=0,2*pi)x=cos(t);y=sin(t);label=1;
int n=80;
mesh Th=buildmesh(gamma(n));
fespace Xh(Th,P1);
Xh uh,vh;

problem Problem1(uh,vh) =
    int2d(Th)(dx(uh)*dx(vh) + dy(uh)*dy(vh)) + int2d(Th)(y*vh)
```

```
   + on(1,uh=1);
Problem1;
plot(uh,ps="circle.eps");
func ue=1+1./8.*(x^2+y^2-1)*y;
func dxue=1./4.*x*y;
func dyue=1./8.*(x^2+y^2-1)+ 1./4.*y^2;
real errL2 = sqrt(int2d(Th)((uh-ue)^2));
real errH1 = sqrt(int2d(Th)((uh-ue)^2)+
   int2d(Th)((dx(uh)-dxue)^2)+int2d(Th)((dy(uh)-dyue)^2));
```

Numerical results

In Fig. 3.21 we report on the left the isolines of the solution for $n = 10$ (left) and $n = 160$ (right). Notice that the circular domain Ω is approximated by a domain Ω_h that is a regular polygon with **n** edges. If we use affine maps, discretization error has order 2 (see Chapter 2 and [EG04]) therefore it does not affect convergence error of linear and quadratic finite elements. Table 3.4 displays the errors for several sizes of the mesh. Even though the number n of nodes on the boundary is just a rough estimate of the size of the mesh, results are in agreement with the theory.   ◇

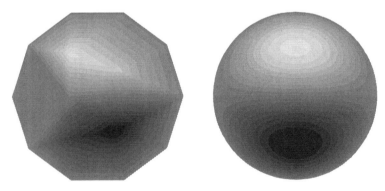

Fig. 3.21 Isolines of two solutions computed with two meshes with different size

Table 3.4 Errors for different sizes of the mesh in Exercise 3.2.2

n	$\|e\|_{L^2}$	$\|e\|_{H^1}$
10	0.0226671	0.110564
20	0.00675168	0.0623377
40	0.00172512	0.0307597
80	0.000466826	0.0165202
160	0.000107896	0.0078256

Exercise 3.2.3. Let Ω be an open bounded subset of $\mathbb{R}^2$ with smooth boundary $\partial\Omega = \Gamma_D \cup \Gamma_N$, $\Gamma_D \cap \Gamma_N = \emptyset$ and denote with $\mathbf{n}$ the outward unit vector normal to $\partial\Omega$. Let $\mathbf{b} : \Omega \to \mathbb{R}^2$ be s.t. $\mathbf{b} \cdot \mathbf{n} \geq 0$ on Γ_N and $\mu : \Omega \to \mathbb{R}$ two continuous functions, $f : \Omega \to \mathbb{R}$ a function of $L^2(\Omega)$ and $g_D : \Gamma_D \to \mathbb{R}$, $g_N : \Gamma_N \to \mathbb{R}$ two regular functions. Let us assume that

$$0 < \mu_0 \leq \mu(\mathbf{x}) \leq \mu_1, \; |\mathbf{b}(\mathbf{x})| \leq b_1, \; \nabla \cdot \mathbf{b} = 0, \; \forall \mathbf{x} \in \Omega.$$

1. Find the weak formulation of the following advection diffusion problem

$$\begin{cases} -\nabla \cdot (\mu\nabla u) + \mathbf{b} \cdot \nabla u = f & \text{in } \Omega, \\ u = g_D & \text{su } \Gamma_D, \\ \mu\nabla u \cdot \mathbf{n} = g_N & \text{su } \Gamma_N. \end{cases} \quad (3.50)$$

2. Show that the weak solution to (3.50) exists and is unique.
3. Assume that $\Omega = (0,1) \times (0,1)$, $\mu = 1$, $\mathbf{b} = [1,1]^T$. Find the portion Γ_N of $\partial\Omega$ such that $\mathbf{b} \cdot \mathbf{n} \geq 0$. Let $\Gamma_D = \partial\Omega \setminus \Gamma_N$, f, g_D and g_N such that the exact solution to (3.50) is $u(x,y) = \sin(2\pi x)\cos(2\pi y)$. Solve the problem with **Freefem++** using triangular finite elements of order 1 and 2. Check the convergence properties.

Solution 3.2.3.

Mathematical analysis

Let us multiply the differential equation in (3.50) by a function $v \in V \equiv H^1_{\Gamma_D}$ and integrate on Ω. We get

$$-\int_\Omega \nabla \cdot (\mu\nabla u)v \, d\omega + \int_\Omega \mathbf{b} \cdot \nabla u v \, d\omega = \int_\Omega fv \, d\omega.$$

By the Green formula, we find

$$\int_\Omega \mu\nabla u \cdot \nabla v \, d\omega + \int_\Omega \mathbf{b} \cdot \nabla u v \, d\omega = \int_\Omega fv \, d\omega + \int_{\Gamma_N} g_N v \, d\gamma.$$

With usual procedure, the weak formulation of (3.50) reads: look for $u \in W_g = \{w \in H^1(\Omega) : w_{|\Gamma_D} = g_D\}$ s.t.

$$a(u,v) = F(v) \qquad \forall v \in V, \quad (3.51)$$

where $a(u,v) \equiv \int_\Omega \mu\nabla u \cdot \nabla v \, d\omega + \int_\Omega \mathbf{b} \cdot \nabla u v \, d\omega$ and $F(v) \equiv \int_\Omega fv \, d\omega + \int_{\Gamma_N} g_N v \, d\gamma$. Let us verify the assumptions of Corollary 3.1.

Form $a(\cdot, \cdot)$ and functional $F(\cdot)$ are bilinear and linear respectively as immediate consequence of integral properties. Moreover

$$|F(v)| \leq ||f||_{L^2(\Omega)} ||v||_{L^2(\Omega)} + ||g_N||_{L^2(\Gamma_N)} ||v||_{L^2(\Gamma_N)}$$

$$\leq \left(||f||_{L^2(\Omega)} + \gamma_T ||g_N||_{L^2(\Gamma_N)} \right) ||v||_{H^1(\Omega)},$$

being γ_T constant in the trace inequality (1.11).

To prove that a is coercive, notice that for any function $v \in V$ we have

$$\int_\Omega v\mathbf{b} \cdot \nabla v = \frac{1}{2} \int_\Omega \mathbf{b} \cdot \nabla v^2 = \frac{1}{2} \int_{\Gamma_N} \mathbf{b} \cdot \mathbf{n} v^2$$

where we exploit the fact that $\nabla \cdot \mathbf{b} = 0$. Consequently,

$$a(u, u) = \int_\Omega \mu |\nabla u|^2 \, d\omega + \frac{1}{2} \int_{\Gamma_N} \mathbf{b} \cdot \mathbf{n} u^2.$$

From $\mathbf{b} \cdot \mathbf{n} \geq 0$ on Γ_N, we conclude

$$a(u, u) \geq \mu_0 ||\nabla u||_{L^2(\Omega)}^2.$$

Since u vanishes on a part of the boundary with positive measure, we can resort to the Poincaré inequality, so that we prove that the bilinear form is coercive with coercivity constant $\alpha = \mu_0 / (1 + C_P^2)$, being C_P the Poincaré constant. We prove also that $a(\cdot, \cdot)$ is continuous. Actually, we have

$$|a(u, v)| \leq \left| \int_\Omega \mu \nabla u \cdot \nabla v \, d\omega \right| + \left| \int_\Omega \mathbf{b} \cdot \nabla u v \, d\omega \right|$$

$$\leq \mu_1 ||\nabla u||_{L^2(\Omega)} ||\nabla v||_{L^2(\Omega)} + b_1 ||\nabla u||_{L^2(\Omega)} ||v||_{L^2(\Omega)}$$

$$\leq M ||u||_V ||v||_V,$$

where $M = \max\{\mu_1, b_1\}$. Solution therefore exists and is unique.

Numerical approximation

Galerkin approximation of (3.51) reads: find $u_h \in W_{g,h}$ s.t.

$$a(u_h, v_h) = F(v_h) \qquad \forall v_h \in V_h,$$

being V_h a finite element subspace of V. In particular we will use subspaces of piecewise polynomial functions of degree 1 and 2.

Numerical results

For the given function to be solution of the problem, we set $f(x,y) = 4\pi^2(\sin(2\pi(x+y))+\sin(2\pi(x-y)))+2\pi\cos(2\pi(x+y))$, $g_D = \sin(2\pi x)\cos(2\pi y)$ and $g_N = -2\pi\cos(2\pi x)\cos(2\pi y)$. Notice that for the given vector $\mathbf{b}$, the portion of the boundary such that $\mathbf{b}\cdot\mathbf{n} \geq 0$ is given by the edges $(x = 1, 0 \leq y \leq 1)$ and $(0 \leq x \leq 1, y = 1)$ which are conventionally labeled by 2 and 3 by FreeFem++.

The problem is defined by the following commands.

```
func gD=sin(2*pi*x)*cos(2*pi*y);
func gN2=2*pi*cos(2*pi*x)*cos(2*pi*y);
func gN3=-2*pi*sin(2*pi*x)*sin(2*pi*y);
func f=8*pi^2*sin(2*pi*x)*cos(2*pi*y) +
           2*pi*cos(2*pi*x)*cos(2*pi*y)-
           2*pi*sin(2*pi*x)*sin(2*pi*y);

problem Problem1(uh,vh) =
     int2d(Th)(dx(uh)*dx(vh) + dy(uh)*dy(vh))
   + int2d(Th)((dx(uh)+dy(uh))*vh)
   - int2d(Th)(f*vh)
   + on(1,2,3,4,uh=gD)- int1d(Th,2)(gN2*vh) - int1d(Th,3)(gN3*vh);
```

Notice the use of command int1d that computes a 1D integral, in our case on the Γ_N sides of the computational domain with mesh Th.

In Table 3.5 we report the error behavior that is in agreement with the theory.   ◇

Table 3.5 Error behavior of the $L^2(\Omega)$ and $H^1(\Omega)$ errors for linear (left) and quadratic (right) finite elements

h	$||e||_{L^2}$	$||e||_{H^1}$	h	$||e||_{L^2}$	$||e||_{H^1}$
0.1	0.0521632	1.36111	0.1	0.00196129	0.1681
0.05	0.0136516	0.693487	0.05	0.000244171	0.0429153
0.025	0.00345268	0.348391	0.025	3.05188e-05	0.0107885
0.0125	0.000865683	0.174403	0.0125	3.81529e-06	0.00270097
0.00625	0.000216578	0.0872272	0.00625	4.76933e-07	0.000675484

Exercise 3.2.4. A simple model used in oceanography is due to Stommel [Sto48]. In this model ocean is assumed to be flat and with uniform depth H, and no vertical movement of the water free boundary it is considered (since it is small compared with the horizontal one). Only Coriolis force, wind action on the surface and friction at the bottom are included.

Incompressibility implies that there exists a function ψ, called *stream function*, related to the velocity components by the equations

$$u = -\frac{\partial \psi}{\partial y}, \quad v = \frac{\partial \psi}{\partial x}.$$

In Stommel model for a rectangular ocean $\Omega = (0, L_x) \times (0, L_y)$ ψ is solution to the following elliptic problem

$$\begin{cases} -\Delta\psi - \alpha\dfrac{\partial \psi}{\partial x} = \gamma \sin(\pi y/L_y) & \text{in } \Omega, \\ \psi = 0 & \text{on } \partial\Omega, \end{cases} \tag{3.52}$$

con $\alpha = \frac{H\beta}{R}$, $\gamma = \frac{W\pi}{RL_y}$, being R the friction coefficient on the bottom, W a coefficient associated with the surface wind, $\beta = df/dy$ where f is the Coriolis parameter, which is in general function only of y (latitude).

1. Give the weak formulation of (3.52).
2. Compute the analytical solution of the model (3.52) for a constant β.
3. Solve the problem numerically with linear finite elements, using the Stommel's parameters: $L_x = 10^7$m, $L_y = 2\pi 10^6$m, $H = 200$m, $W = 0.3 \ 10^{-7}$m^{2}s^{-2}, $R = 0.6 \ 10^{-3}$ms^{-1}. Assume at first $\beta = 0$ and then $\beta = 10^{-10}$m^{-1}s^{-1}. Comment the results.

Solution 3.2.4.

Mathematical analysis

The weak formulation of the problem is obtained with the usual procedure and reads: find $\psi \in V \equiv H_0^1(\Omega)$ such that

$$a(\psi, v) \equiv \int_\Omega \nabla\psi \cdot \nabla v \, d\omega - \alpha \int_\Omega \frac{\partial \psi}{\partial x} v \, d\omega = \gamma \int_\Omega \sin(\pi y/L_y) v \, d\omega \quad \forall v \in V. \tag{3.53}$$

Well posedness analysis relies on the Lax-Milgram Lemma. With arguments similar to the ones used in the previous exercises, we can verify that the bilinear form $a(\cdot, \cdot)$ defined in (3.53) is continuous and coercive. For the coercivity, we just notice that we can write the second term on the left hand side of (3.53) as

$$\alpha \int_\Omega \frac{\partial \psi}{\partial x} v \, d\omega = \int_\Omega \nabla \cdot (\boldsymbol{\beta}\psi) v \, d\omega,$$

with $\boldsymbol{\beta} \equiv (\alpha, 0)^T$. Notice that

$$\alpha \int_\Omega \frac{\partial \psi}{\partial x} \psi \, d\omega = \int_\Omega \nabla \cdot (\boldsymbol{\beta}\psi)\psi \, d\omega = 0.$$

From $\nabla \cdot \boldsymbol{\beta} = 0$, we have in fact

$$\psi \nabla \cdot (\boldsymbol{\beta}\psi) = \psi \boldsymbol{\beta} \cdot \nabla \psi = \frac{1}{2} \boldsymbol{\beta} \cdot \nabla \psi^2 = \frac{1}{2} \nabla \cdot (\boldsymbol{\beta}\psi^2).$$

Therefore, since ψ vanishes on $\partial\Omega$,

$$\int_\Omega \nabla \cdot (\boldsymbol{\beta}\psi)\psi \; d\omega = \frac{1}{2} \int_\Omega \nabla \cdot (\boldsymbol{\beta}\psi^2) \; d\omega = \frac{1}{2} \int_{\partial\Omega} \psi^2 \boldsymbol{\beta} \cdot \mathbf{n} \; d\gamma = 0.$$

Since the functional $F(\cdot)$ defined on the right hand side of (3.53) is linear and continuous, we conclude that the solution exists and is unique.

We can compute this solution following the separation of variables approach. In particular, we assume

$$\psi(x,y) = X(x)Y(y),$$

with $X(0) = X(L_x) = 0$ and $Y(0) = Y(L_y) = 0$. Moreover, the forcing term depends only on y in the form $\gamma \sin(\pi y/L_y)$. Observe that the left hand side of the equation in (3.52) reads now

$$-X''Y - XY'' - \alpha X'Y$$

(where $\prime$ denotes differentiation of each function with respect its independent variable) and that the second derivative of a sinus function is still a sinus function with the same frequency. This suggests to do the following educated guess,

$$Y(y) = \sin\left(\frac{\pi}{L_y}y\right),$$

which fulfills the boundary conditions. We have then

$$\left(-X'' + \frac{\pi^2}{L_y^2}X - \alpha X'\right) \sin\left(\frac{\pi}{L_y}y\right) = \gamma \sin\left(\frac{\pi}{L_y}y\right).$$

The problem is now reduced to the solution of the constant coefficient ordinary differential equation

$$X'' + \alpha X' - \frac{\pi^2}{L_y^2}X = -\gamma.$$

A particular solution is clearly given by

$$X_P(x) = \frac{L_y^2 \gamma}{\pi^2}.$$

The general solution to the homogeneous equation reads

$$X_H(x) = C_1 e^{\lambda_1 x} + C_2 e^{\lambda_2 x}$$

where $\lambda_{1,2}$ are the roots of the algebraic equation

$$\lambda^2 + \alpha\lambda - \frac{\pi^2}{L_y^2} = 0,$$

i.e. $\lambda_{1,2} = -\dfrac{\alpha}{2} \pm \sqrt{\left(\dfrac{\alpha}{2}\right)^2 + \dfrac{\pi^2}{L_y^2}}$. For $X \equiv X_P + X_H$, when we prescribe the boundary conditions $X(0) = X(L_x) = 0$ we find C_1 and C_2, and the final solution reads

$$\psi = -\gamma \frac{L_y^2}{\pi^2}(pe^{\lambda_1 x} + qe^{\lambda_2 x} - 1)\sin(\pi y/L_y)$$

with $p = (1 - e^{\lambda_2 L_x})/(e^{\lambda_1 L_x} - e^{\lambda_2 L_x})$, $q = 1 - p$.

Numerical approximation

Galerkin approximation of (3.53) has the usual form: find $\psi_h \in V_h$ s.t. $a(\psi_h, v_h) = F(v_h)$ for any $v_h \in V_h$, being V_h the subspace of X_h^1 of functions vanishing on the boundary.

Numerical results

Before solving the problem numerically, we *scale* the lengths to avoid working with large numbers. We assume as characteristic length $L = 10^6$m and set $\hat{x} = x/L$, $\hat{y} = y/L$. Problem domain becomes $(0, 10) \times (0, 2\pi)$, and the equation reads

$$-\hat{\Delta}\psi - L\alpha\frac{\partial\psi}{\partial\hat{x}} = \frac{W\pi L}{R2\pi}\sin(\pi\hat{y}/(2\pi)),$$

where $\hat{\Delta}$ is the Laplace operator taken with respect to the adimensional variables $\hat{x}$ and $\hat{y}$. In general the convective coefficient $L\alpha = \dfrac{LH\beta}{R} = \dfrac{10^{10}}{3}\beta$, is large. In our case, β is assumed to be small, so the advection term will be small. In Chapter 4 we will see that if the convective term is large, numerical techniques are required for avoiding numerical oscillations in the solution.

Scaled problem in **Freefem++** syntax reads

```
problem Problem1(uh,vh) =
    int2d(Th)(dx(uh)*dx(vh)+dy(uh)*dy(vh))
  + int2d(Th)(Lalpha*dx(uh)*vh)-int2d(Th)(f*vh)+on(1,2,3,4,uh=0);
```

In Fig. 3.22 we report the solution obtained on a uniform grid for $\beta = 0$ (left) and $\beta = 10^{-10}$ (right). In the latter case, Coriolis force induces an asymmetric ψ. ◇

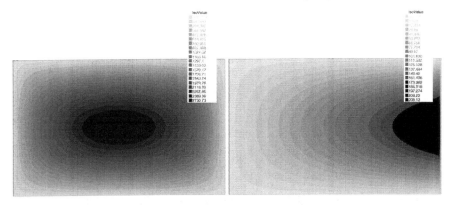

Fig. 3.22 Isolines for ψ computed with linear finite elements for $\beta = 0$ (left), and $\beta = 10^{-10}$ (right)

Exercise 3.2.5. (*) A pancreas cell is grown in gel before its transplantation. The cell is assumed to have a spherical shape with radius R (see Fig. 3.23) and needs Oxygen from the external environment. Oxygen diffuses in the cell and is consumed by the respiration. A realistic assumption is that consumed Oxygen is proportional to the quantity of Oxygen present in the cell. Usually, concentration of gas is assumed to be proportional to its partial pressure.
Oxygen diffusivity is denoted by $\mu > 0$ and the consumption rate is denoted by $K > 0$. They are assumed to be constant. We consider a steady problem with a spherical symmetry. Oxygen (constant) partial pressure in the gel is assumed to be known.
Write a mathematical model that describes Oxygen partial pressure in the cell, analyze it and give a numerical approximation with quadratic finite elements. Numerical data: $R = 0.05\text{mm}$, $\mu = 1.7 \times 10^{-6}\text{nM}/(\text{mm}$ s mmHg), $K = 10^{-2}\text{nM}/(\text{mm}^3$ s mmHg).
Is an external Oxygen partial pressure of 5mmHg enough to guarantee that at the core of the cell pressure is $\geq 1\text{mmHg}$?

Solution 3.2.5.

Mathematical modeling

Let u be the Oxygen partial pressure in the cell. For the mass conservation principle (see e.g. [Sal08], Chapter 2, [Far93]), dynamics of u is driven by the equation

$$\frac{du}{dt} + \nabla \cdot \mathbf{q} = -Ku,$$

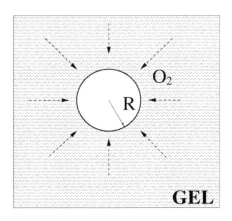

Fig. 3.23 Representation of the spherical cell for Exercise 3.2.5

where **q** is the incoming Oxygen flux. Since the cell is at rest, we do not have convective contributions to the flux. The only contribution is the diffusive one, described by *Fick's Law*,

$$\mathbf{q} = -\mu \nabla u.$$

Steady constant coefficient problem reads

$$-\mu \triangle u + Ku = 0, \quad \text{in} \quad \Omega, \tag{3.54}$$

where Ω is a sphere with radius R. We assume the sphere to be centered in the origin for simplicity. On $\partial\Omega$ we assume that u is known and equal to the external pressure,

$$u = u_{ext} \quad \text{per} \quad \mathbf{x} \in \partial\Omega. \tag{3.55}$$

Equation (3.54) is also known as *Helmholtz equation*.

Let us exploit the spherical symmetry. Let r, θ, ψ be the spherical coordinates. Equation (3.54) becomes (see e.g. [Sal08])

$$-\mu \frac{\partial^2 u}{\partial r^2} - \frac{2\mu}{r}\frac{\partial u}{\partial r} - \frac{\mu}{r^2}\left(\frac{1}{(\sin\psi)^2} + \frac{\partial^2 u}{\partial\theta^2} + \frac{\partial^2 u}{\partial\psi^2} + \cot\psi\frac{\partial u}{\partial\psi}\right) + Ku = 0.$$

Since all the data share the spherical symmetry, we assume that the solution is symmetric, that implies it depends only on the radial coordinate, i.e. $u = u(r)$. This means that the derivatives with respect to θ and ψ vanish and we obtain

$$-\mu \frac{\partial^2 u}{\partial r^2} - \frac{2\mu}{r}\frac{\partial u}{\partial r} + Ku = 0, \quad 0 < r < R.$$

If we multiply both sides by r^2, equation becomes

$$-\frac{\partial}{\partial r}\left(\mu r^2 \frac{\partial u}{\partial r}\right) + Kr^2 u = 0 \tag{3.56}$$

with the boundary condition

$$u(R) = u_{ext}. \tag{3.57}$$

Symmetry condition at the center of the cell requires the radial derivative to be zero,

$$u'(0) = 0. \tag{3.58}$$

Equation (3.56) and boundary conditions (3.57), (3.58) give a possible mathematical model for the problem at hand.

Mathematical analysis

Weak formulation of problem (3.54), (3.55) is standard. Let $G \in H^1(\Omega)$ be a lifting of the boundary data, with $G_{|\partial\Omega} = u_{ext}$. We look for $u \in G + V$ with $V \equiv H_0^1(\Omega)$ s.t.

$$\mu \int_\Omega \nabla u \cdot \nabla v d\omega + K \int_\Omega uv d\omega = 0, \quad \forall v \in V.$$

Well posedness of the problem follows from Corollary 3.1. As a matter of fact, bilinear form $\mu \int_\Omega \nabla w \nabla v d\omega + K \int_\Omega wv d\omega$ is continuous for w and v in V and is coercive in V, since

$$\mu \int_\Omega \nabla v \cdot \nabla v d\omega + K \int_\Omega v^2 d\omega \ge \min(\mu, K) \|v\|_V^2.$$

For the 1D problem (3.56), (3.57), (3.58), we follow similar guidelines, even if now $\Gamma_D = \{R\}$ does not coincide with the whole boundary of the domain. Let us still denote with G a lifting of the boundary data and with V the space $H_R^1(0, R)$ of $H^1(0, R)$ functions vanishing in $x = R$. We have the following problem: find $u \in G + V$ s.t.

$$\int_0^R \left(\mu r^2 \frac{\partial u}{\partial r} \right) \frac{\partial v}{\partial r} dr + \int_0^R K r^2 uv dr = 0 \qquad \forall v \in V. \tag{3.59}$$

Set

$$a(u, v) \equiv \int_0^R \left(\mu r^2 \frac{\partial u}{\partial r} \right) \frac{\partial v}{\partial r} dr + \int_0^R K r^2 uv dr. \tag{3.60}$$

Corollary 3.1 cannot be applied. In fact, the new bilinear form is continuous in V since from the Cauchy-Schwarz inequality we have

$$|a(u, w)| \le R^2 (\mu + K) \|u\|_V \|w\|_V, \quad \forall u, w \in V.$$

However, the bilinear form is not coercive in V since the coefficient of the first term vanishes in the left end point of the domain.

On the other hand, we expect well posedness, since it has been proved in the general 3D case. Let us introduce the following *weighted scalar product*

$$(u, v)_w \equiv \int_0^R r^2 uv\, dr$$

with weight $w \equiv r^2$ and the corresponding norm $||u||_w \equiv \sqrt{(u, u)_w}$. Let $L_w^2(0, R)$ be the space of functions such that $||u||_w < \infty$.

These definitions are formal, and we need to show that $(u, v)_w$ is a scalar product (and consequently $|| \cdot ||_w$ is a norm) on $L_w^2(\Omega)$. To this aim, we have to prove that the following properties hold (see e.g. [Sal08]):

1. bilinearity: for any $\alpha, \beta \in \mathbb{R}$

$$(\alpha u_1 + \beta u_2, v)_w = \int_0^R r^2 (\alpha u_1 + \beta u_2)\, v\, dr = \alpha\, (u_1, v)_w + \beta\, (u_2, v)_w$$

and

$$(u, \alpha v_1 + \beta v_2)_w = \int_0^R r^2 u\, (\alpha v_1 + \beta v_2)\, dr = \alpha\, (u, v_1)_w + \beta\, (u, v_2)_w\,;$$

2. symmetry: $(u, v)_w = \int_0^R r^2 uv\, dr = (v, u)_w$;

3. non-negativity: $(u, u)_w = \int_0^R r^2 u^2 dr \geq 0$. Moreover, since r^2 is positive apart on a measure-null set $(r = 0)$, $(u, u)_w = 0$ if and only if $u = 0$.

Moreover, it is possible to prove that $(u, v)_w \leq ||u||_w ||v||_w$, for any $u, v \in L_w^2(0, R)$, and that $L_w^2(0, R)$ is a Hilbert space [Sal08].

We define therefore the space[5]

$$H_w^1(0, R) \equiv \{w \in L_w^2(0, R) : w' \in L_w^2(0, R)\},$$

with the norm $||u||_{H_w^1(0,R)}^2 = ||u||_w^2 + ||u'||_w^2$. This is a Hibert space too. Moreover, we consider the subspace

$$H_{0,w}^1(0, R) \equiv \{v \in H_w^1(0, R) : v(R) = 0\}\,.$$

Notice that $H^1(0, R) \subset H_w^1(0, R)$, since

$$||u||_{H_w^1(0,R)}^2 \leq R^2 \int_0^R u^2 + (u')^2 dx = R^2 ||u||_{H^1(0,1)}^2$$

so that $||u||_{H^1(0,1)} < \infty \Rightarrow ||u||_{H_w^1(0,R)} < \infty$.

---

[5] Differentiation is understood in distributional sense.

Set $V \equiv H_w^1(0, R)$. Bilinear form $a(u, v)$ is continuous in V (being continuous in $H^1(0, R)$), and is coercive, since

$$a(u, u) \geq \min(\mu, K)\|u\|_{H_w^1}^2, \quad \forall u \in H_w^1(0, R).$$

Now, we can resort to Corollary 3.1 for the problem: find $u \in G + V \equiv \{w \in H_w^1(0, R) : w_{\Gamma_D} = g\}$ s.t.

$$a(u, v) = 0, \quad \forall v \in V.$$

The problem is well posed.

In this linear case, it is possible to compute also the analytic solution (see [Sal08]). For $K = 0$ the general integral has the form

$$u(r) = \frac{C_1}{r} + C_2$$

with C_1 and C_2 constants depending on the prescribed boundary conditions. In our case, since we want to force the first derivative to be 0 for $r = 0$, we have $C_1 = 0, C_2 = u_{ext}$ so the solution is constant. This is expected, because for $K = 0$ there is no activity consuming Oxygen. For $K \neq 0$, solution can be written in terms of Bessel functions.

Numerical approximation

Numerical approximation of (3.59) is obtained in a standard way.

Let $W_h = \{v_h \in X_h^r : v_h(R) = u_{ext}\}$ and $V_h = \{v_h \in X_h^r : v_h(R) = 0\}$, approximate problem reads: find $u_h \in W_h$ s.t.

$$a(u_h, w_h) = F(w_h), \quad \forall v_h \in V_h.$$

This is a standard problem similar to many others considered in the present chapter. We just want to point out some practical aspects. Let us consider the integrals for computation of the entries $a_{ij} = a(\varphi_j, \varphi_i)$, that in our case are given by

$$a_{ij} = \int_0^R \mu r^2 \frac{\partial \varphi_j}{\partial r} \frac{\partial \varphi_i}{\partial r} dr + \int_0^R K r^2 \varphi_j \varphi_i dr.$$

Since integrand functions are polynomials, we can specify a degree of exactness (DOE) for the quadrature formula adopted in the integrals computation, to avoid quadrature errors. Working with finite elements with degree q, the first integrand function in computing a_{ij} is a polynomial of order $2q$, while the second term has degree $2q + 2$. To be exact, a quadrature formula with linear finite elements will require a DOE 4 at least, 6 is required by quadratic finite elements, 8 for cubic finite elements.

Numerical results

Solving the problem with `fem1d` with the given numerical data, quadratic finite elements and an appropriate Gaussian quadrature formula, we find the pressure diagram in Fig. 3.24 (left, solid line). An external partial pressure of 5mmHg does not guarantee a core pressure $\geq$ 1mmHg. This could affect the pancreatic cell. An external pressure of 20mmHg (dashed line) maintains the core pressure to be greater than 3mmHg with better functionality properties. From the simulation we can get remarkable suggestions on the environmental conditions for the conservation of the cell before the transplantation.

In Fig. 3.24 on the right we report the differences (for an external pressure of 5mmHg) between solutions computed with Gauss-Lobatto quadrature formulas featuring a degree of exactness 13 (with 7 nodes) and 5 (with 3 nodes) respectively (see [QSS00]). Second formula is adding a numerical error that is quite evident (even if it is not determinant).   ◇

Remark 3.6 A more reliable model for this problem has been investigated in some research center, such as the Mario Negri Institute in Villa Camozzi, Ranica (BG), Italy. In practice, the cell is not a sphere and the respiration rate is not constant. This is in general a nonlinear function of the Oxygen called *Michaelis Menten law* (see e.g. [AC97]). Moreover, the partial pressure in the gel is not known in general and should be computed by simulating diffusion process in the gel, by assuming a known pressure outside the gel. A 3D example for this model is reported in Fig. 3.25. These results have been obtained with the LifeV code in collaboration with Daniele Di Pietro, currently at IFP, France (see [PV09]).

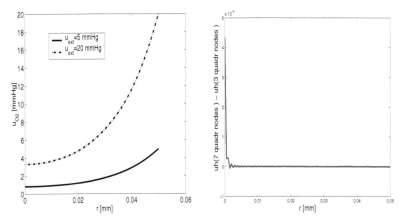

Fig. 3.24 Left: Oxygen partial pressure for the cell problem: $u_{ext} = 5$ (solid line) and $u_{ext} = 20 \, mmHg$ (dashed). Latter solution guarantees a good oxygen concentration to the entire cell. Right: differences between solutions with different quadrature formulas. Three nodes Gauss Lobatto formula introduces a quadrature error

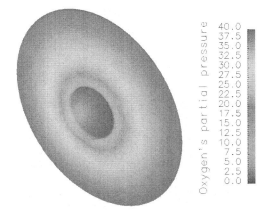

Fig. 3.25 3D simulation of the nonlinear cell model (Michaelis-Menten law). Simulation carried out with **LifeV** (see [lif10])

Exercise 3.2.6. Under appropriate assumptions the study of water filtration in a porous medium occupying a two-dimensional region $\Omega = (0,1)^2$ can be reduced to solving a problem of the form

$$-\nabla \cdot (Ku) = -\frac{\partial}{\partial x}\left(K(x,y)\frac{\partial u}{\partial x}\right) - \frac{\partial}{\partial y}\left(K(x,y)\frac{\partial u}{\partial y}\right) = 0, \quad (3.61)$$

where the function K is the hydraulic conductivity of the medium and u represents the water-level variation. Write a finite-difference discretization of order 2 for the problem (3.61) with (Fig. 3.26 left)

$$K(x,y) = \begin{cases} 101 \text{ if } (x-0.5)^2 + (y-0.5)^2 - 0.04 > 0, \\ 1 \quad \text{otherwise,} \end{cases}$$

when $\Omega = (0,1) \times (0,1)$, and the boundary conditions are

$$\begin{cases} \dfrac{\partial u}{\partial x} = 0 \text{ on } (\{0\} \cup \{1\}) \times (0,1), \\ \\ u = 10 \quad \text{on } (0,1) \times \{0\}, \\ \\ u = 0 \quad \text{on } (0,1) \times \{1\}. \end{cases} \qquad (3.62)$$

Use a 5-node scheme, i.e. each equation for the internal nodes should involve the value of u at the node and that at the 4 adjacent nodes (Fig. 3.26 right).

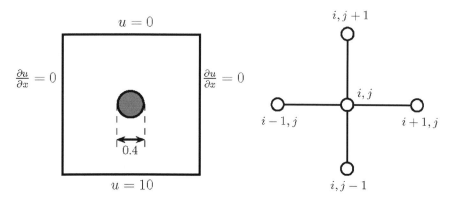

Fig. 3.26 Domain for Exercise 3.2.6 (left) and numeration for the 5-node scheme (right)

Solution 3.2.6.

Problem (3.61) is known as Darcy's problem, and represents a model of fluid filtration in a porous medium. The practical applications of such models spread from irrigation to the use and control of water resources, or the study of blood circulation in the brain. The derivation of the model may be found, for instance, in [Bea88].

Numerical approximation

We introduce in Ω the discretization nodes $\mathbf{x}_{i,j} = (x_i, y_j)$, $i = 0, \dots, n_x$, $j = 0, \dots, n_y$ with $x_i = x_0 + ih_x$, $y_j = y_0 + jh_y$, $h_x = 1/n_x$, $h_y = 1/n_y$, having set $x_0 = 0$, $y_0 = 0$. A priori h_x may not coincide with h_y, yet we assume for simplicity $h = h_x = h_y$, as well as $n = n_x = n_y$.

The presence of a varying coefficient K compels us to rethink the finite-difference scheme (for otherwise it would be enough to consider an approximation like (2.45) both in x and y, thus producing the classical *five-point-scheme* addressed for instance in [QSS00, Ch. 12]).

To keep track of the coefficient we begin by approximating the x-derivative at the generic node $\mathbf{x}_{i,j}$. The aim is to end up with a centered scheme of order 2 having a *stencil* formed by nodes $\mathbf{x}_{i-1,j}$, $\mathbf{x}_{i,j}$, $\mathbf{x}_{i+1,j}$ only. Therefore, we consider the following approximation (see Fig. 3.27)

$$\frac{\partial}{\partial x}\left(K\frac{\partial u}{\partial x}(\mathbf{x}_{i,j})\right) \simeq \frac{K(\mathbf{x}_{i+1/2,j})\dfrac{\partial u}{\partial x}(\mathbf{x}_{i+1/2,j}) - K(\mathbf{x}_{i-1/2,j})\dfrac{\partial u}{\partial x}(\mathbf{x}_{i-1/2,j})}{h}$$

where we have set $\mathbf{x}_{i\pm1/2,j} = (x_i \pm h/2, y_j)$. This is accurate to order 2 in h, if both u and K are regular enough. At this point we approximate the derivatives at the nodes by applying again a centered difference quotient (still accurate to order 2 in h) to arrive at the approximation

$$\frac{\partial}{\partial x}\left(K\frac{\partial u}{\partial x}(\mathbf{x}_{i,j})\right)$$

$$\simeq \frac{K(\mathbf{x}_{i+1/2,j})(u_{i+1,j}-u_{i,j})-K(\mathbf{x}_{i-1/2,j})(u_{i,j}-u_{i-1,j})}{h^2}.$$

(3.63)

The right-hand side of (3.63) furnishes an accurate approximation to order 2. Indeed, substituting the Taylor series expansions

$$K(\mathbf{x}_{i\pm1/2,j}) = K(\mathbf{x}_{i,j}) \pm \frac{h}{2}\frac{\partial K}{\partial x}(\mathbf{x}_{i,j}) + \frac{h^2}{8}\frac{\partial K^2}{\partial x^2}(\mathbf{x}_{i,j}) \pm \frac{h^3}{48}\frac{\partial K^3}{\partial x^3}(\mathbf{x}_{i,j})$$
$$+O(h^4),$$

$$u(\mathbf{x}_{i\pm1,j}) = u(\mathbf{x}_{i,j}) \pm h\frac{\partial u}{\partial x}(\mathbf{x}_{i,j}) + \frac{h^2}{2}\frac{\partial u^2}{\partial x^2}(\mathbf{x}_{i,j}) \pm \frac{h^3}{6}\frac{\partial u^3}{\partial x^3}(\mathbf{x}_{i,j})$$
$$+O(h^4)$$

to the right-hand-side of (3.63) we obtain

$$\frac{K(\mathbf{x}_{i+1/2,j})(u_{i+1,j}-u_{i,j})-K(\mathbf{x}_{i-1/2,j})(u_{i,j}-u_{i-1,j})}{h^2}$$

$$= K(\mathbf{x}_{i,j})\frac{\partial u^2}{\partial x^2}(\mathbf{x}_{i,j}) + \frac{\partial K}{\partial x}(\mathbf{x}_{i,j})\frac{\partial u}{\partial x}(\mathbf{x}_{i,j}) + O(h^2).$$

To get the second derivative's discretization in y we proceed similarly, obtaining,

$$\frac{K(\mathbf{x}_{i,j+1/2})(u_{i,j+1}-u_{i,j})-K(\mathbf{x}_{i,j-1/2})(u_{i,j}-u_{i,j-1})}{h^2}.$$

The finite-difference scheme eventually obtained has a *stencil* with 5 nodes, but the coefficients depend upon K.

As for the boundary conditions, on Dirichlet boundaries they are easy to prescribe, where one imposes u directly. The case $\partial u/\partial x = 0$ can be dealt with by a decentred (yet 2nd order) finite difference, like

$$\frac{\partial u}{\partial x}(0,y_i) \simeq \frac{3}{2h}u_{0,i} + \frac{2}{h}u_{1,i} - \frac{1}{2h}u_{2,i}.$$

Now it is a matter of writing the MATLAB instructions needed to build the matrix and the source of the associated linear system. We need, however, to renumber the nodes first. More precisely, we need to construct a one-to-one map $(i,j) \rightarrow k$, where $i,j = 0,\ldots,n$ and $k = 0,\ldots,nm$. For instance, we can do as indicated in Fig. 3.27, supposing to number nodes proceeding from $(0,0)$ left to right, top to bottom till $(1,1)$.

Due to the varying coefficients, the matrix can now be constructed using a suitable for-loop on all internal nodes and then imposing the boundary conditions (which is trivial since the problem has homogeneous Dirichlet conditions).

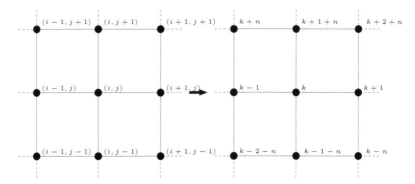

Fig. 3.27 A renumbering of the grid nodes that simplifies the implementation of the finite-difference method in dimension two

---

Numerical results

Program 12 implements the solution of the exercise. It is in fact a little more general since it allows different selection of boundary conditions.

Program 12 - FDnonconstantcoeff : Finite differences solution of Laplace equation on a square with non linear coefficients

```
function u=FDnonconstantcoeff(nh,fun,funK,bcv,bct)
%
%  It evaluates the solution of
%
%  - @x K(x,y) @x u- @y K(x,y) @y u=f
%  in Omega=(0,1)^2
%  with  b(u)=g1 on x=0, b(u)=g2 on x=1, b(u)=g3 on y=0 b(u)=g4
%  on y=1
%  where b(u)= u or b(u) = @n u depending on the type of bc applied
%  (Dirichlet or Neumann).
%
%  We use a 5-point scheme on a regular grid of nh+1 X nh+1 nodes
%
%  nh    Number of intervals (nodes-1) on each side
%  fun   Function name or function handler (x,y)-> R which returns the forcing
%        term
%  funK  Function name or function handler (x,y)->R returning the value of the
%        coefficient K
%  bcv   Vector of boundary condition values: bcv(i)=gi i=1,2,3,4
%  bct   Vector of boundary condition type: 0 Diriclet, 1 Neumann. 4 values
%        one for side according to the given scheme
%  u     Approximated solution. It is a matrix nh+1 X nh+1 giving the values
%        in the lattice (i,j)
%
%  The global node numbering goes from bottom to top, left to right
%
```

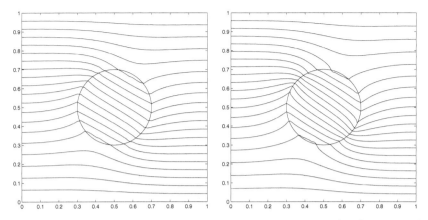

Fig. 3.28 Isolines of the finite-difference approximation of problem (3.61) for $h = 1/50$ (left) and $h = 1/100$ (right). Inside the circle the conductivity is 1, outside 101

Fig. 3.28 shows the isolines (values ranging from 0 to 10 with size 0.5) of the solutions found for two different discretization steps h. As the analytical solution is not available, using two grid spacings allows to qualitatively appreciate, the accuracy of the solution. The presence of an area of low hydraulic conductivity affects, as it should, the solution's behavior (with constant conductivity the solution would be a plane). ◇

Exercise 3.2.7. Determine a centered scheme by compact finite differences of order 4 on a grid of uniform step h to approximate the problem $-\Delta u = -2e^{x-y}$ on $\Omega = (0,1) \times (0,1)$, with $u = e^{x-y}$ on $\partial\Omega$. Check the order of accuracy by first computing the exact solution.

Solution 3.2.7.

Mathematical analysis

To determine the exact solution we perform the following change of variables: $\eta = x - y$ and $\xi = x + y$ Using the fact that

$$\frac{\partial^2 u}{\partial x^2} = \frac{\partial^2 u}{\partial \xi^2} + 2\frac{\partial^2 u}{\partial \xi \partial \eta} + \frac{\partial^2 u}{\partial \eta^2},$$

$$\frac{\partial^2 u}{\partial y^2} = \frac{\partial^2 u}{\partial \xi^2} - 2\frac{\partial^2 u}{\partial \xi \partial \eta} + \frac{\partial^2 u}{\partial \eta^2},$$

we derive that the given differential equation is equivalent to

$$\frac{\partial^2 u}{\partial \xi^2} + \frac{\partial^2 u}{\partial \eta^2} = -2e^{\eta},$$

while the boundary conditions read $u = e^\eta$ on $\partial\Omega$. The left hand side does not depend on ξ, thus $u(\eta, \xi) = \psi(\eta) + c_0\xi + c_1$ for some function ψ and constants c_0 and c_1. Considering the boundary condition, one derives immediately that $c_0 = c_1 = 0$. We are then left with the ordinary differential equation $\psi''(\eta) = e^\eta$, by which (exploiting again the boundary conditions) $\psi = e^\eta$ and thus $u = e^{x-y}$.

Numerical approximation

Let us define the discretization nodes $\mathbf{x}_{i,j} = (x_i, y_j)$ on Ω as in Exercise 3.2.6.

We can solve this exercise by recalling the symbolic expression (3.34) and using it to approximate both the second derivative in x and the one in y. We define $\delta_x^2 u_{i,j} \equiv h^{-2}(u_{i+1,j} - 2u_{i,j} + u_{i-1,j})$ and $\delta_y^2 u_{i,j} \equiv h^{-2}(u_{i,j+1} - 2u_{i,j} + u_{i,j-1})$. The scheme we are seeking will then have the form

$$-\left(1 + \frac{h^2}{12}\delta_x^2\right)^{-1}\delta_x^2 u_{i,j} - \left(1 + \frac{h^2}{12}\delta_y^2\right)^{-1}\delta_y^2 u_{i,j} = f_{i,j},$$

which can be rewritten as

$$-\left(1 + \frac{h^2}{12}\delta_y^2\right)\delta_x^2 u_{i,j} - \left(1 + \frac{h^2}{12}\delta_x^2\right)\delta_y^2 u_{i,j} = \left(1 + \frac{h^2}{12}\delta_x^2\right)\left(1 + \frac{h^2}{12}\delta_y^2\right)f_{i,j}.$$

Since operators δ_x^2 and δ_y^2 commute, we have

$$-(\delta_x^2 + \delta_y^2)u_{i,j} - \frac{h^2}{6}\delta_x^2\delta_y^2 u_{i,j} = f + \frac{h^2}{12}(\delta_x^2 + \delta_y^2)f_{i,j} + \frac{h^4}{144}\delta_x^2\delta_y^2 f_{i,j}. \quad (3.64)$$

Let us highlight in (3.64) the dependency on the $u_{i,j}$ by noting

$$\begin{aligned}
\delta_x^2\delta_y^2 u_{i,j} &= h^{-2}\left(\delta_y^2 u_{i-1,j} - 2\delta_y^2 u_{i,j} + \delta_y^2 u_{i+1,j}\right)\\
&= h^{-4}\left(u_{i-1,j-1} - 2u_{i-1,j} + u_{i-1,j+1} - 2(u_{i,j-1} - 2u_{i,j} + u_{i,j+1})\right.\\
&\quad \left. + u_{i+1,j-1} - 2u_{i+1,j} + u_{i+1,j+1}\right),
\end{aligned}$$

so that, after a few algebraic manipulations, the generic equation of the scheme becomes

$$\begin{aligned}
&-\frac{4}{h^2}(u_{i-1,j} + u_{i,j-1} + u_{i+1,j} + u_{i,j+1}) + \frac{20}{h^2}u_{i,j}\\
&-\frac{1}{h^2}(u_{i-1,j-1} + u_{i-1,j+1} + u_{i+1,j-1} + u_{i+1,j+1}) = \frac{25}{6}f_{i,j}\\
&+\frac{5}{12}(f_{i,j-1} + f_{i-1,j} + f_{i,j+1} + f_{i+1,j})\\
&+\frac{1}{24}(f_{i-1,j-1} + f_{i-1,j+1} + f_{i+1,j-1} + f_{i+1,j+1}).
\end{aligned} \quad (3.65)$$

The matrix governing the linear system (3.65) is sparse, so we use the **sparse** MATLAB instructions to store it appropriately and save memory (see Appendix A for a discussion on sparse matrices). This we did in Program 13, which asks as input parameters the number **nh** of discretization intervals used to decompose $(0, 1)$, the *inline functions* **fun** to determine the forcing term f, and **gD** for the Dirichlet boundary condition. In output Program 13 gives back the solution **u** stored in a matrix of $(\text{nh}+1)^2$ elements, the corresponding x- and y-coordinates **xx**, **yy** of the nodes. In this way the solution is plotted immediately by the MATLAB command **mesh(xx,yy,u)**.

Program 13 - fdlaplace : Laplace problem on the unit square approximated by compact finite differences

```
function [u,xx,yy]=fdlaplace(nh,fun,gD,varargin)
%FDLAPLACE Laplace problem with compact finite differences
%   [U,XX,YY]=FDLAPLACE(NH,FUN,GD) solves the problem -DELTA U=FUN
%   on the unit square with Dirichlet conditions U=GD on the boundary,
%   using a compact-finite-difference scheme of order 4.
%   The grid is uniform of step H=1/(NH+1).
%   [U,XX,YY]=FDLAPLACE(NH,FUN,GD,P1,P2,...) passes P1,P2,... as
%   optional parameters to the inline functions FUN e GD.
```

With this program one can check that the order of accuracy of the method is indeed 4. Table 3.6 shows the errors in discrete infinity norm and the estimated convergence order (usual technique) for various h.   ◇

Table 3.6 Behavior of the error E in infinity norm for grid size h to approximate the problem in Exercise 3.2.7 and corresponding estimate for the convergence order q with respect to h

h	1/10	1/20	1/40	1/80
E	1.8e-07	1.1e-08	7.0e-10	4.4e-11
q	–	3.99913	3.99408	4.01225

Exercise 3.2.8. Using the spectral-collocation method on the GCL grid with n^2 nodes, solve the problem

$$\begin{cases} -\Delta u + u = -\sinh(x+y) - \cosh(x+y) \text{ on } \Omega, \\ u = \sinh(x+y) + \cosh(x+y) \quad\quad\quad \text{ on } \partial\Omega, \end{cases} \quad (3.66)$$

where $\Omega = (0,1)^2$. Solve the associated linear system by the GMRES method and find the number of iterations needed so that the error (in Euclidean norm) on the residual is less than 10^{-12} for $n = 5, 10, 20, 40$.

Then take as preconditioner the matrix obtained by the finite-difference method applied to the same problem, on the same grid. Comment on the number of iterations.

Solution 3.2.8.

Numerical approximation

The discretization by spectral collocation of the given problem is found by observing both x- and y-second derivatives descend from using twice the matrix of pseudo-spectral differentiation (2.30), multiplied by 2 to account for the fact that it was originally computed on the interval $(-1, 1)$, whereas we apply it on $(0, 1)$ (see Exercise 3.1.11).

Program 14 calculates the solution of a more general problem than (3.66). Namely, it finds the approximated solution by spectral collocation for

$$\begin{cases} -\mu \Delta u + \boldsymbol{\beta} \cdot \nabla u + \sigma u = f \text{ on } \Omega \equiv (a, b) \times (c, d), \\ u = g_D \text{ on } \partial \Omega \end{cases}$$

on the GCL grid. The input parameters are the end points of the domain, the number of discretization nodes in each direction, the constant values of the coefficients and two *inline functions* for f and g_D. Note that also for the *symmetric* problem (3.66) (the associated weak form is symmetric) the collocation method on the GCL grid does not produce a symmetric matrix: for this reason Program 14 solves the linear system with the GMRES method (and not by conjugate gradient). A discussion on iterative methods for linear systems of equations may be found in [QSS00].

Program 14 - diff2Dsp : Approximation of a diffusion-advection-reaction problem by a spectral-collocation method

```
function [us,x,y]=diff2Dsp(xspan,yspan,n,mu,beta,sigma,fun,gD,varargin)
% DIFF2DSP solves an  advection-diffusion-reaction problem
% with constant coefficients.
%  [US,XX,YY]=DIFF2DSP(XSPAN,YSPAN,N,MU,BETA,SIGMA,FUN,GD,P) solves
%  the two-dimensional problem -MU*DELTA U + BETA*GRAD(U)+SIGMA*U=FUN
%  on (XSPAN(1),XSPAN(2))x(YSPAN(1),YSPAN(2)) with Dirichlet
%  conditions U=GD on the boundary, using a spectral-collocation scheme
%  on a Gauss-Lobatto-Chebyshev grid with  N nodes in each
%  direction. MU and SIGMA are constants, BETA a
%  vector with two components. FUN and GD are inline functions.
%  P is a  preconditioner for the  GMRES method
%  used to solve the linear system. Set P=[] to avoid
%  preconditioning.
%  [US,XX,YY]=DIFF2DSP(XSPAN,YSPAN,N,MU,BETA,SIGMA,FUN.GD,P,P1,P2,..)
%  passes the optional parameters P1,.. to the inline functions FUN and GD.
```

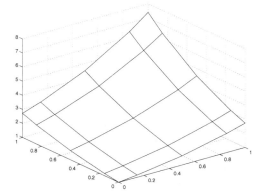

Fig. 3.29 The solution found for $n = 5$ by spectral collocation. The corresponding grid is shown

Let us work on a grid with $n = 5$ nodes in every direction. To solve (3.66) we write

```
>> xspan=[0 1];yspan=[0 1];n=5;mu=1;beta=[0 0];sigma=1;
>> gD=inline('sinh(x+y)+cosh(x+y)','x','y');
>> f=inline('-sinh(x+y)-cosh(x+y)','x','y');
>> [us,x,y]=diff2Dsp(xspan,yspan,n,mu,beta,sigma,gD,f,[]);
Condition number of A 226.295989
gmres(25) converged at iteration 1(7) to a solution with
relative residual 1.4e-14
```

The solution us, shown in Fig. 3.29, is found by the command mesh(x,y,us)).

Let us comment on the messages appearing at the end of the execution of the program. The first concerns the condition number in norm 2 of the matrix of the linear system associated to the discretization. The second is printed by the *function* gmres that the program uses to solve the system. This is the so-called GMRES method with *restart*. It is known that GMRES ([QSS00]) requires to store a basis for Krylov spaces whose dimension increases at each iteration of one vector of dimension N equal to the number of equations of the linear system. With exact arithmetic, convergence is assured in at most N iterations, yet we expect to reach the specified tolerance much earlier.

However, if convergence is slow or the problem is of large size we may end up storing a large set of vectors, clogging up memory. For this reason one often uses the version with *restart*, normally called GMRES(M), in which every M iterations the method starts anew. The price to pay is the possible loss of convergence if M is too small.

In the message popping up, gmres(25) refers to the fact we used the GMRES(25), which in our case it means that we are not using a restart, since 25 is the size of our linear system and (at least with in exact arithmetic) we should converge before reaching that number of iterations. In MATLAB, a counter keeps track of the number of *restarts* used, consequently 1(7) should

Table 3.7 Behavior of the condition number (in the 2-norm) for the matrix associated to the spectral discretization without preconditioning (column 2) and with preconditioning (column 3), as n varies

n	$K(A)$	$K(\mathrm{P}_{FD}^{-1}A)$
5	226.295989	1.810680
10	5524.505302	2.276853
20	1.4479e+05	2.460154
40	3.5706e+06	2.562154

be interpreted as "iteration nr. 7 of the first *restart* cycle", in other words, overall iteration 7.

But if we try to solve the system with $n = 10$ (thus producing a linear system of size 100) the message appearing is different. We have in fact

```
Condition number of A 5524.505302
gmres(100) stopped at iteration 1(40) without converging to
the desired tolerance 1.e-13 because the method stagnated
The iterate returned (number 1(39)) has relative residual 1.e-12
```

Hence, the method fails to converge and it stagnates: it reaches a point where any further iteration would not reduce the error. That is because the matrix is ill conditioned. Indeed by doubling n the condition number has increased considerably. Even if in exact arithmetic GMRES(100) is guaranteed to converge for a system of size up to 100, the round-off errors are amplified by the bad conditioning and hinder convergence. As the second column of Table 3.7 shows, this behavior is confirmed for larger n as well. A *preconditioner* is then clearly necessary.

The text suggests using as preconditioner the matrix P_{FD} associated to the discretization of the same differential problem with finite differences on the Gauss-Lobatto-Chebyshev grid. The point is thus to generate a discretization by finite differences on a non-uniform grid, exactly as in the one-dimensional case of Exercise 3.1.10. Such construction is carried out in Program 15, by assigning to the input parameter `fgrid` a non-zero value and defining the remaining parameters just as in Program 14.

Program 15 - diff2Dfd : Approximation of a diffusion-advection-reaction problem by a finite-difference method

```
function [us,x,y,AFD]=diff2Dfd(xspan,yspan,n,mu,beta,sigma,gD,...
 fun,fgrid,varargin)
%DIFF2DFD solves a diffusion-advection-reaction problem with
%constant coefficients.
%   [UFD,XX,YY,AFD]=DIFF2DFD(XSPAN,YSPAN,N,MU,BETA,SIGMA,GD,FUN,FGRID)
%    solves the problem 2D -MU*DELTA U + BETA*GRAD(U)+SIGMA*U=FUN
%   on (XSPAN(1),XSPAN(2))x(YSPAN(1),YSPAN(2)) with Dirichlet
%   conditions U=GD on the boundary, using a centred finite-difference
%   scheme of order 2 on a uniform grid  (FGRID=0)
```

```
%   or a Gauss-Lobatto-Chebyshev grid with  N nodes
%   in every direction (FGRID=1). MU and SIGMA are constants,
%   BETA a vector of 2 components.FUN and GD are inline
%   functions. AFD is the matrix associated to the discretization.
%   [US,XX,YY,AFD]=DIFF2DFD(XSPAN,YSPAN,N,MU,BETA,SIGMA,GD,FUN,...
%   FGRID,P1,P2,..) passes the optional parameters P1,.. to the inline
%   functions FUN and GD.
```

In this way we obtain for the condition number of the preconditioned matrix the result shown in column three of the table: the condition number is practically constant with n. Indeed, one can prove that such preconditioner is *optimal* for the given problem, in the sense that, beyond being relatively easy to solve, the condition number of the preconditioned matrix is independent of the matrix size.   ◇

3.3 Domain decomposition methods for 1D elliptic problems

In this section we consider some elementary notions of *domain decomposition methods*. We will limit ourselves to 1D elliptic problems on 2 domains, so to give a basic idea of the simplest methods and the possible numerical problems encountered when solving a problem by domain decomposition. We refer the interested reader to [Qua09], Chapter 17 and e.g. to the monographs [QV99, TW05] for an extensive study of these methods.

Here we consider the simple problem: for $f(x) \in L^2(0,1)$ find $u(x)$ s.t.

$$-\frac{d^2u}{dx^2} = f \qquad x \in (0,1)$$
$$u(0) = u(1) = 0. \tag{3.67}$$

Let us introduce two points $0 < x_l \leq x_r < 1$ in the interval $(0,1)$ and the corresponding domains $[0, x_r]$ and $[x_l, 1]$.

3.3.1 Overlapping methods

Let us assume $x_l < x_r$. A possible method for solving (3.67) by subdomains is given by the following *Schwarz algorithm*:

Algorithm 1

1. Let $\lambda^{(0)} \in \mathbb{R}$ be an arbitrary number.
2. Loop: up to convergence, for $i = 1, 2, \ldots$ solve:

   a.
   $$-\frac{d^2u_1^{(i+1)}}{dx^2} = f \; x \in (0, x_r)$$
   $$u_1^{(i+1)}(0) = 0, \quad u_1^{(i+1)}(x_r) = \lambda^{(i)}. \tag{3.68}$$

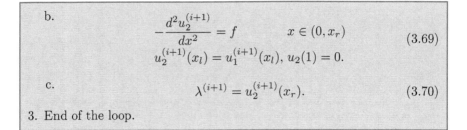

b.
$$-\frac{d^2 u_2^{(i+1)}}{dx^2} = f \qquad x \in (0, x_r)$$
$$u_2^{(i+1)}(x_l) = u_1^{(i+1)}(x_l), \ u_2(1) = 0. \tag{3.69}$$

c.
$$\lambda^{(i+1)} = u_2^{(i+1)}(x_r). \tag{3.70}$$

3. End of the loop.

Another option is the *additive overlapping Schwarz algorithm*.

Algorithm 2

1. Let $\lambda^{(0)} \in \mathbb{R}$ and $\mu^{(0)} \in \mathbb{R}$ be two arbitrary numbers.
2. Loop: up to convergence, for $i = 1, 2, \ldots$ solve:

a.
$$-\frac{d^2 u_1^{(i+1)}}{dx^2} = f \ x \in (0, x_r)$$
$$u_1^{(i+1)}(0) = 0, \ u_1^{(i+1)}(x_r) = \lambda^{(i)}. \tag{3.71}$$

b.
$$-\frac{d^2 u_2^{(i+1)}}{dx^2} = f \qquad x \in (0, x_r)$$
$$u_2^{(i+1)}(x_l) = \mu^{(i)}, \ u_2(1) = 0. \tag{3.72}$$

c.
$$\lambda^{(i+1)} = u_2^{(i+1)}(x_r), \quad \mu^{(i+1)} = u_1^{(i+1)}(x_l). \tag{3.73}$$

3. End of the loop.

The latter method has the advantage that the computation of u_1 and u_2 can be carried out in parallel, since the solution of u_2 does not require data from u_1 computed at the same iteration (as the first method does).

When solving a problem with a domain decomposition approach a crucial issue is obviously the *convergence* of the couple of solutions (u_1, u_2) to the "exact" solution u. We will investigate this aspect in the exercises.

Another possible approach consists of the introduction of Robin conditions replacing the Dirichlet ones.

Algorithm 3

1. Let $\lambda^{(0)} \in \mathbb{R}$ be an arbitrary number.
2. Loop: up to convergence, for $i = 1, 2, \ldots$ solve:

   a.

$$-\frac{d^2 u_1^{(i+1)}}{dx^2} = f \ x \in (0, x_r)$$

$$u_1(0) = 0, \qquad \frac{du_1^{(i+1)}}{dx}(x_r) + \gamma_1 u_1^{(i+1)}(x_r) = \lambda^{(i)}.$$

    (3.74)

   b.

$$-\frac{d^2 u_2^{(i+1)}}{dx^2} = f \qquad\qquad x \in (0, x_r) \qquad\qquad (3.75)$$

$$\frac{du_2^{(i+1)}}{dx}(x_l) + \gamma_2 u_2^{(i+1)}(x_l) = \frac{du_1^{(i+1)}}{dx}(x_l) + \gamma_2 u_1^{(i+1)}(x_l), \ u_2(1) = 0.$$

   c.

$$\lambda^{(i+1)} = \frac{du_2^{(i+1)}}{dx}(x_r) + \gamma_2 u_2^{(i+1)}(x_r). \qquad (3.76)$$

3. End of the loop.

We will see in the exercises how the Robin conditions, and in particular the parameters γ_1 and γ_2 affect the convergence.

Exercise 3.3.1. Consider the problem (3.67) and Algorithm 1.

1. Prove that the method of Algorithm 1 gives a convergent solution, the convergence being faster when the size of the overlap increases.
2. Write a code that implement the method and verify the convergence properties proved previously.
3. Repeat the two previous points for the Algorithm 2.

Solution 3.3.1.

Mathematical analysis

Let us consider the two error equations. More precisely, we set

$$e_1^{(i)} \equiv u_1^{(i)} - u, \quad e_2^{(i)} \equiv u_2^{(i)} - u, \quad \varepsilon^{(i)} \equiv \lambda^{(i)} - u(x_r).$$

By component-wise subtraction of the corresponding equations, errors $e_1^{(i+1)}$ and $e_2^{(i+1)}$ fulfill the equations

$$
\begin{cases}
-\dfrac{d^2 e_1^{(i+1)}}{dx^2} = 0, \ x \in (0, x_r) \\[2mm]
e_1^{(i+1)}(0) = 0 \quad e_1^{(i+1)}(x_r) = \varepsilon^{(i)}
\end{cases}
\quad,\quad
\begin{cases}
-\dfrac{d^2 e_2^{(i+1)}}{dx^2} = 0, \ x \in (x_l, 1) \\[2mm]
e_2^{(i+1)}(x_l) = \quad e_1^{(i+1)}(x_r)
\end{cases}.
$$

The error equations prescribe the error to be a linear function in space (being the second derivative $=0$). After including the boundary conditions, we have therefore

$$
e_1^{(i+1)}(x) = \frac{\varepsilon^{(i)}}{x_r} x \tag{3.77}
$$

and

$$
e_2^{(i+1)}(x) = \frac{e_1^{(i+1)}(x_l)}{1 - x_l}(1 - x) = \frac{\varepsilon^{(i)} x_l}{x_r(1 - x_l)}(1 - x). \tag{3.78}
$$

By definition

$$
\varepsilon^{(i+1)} = e_2^{(i+1)}(x_r) = \frac{1 - x_r}{1 - x_l}\frac{x_l}{x_r}\varepsilon^{(i)}.
$$

If we regard the method as a fixed point iteration with respect to the value $\lambda^{(i)}$, this is a contraction provided that

$$
\left| \frac{1 - x_r}{1 - x_l}\frac{x_l}{x_r} \right| < 1 \Rightarrow -1 < \frac{1 - x_r}{1 - x_l}\frac{x_l}{x_r} < 1. \tag{3.79}
$$

The left hand inequality is trivially true since by construction $(1 - x_r)/(1 - x_l)x_l/x_r$ is positive. The right hand inequality leads to $x_l < x_r$ which is true by construction too. This proves that, for any $x_l < x_r$, $\lim_{i \to \infty} \varepsilon^{(i+1)} = 0$, i.e. the method is convergent in x_r. From the convergence in this point we prove consequently the convergence over the entire interval, as follows straightforwardly from (3.77) and (3.78). We conclude therefore that $\lim_{i \to \infty} e_1^{(i+1)} = 0$ and $\lim_{i \to \infty} e_2^{(i+1)} = 0$.

Moreover, notice that:

1. for either $x_l = 0$ or $x_r = 1$ the iteration factor $(1 - x_r)/(1 - x_l)x_l/x_r = 0$, that guarantees that the method converges in one iteration; this is trivially expected, since in this case one of the two subdomains coincide with the entire domain;
2. for $x_l = x_r$ the method is not convergent, being the reduction factor $= 1$;
3. if we write $x_r = x_l + \delta$ being $\delta > 0$ the size of the overlapping, the iteration factor reads

$$
\frac{1 - x_r}{1 - x_l}\frac{x_l}{x_r} = 1 - \frac{\delta}{(x_l + \delta)(1 - x_l)}.
$$

Notice that the function

$$g(\delta) = 1 - \frac{\delta}{(x_l + \delta)(1 - x_l)}$$

is such that $g(0) = 1$, $g(1 - x_l) = 0$ and (since $x_l < 1$)

$$\frac{dg}{d\delta} = -\frac{x_l - x_l^2}{(x + \delta)^2(1 - x_l)^2} < 0.$$

The iteration factor is therefore monotonically decreasing when the overlapping region increases. This proves that larger is the overlapping and faster is the convergence of the iterative scheme.
Let us consider now the Algorithm n.2. Set

$$\varepsilon^{(i)} = \lambda^{(i)} - u(x_r), \quad \eta^{(i)} = \mu^{(i)} - u(x_l).$$

Notice that with similar arguments as for Algorithm n.1 we have

$$\varepsilon^{(i+1)} = \frac{\eta^i}{1 - x_l}(1 - x_r), \quad \eta^{(i+1)} = \frac{\varepsilon^{(i)}}{x_r}x_l$$

so that, recursively

$$\varepsilon^{(i+2)} = \frac{x_l(1 - x_r)}{x_r(1 - x_l)}\varepsilon^{(i)}, \quad \eta^{(i+2)} = \frac{x_l(1 - x_r)}{x_r(1 - x_l)}\eta^{(i)}.$$

From the latter equation, we notice that the fixed point iteration is still a contraction, with the same factor as for the previous scheme (also called "multiplicative" Schwarz method as opposed to the present one, called "additive"). However, in this case, the reduction obtained by one iteration of the multiplicative version requires two iterations. In other terms, the intrinsic parallelism of the additive method is obtained by reducing the convergence rate. The expected number of iterations required by the additive method will therefore be about twice the number of iteration of the multiplicative version.

Numerical approximation

A simple implementation of the code is in the code `overlap_schwarz` partially reported in Program 16. The code is written in MATLAB, and has been tested also in `Octave`.

After the input phase (where the consistency of the data is checked and exceptions are handled), the matrix of each subdomain is assembled, by using linear finite elements. The right hand side of weak formulation of equation (3.67) is approximated with the trapezoidal rule.

The convergence is tested by evaluating the maximum of the difference of u_1 and u_2 on the overlapping region. Another possible test considers the difference $\lambda^{(i+1)} - \lambda^{(i)}$.

The code is available at the book website.

Program 16 - overlap_schwarz : Schwarz method with overlapping

```
function [u1,u2,x1,x2,it]=overlap_schwarz(f,xl,xr,ig,h1,h2,tol,Nmax)
%
% [u1,u2,x1,x2,it]=overlap_schwarz(f,xl,xr,ig,h1,h2,tol,Nmax)
%
% MULTIPLICATIVE Schwarz method with overlapping
%   Linear finite elements + Trapezoidal rule for the rhs
%
% Problem: -u"=f in (0,1), u(0)=u(1)=0
%
%INPUT
% f = STRING with the forcing term (e.g. '0*x+0')
%xl,xr = interfaces positions (xl<xr):
%   domain 1 is [0:xr], domain 2 is [xl:0]
%ig = initial guess for the solutions in the interfaces (in xr)
%h1,h2 = mesh size in the two subdomains
%tol= tolerance for the convergence
%Nmax = max number of iterations
%%%%%%%%%%%%%%%%%%%%%%%%%%%%%%%%%%%%%%%%
%OUTPUT
%u1,u2 = solution in the two subdomains
%x1,x2 = mesh of the two subdomains
%it= number of iteration
```

The signature of the code `overlap_schwarz_add` implementing the additive Schwarz method is reported in Program 17. In this case, two initial guesses are required and the two solution $u_1^{(i)}$ and $u_2^{(i)}$ at each iteration can be computed at the same time by two processors independently.

Program 17 - overlap_schwarz_add : Additive Schwarz method

```
function [u1,u2,x1,x2,it]=overlap_schwarz_add(f,xl,xr,ig1,ig2,h1,h2,
                                              tol,Nmax)
%
% [u1,u2,x1,x2,it]=overlap_schwarz_add(f,xl,xr,ig1,ig2,h1,h2,tol,Nmax)
%
% ADDITIVE Schwarz method with overlapping
% with Linear finite elements + Trapezoidal rule for the rhs
%
% Problem: -u"=f in (0,1), u(0)=u(1)=0
%
%INPUT
% f = STRING with the forcing term (e.g. '0*x+0')
%xl,xr = interfaces positions (xl<xr): domain 1=[0:xr], domain 2=[xl:0]
%ig1,ig2 = initial guesses for the solutions in the interfaces
%        (ig1 is in xr, ig2 is in xl)
%h1,h2 = mesh size in the two subdomains
%tol= tolerance for the convergence
%Nmax = max number of iterations
```

```
%%%%%%%%%%%%%%%%%%%%%%%%%%%%%%%%%%%%%%%%%%
%OUTPUT
%u1,u2 = solution in the two subdomains
%x1,x2 = mesh of the two subdomains
%it= number of iteration
```

Numerical results

In Table 3.8 we report the number of iterations obtained with the multiplicative (it_M) and the additive (it_A) Schwarz methods when solving Problem 3.67 with $f = -1$.

Table 3.8 Number of iterations required by the Multiplicative Schwarz method and the Additive one for different sizes of the overlap

(x_l, x_r)	(0.5, 0.55)	(0.4, 0.6)	(0.3, 0.7)	(0.2, 0.8)	(0, 0.5)	(0.5, 1)
it_M	38	11	6	4	1	1
it_A	70	21	11	7	2	2

We consider different positions for the points x_l and x_r. Tolerance is set to $1.e - 4$ and the maximum number of iterations is 1000. The initial guess is $\lambda^{(0)} = 1$ (and $\mu^{(0)} = 1$ for the additive method). Mesh size is taken to be $h = 0.01$ for both the sub-domains.

Results confirm the theoretical analysis, even if the latter is carried out on the continuous problem (i.e. discretization errors are not included). When the overlap gets larger, the acceleration is faster. Convergence of the multiplicative method is immediate if one of the two domains covers the entire interval.

The multiplicative method is as fast as twice the additive one. The latter is however easily implemented on parallel architectures.

Notice that the mesh size in this case is not significantly affecting the convergence rate, as can be verified by running the code for instance with $h = .0.005, 0.001$. In general, however, the convergence rate can significantly depend on the mesh size.

In Fig. 3.30 we report the snapshots of the solutions on the two subdomains for the multiplicative Schwarz method in the case $x_l = 0.4, x_r = 0.6$.   ◊

Exercise 3.3.2. Consider the problem (3.67) and Algorithm 3.

1. Prove that for a suitable choice of the parameters γ_1 and γ_2 the method of Algorithm 3 gives a convergent solution and find the values of the parameter that maximize the convergence rate.
2. Write a code that implements the method and verify the convergence properties proved previously.

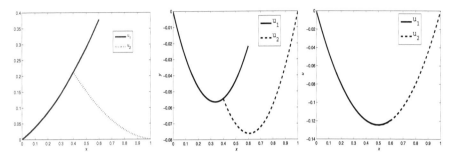

Fig. 3.30 Snapshots of the subdomain solutions for the Multiplicative Schwarz case with $x_l = 0.4, x_r = 0.6$. Iterations 1 (left) 3 (center) and 9 (right)

Solution 3.3.2.

Mathematical analysis

We follow similar arguments of the previous exercise. Let us define

$$\varepsilon^{(k)} \equiv \lambda^{(k)} - \frac{du}{dx}(x_r) + \gamma_1 u(x_r).$$

Then, the error equations read

$$
\begin{cases}
\dfrac{d^2 e_1^{(i+1)}}{dx^2} = 0 \\[2mm]
e_1^{(i+1)}(0) = 0, \quad \dfrac{de_1^{(i+1)}}{dx}(x_r) + \gamma_1 e_1^{(i+1)}(x_r) = \varepsilon^{(i)}
\end{cases},
$$

$$(3.80)$$

$$
\begin{cases}
\dfrac{d^2 e_2^{(i+1)}}{dx^2} = 0 \\[2mm]
\dfrac{de_2^{(i+1)}}{dx}(x_l) + \gamma_2 e_2^{(i+1)}(x_l) = \dfrac{de_1^{(i+1)}}{dx}(x_l) + \gamma_2 e_1^{(i+1)}(x_l), \quad e_2^{(i+1)}(1) = 0
\end{cases},
$$

with $\varepsilon^{(i+1)} = \dfrac{de_2^{(i+1)}}{dx}(x_r) + \gamma_1 e_2^{(i+1)}(x_r)$.

As for the previous exercise, we can solve directly the error equations and obtain

$$e_1^{(i+1)} = \frac{\varepsilon^{(i)} x}{1 + \gamma_1 x_r}, \quad e_2^{(i+1)} = \left(\frac{1 + \gamma_2 x_l}{1 + \gamma_1 x_r}\right) \frac{(1 - x)\varepsilon^{(i)}}{(-1 + \gamma_1(1 - x_l))} \qquad (3.81)$$

and this leads to the recursive relation

$$\varepsilon^{(i+1)} = \left(\frac{-1 + \gamma_1(1 - x_r)}{-1 + \gamma_2(1 - x_l)}\right)\left(\frac{1 + \gamma_2 x_l}{1 + \gamma_1 x_r}\right)\varepsilon^{(i+1)}.$$

The convergence factor in this case is

$$\varphi = \left(\frac{-1 + \gamma_1(1 - x_r)}{-1 + \gamma_2(1 - x_l)}\right)\left(\frac{1 + \gamma_2 x_l}{1 + \gamma_1 x_r}\right).$$

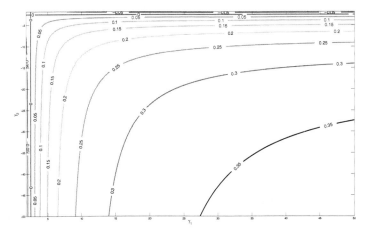

Fig. 3.31 Contour lines of the convergence factor φ for a Robin-Robin Schwarz method as a function of γ_1 and γ_2. The optimal values correspond to the isolines $\varphi = 0$

From the form of φ we infer that:

1. for either $\gamma_2 = -x_l^{-1}$ or $\gamma_1 = (1 - x_r)^{-1}$, convergence is reached within one iteration, since the convergence factor is $\varphi = 0$;
2. different combinations of the coefficients γ_1 and γ_2 attain the convergence, for different size of the overlapping; for instance, in Fig. 3.31 we report the contour lines of φ for $x_l = 0.4$ and $x_r = 0.6$ in a region of the γ_1, γ_2 plane where the convergence is guaranteed;
3. an appropriate choice of γ_1 and γ_2 guarantees convergence also when the overlapping is null; in other terms, *overlapping in this case is not necessary for the convergence*;
4. the values $\gamma_1 = -x_r^{-1}$ or $\gamma_2 = (1 - x_l)^{-1}$ entail a blow up of φ.

Numerical approximation

We report in Program 18 the signature of the code **schwarz_robin**. The complete code is downloadable at the book website. In the present approximation, the piecewise constant derivative of the solution is computed simply with the instruction **diff**. The derivative in the domain is needed only if the measure of the overlapping region is positive.

Should $x_l = x_r \equiv \hat{x}$, the boundary derivatives $\dfrac{du_j}{dx}\phi(\hat{x}), j = 1, 2$, where ϕ is a generic Lagrangian shape function, can be computed directly from the residual of the subdomain computation, as we illustrate in the Exercise 3.3.3.

Program 18 - schwarz_robin : Schwarz with Robin conditions

```
%
% [u1,u2,x1,x2,it]=schwarz_robin(f,xl,xr,ig,g1,g2,h1,h2,tol,Nmax)
%
% MULTIPLICATIVE Schwarz method with overlapping
% Robin conditions
%  Linear finite elements + Trapezoidal rule for the rhs
%
% Problem: -u"=f in (0,1), u(0)=u(1)=0
%
%INPUT
% f = STRING with the forcing term (e.g. '0*x+0')
%xl,xr = interfaces positions (xl<xr): domain 1=[0:xr], domain 2=[xl:0]
%ig = initial guess for the solutions in the interfaces (in xr)
%g1,g2 = parameters for the Robin conditions
%h1,h2 = mesh size in the two subdomains
%tol= tolerance for the convergence
%Nmax = max number of iterations
%%%%%%%%%%%%%%%%%%%%%%%%%%%%%%%%%%%%%%%%%%%
%OUTPUT
%u1,u2 = solution in the two subdomains
%x1,x2 = mesh of the two subdomains
%it= number of iteration
```

Numerical results

We solve the problem (3.67) with $f = 1$. We set the tolerance to 10^{-4} and the mesh size is $h_1 = h_2 = 0.001$. We start from $\lambda_1^{(0)} = 1$.

Results in Table 3.9 show the expected behavior. For the Robin Schwarz algorithm the size of the overlapping is not decisive for the convergence provided that parameters γ_1 and γ_2 are selected properly. In particular, for the predicted optimal values of the parameters (denoted by a $*$ in the table), the

Table 3.9 Results of problem (3.67) obtained using Algorithm 3, with different values of the parameters. The symbol * here denotes the optimal value

x_l	x_r	γ_1	γ_2	it
0.2	0.5	1	-1	5
0.2	0.4	1	-1	5
0.2	0.3	1	-1	4
0.2	0.2	1	-1	4
0.2	0.5	$\dfrac{1}{0.8}$ *	-1	1
0.2	0.5	1	$-\dfrac{1}{0.5}$ *	1
0.2	0.5	-1	1	$-$

convergence is immediate. The last line of the table presents a case with a "wrong" selection of the parameters, which prevents convergence.

In practical cases with 2D or 3D problems, the analytical identification of the optimal parameters is not possible or practically unaffordable. Different approximation techniques are necessary, depending on the specific problem at hand. ◇

3.3.2 Non-overlapping methods

A completely different approach considers *non-overlapping subdomains*. Let $x_\Gamma(= x_l = x_r)$ be a point in $(0,1)$.

Algorithm 4

1. Let $\lambda^{(0)} \in \mathbb{R}$ be an arbitrary number and ϑ a relaxation parameter.
2. Loop: up to convergence, for $i = 1, 2, \ldots$ solve

a.
$$-\frac{d^2 u_1^{(i+1)}}{dx^2} = f \ x \in (0, x_\Gamma) \tag{3.82}$$

$$u_1^{(i+1)}(0) = 0, \quad u_1^{(i+1)}(x_\Gamma) = \lambda^{(i)}.$$

b.
$$-\frac{d^2 u_2^{(i+1)}}{dx^2} = f \qquad\qquad x \in (x_\Gamma, 1) \tag{3.83}$$

$$\frac{du_2^{(i+1)}}{dx}(x_\Gamma) = \frac{du_1^{(i+1)}}{dx}(x_\Gamma), \quad u_2^{(i+1)}(1) = 0.$$

c.
$$\lambda^{(i+1)} = \vartheta u_2^{(i+1)}(x_\Gamma) + (1 - \vartheta)\lambda^{(i)}. \tag{3.84}$$

3. End of the loop.

This algorithm relies upon the equivalence of Problem (3.67) with the decomposed problem

$$-\frac{d^2 u_1}{dx^2} = f \qquad\qquad x \in (0, x_\Gamma)$$

$$-\frac{d^2 u_2}{dx^2} = f \qquad\qquad x \in (x_\Gamma, 1)$$

$$u_1(x_\Gamma) = u_2(x_\Gamma), \quad \frac{du_1}{dx}(x_\Gamma) = \frac{du_2}{dx}(x_\Gamma)$$

(see e.g. [QV99]). The convergence of this method and the role of the relaxation parameter ϑ will be investigated in the exercises.

Exercise 3.3.3. Consider the problem (3.67) and Algorithm 4.

1. Set $\vartheta = 1$ and investigate the conditions that guarantee the convergence.
2. Identify the optimal value of ϑ that maximizes the convergence rate.
3. Write a code that implements the method and verify the convergence properties.

Solution 3.3.3.

Mathematical analysis

By proceeding as in the previous exercises, we write the equations for the error $e_1^{(i)}$ and $e_2^{(i)}$. In this case, for $\varepsilon^{(i)} \equiv \lambda^{(i)} - u(x_\Gamma)$, the equations read

$$
\begin{cases}
\dfrac{d^2 e_1^{(i+1)}}{dx^2} = 0 \quad x \in (0, x_\Gamma) \\[2mm]
e_1^{(i+1)}(0) = 0, e_1^{(i+1)}(x_\Gamma) = \varepsilon^{(i)},
\end{cases}
\begin{cases}
\dfrac{d^2 e_2^{(i+1)}}{dx^2} = 0 \quad x \in (x_\Gamma, 1) \\[2mm]
\dfrac{de_2^{(i+1)}}{dx}(x_\Gamma) = \dfrac{de_1^{(i+1)}}{dx}(x_\Gamma), e_2^{(i+1)}(1) = 0,
\end{cases}
\tag{3.85}
$$

with

$$
\varepsilon^{(i+1)} = \vartheta e_2^{(i)}(x_\Gamma) + (1 - \vartheta)\varepsilon^{(i)}.
$$

By solving these equations, we obtain

$$
e_1^{(i+1)}(x) = \frac{\varepsilon^{(i)}}{x_\Gamma} x, \qquad e_2^{(i+1)}(x) = -\frac{\varepsilon^{(i)}}{x_\Gamma}(1 - x)
$$

so that for $\vartheta = 1$,

$$
\varepsilon^{(i+1)} = -\frac{1 - x_\Gamma}{x_\Gamma} \varepsilon^{(i)}.
$$

The reduction factor $\varphi = -\dfrac{1 - x_\Gamma}{x_\Gamma}$ drives the convergence of the method. In particular, convergence is guaranteed if $|\varphi| < 1$, i.e.

$$
x_\Gamma > \frac{1}{2}.
$$

In other terms, the measure of the domain where a Dirichlet problem at the interface is solved must be greater than the measure of the domain where a Neumann condition in x_Γ is prescribed.

In the relaxed case, we have

$$
\varepsilon^{(i+1)} = -\vartheta \frac{1 - x_\Gamma}{x_\Gamma} \varepsilon^{(i)} + (1 - \vartheta)\varepsilon^{(i)} = \left(1 - \frac{\vartheta}{x_\Gamma}\right)\varepsilon^{(i)}.
$$

The relaxed reduction factor reads

$$
\varphi(x_\Gamma, \vartheta) = 1 - \frac{\vartheta}{x_\Gamma}.
$$

By forcing $|\varphi| < 1$ we deduce that:

1. for any $x_\Gamma \in (0,1)$ and for $0 < \theta < 2x_\Gamma$ the method is convergent;
2. for $\theta = x_\Gamma$ in exact arithmetic the convergence is attained in one iteration, since $\varphi(x_\Gamma, x_\Gamma) = 0$.

Numerical approximation

We report a possible implementation of the method is the `dirichlet_neumann` code whose signature is in Program 19. The complete code can be downloaded at the book website.

The method basically couples a Dirichlet step for the domain 1 and a Neumann problem for the domain 2. However, it is worth addressing in detail how the Neumann data are prescribed to the second domain,

$$\frac{du_2^{(i+1)}}{dx}(x_\Gamma) = \frac{du_1^{(i+1)}}{dx}(x_\Gamma)$$

where $u_1^{(i+1)}$ is the solution to the Dirichlet problem in the domain 1. To this aim, let us consider the weak form of each subproblem.

By proceeding in the usual way, we take a function $v \in H^1(0, x_\Gamma)$, we get in the first domain

$$-[\frac{du_1^{(i+1)}}{dx}v]_0^{x_\Gamma} + \int_0^{x_\Gamma} \frac{du_1^{(i+1)}}{dx}\frac{dv}{dx} = \int_0^{x_\Gamma} fv. \tag{3.86}$$

Let $\mathcal{L}_{\lambda^{(i)}}$ be a lifting in the domain 1 of $\lambda^{(i)}$ and take in particular a function v vanishing on the boundary, i.e. $v \in H_0^1(0, x_\Gamma)$. In the first domain, we solve the problem: find $u_1^{(i+1)} \equiv \hat{u}_1^{(i+1)} + \mathcal{L}_{\lambda^{(i)}}$ s.t. for any function $v \in H_0^1(0, x_\Gamma)$

$$\int_0^{x_\Gamma} \frac{du_1^{(i+1)}}{dx}\frac{dv}{dx} = \int_0^{x_\Gamma} fv. \tag{3.87}$$

The same relation holds at the discrete level when we take a test function v_h in a finite dimensional subspace of $H_0^1(0, x_\Gamma)$. Notice however that if we take a function v that belong to H^1 but not to H_0^1 and in particular $v(0) = 0$, but $v(x_\Gamma) \neq 0$, from (3.86) we get for the solution $u_1^{(i+1)}$

$$(\frac{du_1^{(i+1)}}{dx}v)(x_\Gamma) = \int_0^{x_\Gamma} \frac{du_1^{(i+1)}}{dx}\frac{dv}{dx} - \int_0^{x_\Gamma} fv. \tag{3.88}$$

In the discrete setting, this is in particular useful for $v(x) = \varphi_{x_\Gamma}(x)$ the Lagrangian finite element basis function in the domain 1, corresponding to the rightmost degree of freedom placed at the interface x_Γ. Since $\varphi_{x_\Gamma}(x_\Gamma) = 1$, it follows that

$$\frac{du_1^{(i+1)}}{dx}(x_\Gamma) = \int\limits_0^{x_\Gamma} \frac{du_1^{(i+1)}}{dx} \frac{d\varphi_{x_\Gamma}}{dx} - \int\limits_0^{x_\Gamma} f\varphi_{x_\Gamma}. \qquad (3.89)$$

This relation shows that the Neumann data for the second domain can be calculated without any further approximation of the derivative of $u_1^{(i+1)}$. At the algebraic level, notice that this operation can be accomplished following the steps:

1. assembly the matrix A_{nbc} of the system without prescribing the boundary conditions (as it is usual done) and store it;
2. modify the matrix by applying the boundary conditions; in 1D problems the usual strategy is to eliminate the first (last) row and column of the matrix when Dirichlet conditions are prescribed at the left (right) end point; modify the right hand side of the system accordingly;
3. solve the system including the boundary conditions $A_{wbc}u = rhs$;
4. compute the residual $r = A_{nbc}u - rhs$.

In the entries corresponding to Dirichlet boundary data the residual r contains the derivative of the solution, as indicated by equation (3.89).

It is worth stressing that the pay-off of this approach is limited for 1D problems, where the direct computation of the (piecewise constant) derivative of the linear finite element solution is immediate as well. However, the strength of the residual method is evident in multidimensional problems where the term to be computed reads

$$\int\limits_\Gamma \frac{du_1^{(i+1)}}{d\mathbf{n}} \varphi.$$

If the computational grid is conformal on the interface, then the residual method is promptly usable, independently of the geometry of the interface.

Program 19 - dirichlet _ neumann : Relaxed Dirichlet-Neumann method

```
% [u1,u2,x1,x2,it]=dirichlet_neumann(f,xi,ig,theta,h1,h2,tol,Nmax)
%
% Dirichlet Neumann
%   Linear finite elements + Trapezoidal rule for the rhs
%
% Problem: -u"=f in (0,1), u(0)=u(1)=0
%
% |-----------------------------------------------------------------|
%
% |----------------------------| xi (DIRICHLET)
%
%    (NEUMANN)     xi  |--------------------------------------------|
%
%
%INPUT
```

```
% f = STRING with the forcing term (e.g. '0*x+0')
%xi = interface position: domain 1 is [0:xi], domain 2 is [xi:0]
%ig = initial guess for the solutions in the interface
%theta = relaxation parameter
%h1,h2 = mesh size in the two subdomains
%tol= tolerance for the convergence
%Nmax = max number of iterations
%%%%%%%%%%%%%%%%%%%%%%%%%%%%%%%%%%%%%%%%%
%OUTPUT
%u1,u2 = solution in the two subdomains
%x1,x2 = mesh of the two subdomains
%it= number of iteration
%
```

Numerical results

In Table 3.10 we report some results obtained by the relaxed Dirichlet Neumann method.

Table 3.10 Number of iterations for the relaxed Dirichlet-Neumann method with different combinations of the position of the interface and the relaxation parameter. The optimal choice for ϑ predicted by the theory is denoted by a *. Symbol '-' denotes convergence failure

x_Γ	0.5	0.6	0.7	0.8	0.5	0.5	0.5	0.3	0.3	0.3	0.3	0.7	0.7	0.7
ϑ	1	1	1	1	0.1	0.3	0.5*	0.1	0.2	0.3*	0.5	0.1	0.5	0.7*
it	-	25	12	7	35	10	1	21	9	1	25	48	8	1

The results confirm the theoretical findings. In the non-relaxed case ($\vartheta = 1$) the convergence is guaranteed only when the Dirichlet domain is larger than the Neumann one. Should this assumption be fulfilled, the convergence is faster when x_Γ approaches 1.

In the relaxed case, convergence can be guaranteed for a proper choice of ϑ. In particular, for $\vartheta = x_\Gamma$ the convergence is reached in 1 iteration (starred cases in the table).

In Fig. 3.32 we report some snapshots of the two solutions for $x_\Gamma = 0.3$, $\vartheta = 0.2$. We display the solution after iterations 1, 3 and 6.

The identification of the optimal parameter is obviously possible because the problem at hand is simple. In practical problems, it requires specific investigations.

Dirichlet-Neumann method does not feature overlapping and can be used therefore when the two subdomains actually correspond to two different regions of a physical problem, such as

$$-\frac{d}{dx}\left(\chi\frac{du}{dx}\right) = f, \quad x \in (0,1)$$

with

$$\chi = \begin{cases} \chi_1 \ x \in (0, x_\Gamma) \\ \chi_2 \ x \in (x_\Gamma, 1), \end{cases}$$

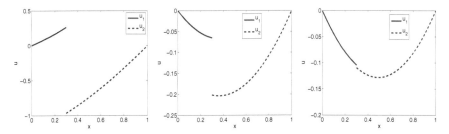

Fig. 3.32 Results of the relaxed Dirichlet Neumann method, for $x_\Gamma = 0.3$, $\vartheta = 0.2$. Iterations 1 (left), 3(center), 6 (right). Notice that the unrelaxed method in this case does not converge

with $\chi_1 \neq \chi_2$ positive coefficients, as in Exercise 3.1.7. The subdomain approach can be a viable alternative to a monolithic solver, the subproblems being in general better conditioned than the one domain problem. The discontinuity of the coefficients is in fact anticipated to introduce an increment of the condition number of the matrix associated to the problem.

We leave to the reader the extension of the Program 19 to this case.   ◇

4

Advection-diffusion-reaction (ADR) problems

We have seen in Chapter 3 that Galerkin method applied to elliptic problems in the form: find $u \in V \subseteq H^1(\Omega)$ such that

$$a(u, v) = F(v) \quad \forall v \in V,$$

provides a *convergent solution* in the $H^1(\Omega)$ norm that satisfies

$$||u_h||_V \leq \frac{M}{\alpha}, \quad ||u - u_h||_{H^1(\Omega)} \leq \frac{\gamma}{\alpha} \inf_{v_h \in V_h} ||u - v_h||_{H^1(\Omega)},$$

where M is the continuity constant of $F(\cdot)$, α and γ coercivity and continuity constants of $a(\cdot, \cdot)$ respectively. In practice, these inequalities can be meaningless when the constants involved are large. In particular if $\gamma \gg \alpha$ the second inequality is an effective bound for the error only if $\inf_{v_h \in V_h} ||u - v_h||_{H^1(\Omega)}$ is small. For a finite element discretization, this corresponds to a small value of the mesh size h. The associated discretized problem can be therefore computationally expensive or even not affordable.

Advection-dominated as well as *reaction dominated* advection-diffusion-reaction (ADR) problems feature a continuity constant γ (dependent on the advection/reaction terms) significantly larger than the coercivity one (which depends on the diffusivity term). These problems are frequent in engineering applications. The selection of a small enough h is often unfeasible. In these cases, specific numerical techniques are required to address this issue. These are called *stabilization techniques* (even if - as we have seen - Galerkin solution *is* stable in the usual mathematical meaning). A general formulation of the stabilized problem reads: find $u_h \in V_h \subseteq H^1(\Omega)$ such that

$$a_h(u_h, v_h) = F_h(v_h) \quad \forall v_h \in V_h,$$

where index h added to the bilinear form and the functional indicates the perturbation due to the stabilization method. Error analysis for this kind of problem relies upon a fundamental results, the Strang Lemma (see e.g. [QV94]). Under the assumption that $a_h(\cdot, \cdot)$ is uniformly coercive in V, this

Formaggia L., Saleri F., Veneziani A.: Solving Numerical PDEs: Problems, Applications, Exercises. DOI 10.1007/978-88-470-2412-0_4, © Springer-Verlag Italia 2012

Lemma states that

$$||u - u_h||_V \leq C \left(\inf_{v_h \in V_h} ||v - v_h||_V + \inf_{v_h \in V_h} \sup_{w_h \in V_h \backslash 0} \frac{|a(v_h, w_h) - a_h(v_h, w_h)|}{||w_h||_V} \right.$$

$$\left. + \sup_{w_h \in V_h \backslash 0} \frac{|F(v_h, w_h) - F_h(v_h, w_h)|}{||w_h||_V} \right). \tag{4.1}$$

In this chapter we solve some exercises for this class of problems. Hereafter we introduce some preliminary exercises aimed at introducing basic tools for stabilization methods. Then we consider advection-dominated problems, and finally reaction-dominated ones.

Mass matrix and mass matrix lumping

We mention here a technique that is relevant for the stabilization of reaction dominated problems, but also in time dependent problems, as we will see. The discretization by finite element of terms homogeneous in u leads to the computation of the so call *mass matrix* M, whose components are

$$M_{ij} = \int_\Omega \varphi_j \varphi_i d\Omega, \quad 1 \leq i, j \leq N_h$$

where $\{\varphi_i, \quad i = 1, \ldots N_h\}$ is the set of finite element basis function. It is immediate to verify that M is symmetric positive definite. As usual in finite elements, the mass matrix is assembled from the contribution coming from each mesh element K, given by the *local mass matrix* M_K of elements

$$M_{K,sl} = \int_K \varphi_{\nu_{K,l}} \varphi_{\nu_{K,l}} \, dK = \int_{\widehat{K}} |J_K| \widehat{\varphi}_l \widehat{\varphi}_s \, d\widehat{K},$$

for $1 \leq s, l \leq n_l$. Here, $\nu_{K,l}$ indicates the connectivity matrix, J_K the Jacobian of the transformation between reference and current element, n_l the number of local degrees of freedom, and the quantities with a hat are related to the reference element, see Chapter 2. For affine finite elements (see Section 2.2.3), the Jacobian is constant and $M_K = |J_K| M_{\widehat{K}}$, where the local mass matrix on the reference element $M_{\widehat{K}}$ is given by

$$M_{\widehat{K},sl} = \int_{\widehat{K}} \widehat{\varphi}_l \widehat{\varphi}_s \, d\widehat{K}. \tag{4.2}$$

In certain situations, which will be detailed in the pertinent sections of the book, it is convenient to replace M with a diagonal matrix M^L, called *lumped mass matrix*. A common strategy is summing the rows, i.e. we set for all i and j,

$$M_{K,ll}^L = \sum_{s=1}^{n_l} M_{K,ls}, \quad M_{K,ls}^L = 0 \quad \text{for } l \neq s. \tag{4.3}$$

For higher order elements in more than one dimension, this strategy may lead to a singular matrix and other techniques are at hand, one of which is to set on each element

$$M_{K,ll}^L = |K| \frac{M_{K,ll}}{\sum_s M_{K,ss}}, \quad M_{K,ls}^L = 0 \quad \text{for } l \neq s, \tag{4.4}$$

and assemble the global matrix as usual. Both formulas ensure that the fundamental property of the local mass matrix $\sum_{sl} M_{K,sl} = |K|$ is maintained by its lumped form as well.

4.1 Preliminary problems

Exercise 4.1.1. Let $u \in V \equiv H_0^1(0,1)$, μ and σ be given constants, and b be a given function. Split the 1D ADR operator

$$\mathcal{L}u = -\mu u'' + bu' + \sigma u,$$

into its *symmetric* and *skew-symmetric* parts.

Solution 4.1.1. An operator $\mathcal{L} : V \to V'$, associated with the corresponding bilinear form $a(\cdot, \cdot)$, is *symmetric* if

$$_{V'} < \mathcal{L}u, v >_V = a(u,v) = a(v,u) =_V < u, \mathcal{L}v >_{V'}$$

and *skew-symmetric* if

$$_{V'} < \mathcal{L}u, v >_V = a(u,v) = -a(v,u) = -_V < u, \mathcal{L}v >_{V'}.$$

Similarly to what can be done for matrices, a generic linear operator $\mathcal{L}$ can be decomposed into its symmetric part, denoted by $\mathcal{L}_S$, and the skew-symmetric one $\mathcal{L}_{SS}$. If $\mathcal{L}^*$ denotes the *adjoint operator* such that

$$_{V'} < \mathcal{L}^*u, v >_V \equiv a(v,u) =_V < u, \mathcal{L}v >_{V'},$$

we have

$$\mathcal{L} = \mathcal{L}_S + \mathcal{L}_{SS}, \quad \text{with} \quad \begin{cases} \mathcal{L}_S = \dfrac{1}{2} (\mathcal{L} + \mathcal{L}^*), \\[2mm] \mathcal{L}_{SS} = \dfrac{1}{2} (\mathcal{L} - \mathcal{L}^*). \end{cases}$$

We refer to the following bilinear form with arguments in $H_0^1(0,1)$

$$\int_0^1 (-\mu u'' + \sigma u)\, v\, dx = \mu \int_0^1 u'v'\, dx + \int_0^1 \sigma u v\, dx \equiv a_{\mathrm{DR}}(u,v).$$

Notice that in the previous expression with $\mathcal{L}u = -\mu u'' + \sigma u$, the first integral is a formal way for denoting $_{V'} < \mathcal{L}u, v >_V$. Since the product of u', v' and u, v is commutative, we have that $a_{\mathrm{DR}}(\cdot, \cdot)$ is symmetric.

Let us consider the advection term. At this stage, for the sake of simplicity, let us assume that b is constant. The associated bilinear form is such that

$$a_A(u,v) \equiv b \int_0^1 u'v dx = -b \int_0^1 uv' dx = -a_A(v,u). \tag{4.5}$$

We conclude that if b is constant the advection operator is skew-symmetric, so that

$$\mathcal{L}_S u = -\mu u'' + \sigma u, \quad \mathcal{L}_{SS} u = bu'. \tag{4.6}$$

In the general case of a non-constant advection coefficient b, we have

$$a_A(u,v) = \int_0^1 bu'v dx = \frac{1}{2}\int_0^1 bu'v dx + \frac{1}{2}\int_0^1 (bu)' v dx - \frac{1}{2}\int_0^1 b'uv dx.$$

The last term can be assimilated to a reactive term and is associated with the symmetric part of the operator. For the other two previous terms, integrating by parts we have

$$\frac{1}{2}\int_0^1 (bu' + (bu)') v dx = -\frac{1}{2}\int_0^1 ((bv)' + bv') u dx.$$

We conclude that

$$\mathcal{L}_S = -\mu u'' + \left(\sigma - \frac{1}{2}b'\right)u, \quad \mathcal{L}_{SS} = \frac{1}{2}bu' + \frac{1}{2}(bu)' = bu' + \frac{1}{2}b'u. \tag{4.7}$$

This clearly reduces to (4.6) for a constant b.   ◇

Exercise 4.1.2. Split the ADR operator in non-divergence form

$$\mathcal{L}u = -\mu\triangle u + \mathbf{b}\cdot\nabla u + \sigma u$$

where μ and σ are given constants, $\mathbf{b}$ is a given advective field and $u \in H_0^1(\Omega)$, into its symmetric and skew-symmetric parts.

Solution 4.1.2.   Notice that

$$\mathbf{b}\cdot\nabla u = \frac{1}{2}\mathbf{b}\cdot\nabla u + \frac{1}{2}\nabla\cdot(\mathbf{b}u) - \frac{1}{2}(\nabla\cdot\mathbf{b})u.$$

We have therefore

$$_{V'} < \mathcal{L}u, v >_V \equiv \int_\Omega (-\mu \triangle u + \mathbf{b} \cdot \nabla u + \sigma u) \, v d\omega =$$

$$\int_\Omega \mu \nabla u \cdot \nabla v d\omega + \frac{1}{2} \int_\Omega (\mathbf{b} \cdot \nabla u) \, v d\omega + \frac{1}{2} \int_\Omega \nabla \cdot (\mathbf{b}u) v d\omega -$$

$$\frac{1}{2} \int_\Omega (\nabla \cdot \mathbf{b}) \, uv d\omega + \int_\Omega \sigma uv d\omega \equiv a(u, v).$$

Bilinear form $a_S(u, v) \equiv \int_\Omega \mu \nabla u \cdot \nabla v d\omega + \int_\Omega (\sigma - \frac{1}{2}\nabla \cdot \mathbf{b}) uv d\omega$ is symmetric for the commutativity of the scalar product and function product. Moreover, upon application of the Green formula

$$a_{SS}(u, v) \equiv \frac{1}{2} \int_\Omega (\mathbf{b} \cdot \nabla u) \, v d\omega + \frac{1}{2} \int_\Omega \nabla \cdot (\mathbf{b}u) v d\omega =$$

$$-\frac{1}{2} \int_\Omega \nabla \cdot (\mathbf{b}v) u d\omega - \frac{1}{2} \int_\Omega (\mathbf{b} \cdot \nabla v) \, u d\omega = -a_{SS}(v, u).$$

The symmetric part of the operator reads therefore $\mathcal{L}_S u = -\mu \triangle u - \frac{1}{2}(\nabla \cdot \mathbf{b})u + \sigma u$ and the skew-symmetric one $\mathcal{L}_{SS} u = \frac{1}{2}(\mathbf{b} \cdot \nabla u) + \frac{1}{2}\nabla \cdot (\mathbf{b}u)$. $\diamond$

Exercise 4.1.3. Given the problem

$$\begin{cases} -\sum_{i,j=1}^{2} \frac{\partial^2 u}{\partial x_i \partial x_j} + \beta \frac{\partial^2 u}{\partial x_1^2} + \gamma \frac{\partial^2 u}{\partial x_1 \partial x_2} + \delta \frac{\partial^2 u}{\partial x_2^2} + \eta \frac{\partial u}{\partial x_1} = f \text{ in } \Omega, \\ u = 0 \hspace{6cm} \text{on } \partial\Omega, \end{cases} \quad (4.8)$$

where β, γ, δ e η are given constant coefficients and f is a given function of $\mathbf{x} = (x_1, x_2) \in \Omega$:

1. Find sufficient conditions on the data that guarantee well posedness of the weak problem.
2. Write a Galerkin finite element discretization and analyze its convergence.
3. Specify conditions that guarantee that the problem is symmetric and appropriate methods for solving the associated linear system.

Solution 4.1.3.

> *Mathematical analysis*

Equation (4.8) can be rewritten

$$-\nabla \cdot (\mathcal{K}\nabla u) + \mathcal{B} \cdot \nabla u = f \tag{4.9}$$

in Ω, where $\mathcal{K}$ and $\mathcal{B}$ are a tensor and a vector respectively given by

$$\mathcal{K} \equiv \begin{bmatrix} 1-\beta & 1-\dfrac{\gamma}{2} \\[2mm] 1-\dfrac{\gamma}{2} & 1-\delta \end{bmatrix}, \quad \mathcal{B} \equiv \begin{bmatrix} \eta \\ 0 \end{bmatrix}.$$

To (4.9) we associate the boundary condition $u = 0$ on $\partial\Omega$. The differential operator of this problem (following the definition of e.g. [QV94], Chapter 6) is elliptic if there exists a constant $\mu_0 > 0$ such that for any vector χ in $\mathbb{R}^2$ we have

$$\sum_{i,j=1}^{2} \mathcal{K}_{ij}\chi_i\chi_j \geq \mu_0 ||\chi||^2. \tag{4.10}$$

In other terms, we require that the tensor $\mathcal{K}$ by symmetric positive definite. The Sylvester criterion (see e.g. [GV96]) states that a real symmetric matrix is p.d. if and only if all the principal minors are positive. We prescribe these conditions on the data

$$1 - \beta > 0, \quad (1-\beta)(1-\delta) - \frac{1}{4}(1-\gamma)^2 > 0. \tag{4.11}$$

Thanks to the first of (4.11), the second one reads

$$\delta < 1 - \frac{(1-\gamma)^2}{4(1-\beta)}. \tag{4.12}$$

Under these conditions, the eigenvalues of $\mathcal{K}$

$$\lambda_{0,1} = (2 - \beta - \delta) \pm \sqrt{(\beta-\delta)^2 + (1-\gamma)^2} \tag{4.13}$$

are both positive. Hereafter we assume (4.11) to hold. Constant μ_0 in (4.10) is the minimal eigenvalue (corresponding to the negative sign in (4.13)). Let μ_1 be the maximum eigenvalue. For $V = H_0^1(\Omega)$, the weak formulation of the problem reads: find $u \in V$ such that for every $v \in V$

$$\int_\Omega (\mathcal{K}\nabla u) \cdot \nabla v \, d\omega + \int_\Omega (\mathcal{B} \cdot \nabla u)\, v \, d\omega = \int_\Omega f v \, d\omega.$$

The selected functional space makes the bilinear form

$$a(u,v) \equiv \int_\Omega (\mathcal{K}\nabla u) \cdot \nabla v d\omega + \int_\Omega \mathcal{B} \cdot \nabla u v d\omega$$

continuous, since $|a(u,v)| \le (\mu_1 + |\eta|)||u||_V||v||_V$. Moreover, we assume that f is a linear and continuous functional on V, which is equivalent to say that f belongs to the dual space of V, usually denoted by V'. In this case, the expression $\int_\Omega fv d\omega$ is a formal way for denoting the duality $\mathcal{F}(v) = {}_{V'} < f, v >_V$.

Let us show that the bilinear form is coercive. From the boundary conditions, the identity $(\nabla u)u = \frac{1}{2}\nabla |u|^2$ and the Poincaré inequality, we have

$$a(u,u) = \int_\Omega (\mathcal{K}\nabla u) \cdot \nabla u d\omega + \int_\Omega \mathcal{B} \cdot (\nabla u)u d\omega \ge$$

$$\mu_0 ||\nabla u||_{L^2}^2 + \frac{1}{2}\int_\Omega \nabla \cdot (\mathcal{B}u^2)\, d\omega \ge \alpha ||u||_V^2 + \frac{1}{2}\int_{\partial\Omega} \mathcal{B} \cdot \mathbf{n} u^2 d\gamma = \alpha ||u||_V^2$$

where $\alpha = \mu_0/(1 + C_\Omega^2)$ e C_Ω is the constant of the Poincaré inequality. Coercivity is proven and well posedness (under (4.11) and (4.12)) stems from the Lax-Milgram lemma.

Numerical approximation

Let us select a subspace of V, $V_h \subset X_h^r$, where X_h^r is the space introduced in (2.22) of Chapter 2. In particular, V_h is the space of functions of X_h^r vanishing at the boundary.

Discrete formulation of the problem reads: find $u_h \in V_h$ such that for any $v_h \in V_h$ we have $a(u_h, v_h) = \mathcal{F}(v_h)$. The discrete problem is well posed as an immediate consequence of the analysis performed on the continuous one. We remind moreover that for $u \in H^{s+1}(\Omega)$ (see (2.27))

$$||u - u_h||_{H^1} \le \frac{M}{\alpha} C h^q |u|_{H^{q+1}},$$

where M is the continuity constant of the bilinear form, C is a constant associated with the interpolation error and $q = \min(r, s)$. In particular, notice that in our case $M = \mu_1 + |\eta|$. When $|\eta| \gg \mu_0$ so that $M/\alpha \gg 1$, the error estimate is practically meaningless. In this case, the problem is *advection dominated* and it needs a suitable "stabilization".

Finally, notice that if $\eta = 0$ the bilinear form $a(u, v)$ is symmetric (see Exercise 4.1.2). Actually, since $\mathcal{K}$ is symmetric, we have

$$a(u, v) = \int_\Omega (\mathcal{K}\nabla u) \cdot \nabla v \, d\omega = \int_\Omega \nabla u \cdot (\mathcal{K}\nabla v) \, d\omega = a(v, u).$$

The discrete problem leads to a linear system in the stiffness matrix $A_{ij} = a(\varphi_j, \varphi_i)$, which for $\eta = 0$ is symmetric (and positive definite), as a consequence of the symmetry and coercivity of the bilinear form. Suitable methods for solving the associated linear system are (see e.g. [QSS00]):

1. *Direct methods*: Cholesky factorization requires preliminary reordering of the matrix for limiting the *fill-in*. Actually A is sparse, whilst in general the Cholesky coefficients are not.
2. *Iterative methods*: the *conjugate gradient* method, suitably preconditioned (possibly with an incomplete Cholesky factorization, see e.g. [Saa03]). $\diamond$

Exercise 4.1.4. Verify that for 1D linear, quadratic and cubic finite elements lumping the mass matrices using (4.3) gives the following results for the local lumped mass matrix on the reference interval $\widehat{K} = [0, 1]$,

$$r = 1: \quad \mathrm{M}^L_{\widehat{K}} = \frac{1}{2} \begin{bmatrix} 1 & 0 \\ 0 & 1 \end{bmatrix},$$

$$r = 2: \quad \mathrm{M}^L_{\widehat{K}} = \frac{1}{6} \begin{bmatrix} 1 & 0 & 0 \\ 0 & 4 & 0 \\ 0 & 0 & 1 \end{bmatrix},$$

$$r = 3: \quad \mathrm{M}^L_{\widehat{K}} = \frac{1}{8} \begin{bmatrix} 1 & 0 & 0 & 0 \\ 0 & 3 & 0 & 0 \\ 0 & 0 & 3 & 0 \\ 0 & 0 & 0 & 1 \end{bmatrix}. \tag{4.14}$$

Verify moreover that the same result is obtained for $r = 1, 2$ by adopting formula (4.4) instead, while for $r = 3$ the latter formula would give

$$\mathrm{M}^L_{\widehat{K}} = \frac{1}{1552} \begin{bmatrix} 128 & 0 & 0 & 0 \\ 0 & 648 & 0 & 0 \\ 0 & 0 & 648 & 0 \\ 0 & 0 & 0 & 128 \end{bmatrix} = \begin{bmatrix} \frac{8}{97} & 0 & 0 & 0 \\ 0 & \frac{81}{194} & 0 & 0 \\ 0 & 0 & \frac{81}{194} & 0 \\ 0 & 0 & 0 & \frac{8}{97} \end{bmatrix}. \tag{4.15}$$

Observe, moreover, that for $r = 1$ the mass lumping is equivalent to apply the trapezoidal formula to approximate the integration over the element when computing the local mass matrix, while for $r = 2$ the same result is obtained using the Simpson-Cavalieri formula.

We remark that the equivalence with quadrature formulas allows to analyze of the error introduced by lumping on the solution of the differential problem by resorting to the Strang Lemma (4.1).

Solution 4.1.4.   The application of (4.2) to the one dimensional case and the reference interval gives $\mathrm{M}_{\widehat{K},sl} = \int_0^1 \widehat{\varphi}_s \widehat{\varphi}_l d\widehat{x}$. From Ch. 2 we know that for $r = 1$ the basis functions on the reference interval $(0,1)$ are given by $\widehat{\varphi}_0(\widehat{x}) = 1 - \widehat{x}$, $\widehat{\varphi}_1(\widehat{x}) = \widehat{x}$ and the local nodes are $\widehat{x}_1 = 0$ and $\widehat{x}_2 = 1$. Performing the integral we obtain

$$\mathrm{M}_{\widehat{K}} = \frac{1}{6} \begin{bmatrix} 2 & 1 \\ 1 & 2 \end{bmatrix}. \tag{4.16}$$

Lumping by row sum we obtain the first expression in (4.14), and by direct inspection, we notice the same result would have been obtained by adopting (4.4). The application of the trapezoidal rule is equivalent to the following approximation

$$\int_0^1 \widehat{\varphi}_s \widehat{\varphi}_l \, d\widehat{x} \simeq \frac{1}{2} \sum_{k=0}^1 (\widehat{\varphi}_s(0)\widehat{\varphi}_l(0) + \widehat{\varphi}_s(1)\widehat{\varphi}_l(1)),$$

for l and s in $1, 2$. Exploiting property $\widehat{\varphi}_l(\widehat{x}_s) = \delta_{ls}$ and the fact that the finite element nodes on the reference element coincide with the quadrature points, we note that the resulting approximation is diagonal. The diagonal term is given by

$$\int_0^1 \widehat{\varphi}_l \widehat{\varphi}_l \, d\widehat{x} \simeq \frac{1}{2} \widehat{\varphi}_l(\widehat{x}_l)\widehat{\varphi}_l(\widehat{x}_l) = \frac{1}{2}, \quad \text{for } l = 1, 2.$$

The approximation thus coincides with the one found previously.

In the case $r = 2$, the basis functions on the reference interval are

$$\widehat{\varphi}_0 = 2(1-x)\left(\frac{1}{2} - x\right), \quad \widehat{\varphi}_1 = 4x(1-x), \quad \widehat{\varphi}_2 = 2x\left(x - \frac{1}{2}\right),$$

and the nodes are given by $\widehat{x}_1 = 0$, $\widehat{x}_2 = 0.5$ and $\widehat{x}_3 = 1$ Upon direct computation, we get

$$\mathrm{M}_{\widehat{K}} = \begin{bmatrix} \dfrac{2}{15} & \dfrac{1}{15} & -\dfrac{1}{30} \\[2mm] \dfrac{1}{15} & \dfrac{8}{15} & \dfrac{1}{15} \\[2mm] -\dfrac{1}{30} & \dfrac{1}{15} & \dfrac{2}{15} \end{bmatrix}.$$

Summing by row we find the matrix $M_{\widehat{K}}^{L}$ suggested in the exercise and we may easily verify that the same result is obtained by formula (4.4). Simpson rule provides the approximation

$$\int_0^1 \widehat{\varphi}_s\widehat{\varphi}_l \, d\widehat{x} \simeq \frac{1}{6}\sum_{k=0}^{1} \left(\widehat{\varphi}_s(0)\widehat{\varphi}_l(0) + 4\widehat{\varphi}_s(0.5)\widehat{\varphi}_l(0.5) + \widehat{\varphi}_s(1)\widehat{\varphi}_l(1) \right).$$

Again, we get the desired result by noting that the quadrature points of the Simpson formula coincide with the finite element nodes and using the property of finite element basis functions.

For the case $r = 3$ we could work in a similar way. Since computations are quite involved, as an alternative we report the snapshot of MATLAB code (based on the symbolic toolbox) that performs the required computation.

```
>> format rat; syms x
>> phi=[9/2*(1/3-x)*(2/3-x)*(1-x),27/2*x*(2/3-x)*(1-x),...
           27/2*x*(x-1/3)*(1-x),9/2*x*(x-1/3)*(x-2/3)'];
>> for i=1:4,
        for j=1:4, M(i,j)=int(phi(i)*phi(j),0,1); end;
    end;
>> for i=1:4,
        for j=1:4,
            ML(i,j)=sum(M(i,:))*(i==j); %implements row sum
        end;
    end;
>> Mhat=zeros(4);
>> T=trace(M);
>> for i=1:4,
        Mhat(i,i)=M(i,i)/T; %implements alternate formula
    end;
>> M


M =
[   8/105,   33/560,   -3/140,  19/1680]
[  33/560,    27/70,  -27/560,   -3/140]
[  -3/140,  -27/560,    27/70,   33/560]
[ 19/1680,   -3/140,   33/560,    8/105]


>> Mhat


Mhat =
        8/97            0            0            0
           0       81/194            0            0
           0            0       81/194            0
           0            0            0         8/97


>> ML
```

```
ML =
[ 1/8,    0,    0,    0]
[   0,  3/8,    0,    0]
[   0,    0,  3/8,    0]
[   0,    0,    0,  1/8]
```

In this case (4.3) and (4.4) are not equivalent.   ◇

4.2 Advection dominated problems

Exercise 4.2.1. Let us consider the problem

$$
\begin{cases}
-\varepsilon u''(x) + bu'(x) = 1, \ 0 < x < 1, \\
u(0) = \alpha, \qquad\qquad u(1) = \beta,
\end{cases}
\tag{4.17}
$$

where $\varepsilon > 0$ and $\alpha, \beta, b \in \mathbb{R}$ are given. Write the weak formulation of the problem and its standard finite element approximation. Write the approximation using the *artificial viscosity* techniques based on the standard *Upwind* and the Scharfetter-Gummel methods (refer, for instance, to [Qua09] for details). Analyze the stability and convergence properties of the results obtained.
After computing the analytical solution, compute the numerical approximation with **fem1d**, by using quadratic finite elements and assuming $b = -1, \varepsilon = 10^{-3}, \alpha = 0, \beta = 1$.
Identify a value for h by which the standard finite elements solution does not oscillate.

Solution 4.2.1.

Mathematical analysis

A weak formulation of the problem is obtained formally by multiplying the equation (4.17) by a function $v \in H_0^1(0,1)$ (this space will be denoted hereafter by V), by integrating on the interval $(0,1)$ and applying the integration by parts formula. We have the problem: find $u \in H^1(0,1)$ such that for any $v \in V$

$$
\varepsilon \int_0^1 u'v'dx + b \int_0^1 u'vdx = \int_0^1 vdx
$$

with $u(0) = \alpha$ and $u(1) = \beta$. Working as done in Chapter 3, if $G \in H^1(0,1)$ is a lifting of the boundary data, the problem reads: find $u \in G + V$ such that $a(u, v) = \mathcal{F}(v)$ for any $v \in V$, where

$$a(w, v) \equiv \varepsilon \int_0^1 w'v'dx + b \int_0^1 w'vdx, \quad \mathcal{F}(v) = \int_0^1 vdx.$$

Using Cauchy-Schwarz inequality one easily verifies that functional $\mathcal{F}(v)$ is continuous as well as the bilinear form $a(\cdot, \cdot)$, being the continuity constant of the latter given by $\gamma = \varepsilon + |b|$.

Moreover, $a(\cdot, \cdot)$ is coercive in $V \times V$. Indeed, for any $w \in V$

$$a(w, w) = \varepsilon \int_0^1 (w')^2 dx + b \int_0^1 w'wdx = \varepsilon||w'||_{L^2}^2 + \frac{b}{2}\left[w^2\right]_0^1 = \varepsilon||w'||_{L^2}^2,$$

since w vanishes on the boundary. From the Poincaré inequality, it follows that $a(w, w) \geq \alpha||w||_V^2$ with constant $\alpha = \varepsilon/(1 + C_\Omega^2)$ where C_Ω is the constant of the Poincaré inequality. The well posedness of the problem follows from Corollary 3.1.

In this special case, we can find the analytical solution by finding the general integral and prescribing the boundary conditions. A particular integral of $(4.17)_1$ is given by $u_P = x/b$. The general integral of the equation is given by [BD70] $u_G = C_1 + C_2 e^{bx/\varepsilon} + x/b$. C_1 and C_2 are found by prescribing the boundary conditions,

$$\begin{cases} C_1 + C_2 = \alpha, \\ \\ C_1 + C_2 e^{b/\varepsilon} + \dfrac{1}{b} = \beta. \end{cases}$$

The analytical solution to the problem reads therefore $u(x) = \dfrac{2(e^{10^5 x} - 1)}{e^{10^5} - 1} - x$.

Numerical approximation

To carry out a finite element discretization of the problem at hand, we introduce a subdivision of the interval $(0, 1)$. For the sake of simplicity, we assume this mesh to be uniform, with size h. Let us introduce the subspace $V_h \subset X_h^r(0, 1)$ of V of piecewise polynomial functions on each subinterval, with degree r and vanishing on the boundary. If u_h denotes the numerical solution, the discretized problem reads: find $u_h \in V_h$ such that for any $v_h \in V_h$

$$a(u_h, v_h) = \mathcal{F}(v_h).$$

Convergence theory of finite elements (see Chapter 2) states that

$$||u - u_h||_V \leq (1 + C_\Omega^2)\frac{\varepsilon + |b|}{\varepsilon}h^q|u|_{H^{q+1}}, \tag{4.18}$$

holds for $u \in H^{s+1}(0,1)$ and $q = \min(r,s)$. If $|b| \gg \varepsilon$, then $\dfrac{\varepsilon + |b|}{\varepsilon} \gg 1$, so that the convergence estimate is useless unless h is small enough. In practice (see e.g. [Qua09]), standard linear finite element approximation features unphysical oscillations when $\mathbb{Pe} > 1$, where $\mathbb{Pe}$ is the Péclet number defined as $\dfrac{|b|h}{2\varepsilon}$. Oscillations can be avoided with the standard Galerkin method if the grid size is small enough to have $\mathbb{Pe} < 1$. However, this approach can be unaffordable in practice. With the *upwind* method we augment the viscosity of the problem by adding a numerical diffusive term in the form $-\dfrac{|b|h}{2}u'' = -\varepsilon\mathbb{Pe}u''$.

In this way, the actual viscosity of the numerical problem is $\varepsilon^* = \varepsilon\,(1 + \mathbb{Pe})$, and the actual Péclet number is

$$\mathbb{Pe}^* = \frac{|b|h}{2\varepsilon^*} = \frac{\mathbb{Pe}}{1 + \mathbb{Pe}}.$$

Clearly $\mathbb{Pe}^* < 1$ independently of h. Upwind approximation does not introduce spurious oscillations for any h, at the price of solving a modified, more diffusive, problem.

For the accuracy analysis, the Strang Lemma leads to the conclusion

$$\|u - u_h\|_V \leq K h^s |u|_{q+1} + \frac{|b|h}{2}\|u_h\|_V,$$

where q is the index in (4.18), and $K = (1 + C_\Omega^2)\dfrac{\varepsilon + |b|}{\varepsilon}$. This estimate shows that the perturbation to the bilinear form caused by the numerical viscosity reduces the order of accuracy to 1, independently of the finite element degree. This is the price to pay to get non-oscillating solutions using this technique.

Numerical results

We solve the problem with `fem1d` with quadratic finite elements and the standard Galerkin scheme with $h = 0.1$ and $h = 0.05$. As Fig. 4.1 shows, solution is affected by numerical oscillations that make it meaningless. Oscillations reduce when the mesh size h gets smaller (see Fig. 4.2) and are eliminated for $\mathbb{Pe} < 1$, more precisely for

$$\frac{|b|h}{2\varepsilon} < 1 \Rightarrow h < \frac{2\varepsilon}{|b|} = 0.002.$$

Mesh size $h = 0.001$ yields an accurate solution (see Fig. 4.3), even if relatively to the problem at hand the computational cost is high.

Fig. 4.4 shows the solution obtained with the Upwind method for $\mathbb{Pe} > 1$. It is evident that oscillations have been eliminated, even if the boundary layer (i.e. the region close to the boundary where the exact solution features a steep gradient) has widened as a consequence of the numerical over-diffusion.

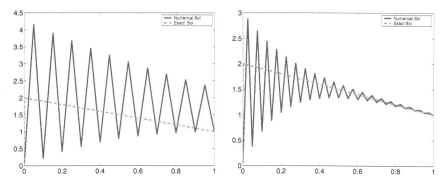

Fig. 4.1 Solution of Exercise 4.2.1: standard Galerkin method with mesh size $h = 0.1$ and $h = 0.05$. The gray line is the exact solution

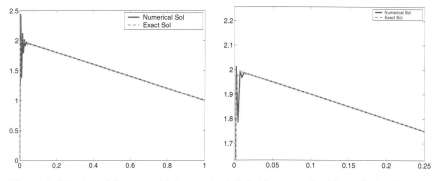

Fig. 4.2 Solution of Exercise 4.2.1: standard Galerkin method with mesh size $h = 0.01$ and $h = 0.005$ (in the latter case, a zoom of the solution is showed)

In Table 4.1 we report the errors computed with respect to the $L^2(0, 1)$ and $H^1(0, 1)$ norms with both the Galerkin and the Upwind methods. For h getting smaller the expected asymptotic orders for the standard Galerkin method of 3 and 2 respectively are found. For the Upwind method we have order 1.

Table 4.1 Errors of both Galerkin and Upwind methods for different mesh sizes and L^2 and H^1 norms, in solving Exercise 4.2.1

h	Galerkin $L^2(0, 1)$	Galerkin $H^1(0, 1)$	Upwind $L^2(0, 1)$	Upwind $H^1(0, 1)$
0.01	0.042027	34.7604	0.084734	35.4435
0.005	0.013108	19.7796	0.052634	28.9093
0.0025	0.0028641	7.9463	0.030937	20.9863
0.001	0.00024082	1.5831	0.014131	11.5944
0.0005	3.1605e-05	0.41118	0.0074522	6.6747

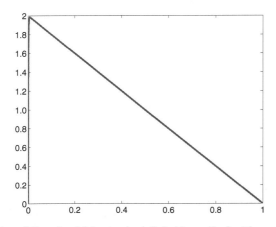

Fig. 4.3 Solution of Exercise 4.2.1: standard Galerkin method with mesh size $h = 0.001$

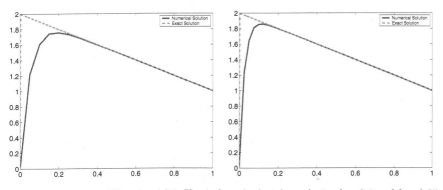

Fig. 4.4 Solution of Exercise 4.2.1: Upwind method with mesh size $h = 0.1$ and $h = 0.05$

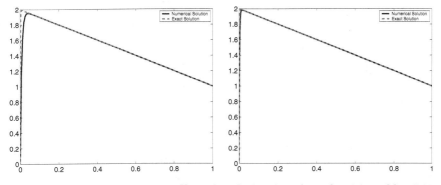

Fig. 4.5 Solution of Exercise 4.2.1: Upwind method with mesh size $h = 0.01$ and $h = 0.005$

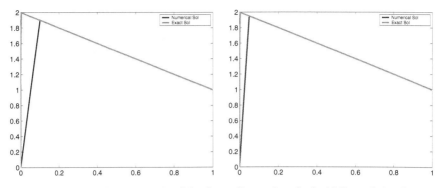

Fig. 4.6 Solution of Exercise 4.2.1: Scharfetter-Gummel method with linear finite elements and mesh size $h = 0.1$ and $h = 0.05$

A more accurate stabilization technique is the Scharfetter-Gummel method (see [Qua09] and Exercise 4.2.2). Since the forcing term and coefficients are constant, for linear finite elements this method provides the exact solution at the discretization nodes. In Fig. 4.6 we report the relative results, which highlight the reduced effects of numerical diffusion with respect to standard Upwind and the nodal super-convergence.   ◇

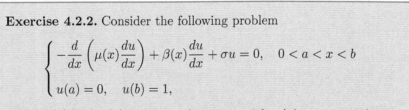

Exercise 4.2.2. Consider the following problem

$$\begin{cases} -\dfrac{d}{dx}\left(\mu(x)\dfrac{du}{dx}\right) + \beta(x)\dfrac{du}{dx} + \sigma u = 0, & 0 < a < x < b \\[2mm] u(a) = 0, \quad u(b) = 1, \end{cases}$$

where $\beta(x)$ and $\mu(x)$ are given functions with $\mu(x) \geq \mu_0 > 0$ for any $x \in [a,b]$ and σ is a positive constant.

1. Write the weak formulation of the problem and find suitable assumptions on the data that guarantee well posedness.
2. For $\beta = 10^3, \sigma = 0, \mu(x) = x$ compute the exact solution to the problem as a function of a and b.
3. Analyze the accuracy of the numerical solution obtained with the standard Galerkin method for $a = 1, b = 2$ and $a = 10^3, b = 10^3 + 1$. Identify possible remedies to inaccuracies of the numerical solution.
4. Verify the answer to the previous point, by solving the problem with linear finite elements. Use the Upwind and the Sharfetter-Gummel methods ($h = 0.05$), and comment on the solutions.

Solution 4.2.2.

Mathematical analysis

We introduce the functional space $V \equiv H_0^1(a,b)$ of $H^1(a,b)$ functions vanishing at the end points. Moreover, we have the bilinear form

$$a(w,v) \equiv \int_a^b (\mu w'v' + \beta w'v + \sigma wv)\, dx,$$

and a lifting function $G \in H^1(a,b)$ such that $G(a) = 0$ and $G(b) = 1$. The weak form of the problem reads: find $u \in G+V$ such that $a(u,v) = 0$ for any $v \in V$.

Well posedness is guaranteed by the continuity of the bilinear form in $H^1(a,b)$ and coercivity in V. We assume[1] $\mu \in L^\infty(a,b)$, $\beta \in L^\infty(a,b)$. Under these hypotheses, we have $|a(w,v)| \leq \max(\|\mu\|_{L^\infty}, \|\beta\|_{L^\infty}, \sigma)\|w\|_V \|v\|_V$.

For the coercivity, if w is a generic function of V, we have

$$a(w,w) = \int_a^b \left(\mu\,(w')^2 + \beta w'w + \sigma w^2\right) dx \geq$$

$$\min(\mu_0, \sigma)\|w\|_V^2 + \int_a^b \beta w'w\, dx.$$

On the other hand

$$\int_a^b \beta w'w\, dx = \frac{1}{2}\int_a^b \beta \frac{d\,(w^2)}{dx} dx = \frac{1}{2}[\beta w]_a^b - \frac{1}{2}\int_a^b \beta' w^2\, dx = -\frac{1}{2}\int_a^b \beta' w^2\, dx.$$

The last equality follows from $w(a) = w(b) = 0$. If we assume that $\beta' \leq 0$, we have that the bilinear form is coercive and the problem is well posed.

Another possible proof of coercivity can be found without exploiting the assumption $\sigma > 0$, but the Poincaré inequality, which so far has not been used. The coercivity constant α is in this case

$$\alpha = \max\left(\frac{\mu_0}{1+C_\Omega^2}, \min(\mu_0, \sigma)\right).$$

Coercivity thus holds also for $\sigma = 0$. Well posedness is a consequence of Corollary 3.1.

---

[1] These are not the most general assumptions for ensuring continuity. For instance, since Sobolev Theorem states that in 1D $H^1(a,b)$ functions are bounded, we could assume $\beta(x) \in L^2(a,b)$. Here we refer here to somehow more restrictive assumptions, yet reasonable for the applications.

Let us set $\mu = x$, $\beta = 10^3$ and $\sigma = 0$. The differential equation of our problem reads

$$-(xu')' + \beta u' = 0.$$

Exact solution can be computed for instance by setting $v = u'$ and solving the auxiliary equation

$$-(xv)' + \beta v = 0 \Rightarrow \frac{v'}{v} = (\beta - 1)\frac{1}{x}.$$

By direct integration we get $v = C_0 x^{\beta-1}$ where C_0 is constant. With a second integration, we get $u = \frac{C_0}{\beta}x^\beta + C_1$ where C_1 is a constant. By prescribing the boundary condition, we get the solution

$$u = \frac{x^\beta - a^\beta}{b^\beta - a^\beta}.$$

Numerical approximation

For the third question, notice that for $a = 1, b = 2$ all the hypotheses required by the well posedness are fulfilled. Similarly, hypotheses are fulfilled for $a = 10^3, b = 10^3 + 1$ with $\mu_0 = 10^3$. Péclet number in the first case is $\mathbb{Pe} = \dfrac{\beta h}{2\mu_0} = 500h$ so that finite elements with no stabilization can be used for $h < 1/500$.

In the second case we have $\mathbb{Pe} = 1000h/2000 \Rightarrow h < 2$, which certainly holds since $b - a = 1$.

In the first case large discretization steps are therefore allowed only with some stabilization method. For instance, if we use the *Upwind* method, as done in the previous exercise, we modify the viscosity μ in $\mu_{upwind} = \mu(1 + \mathbb{Pe})$. More in general, viscosity can be modified by setting $\mu^* = \mu(1 + \phi(\mathbb{Pe}))$, where $\phi(\cdot)$ is such that

$$\phi(x) > 0 \quad \forall x > 0, \qquad \phi(0) = 0.$$

The first requirement states that the scheme should add (and not subtract!) numerical viscosity so to force the stabilization for large values of h. The second requirement guarantees that the modified numerical problem is still consistent with the original one, so that for $h \to 0$ (which implies for $\mathbb{Pe} \to 0$), the perturbation due to the stabilization vanishes. Another possible choice relies on the so called *exponential fitting* given by

$$\phi(\tau) = \tau - 1 + \frac{2\tau}{e^{2\tau} - 1}, \quad \text{for} \quad \tau > 0.$$

This is the *Scharfetter-Gummel* method.

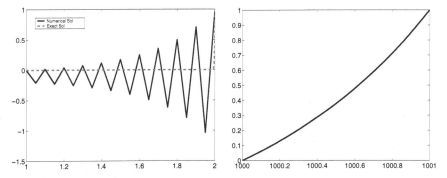

Fig. 4.7 Solution of Exercise 4.2.2 with standard Galerkin finite element method. On the left: $a = 1$ and $b = 2$, on the right $a = 1000$, $b = 1001$. Mesh grid size is $h = 0.05$ in both cases. As expected, numerical oscillations affect the first case and are absent in the second one

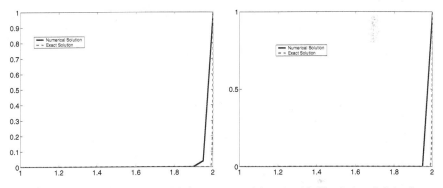

Fig. 4.8 Solution of Exercise 4.2.2 for $a = 1$ and $b = 2$ with Upwind and Scharfetter-Gummel methods. As the comparison with the exact solution (dashed line) points out, the second method is less diffusive

Numerical results

Fig. 4.7 illustrates the numerical solution with the standard (non stabilized) Galerkin method and mesh size $h = 0.05$ for $a = 1$, $b = 2$ (left) and $a = 1000$, $b = 1001$ (right). As expected, numerical oscillations are present in the first case, not in the second one. In Fig. 4.8 we report the solutions for the case $a = 1$, $b = 2$ with $h = 0.05$ and Upwind (left) and Scharfetter-Gummel (right) stabilization methods.

The second approach is clearly less dissipative. The *boundary layer*, i.e. the interval along the x axis where the solution changes abruptly from 0 to the prescribed value of 1, is small for the exact solution ($\mathcal{O}(\mu_0/\beta)$). On the contrary, it is larger for both the numerical solutions. This is the side-effect of the increment of the viscosity induced by the stabilization. Nevertheless, the Scharfetter-Gummel method introduces a smaller numerical diffusion and the associated boundary layer is closer to the exact one. Notice that the numerical solution is piecewise linear and the internal node closest to the end point $x = 2$ is in $x = 1.95$, so the Sharfetter-Gummel approximation of the boundary layer with this method is excellent. $\diamond$

Exercise 4.2.3. Let us consider the problem

$$\begin{cases} -\varepsilon u''(x) + u'(x) = 1, \, 0 < x < 1, \\ \\ u(0) = 0, \qquad\qquad u'(1) = 1, \end{cases} \qquad (4.19)$$

with $\varepsilon > 0$ a given constant. Write the weak formulation and its Galerkin finite element approximation. Compute the numerical solution with **femld** for $\varepsilon = 10^{-6}$. Does the standard Galerkin method produce spurious oscillation? Why?

Solution 4.2.3.

Mathematical analysis

Weak formulation is obtained in the usual way. Solution u and test function v are assumed to belong to $V \equiv \{v \in H^1(0,1), v(0) = 0\}$. Weak formulation reads: find $u \in V$ such that for any $v \in V$

$$\varepsilon \int_0^1 u'v'dx + \int_0^1 u'vdx = \int_0^1 vdx + \varepsilon v(1).$$

Functional $\mathcal{F}(v) \equiv \int_0^1 vdx + \varepsilon v(1)$ is continuous since

$$\left| \int_0^1 vdx + \varepsilon v(1) \right| \leq (1 + \varepsilon \gamma_T) \|v\|_V,$$

where γ_T is the continuity constant of the trace operator (see (1.11)).

Bilinear form $a(u, v) \equiv \varepsilon \int_0^1 u'v'dx + \int_0^1 u'vdx$ is continuous and coercive. Continuity descends from the functional space selected for u and v. Moreover, Poincaré inequality yields

$$a(u, u) = \varepsilon \|u'\|_{L^2}^2 + \frac{1}{2}u^2(1) \geq \varepsilon \|u'\|_{L^2}^2 \geq \varepsilon C\|u\|_V^2,$$

where $C = 1/(1 + C_\Omega^2)$, C_Ω being the Poincaré inequality constant.

Analytical solution of (4.19) is computed as follows. Working as in the previous exercise, we find the general integral of the differential equation $u_G = C_1 + C_2 e^{x/\varepsilon} + x$, where constants C_1 and C_2 depend on the boundary conditions.

By direct computation we find that the exact solution is $u(x) = x$.

Numerical approximation

Standard linear finite element approximation is obtained in the usual way, with a (uniform) mesh with size h and the space V_h of functions of X_h^1 vanishing in $h = 0$. With this notation, Galerkin finite elements discretization of the problem reads: find $u_h \in V_h$ such that for any $v_h \in V_h$ $a(u_h, v_h) = \mathcal{F}(v_h)$.

Numerical results

Since the exact solution is linear, it belongs to the finite element subspace of piecewise polynomial functions (with degree $q \geq 1$). In this case that the numerical solution coincides with the exact one. This explains why spurious oscillations are absent for any value of $\mathbb{Pe}$.

In Fig. 4.9 we illustrate the solution obtained with linear finite elements (standard Galerkin method) and mesh size $h = 0.1$. Numerical solution coincides with the exact one up to the machine epsilon, even for $\mathbb{Pe} = 0.5 \times 10^6$.   ◊

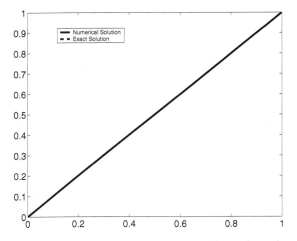

Fig. 4.9 Solution of Exercise 4.2.3: Galerkin method with linear finite elements with mesh size $h = 0.1$. Numerical error is within 10^{-15}

Exercise 4.2.4. Find a *finite difference* approximation accurate to order 2, and suitably decentred, of the boundary-value problem: find $u : (0, 1) \to \mathbb{R}$ such that

$$
\begin{cases}
-0.01 \dfrac{d^2 u}{dx^2} - \dfrac{du}{dx} = 0 \ \text{ for } x \in (0, 1), \\[2mm]
u(0) = 0, \qquad\qquad u(1) = 1.
\end{cases}
\tag{4.20}
$$

Consider a grid with uniform spacing h formed by $n + 1$ nodes. Finally, compute the error in the $L^\infty(0, 1)$ norm, recalling that the exact solution is $u = \left(e^{-100x} - 1 \right) / \left(e^{-100} - 1 \right)$.

Solution 4.2.4.

Numerical approximation

This advection-diffusion problem has dominant advective term: the global Péclet number (computed on the domain's length) is in fact 50. To have solutions without spurious oscillations the constraint $\mathbb{Pe} = 50h < 1$ must be satisfied. Alternatively, as mentioned in the text, the scheme must be decentred. Since standard *upwind*-type centering stops the accuracy of the scheme to the first order, to guarantee order 2 we make use of the decentred approximation $(2.31)_1$, which we have already proven to have order two. The finite-difference scheme will thus be

$$
\begin{cases}
-0.01 \dfrac{u_{i+1} - 2u_i + u_{i-1}}{h^2} - \dfrac{-u_{i+2} + 4u_{i+1} - 3u_i}{2h} = 0 \\[2mm]
\qquad\qquad\qquad\qquad\qquad \text{for } i = 1, \ldots, n - 2, \\[2mm]
u_0 = 0, \qquad\qquad\qquad\qquad\qquad u_n = 1.
\end{cases}
\tag{4.21}
$$

There remains to determine the equation for the node x_{n-1}: approximation $(2.31)_1$ cannot be used, for it would involve the value of u at the node x_{n+1} lying outside the domain. A first possibility is to use a centered difference quotient accurate to order 2 (like $(u_n - u_{n-2})/(2h)$, for instance), though shedding the centering at x_{n-1}. A second option consists in creating a the fake unknown u_{n+1} in the scheme so to use, at x_{n-1} as well, the difference quotient $(2.31)_1$. This technique is akin to the ghost node adopted in hyperbolic problems, see Ch. 6. The problem lies in finding an appropriate value for u_{n+1}. A possible solution is to extrapolate with a suitable order of approximation starting from $u_n, u_{n-1}, u_{n-2}, \ldots$ Evidently, order 2 will not be enough, because this will then be used to approximate the derivative, which entails dividing by h and thus reducing the error to order one. We need order 3. For example, setting $u_{n+1} \simeq 3u_n - 3u_{n-1} + u_{n-2}$, scheme (4.21) for $i = n - 1$ will become

$$-0.01 \frac{u_n - 2u_{n-1} + u_{n-2}}{h^2} - \frac{u_n - u_{n-2}}{2h} = 0.$$

We still obtain a second-order centered discretization of the transport term! However, this does not inhibit the stabilizing action of decentred derivatives as we adopt a centered scheme at just one node.

Numerical results

Program 20 implements the scheme on a uniform grid of step h to solve the more general problem

$$\begin{cases} -\mu u''(x) + \beta u'(x) + \sigma u(x) = f(x) & x \in (a, b), \\ u(a) = g_D(a), \quad u(b) = g_D(b). \end{cases}$$

Program 20 - diff1Dfd : Approximation by finite differences of order 2 for a quadratic boundary-value problem

```
%DIFF1DFD Finite differences for the boundary-value problem of order 2.
%   [UH,X]=DIFF1DFD(XSPAN,NH,MU,BETA,SIGMA,GD,F) solves the   problem
%   - MU*U" + BETA*U'+SIGMA*U = F  in (XSPAN(1),XSPAN(2))
%     U(XSPAN(1))=GD(XSPAN(1)), U(XSPAN(2))=GD(XSPAN(2))
% on an uniform grid of step H=(XSPAN(2)-XSPAN(1))/NH using
% finite differences of order 2 where MU, BETA and SIGMA are constants
% and F, GD inline functions.
% If PE=0.5*ABS(BETA)*H/MU is greater than 1 we use a decentred scheme
% again of order 2.
```

In Fig. 4.10, left, we have the solution found by the decentred scheme with $h = 1/10$ (solid curve) and a centered scheme with $h = 1/100$ (dashed curve). As expected, having chosen a decentred scheme has allowed to eliminate spurious oscillations for "large" h as well. On the right we can read the behavior of the error in norm $L^\infty(0, 1)$ (log scale) as h varies. For h sufficiently small the solution is accurate to order 2. $\diamondsuit$

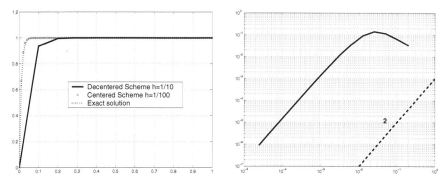

Fig. 4.10 Left: solution obtained by scheme (4.21) with $h = 1/10$, solid curve. The dashed curve is the exact solution, while the values marked with $\times$ refer to the solution with $h = 1/100$ and centered scheme. Right: graph, in log scale, of the error in norm $L^\infty(0, 1)$ for the decentred scheme, as h varies (horizontal axis). The dashed curve provides the reference slope corresponding to order 2. Note that smaller values of h would make the effect of round-off errors manifest

Exercise 4.2.5. Let us consider the following 1D advection diffusion problem

$$\begin{cases} -(\mu u' - \psi' u)' = 1, & 0 < x < 1, \\ u(0) = u(1) = 0, \end{cases} \tag{4.22}$$

where μ is a positive constant and ψ is a given function.

1. Prove existence and uniqueness of the (weak) solution of problem (4.22) by introducing suitable assumptions on the function ψ.
2. Set $u = \rho e^{\psi/\mu}$, where ρ is an unknown auxiliary function. Analyze existence and uniqueness of the solution for the new formulation. Compute the analytic solution for $\psi = \alpha x$, where α is a real constant.
3. For both the formulations (in u and ρ) write the finite elements discretization.
4. Compare the two approaches in 1 and 2, both from the theoretical and numerical viewpoints. For $\psi = \alpha x$ compute the numerical solution with $\mu = 0.1, \mu = 0.01$, $\alpha = 1$ e $\alpha = -1$ and grid size $h = 0.1$ and $h = 0.01$.

Solution 4.2.5.

Mathematical analysis

If ψ is regular enough, we have the equivalent problem

$$-(\mu u')' + \psi' u' + \psi'' u = 1,$$

for $x \in (0, 1)$, with boundary conditions $u(0) = u(1) = 0$.

The weak form reads: find $u \in V \equiv H_0^1(0,1)$ such that for any $v \in V$

$$a(u,v) \equiv \int_0^1 \mu u'v'dx + \int_0^1 \psi'u'vdx + \int_0^1 \psi''uvdx = \int_0^1 vdx.$$

Let us assume $\psi \in H^2(0,1)$. Sobolev embedding Theorem states that $v, \psi, \psi' \in L^\infty(0,1)$ and (C is here the constant of the embedding Theorem)

$$\left| \int_0^1 \psi'u'vdx \right| \leq ||\psi'||_{L^\infty}||u'||_{L^2}||v||_{L^2} \leq ||\psi'||_{L^\infty}||u||_V||v||_V,$$

$$\left| \int_0^1 \psi''uvdx \right| \leq ||\psi''||_{L^2}||u||_{L^\infty}||v||_{L^\infty} \leq C^2||\psi''||_{L^2}||u||_V||v||_V.$$

The bilinear form is therefore continuous. Let us investigate the coercivity. By direct computation we have

$$a(u,u) = \int_0^1 \mu\,(u')^2\,dx + \int_0^1 \psi'u'udx + \int_0^1 \psi''u^2dx = \int_0^1 \mu\,(u')^2dx+$$

$$\frac{1}{2}\int_0^1 \psi'\frac{d(u^2)}{dx}dx + \int_0^1 \psi''u^2dx = \int_0^1 \mu\,(u')^2\,dx - \frac{1}{2}\int_0^1 \psi''u^2dx + \int_0^1 \psi''u^2dx =$$

$$\int_0^1 \mu\,(u')^2\,dx + \frac{1}{2}\int_0^1 \psi''u^2dx \geq \frac{\mu_0}{1+C_\Omega^2}||u||_V^2 + \frac{1}{2}\int_0^1 \psi''u^2dx,$$

where C_Ω is the Poincaré constant. The bilinear form is coercive provided that $\psi'' \geq 0$. More in general, let us set $\psi''_{min} = \min(\psi'')$. From the previous inequality we get

$$a(u,u) \geq \left(\frac{\mu_0}{1+C_\Omega^2} + \psi''_{min} \right)||u||_V^2,$$

that implies the coercivity of the bilinear form under the condition

$$\frac{\mu_0}{1+C_\Omega^2} - \min(0, |\psi''_{min}|) > 0.$$

From now on we will assume this assumption to hold.

Let us formulate the problem with the auxiliary variable ρ, by setting $u = \rho e^{\psi/\mu}$. We have

$$u' = \rho'e^{\psi/\mu} + \rho e^{\psi/\mu}\frac{\psi'}{\mu} \Rightarrow \mu u' = \mu\rho'e^{\psi/\mu} + \psi'u.$$

Thanks to the boundary conditions for u, we have

$$u(0) = u(1) = 0 \Rightarrow \rho(0) = \rho(1) = 0.$$

When formulated with ρ, the problems becomes purely diffusive

$$\begin{cases} -\left(\mu e^{\psi/\mu} \rho'\right)' = 1 & x \in (0,1), \\ \rho(0) = 0, & \rho(1) = 0. \end{cases}$$

The weak formulation reads: find $\rho \in V$ such that

$$\mu \int_0^1 e^{\psi/\mu} \rho' v' dx = \int_0^1 v dx \quad \forall v \in V.$$

The bilinear form is continuous provided that $\psi \in L^\infty(0,1)$, that is less restrictive than the assumptions introduced with the previous formulation. On the other hand, coercivity derives straightforwardly from $\mu > 0$ since, thanks to the Dirichlet boundary conditions, Poincaré inequality holds.

Let us compute the analytical solution for $\psi = \alpha x$. If we refer to the second (strong) formulation, we have

$$-\mu \left(e^{\alpha x/\mu} \rho'\right)' = 1 \quad \text{so that} \quad \rho' = \left(C_1 - \frac{x}{\mu}\right) e^{-\alpha x/\mu},$$

where C_1 is an integration constant. After a second integration we get

$$\rho = -C_1 \frac{\mu}{\alpha} e^{-\alpha x/\mu} + \frac{x}{\alpha} e^{-\alpha x/\mu} + \frac{\mu}{\alpha^2} e^{-\alpha x/\mu} + C_2,$$

where C_2 is a constant. Constants C_1 and C_2 are computed by forcing the boundary conditions. We get therefore the solution

$$\rho = \frac{1}{\alpha} \left[\frac{1 - e^{-\alpha x/\mu}}{1 - e^{\alpha/\mu}} + x e^{-\alpha x/\mu} \right] \Rightarrow u = \frac{1}{\alpha} \left[x - \frac{e^{\alpha x/\mu} - 1}{e^{\alpha/\mu} - 1} \right].$$

Numerical approximation

Let us consider the adimensional numbers (see also Section 4.3):

$$\frac{\|\psi'\|_{L^\infty(0,1)} h}{2\mu} \quad \text{and} \quad \frac{\|\psi''\|_{L^\infty(0,1)} h^2}{6\mu}.$$

If the following conditions [Qua09]

$$h < \frac{2\mu}{\|\psi'\|_{L^\infty(0,1)}}, \quad h < \sqrt{\frac{6\mu}{\|\psi''\|_{L^\infty(0,1)}}} \tag{4.23}$$

hold, the linear finite element approximation is not affected by spurious oscillations. When such conditions do not hold, we can stabilize the problem as follows.

1. *Upwind* when the first of (4.23) does not hold. This implies an increment of the viscosity of the problem

$$\mu \to \mu^* = \mu \left(1 + \frac{|\psi'|h}{2\mu} \right).$$

This cancels the numerical oscillations, even if reduces the accuracy to order 1 independently of the degree of finite elements used (see Exercise 4.2.1).

2. *Mass lumping* for the reactive term when the second of (4.23) does not hold. *If the mass-lumping is applied properly*, the order of accuracy of the solution is not reduced. For instance, using linear finite elements and the matrix introduced in the Exercise 4.1.4, the accuracy of the numerical solution (in norm H^1) is still of order 1. However, in general mass lumping can affect the accuracy.

If we change the variable, computation of ρ is simpler at the numerical level, since the reformulated problem is purely diffusive. However, notice that upon discretization of the problem, the associated linear system features a matrix A with entries $A_{ij} = \mu \int_0^1 e^{\psi/\mu} \varphi'_j \varphi'_i dx$. These entries can span a large numerical range, depending on function ψ. This can induce a significant ill conditioning of the matrix and consequent difficulties in obtaining an accurate computation of ρ.

In addition, observe that the variable we are interested in is u (not ρ). Even when ρ is well approximated by the numerical solution ρ_h, for computing u_h — approximation of u — we have to compute $u_h = \rho_h e^{\psi/\mu}$. Now, if $\psi \gg \mu$, then $e^{\psi/\mu} \gg 1$ and the multiplicative term can amplify numerical errors, so to make u_h inaccurate. More precisely, if $\delta\rho = \rho - \rho_h$, we have (neglecting other source of error)

$$u_h = \rho_h e^{\psi/\mu} = (\rho + \delta\rho)e^{\psi/\mu} = \rho e^{\psi/\mu} + \delta\rho e^{\psi/\mu} = u + \delta u,$$

with $\delta u \equiv \delta\rho e^{\psi/\mu}$. Should $\psi \gg \mu$ then $e^{\psi/\mu}$ will amplify the absolute error $\delta\rho$. Relative errors on u and ρ are however of the same order, since

$$\frac{\delta u}{u} = \frac{\delta\rho e^{\psi/\mu}}{\rho e^{\psi/\mu}} = \frac{\delta\rho}{\rho}.$$

In summary, the second formulation requires less regularity on ψ and in principle is easier to be solved numerically. However, it has two drawbacks in particular if ψ/μ is large.

1. *Ill-conditioning of the associated linear system*;
2. *Amplification of numerical errors in computing u_h from ρ_h.*

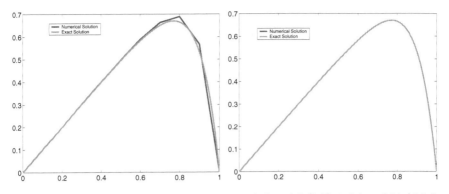

Fig. 4.11 Computation of u with $\mu = 0.1, \alpha = 1$, $h = 0.1$ (left) and $h = 0.01$ (right). Exact reference solution is illustrated in light gray

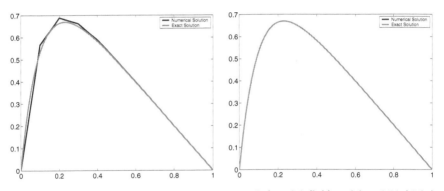

Fig. 4.12 Computation of u with $\mu = 0.1, \alpha = -1$, $h = 0.1$ (left) and $h = 0.01$ (right). Exact reference solution is drawn in light gray

Numerical results

Let us set $\mu = 0.1$, $\psi = \alpha x$ with $\alpha = 1$, and consider the first formulation. In Fig. 4.11 we illustrate the case $\mu = 0.1$ with two different values of the grid size ($h = 0.1$ and $h = 0.01$). Since the Péclet number is $5h$, upper bounds for h are 0.5 and 0.05 respectively. These values yield non-oscillating solutions. Condition numbers in the two cases are 20.4081 and 2014.9633 respectively, while:

1. errors in norm L^2 are 0.015137 and 0.00015581, with a reduction factor of about 100;
2. errors in norm H^1 are 0.62769 and 0.064534, with a reduction factor of about 10.

All these results are coherent with the theory.

In the case $\alpha = -1$ we do not find relevant differences in the accuracy of the numerical solution (see Fig. 4.12). For $\mu = 0.01$, Péclet number is $50h$.

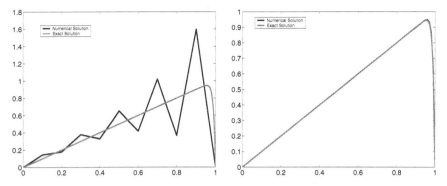

Fig. 4.13 Computation of u with $\mu = 0.01, \alpha = 1$, $h = 0.1$ (on the left) and $h = 0.01$ (on the right). Exact solution is reported in light gray as a reference curve

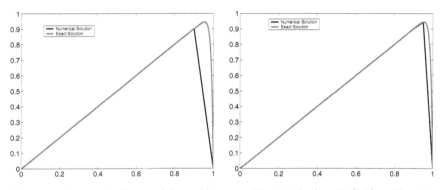

Fig. 4.14 Numerical solution of the problem as in Fig. 4.13, for $h = 0.1$ (left) and $h = 0.05$ (right) and Scharfetter-Gummel stabilization

For $h = 0.1$ $\mathbb{P}e \geq 1$ and the numerical solution oscillates. (see Fig. 4.13 on the left). For $h = 0.01$, on the contrary, numerical solution does not oscillate (see Fig. 4.13 on the right). Larger mesh sizes are allowed when a stabilization technique is adopted. For instance, in Fig. 4.14 we report the results obtained with the Scharfetter-Gummel method with $h = 0.1$ and $h = 0.05$. For $\alpha = -1$ we obtain similar results.

Let us consider the second formulation in ρ. In Fig. 4.15 we report the case $\mu = 0.1$, $\alpha = 1$, $h = 0.1$ and 0.01. The condition number in this case ranges from about 5585 for $h = 0.1$ to about 1.7×10^6 for $h = 0.01$.

Computation of u from ρ gives acceptable results (Fig. 4.16) and the error in norm L^2 is approximately 5×10^{-4}.

For $\alpha = -1$ (Fig. 4.17) the exact ρ takes values over a larger range. Even if the condition number of the associated matrix is the same of the case $\alpha = 1$, the absolute errors are significantly larger (see Table 4.2), although they follow the theoretical convergence estimates.

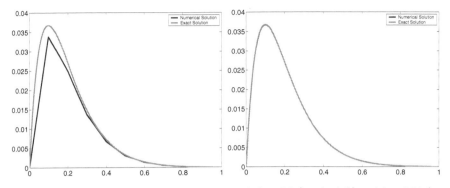

Fig. 4.15 Computation of ρ with $\mu = 0.1, \alpha = 1$, $h = 0.1$ (on the left) and $h = 0.01$ (on the right). Exact solution is reported in light gray as reference curve

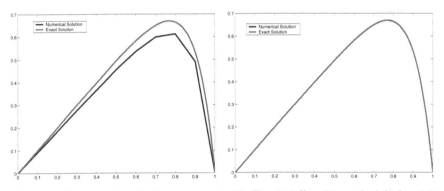

Fig. 4.16 Computation of u from ρ computed in Fig. 4.15 ($h = 0.1$ on the left, $h = 0.01$ on the right)

Table 4.2 Condition number and error in computing ρ for $\mu = 0.1$, $\alpha = 1$ (second and third columns) and $\alpha = -1$ (columns 5-7)

$\alpha = 1$	$h = 0.1$	$h = 0.01$	$\alpha = -1$	$h = 0.1$	$h = 0.01$	$h = 0.001$
$K(A)$	5.5e+3	1.7e+6		5.5e+3	1.7e+6	0.2e+9
e_{L^2}	0.0034376	3.9454e-05		75.7181	0.86903	0.0087028
e_{H^1}	0.091554	0.010192		2016.603	224.4889	22.4743

For $\mu = 0.01$ and $\alpha = 1$ notice that the computation of ρ is completely inaccurate for $h = 0.1$. In the latter case, the numerical solution is unable of capturing the exact one, since we do not have enough nodes in the mesh. Solution becomes acceptable for $h = 0.01$ (see Fig. 4.18).

The situation is qualitatively similar for $\alpha = -1$, even if - again - the solution ranges over a large interval of values so that the absolute errors are large (Table 4.3 and Fig. 4.19). As expected the condition number is

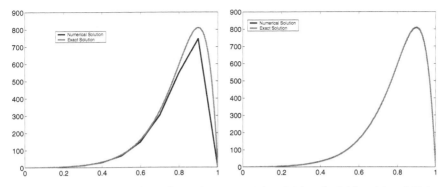

Fig. 4.17 Computation of ρ with $\mu = 0.1, \alpha = -1$, $h = 0.1$ (on the left) and $h = 0.01$ (on the right). For $h = 0.01$ the numerical solution is graphically overlapping the exact one

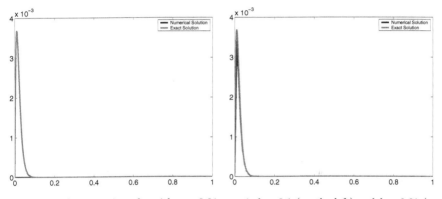

Fig. 4.18 Computation of ρ with $\mu = 0.01, \alpha = 1$, $h = 0.1$ (on the left) and $h = 0.01$ (on the right). Notice that for $h = 0.1$ the numerical solution is very close to be zero

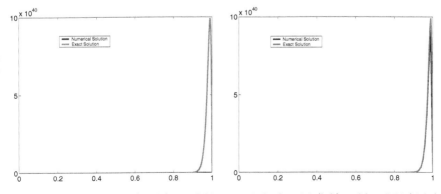

Fig. 4.19 Computation of ρ with $\mu = 0.01, \alpha = -1$, for $h = 0.1$ (left) and $h = 0.01$ (right)

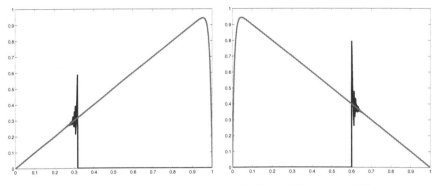

Fig. 4.20 Computation of u form ρ for $\mu = 0.01$, $h = 0.001$, $\alpha = 1$ (left) and $\alpha = -1$ (right). The exact solution is drawn in gray, the numerical one is in black

Table 4.3 Condition number and error in computing ρ for $\mu = 0.01$, $\alpha = 1$ (columns 2-4) and $\alpha = -1$ (columns 6-8)

$\alpha = 1$	$h = 0.1$	$h = 0.01$	$h = 0.001$	$\alpha = -1$	$h = 0.1$	$h = 0.01$	$h = 0.001$
$K(A)$	$3.9e + 34$	$3.5e + 42$	$1.3e + 44$		$3.9e + 34$	$6.8e + 42$	$4.3e + 46$
e_{L^2}	0.00049809	0.00010876	$1.2e - 06$		$1.3e + 40$	$2.9e + 39$	$2.2e + 33$
e_{H^1}	0.049641	0.02894	0.0032238		$1.3e + 42$	$7.7e + 41$	$2.1e + 37$

extremely large. Errors depend on h, even if the expected order is attained with small values of the mesh size.

When we recompute u form ρ for $\mu = 0.01$, that is when ψ/μ has a large absolute value, matrix for computing ρ is so ill-conditioned that the computation of u is completely inaccurate. When $\alpha > 1$ the errors on ρ are small, but they are amplified by the computation of u. On the contrary, when $\alpha = -1$ the errors on ρ are larger and the computation of u is not accurate at all. In both cases, numerical cancellation makes u vanish on a part of the domain with large oscillations (see Fig. 4.20). ◇

Remark 4.1 In the investigation of simplified models for semi-conductor devices (*drift-diffusion* equations) we find problems similar to the ones considered in this exercise (usually non-linear). In this context, ρ is called *Slotboom variable*, see [Slo73]. S. Selberherr in [Sel84] Section 5.2 notices that the major drawback of these variables is the huge interval of values required for representing them in real computations. In other words, using the Slotboom variable is useful for analytical more than numerical purposes, as the previous exercise points out.

Exercise 4.2.6. Consider the advection-diffusion-reaction problem

$$\begin{cases} -\Delta u + \nabla \cdot (\boldsymbol{\beta} u) + u = 0 & \text{in} \quad \Omega \subset \mathbb{R}^2, \\ u = \varphi & \text{on} \quad \Gamma_D, \\ \nabla u \cdot \mathbf{n} = \boldsymbol{\beta} \cdot \mathbf{n} u & \text{on} \quad \Gamma_N, \end{cases} \tag{4.24}$$

where Ω is an open domain, $\partial\Omega = \Gamma_D \cup \Gamma_N$, $\Gamma_D \cap \Gamma_N = \emptyset$, $\Gamma_D \neq \emptyset$. Prove existence and uniqueness of the weak solution of the problem under suitable assumptions on the data $\boldsymbol{\beta} = (\beta_1(x,y), \beta_2(x,y))^T$ and $\varphi = \varphi(x,y)$. Identify when it is appropriate to stabilize the problem to avoid spurious oscillations in the solution. To this aim, introduce finite element discretizations based on the artificial viscosity method, and the SUPG stabilization, pointing out pros and cons in comparison with the non-stabilized Galerkin finite element method.

Finally, assume $\Omega = (0,1) \times (0,1)$, $\boldsymbol{\beta} = (10^3, 10^3)^T$, $\Gamma_D = \partial\Omega$ and

$$\varphi = \begin{cases} 1 & \text{for} & x = 0 \quad 0 < y < 1, \\ 1 & \text{for} & y = 0 \quad 0 < x < 1, \\ 0 \ \text{elsewhere}. \end{cases}$$

Check that the assumptions that make the problem well posed are fulfilled and solve it numerically with the different stabilization techniques proposed.

Solution 4.2.6.

Mathematical analysis

For the well posedness analysis we follow the usual approach. We multiply the equation by a test function $v \in V \equiv H^1_{\Gamma_D}(\Omega)$. After integrating by parts both the second and first order terms, we get

$$\int_\Omega \nabla u \cdot \nabla v \, d\omega - \int_{\partial\Omega} \nabla u \cdot \mathbf{n} v \, d\gamma + \int_{\partial\Omega} \boldsymbol{\beta} \cdot \mathbf{n} u v \, d\gamma - \int_\Omega \boldsymbol{\beta} \cdot \nabla v u \, d\omega + \int_\Omega u v \, d\omega = 0.$$

We consider in particular the boundary integral on Γ_D and Γ_N separately

$$-\int_{\partial\Omega} \nabla u \cdot \mathbf{n} v \, d\gamma + \int_{\partial\Omega} \boldsymbol{\beta} \cdot \mathbf{n} u v \, d\gamma =$$
$$-\int_{\Gamma_D} (\nabla u \cdot \mathbf{n} - \boldsymbol{\beta} \cdot \mathbf{n} u) \, v \, d\gamma - \int_{\Gamma_N} (\nabla u \cdot \mathbf{n} - \boldsymbol{\beta} \cdot \mathbf{n} u) \, v \, d\gamma.$$

Integral on Γ_N vanishes, since $\nabla u \cdot \mathbf{n} - \boldsymbol{\beta} \cdot \mathbf{n} u = 0$ on Γ_N. This justifies the integration by parts of the convective term, so that the given boundary condition on Γ_N is treated in "natural" way. On Γ_D the boundary integral vanishes, being $v = 0$. Moreover, let $G(x,y)$ be a function in $H^1(\Omega)$ such that $G(x,y) = \varphi(x,y)$ on Γ_D. Existence of G requires that $\varphi \in H^{1/2}(\Gamma_D)$ and that the domain Ω is regular enough (for instance a polygonal domain). Under these assumptions, the weak formulation of the problem reads: find

$u \in G + V$ such that for any $v \in V$

$$a(u, v) = 0,$$

where

$$a(u, v) \equiv \int_{\Omega} \nabla u \cdot \nabla v d\omega - \int_{\Omega} \boldsymbol{\beta} \cdot \nabla v u d\omega + \int_{\Omega} u v d\omega.$$

For the well posedness analysis we refer to Lemma 3.1. Let us assume that the components of $\boldsymbol{\beta}$ are functions of $L^{\infty}(\Omega)$. With this assumption, as we have seen in the previous exercises, we can prove that the bilinear form is continuous in $H^1(\Omega)$. Furthermore, for a generic function $w \in V$,

$$a(w, w) = ||\nabla w||^2_{L^2(\Omega)} - \frac{1}{2} \int_{\Omega} \boldsymbol{\beta} \cdot \nabla w^2 d\omega + ||w||^2_{L^2(\Omega)}.$$

The first and the third terms summed up together gives $||w||^2_V$. For the second term, by the Green formula and the boundary conditions, we have

$$-\frac{1}{2} \int_{\Omega} \boldsymbol{\beta} \cdot \nabla w^2 d\omega = -\frac{1}{2} \int_{\Gamma_N} \boldsymbol{\beta} \cdot \mathbf{n} w^2 d\gamma + \frac{1}{2} \int_{\Omega} \nabla \cdot \boldsymbol{\beta} w^2 d\omega.$$

If we assume that

$$\nabla \cdot \boldsymbol{\beta} \geq 0 \quad \text{in} \quad \Omega \quad \text{and} \quad \boldsymbol{\beta} \cdot \mathbf{n} \leq 0 \quad \text{on} \quad \Gamma_N, \qquad (4.25)$$

we conclude that the bilinear form is coercive and the problem is well posed. Notice that if $\boldsymbol{\beta}$ is the velocity field of an incompressible fluid, then $\nabla \cdot \boldsymbol{\beta} = 0$, so that the first of (4.25) is verified. The second of (4.25) corresponds to requiring that the Neumann condition is applied to an inflow portion of the boundary. Note that this condition is different from the one that is obtained when considering a problem not in conservation form, where the transport term is $\boldsymbol{\beta} \cdot \nabla u$, like in Exercise 4.2.7.

Numerical approximation

Finite element discretization requires the introduction of a finite-dimensional subspace $V_h \subset X_h^r$ (see (2.22)) of functions vanishing on Γ_D. We want to find the numerical solution u_h such that for any $v_h \in V_h$

$$a(u_h, v_h) = \mathcal{F}(v_h).$$

The local Péclet number on each element K of the mesh $\mathcal{T}_h$ is given by $\mathbb{Pe}_K = \dfrac{||\boldsymbol{\beta}||_{L^{\infty}(K)} h_K}{2\mu} = \dfrac{||\boldsymbol{\beta}||_{L^{\infty}(K)} h_K}{2}$. In order to avoid spurious oscillations we have to force $\mathbb{P}e_K < 1$ for all $K \in \mathcal{T}_h$. This is guaranteed if $h = \min_{K \in \mathcal{T}_h} h_K$ satisfies[2] $\dfrac{||\boldsymbol{\beta}||_{L^\infty(\Omega)} h}{2} < 1$.

Alternatively, we resort to a "stabilized" method, in the class of *Generalized Galerkin* methods. A first option is to introduce an isotropic numerical viscosity (that is uniform in all directions), as an immediate extension of the 1D case (see e.g. [Qua09], Chapter 5).

Accounting for non constant h_k, we modify the bilinear form as follows

$$a_h(u_h, v_h) = a(u_h, v_h) + \sum_{K \in \mathcal{T}_h} \frac{h_K}{||\boldsymbol{\beta}||_{L^\infty(K)}} \int_K \nabla u_h \cdot \nabla v_h \, dK.$$

The drawback of this stabilization is to introduce viscosity in all the directions, including the *crosswind* one, which is orthogonal to the convective direction. Along the crosswind direction, the problem is diffusive and no stabilization is needed. If we apply the Strang Lemma, on the modified formulation, we find that accuracy will be of order 1 independently of the finite element degree adopted, as for the 1D case (Exercise 4.2.1).

A more accurate method is the *Streamline Upwind-Petrov Galerkin (SUPG)*. This introduces artificial viscosity only in the direction of convective term and moreover preserves the strong consistency of Galerkin methods Let us denote with $\mathcal{L}$ the differential operator associated with the problem

$$\mathcal{L}u = -\triangle u + \nabla \cdot (\boldsymbol{\beta} u) + u,$$

whose skew-symmetric part is given by (see Exercise 4.1.2)

$$\mathcal{L}_{SS}u = \frac{1}{2}(\boldsymbol{\beta} \cdot \nabla u) + \frac{1}{2}\nabla \cdot (\boldsymbol{\beta} u).$$

When the forcing term is null, the SUPG stabilized bilinear form reads

$$a_h(u_h, v_h) = a(u_h, v_h) + \sum_{K \in \mathcal{T}_h} \delta_K \left(\mathcal{L}u_h, \frac{h_K}{||\boldsymbol{\beta}||} \mathcal{L}_{SS} v_h \right),$$

where $||\boldsymbol{\beta}||$ is the Euclidean norm of $\boldsymbol{\beta}$, and δ_K are parameters to be suitably tuned. The scalar products in the summation are well defined even if differential operators are in their "strong form", since they are applied element-wise to members of V_h (which, restricted to K, are polynomials and thus infinitely regular functions).

This stabilization method is strongly consistent. Indeed the "perturbation" introduced by the method vanishes when applied to the exact solution since $\mathcal{L}u = 0$. Moreover, if the exact solution is regular enough the discretization error measured in a suitable norm is proportional to $h^{r+1/2}$, where r is the chosen finite element degree, with a constant of proportionality which is in-

---

[2] This is however "pessimistic", normally in multidimensional problems the element-wise definition is used and one may adopt grid adaptivity to satisfy the condition.

dependent of $\mathbb{P}e_K$ [Qua09, QV99]. This estimate is valid as long as $\mathbb{P}e_K \geq 1$ and tells us that by using this stabilization the error is well controlled also when $\mathbb{P}e_K > 1$.

The proposed problem is well posed, since $\nabla \cdot \boldsymbol{\beta} = 0$ and meas(Γ_N) = 0, so that the assumptions (4.25) are fulfilled.

In Program 21 we report the **FreeFem** code of the problem stabilized with artificial viscosity, while the SUPG-stabilization code is reported in Program 22. Notice that **hTriangle** is a built-in function of **FreeFem++**, returning the size of the current element.

Program 21 - adv-diff2d-va : 2D advection-diffusion problem stabilized with artificial viscosity

```
mesh Th=square(50,50);
fespace Vh(Th,P1);
Vh u=0,v;
real mu=1;
int i=0;
real betax=-1000;
real betay=-1000;
real sigma=1;
real modbeta = sqrt(betax^2+betay^2);
real betalinfty = 1000;
problem adv_diff2D(u,v,solver=GMRES,init=i,eps=-1.0e-6) =
    int2d(Th)( mu*(dx(u)*dx(v) + dy(u)*dy(v)))
//stabilization:
  + int2d(Th)( betalinfty*hTriangle*(dx(u)*dx(v) + dy(u)*dy(v)))
  + int2d(Th) ( betax*dx(u)*v + betay*dy(u)*v )
  + int2d(Th) (sigma*u*v)
  + on(1,4,u=1)
  + on(2,3,u=0)   ;
adv_diff2D;
```

Program 22 - adv-diff2d-supg : 2D advection-diffusion problem stabilized with SUPG

```
real delta = 1.0;
fespace Nh(Th,P0);
real modbeta = sqrt(betax^2+betay^2);
Nh tau = hTriangle/sqrt(betax^2+betay^2);
problem adv_diff2Dsupg(u,v,solver=GMRES,init=i,eps=-1.0e-6) =
    int2d(Th)( mu*dx(u)*dx(v) + mu*dy(u)*dy(v))
  + int2d(Th) ( betax*dx(u)*v + betay*dy(u)*v ) + int2d(Th) (sigma*u*v)
  + int2d(Th)( delta*(-mu*(dxx(u)+dyy(u))+betax*dx(u)
    +betay*dy(u)+sigma*u)*tau*(betax*dx(v) + betay*dy(v)) )
  + on(1,4,u=1) + on(2,3,u=0)   ;
adv_diff2Dsupg;
```

Fig. 4.21 shows the iso-lines of the numerical solution with the standard Galerkin linear finite element method and a mesh size $h = 1/50$, so that Péclet number is 10. Numerical solution is completely wrong because of the spurious oscillations.

In Fig. 4.22 (on the left) we report the numerical solution computed with the artificial viscosity method on a grid with size $h = 0.1$. No oscillations are present.

Finally in Fig. 4.22 on the right we report the solution computed with the SUPG method. Notice how the boundary layer is steeper, since the method is less diffusive.   ◇

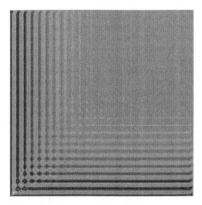

Fig. 4.21 Iso-lines of the numerical solution for Exercise 4.2.6 with standard Galerkin method. Mesh size is $h = 1/50$ and $\mathbb{P}e = 10$. Spurious numerical oscillations are evident

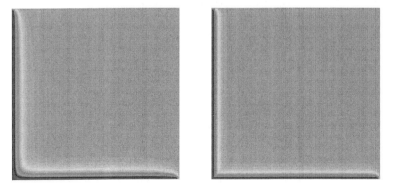

Fig. 4.22 Iso-lines of the numerical solution for Exercise 4.2.6 with the artificial viscosity (left) and SUPG (right). Mesh size is $h = 1/50$. Action of stabilization is evident. The steeper (and more accurate) boundary layer in the SUPG solution is a consequence of the less artificial diffusion introduced by the method

Exercise 4.2.7. Let us consider the following problem

$$
\begin{cases}
-\Delta u + \dfrac{\partial}{\partial x}\left(\dfrac{1}{2}x^2 y^2 u\right) - \dfrac{\partial}{\partial y}\left(\dfrac{1}{3}xy^3 u\right) = f & \text{in} \quad \Omega \subset \mathbb{R}^2, \\[2mm]
u = 0 & \text{on} \quad \Gamma_D, \\[2mm]
\dfrac{\partial u}{\partial \mathbf{n}} + u = 0 & \text{on} \quad \Gamma_N,
\end{cases}
\tag{4.26}
$$

where $\partial\Omega = \Gamma_D \cup \Gamma_N$, $\Gamma_D \cap \Gamma_N = \emptyset$. Domain Ω is a circle with center in the generic point $\widetilde{x}, \widetilde{y}$ and radius $r = 1$.

1. Write the weak formulation of the problem (4.26) after having introduced suitable functional spaces.
2. Find assumptions on the data that guarantee well posedness to the solution; in particular, give geometrical constraints on Γ_N as sufficient conditions for the well posedness; compute explicitly these constraints for $\widetilde{x} = \widetilde{y} = 0$.
3. Discretize the problem by the Galerkin finite element method.
4. Discuss the accuracy of the numerical solution when $\widetilde{x} = \widetilde{y} = 0$ and $\widetilde{x} = \widetilde{y} = 1000$, identifying possible remedies to numerical inaccuracies. Verify the answers with **FreeFem** for $f(x,y) = \sqrt{\left(\dfrac{1}{2}x^2 y^2\right)^2 + \left(-\dfrac{1}{3}xy^3\right)^2}$.

Solution 4.2.7.

Mathematical analysis

Notice that for $\boldsymbol{\beta} = \left[\dfrac{1}{2}x^2 y^2, -\dfrac{1}{3}xy^3\right]^T$, the convective term in (4.26) can be written as

$$
\frac{\partial}{\partial x}\left(\frac{1}{2}x^2 y^2 u\right) - \frac{\partial}{\partial y}\left(\frac{1}{3}xy^3 u\right) = \nabla \cdot (\boldsymbol{\beta} u) = \boldsymbol{\beta} \cdot \nabla u,
$$

since $\boldsymbol{\beta}$ is a divergence-free vector.

In order to derive the weak formulation of the problem, let us multiply the equation by a test function v in the space $V \equiv H^1_{\Gamma_D}(\Omega)$ of $H^1(\Omega)$ functions with null trace on Γ_D. After integration over Ω, and application of the Green formula

$$
\int_{\Omega} \nabla u \cdot \nabla v\, d\omega - \int_{\Gamma_N} \nabla u \cdot \mathbf{n}v\, d\gamma + \int_{\Omega} \boldsymbol{\beta} \cdot \nabla u v\, d\omega = \int_{\Omega} f v\, d\omega.
$$

Thanks to boundary conditions on Γ_N, we obtain the weak formulation: find $u \in V$ s.t. for any $v \in V$,

$$\int_\Omega \nabla u \cdot \nabla v d\omega + \int_{\Gamma_N} uv d\gamma + \int_\Omega \nabla \cdot (\boldsymbol{\beta} u)\, v d\omega = \int_\Omega fv d\omega. \tag{4.27}$$

We prove that the bilinear form

$$a(u,v) \equiv \int_\Omega \nabla u \cdot \nabla v d\omega + \int_{\Gamma_N} uv d\gamma + \int_\Omega \boldsymbol{\beta} \cdot (\nabla u) v d\omega$$

is continuous and coercive. Continuity follows from:

1. boundedness of the integrals on Ω since u and v belong to V;
2. boundedness of the integrals in Γ_N since the trace of u and v (functions of V) on Γ_N is in $L^2(\Gamma_N)$ (see Chapter 1).

Moreover we have the inequality

$$a(u,u) = \|\nabla u\|_{L^2}^2 + \int_{\Gamma_N} u^2 d\gamma + \frac{1}{2}\int_\Omega \boldsymbol{\beta}\cdot\nabla u^2 d\omega \geq \|\nabla u\|_{L^2}^2 + \frac{1}{2}\int_{\Gamma_N}(\boldsymbol{\beta}\cdot\mathbf{n})\,u^2 d\gamma.$$

Assumption that Γ_N is an *outlet* for Ω,

$$\boldsymbol{\beta} \cdot \mathbf{n} \geq 0 \qquad \text{on} \qquad \Gamma_N, \tag{4.28}$$

is enough for the coercivity of the bilinear form. Should $\text{meas}(\Gamma_N) = 0$ (only Dirichlet conditions), bilinear form will be automatically coercive.

Finally, notice that the right hand side of (4.27) is a linear and continuous functional in V provided $f \in V'$, the dual space of V. In particular this holds when $f \in L^2(\Omega)$.

Assumption (4.28) in this problem can be formulated as a constraint on Γ_N. The outward normal unit vector for the boundary $(x - \widetilde{x})^2 + (y - \widetilde{y})^2 = 1$ is

$$\mathbf{n} = [n_1, n_2]^T \quad \text{con} \quad n_1 = x - \widetilde{x}, \quad n_2 = y - \widetilde{y}.$$

We have therefore the following constraint

$$\frac{1}{2}x^2y^2(x - \widetilde{x}) - \frac{1}{3}xy^3(y - \widetilde{y}) \geq 0.$$

For $\widetilde{x} = \widetilde{y} = 0$, we have the conditions

$$|y| \leq \sqrt{\frac{3}{2}}x \quad \text{for} \quad x \geq 0, \qquad |y| \geq \sqrt{\frac{3}{2}}|x| \quad \text{for} \quad x \leq 0,$$

illustrated with the solid lines in Fig. 4.23.

Condition (4.28) requires that Γ_N is a subset of this region of $\partial\Omega$.

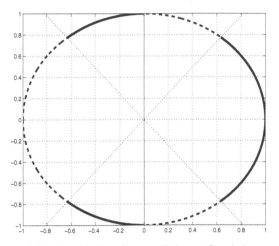

Fig. 4.23 Exercise 4.2.7: Ω with $\widetilde{x} = \widetilde{y} = 0$. If the outflow belongs to the portion of $\partial\Omega$ denoted by the solid lines, well posedness is guaranteed. Dotted lines correspond to equations $y = \pm\sqrt{3/2}x$

Numerical approximation

For a finite dimensional subspace V_h of V, the finite element approximation reads: find $u_h \in V_h$ such that, for any $v_h \in V_h$, we have

$$\int_\Omega \nabla u_h \cdot \nabla v_h d\omega + \int_{\Gamma_N} u_h v_h d\gamma + \int_\Omega \nabla \cdot (\boldsymbol{\beta} u_h)\, v_h d\omega = \int_\Omega f v_h d\omega. \qquad (4.29)$$

After introducing a conformal triangular grid of Ω, possible candidates for V_h are the spaces X_h^r in (2.22).

Local Peclét number for this problem is $\mathbb{P}e_K = \dfrac{\|\beta\|_{L^\infty(K)} h_K}{2}$, which can be bound by $\dfrac{\|\beta\|_{L^\infty(\Omega)} h}{2}$. If we take, $\widetilde{x} = \widetilde{y} = 0$ we have $\|\beta\|_{L^\infty(\Omega)} < \frac{1}{2}$. Therefore, in this case, non-oscillating solution are obtained even with large values of h. For $\widetilde{x} = \widetilde{y} = 1000$, on the contrary, $\|\beta\|_{L^\infty(\Omega)} \approx \frac{1}{2} \times 10^{12}$, and the Galerkin method computes non-oscillating solutions only for truly small values of h (order of 10^{-11}). Strongly consistent stabilization methods yield non-oscillating solutions for large h, maintaining the strong consistency of the Galerkin method.

Numerical results

FreeFem code is reported in Program 23.

Program 23 - ad-circle : Advection-diffusion problem on a circular domain

```
real xbar=1000,ybar=1000;
border a(t=0, 2*pi)x = cos(t)+xbar; y = sin(t)+ybar;   ;
mesh th = buildmesh(a(100));
fespace Vh(th,P1);
Vh u=0,v;
real mu=1;
Vh betax=1./2*x^2*y^2;
Vh betay=-1./3*x*y^3;
Vh f=sqrt(betax*betax+betay*betay);
int i=0;
real betalinfty=0.5*1001.^2;
problem Circle(u,v,solver=GMRES,init=i,eps=-1.0e-8) =
    int2d(th)( (mu)*(dx(u)*dx(v) + dy(u)*dy(v)))
  + int2d(th) ( betax*dx(u)*v + betay*dy(u)*v )
  + int2d(th)(-f*v)   + on(1,u=0)   ;
Circle;
```

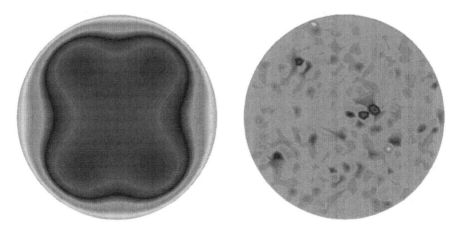

Fig. 4.24 Numerical solution of Exercise 4.2.7. On the left solution isolines for the domain centered in $(0,0)$ and $h \approx 0.05$. Diffusion is dominating. On the right, the domain centered in $(1000, 1000)$ and $h \approx 0.05$. Convection is dominating and the solution oscillates

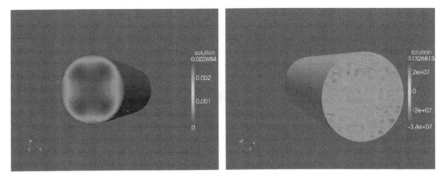

Fig. 4.25 Extension in 3D of Exercise 4.2.7. On the left the case where diffusion is dominating, on the right the case of advection dominated problem

Results in Fig. 4.24, computed with domain centered in $(0,0)$ (left) and $(1000, 1000)$ (right) confirm the expected behavior. In the second case, numerical solution strongly oscillates and it is meaningless.   ◇

Remark 4.2 We could consider the 3D extension of the previous exercise, assuming a cylindrical domain with convective field $\beta = \left[\frac{1}{2}x^2y^2, -\frac{1}{3}xy^3, 0\right]^T$. The cylinder is aligned with the z axis, with radius $R = 1$. A point $[x_0, y_0, z_0]$ on the axis is located in $[0, 0, 0]$ in one case and $[1000, 1000, 1000]$ in the second case. The axial coordinate ranges over the interval $[z_0, z_0 + L]$ with $L = 3$.

In Fig. 4.25 we report two snapshots of the solutions in the two cases. As expected, in the case $[x_0, y_0, z_0] = [1000, 1000, 1000]$ the solution is meaningless.

These results have been obtained with the LifeV library by Alessio Fumagalli (MOX, Politecnico di Milano, Milan, Italy). We report a part of the code used in Program 24. The entire code is available at www.lifev.org.

Program 24 - 3D Extension of Exercise 4.2.7 :

```
Real betaFctCylinder( const Real& /* t */, const Real& x,
        const Real& y, const Real& /* z */, const ID& i )
{
    switch (i){
    case 0:
        return 0.5 * std::pow(x, 2) * std::pow(y, 2);
        break;
    case 1:
        return - 1. / 3. * x * std::pow(y, 3);
        break;
    case 2:
        return 0.;
        break;
    default:
        return 0.;
    }
}
Real fRhs( const Real& /* t */, const Real& x, const Real& y,
```

```
          const Real& /* z */ , const ID& /* i */ )
{ return   std::sqrt( std::pow( 0.5 * std::pow( x, 2 )
                  * std::pow( y, 2 ), 2 ) +
             std::pow( - 1. / 3. * x * std::pow( y, 3 ), 2 ) );}


typedef RegionMesh3D<LinearTetra> mesh_type;
typedef MatrixEpetra<Real> matrix_type;
typedef VectorEpetra vector_type;


int
main( int argc, char** argv )
{
(...)
// Perform the assembly of the matrix
    adrAssembler.addDiffusion(systemMatrix, diffusionCoeff );
    vector_type beta(betaFESpace->map(),Repeated);
    betaFESpace->interpolate(betaFctCylinder,beta,0.0);
    adrAssembler.addAdvection(systemMatrix,beta);\
    systemMatrix->globalAssemble();
// Definition and assembly of the RHS
    vector_type rhs(uFESpace->map(),Repeated);
    rhs*=0.0;
    vector_type fInterpolated(uFESpace->map(),Repeated);
    uFESpace->interpolate(fRhs,fInterpolated,0.0);
    adrAssembler.addMassRhs(rhs,fInterpolated);
    rhs.globalAssemble();
// Definition and application of the BCs
    BCHandler bchandler;
    BCFunctionBase BCd ( zeroFct );
    BCFunctionBase BCn ( zeroFct );
    bchandler.addBC("Neumann", 2, Natural, Full,BCn,1);
    bchandler.addBC("Neumann", 3, Natural, Full,BCn,1);
    bchandler.addBC("Dirichlet", 1, Essential, Scalar, BCd);
    bchandler.bcUpdate(*uFESpace->mesh(),uFESpace->feBd(),uFESpace->dof());
    vector_type rhsBC(rhs,Unique);
    bcManage(*systemMatrix,rhsBC,*uFESpace->mesh(),
        uFESpace->dof(),bchandler,uFESpace->feBd(),1.0,0.0);
    rhs = rhsBC;
// Definition of the solver
    SolverAztecOO linearSolver;
    linearSolver.setMatrix(*systemMatrix);
// Definition of the solution
    vector_type solution(uFESpace->map(),Unique);
    solution*=0.0;
    linearSolver.solveSystem(rhsBC,solution,systemMatrix);

(...)

    return( EXIT_SUCCESS );
}
```

4.3 Reaction dominated problems

Exercise 4.3.1. Consider the diffusion-reaction problem

$$-\mu u'' + \sigma u = 0 \qquad x \in (0,1)$$

with $u(0) = 0$, $u(1) = 1$, and μ, σ positive constants.

1. Prove existence and uniqueness of the weak solution to this problem.
2. Write the linear finite element approximation, in particular the i−th equation of the linear system obtained after the discretization.
3. Give the definition of *reaction dominated* problem and discuss the quality of the solution of the Galerkin method in this case. Specify the conditions on the mesh size such that the linear finite element solution is not oscillating. Compare the condition with the corresponding one for advection dominated problems.
4. Identify possible stabilization techniques for the Galerkin method.
5. Consider the case $\mu = 1$, $\sigma = 10^4$. Compute the analytic solution and numerical solution with `fem1d` with linear finite elements. Compare the results with the answer given at the previous points.

Solution 4.3.1.

Mathematical analysis

Let G be a function of $H^1(0,1)$ such that $G(0) = 0$ and $G(1) = 1$ (e.g. $G(x) = x$). Moreover, for $a(u,v) \equiv \int_0^1 \mu u'v' + \sigma u v dx$ with the usual procedure the weak form of the problem reads: find $u \in G + V$ with $V \equiv H_0^1(0,1)$ such that $a(u,v) = 0$ for any $v \in V$. The bilinear form $a(\cdot,\cdot)$ is continuous as a consequence of the functional spaces selected. Coercivity follows from the positivity of μ and σ, since

$$a(u,u) = \int_0^1 \mu(u')^2 + \sigma u^2 dx \geq \min(\mu, \sigma) \|u\|_V^2.$$

Well posedness follows from Lemma 3.1.

In this simple problem, the analytic solution can be computed by hand. The general solution is indeed $u(x) = C_1 e^{\lambda_1 x} + C_2 e^{\lambda_2 x}$, where λ_1 and λ_2 are roots of the equation $-\mu\lambda^2 + \sigma = 0$, i.e. $\lambda_{1,2} = \pm(\sigma/\mu)^{1/2}$. The particular solution is obtained by prescribing the boundary conditions, obtaining (see Fig. 4.26)

$$u(x) = \frac{e^{\sqrt{\sigma/\mu}x} - e^{-\sqrt{\sigma/\mu}x}}{e^{\sqrt{\sigma/\mu}} - e^{-\sqrt{\sigma/\mu}}} = \frac{\sinh(\sqrt{\sigma/\mu}x)}{\sinh(\sqrt{\sigma/\mu})}.$$

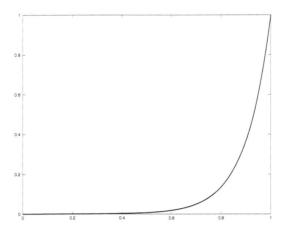

Fig. 4.26 Exact solution of Exercise 4.3.1

Numerical approximation

The Galerkin finite dimensional approximation reads: find $u_h \in V_h$ such that for any $v_h \in V_h$

$$a(u_h, v_h) = \mathcal{F}(v_h). \tag{4.30}$$

In the case of linear finite elements, V_h is the subset of X_h^1 of the functions vanishing at the end points. Basis functions φ_i (for $i = 1, 2, \ldots, N$, being N the dimension of the finite dimension space) are given in Ch. 2 and in (2.18).

The algebraic formulation of (4.30) is obtained by selecting the basis functions as test functions,

$$(\mu K + \sigma M) \mathbf{U} = \mathbf{F},$$

where for $i, j = 1, 2, \ldots N$,

$$K_{ij} = \int_0^1 \varphi_j' \varphi_i' dx, \quad M_{ij} = \int_0^1 \varphi_j \varphi_i dx, \quad F_i = \mathcal{F}(\varphi_i)$$

and $\mathbf{U}$ is the vector of the nodal vectors $U_i = u_h(x_i)$. After some computation one finds that the generic i–th equation reads

$$\mu \frac{-u_{i-1} + 2u_i - u_{i+1}}{h} + \sigma h \left(\frac{1}{6} u_{i-1} + \frac{2}{3} u_i + \frac{1}{6} u_{i+1} \right) = \mathbf{F}_i.$$

This finite difference equation after the inclusion of the boundary conditions (see [Qua09]) has the solution

$$u_i = \frac{\rho_1^i - \rho_2^i}{\rho_1^N - \rho_2^N}, \quad i = 1, 2, \ldots, N$$

where

$$\rho_{\pm} = \frac{1 + 2\mathbb{P}\text{e}_r \pm \sqrt{3\mathbb{P}\text{e}_r(\mathbb{P}\text{e}_r + 2)}}{1 - \mathbb{P}\text{e}_r},$$

where $\mathbb{P}\text{e}_r \equiv \dfrac{\sigma h^2}{6\mu}$ is the so-called *reactive Péclet number.* . The numerator of ρ_+ is always positive, the one of ρ_- is always non-negative ($\rho_- = 0$ for $\mathbb{P}\text{e}_r = 1$). We

$$u_i = \frac{(1 - \mathbb{P}\text{e}_r)^M}{(1 - \mathbb{P}\text{e}_r)^i} \frac{(1 + 2\mathbb{P}\text{e}_r + \sqrt{3\mathbb{P}\text{e}_r(\mathbb{P}\text{e}_r + 2)})^i - (1 + 2\mathbb{P}\text{e}_r - \sqrt{3\mathbb{P}\text{e}_r(\mathbb{P}\text{e}_r + 2)})^i}{(1 + 2\mathbb{P}\text{e}_r + \sqrt{3\mathbb{P}\text{e}_r(\mathbb{P}\text{e}_r + 2)})^M - (1 + 2\mathbb{P}\text{e}_r - \sqrt{3\mathbb{P}\text{e}_r(\mathbb{P}\text{e}_r + 2)})^M}.$$

The sign of u_i is driven by the first factor. For $\mathbb{P}\text{e}_r < 1$, u_i is always positive, as the exact solution is. For $\mathbb{P}\text{e} > 1$, the sign of the solution depends on i, so the numerical solution oscillates along x. Finally, we may stress that the condition $\mathbb{P}\text{e}_r < 1$ is verified for

$$h < \sqrt{\frac{6\mu}{\sigma}}.$$

Should this constraint be too restrictive to be fulfilled, a possible way for the stabilization, borrowed from the Finite Difference method, is *mass lumping*, (see Section 4), which consists in replacing the mass matrix with its lumped approximation. After this approximation, the discretization obtained can be regarded as the application of a *Generalized Galerkin method*. More precisely, we have

$$a_h(u_h, v_h) = \mathcal{F}(v_h),$$

where the approximation of the bilinear form results from the mass matrix approximation. The mathematical analysis of this problem can be carried out thanks to the Strang Lemma. As we have pointed out, for linear finite elements mass lumping can be considered as the consequence of the application of the trapezoidal formula to the computation of the mass matrix entries, so that

$$|a(u_h, v_h) - a_h(u_h, v_h)| = \text{quadrature error of trapez. formula for} \int_{\Omega} u_h v_h.$$

The theory of numerical quadrature states that this error is quadratic with respect to the size of the mesh. Application of the Lemma leads therefore to the conclusion that the linear finite element method with mass lumping is still first order accurate in norm H^1.

Numerical results

We use `femld` for testing the previous considerations. We force the Péclet number to be < 1 by taking

$$h < \sqrt{\frac{6}{10000}} = 0.024494897. \qquad (4.31)$$

Figs. 4.27 and 4.28 show the numerical solutions obtained with different values of h. The numerical solution oscillates apart from the case $h = 0.01$ (Fig. 4.28, left). For $h = 0.025$ (which is close to the bound (4.31)) oscillations are present, but they are extremely small.

Simulations obtained with mass lumping are in Figs. 4.29 and 4.30. Oscillations are clearly cut out by mass lumping. $\Diamond$

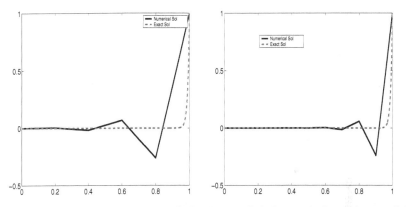

Fig. 4.27 Solution of Exercise 4.3.1 with the classical Galerkin method with $h = 0.2$ (left) and $h = 0.1$ (right)

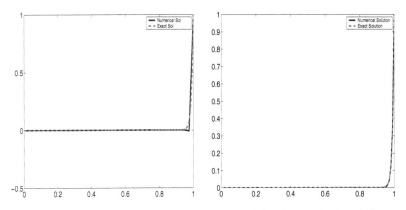

Fig. 4.28 Solution of Exercise 4.3.1 with the classical Galerkin method with $h = 0.025$ (left) and $h = 0.01$ (right)

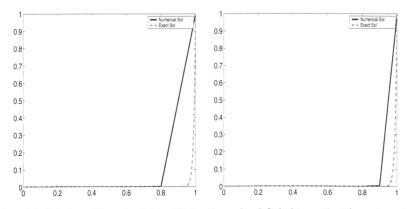

Fig. 4.29 Solution of Exercise 4.3.1 with the Generalized Galerkin method (mass lumping) with $h = 0.2$ (left) and $h = 0.1$ (right)

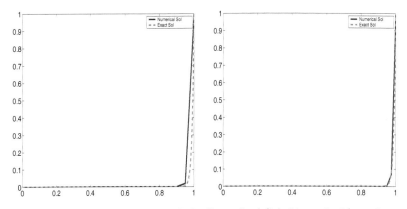

Fig. 4.30 Solution of Exercise 4.3.1 with the Generalized Galerkin method (mass lumping) with $h = 0.05$ (left) and $h = 0.025$ (right)

Exercise 4.3.2. (*) A factory spills a pollutant in the river Ω depicted in Fig. 4.31. The pollutant concentration C_{in} at the inlet Γ_{in} is constant. Pollution does not involve the underwater layers of the river, so we consider this as a 2D problem, with no dependence on the height. Moreover we assume that:

a. in correspondence of the upstream boundary Γ_{up} there is a baseline concentration of pollutant C_p;
b. the downstream section Γ_{down} is far enough to consider that the concentration does not change anymore in the direction of the flux normal to the boundary;

c. the flux of pollutant on the sides is proportional to the difference between the natural concentration C_{dry} and the concentration in the river;
d. diffusivity of the pollutant in the river is isotropic and constant, so it is represented by the scalar number μ;
e. the velocity of the river on the surface is steady and divergence free;
f. a bacterium in the river destroys the pollutant with a rate σ;
g. problem is steady.

1. Write a PDE model for the concentration C in Ω.
2. Analyze its well posedness.
3. Discretize the problem with finite elements; in particular, assuming that the solution $C \in H^2(\Omega)$ (and $C \notin H^3(\Omega)$), select a proper degree for the finite elements.
4. Discuss the accuracy of the approximation as a function of the parameters of the problem.
5. Assume that the river is rectilinear with length 10 m and width 2m. Data: river velocity $\mathbf{u} = [u_1, \quad u_2]^T$ with $u_1 = u_M(2-y)y \ m/s$, $u_2 = 0$, $C_{up} = 10g/m^3$, $C_{in} = 100g/m^3$, $C_{dry} = 1g/m^3$, $\alpha = 0.1$. With a grid with size $h = 0.1$, simulate the following two cases:

   i. $\sigma = 0.5$, $\mu = 10^{-6}$, $u_M = 10$;
   ii. $\mu = 0.1$, $\sigma = 300$, $u_M = 2$;

   with linear generalized Galerkin finite elements-Streamline Diffusion for the advection dominated case and mass lumping for the reaction-dominated case.

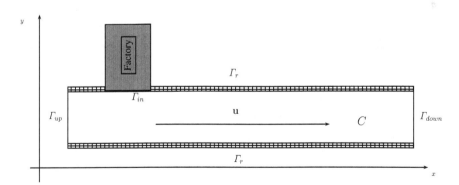

Fig. 4.31 Domain of the river for the Exercise 4.3.2

Solution 4.3.2.

Mathematical modeling

With the notation suggested in the text, we denote with $C = C(x, y)$ the pollutant concentration in the river. Three processes are involved in the pollutant dynamics.

1. *Diffusion* which is assumed to be isotropic.
2. *Convection* induced by the river motion. Let $\mathbf{u}$ be the river velocity, which is assumed to be divergence free ($\nabla \cdot \mathbf{u} = 0$).
3. *Reaction* induced by the consumption of pollutant by the bacterium.

Mass conservation yields (see e.g. [Sal08])

$$\frac{\partial C}{\partial t} = -\nabla \cdot (q_c + q_d) - \sigma C,$$

where:

1. q_t is the mass flux driven by the convection, induced by the river stream, i.e. $q_c = \mathbf{u}C$.
2. q_d is the diffusive flux, that according to the Fick's law can be written $q_d = -\mu\nabla C$, where μ is the pollutant diffusivity.

Here we are assuming a steady problem, so mass balance reads

$$\nabla \cdot (\mathbf{u}C - \mu\nabla C) + \sigma C = 0. \tag{4.32}$$

We split the river boundary as follows:

1. Γ_{in} is the portion where the pollutant emitted by the factory enters with concentration C_{in}; here we assume a Dirichlet condition.
2. Γ_{down} is the outlet portion; for the assumptions given in the text, this side is far from the inlet and we can assume that the variations of C in the flow direction normal to the boundary Γ_{down} are null. We postulate therefore the condition $\nabla C \cdot \mathbf{n} = 0$.
3. Γ_{up} is the upstream boundary, where we prescribe a Dirichlet condition $C = C_f$.
4. The lateral sides Γ_r feature a pollutant flux proportional to the difference of the concentration across the boundary

$$\mu\nabla C \cdot \mathbf{n} = \alpha (C_{dry} - C), \tag{4.33}$$

where $\alpha > 0$ is assumed to be constant.

In summary, the differential problem for the steady dynamics of the pollutant is

$$\begin{cases} -\mu\Delta C + \nabla \cdot (\mathbf{u}C) + \sigma C = 0 & (x,y) \in \Omega, \\ C = C_{in} & \text{on} \quad \Gamma_{in}, \\ C = C_f & \text{on} \quad \Gamma_{up}, \\ \mu\nabla C \cdot \mathbf{n} = 0 & \text{on} \quad \Gamma_{down}, \\ \mu\nabla C \cdot \mathbf{n} + \alpha C = \alpha C_{dry} & \text{on} \quad \Gamma_r. \end{cases} \qquad (4.34)$$

Mathematical analysis

Let $\Gamma_D = \Gamma_{in} \cup \Gamma_{up}$ be the portion of the boundary where we prescribe Dirichlet boundary conditions. Moreover, $V \equiv H^1_{\Gamma_D}(\Omega)$ denotes the space of $H^1(\Omega)$ functions with null trace on Γ_D. Multiply the equation $(4.34)_1$ by a generic function $v \in V$ and integrate over Ω. From the integration by parts, we obtain the problem: find $C \in H^1(\Omega)$ fulfilling the conditions $(4.34)_2$ and $(4.34)_3$, s.t.

$$\mu\int_\Omega \nabla C \nabla v d\omega + \int_{\Gamma_r} \alpha C v d\gamma + \int_\Omega \nabla \cdot (\mathbf{u}C)\, v d\omega + \sigma \int_\Omega C v d\omega = \alpha \int_{\Gamma_r} C_{dry} v d\gamma,$$

where we have exploited the conditions on Γ_{down} e Γ_r and the fact that v has null trace on Γ_D. Let $R(x,y)$ be a function s.t.

$$R_{\Gamma_{in}} = C_{in} \qquad \text{and} \qquad R_{\Gamma_{up}} = C_f.$$

In general, for a rectangular domain, if $C_{in} \in H^{1/2}(\Gamma_{in})$ and $C_f \in H^{1/2}(\Gamma_{up})$, a function $R(x,y) \in H^1(\Omega)$ with these features exists (see Section 1.3). In the present case, C_{in} and C_f are constants, so they satisfy this assumption.

We introduce the bilinear form

$$a(C,v) \equiv \mu\int_\Omega \nabla C \nabla v d\omega + \int_{\Gamma_r} \alpha C v d\gamma + \int_\Omega \nabla \cdot (\mathbf{u}C)\, v d\omega + \sigma \int_\Omega C v d\omega \quad (4.35)$$

and the functional

$$\mathcal{F}(v) \equiv \alpha C_{dry} \int_{\Gamma_r} v d\gamma. \qquad (4.36)$$

The problem reads: find $C \in R + V$ s.t. for any $v \in V$

$$a(C,v) = \mathcal{F}(v).$$

From the Cauchy-Schwarz inequality, we have

$$\left|\mu\int_\Omega \nabla w \nabla v d\omega\right| \le \mu\|w\|_V \|v\|_V.$$

The second term of (4.35) is continuous since the trace operator is continuous from $H^1(\Omega)$ to $H^{1/2}(\Gamma_r)$ so that

$$|\int_{\Gamma_r} \alpha C v| \leq \alpha ||C||_{H^{1/2}(\Gamma_r)} ||v||_{H^{1/2}(\Gamma_r)} \leq \alpha^* ||C||_V ||v||_V.$$

Here α^* includes the trace inequality constant.

Since $\mathbf{u}$ is divergence free, the third term in (4.35) reads $\int_{\Omega} (\mathbf{u} \cdot \nabla C) v d\omega$.

Remind that if $v \in H^1(\Omega)$ the Sobolev embedding Theorem states that in 2D problems $v \in L^6(\Omega)$ and in 3D problems $v \in L^4(\Omega)$. Now, the product of two $L^4(\Omega)$ functions belongs to $L^2(\Omega)$, so that for the generic component u_i of $\mathbf{u}$, $u_i v \in L^2(\Omega)$. Moreover $\nabla C \in L^2(\Omega)$. Therefore, if $\mathbf{u} \in H^1(\Omega)$, the integrand function in the previous term belongs to $L^1(\Omega)$, which guarantees that the integral exists and is finite. Moreover, continuity is obtained. By the way, the assumption $\mathbf{u} \in H^1(\Omega)$ is reasonable since we can think to the velocity field as the solution to the Navier-Stokes equations (see Chapter 7). As such, $\mathbf{u}$ actually belongs to $H^1(\Omega)$. We conclude therefore

$$|\int_{\Omega} (\mathbf{u} \cdot \nabla C) v d\omega| \leq ||\mathbf{u}||_{H^1} ||C||_{H^1} ||v||_V.$$

Continuity of the last term for a constant σ is a trivial consequence of the continuity of the $L^2(\Omega)$ scalar product for functions in V. Continuity of $\mathcal{F}(v)$ follows from the continuity of $\int_{\Gamma_r} \alpha C_{dry} v d\gamma$, which is true, since C_{dry} is constant ($C_{dry} \in H^{1/2}(\Gamma_r)$ would be enough).

For the coercivity of the bilinear form notice that

$$a(w,w) = \mu ||\nabla w||^2_{L^2} + \alpha \int_{\Gamma_r} w^2 d\gamma + \frac{1}{2} \int_{\Omega} \nabla \cdot (\mathbf{u} w^2) d\omega + \sigma \int_{\Omega} w^2 d\omega.$$

Observe that thanks to the divergence Theorem and the boundary conditions

$$\int_{\Omega} \nabla \cdot (\mathbf{u} w^2) d\omega = \int_{\Gamma_r} \mathbf{u} \cdot \mathbf{n} w^2 d\gamma + \int_{\Gamma_{down}} \mathbf{u} \cdot \mathbf{n} w^2 d\gamma.$$

Since the velocity of the river on the beach is zero, $\mathbf{u} \cdot \mathbf{n} = 0$ on Γ_r, whilst we assume $\mathbf{u} \cdot \mathbf{n} \geq 0$ on Γ_{down}, since Γ_{down} is the outlet section. Under these hypotheses

$$a(w,w) \geq \min(\sigma, \mu) ||w||^2_V.$$

Continuity and coercivity arguments guarantee the well posedness thanks to Lemma 3.1.

As a concluding remark, we may notice that it is not possible to find the analytical solution with a variable separation approach, since the convective term couples the two directions and the location of the boundary conditions depend on x.

Numerical approximation

Let V_h be a finite dimensional subspace of V where we look for a solution w_h by solving $a(w_h, v_h) = \mathcal{F}(v_h)$ for any $v_h \in V_h$. In particular, if we introduce a regular triangular mesh $\mathcal{T}_h$ of Ω, and choose V_h to be the space of X_h^r functions vanishing on Γ_D, we get a finite element discretization.

If the solution belongs[3] to $H^2(\Omega)$ we have the estimate

$$||u - u_h||_V \leq Ch|u|_2$$

for any degree $r > 1$. To reduce the computational cost, we select $r = 1$.

In this advection-reaction-diffusion problem, the quality of the solution depends on the weight of the convection or reaction coefficients in comparison with the diffusion one. In particular, spurious oscillations are eliminated if we ensure that on each element K of our mesh h_K is such that

$$\frac{||\mathbf{u}||_{L^\infty(K)} h_K}{2\mu} < 1 \quad \text{and} \quad \frac{|\sigma| h_K^2}{6\mu} < 1,$$

or by resorting to stabilization techniques. The latter are mandatory if the previous constraints on the mesh spacing are too restrictive.

Let us consider the Streamline Diffusion (SD) stabilization . The river features a non-zero velocity component only in the direction x. Only in this direction we have a convective effect, therefore the stabilization is needed only along x. The stabilized bilinear form thus reads

$$a_h(C, v) = a(C, v) + \sum_{k \in \mathcal{T}_h} \frac{||\mathbf{u}||_{L^\infty(K)} h_K}{2} \int_K \frac{\partial C}{\partial x} \frac{\partial v}{\partial x} dK.$$

In general, if the convection is not aligned with one of the axes, the SD stabilizing term on each element reads (see e.g. [Qua09]) $-\frac{h_K}{2||\mathbf{u}||_{L^\infty(K)}} \nabla \cdot [(\mathbf{u} \cdot \nabla C) \mathbf{u}]$ leading to the bilinear form

$$a_h(C, v) = a(C, v) + \sum_{K \in \mathcal{T}_h} \frac{1}{2||\mathbf{u}||_{L^\infty(K)} h_K} \int_K (\mathbf{u} \cdot \nabla C)(\mathbf{u} \cdot \nabla v) dK.$$

FreeFem code with the standard Galerkin method is given in Program 25. Program 26 reports the code with the SD stabilization and mass lumping for the reactive term. For simplicity we have used $||\mathbf{u}||_{L^\infty(\Omega)}$ instead of $||\mathbf{u}||_{L^\infty(K)}$ in the stabilization term (this is of course more diffusive but very practical).

---

[3] In general it is not trivial to prove regularity of the solution to problems with different boundary conditions on different portions of the boundary.

Program 25 - river1 : River pollution problem solved with standard Finite Elements

```
// Mesh
border floor(t=0,10){ x=t; y=0; label=10;}; //lato y=0 (Gamma_r)
border right(t=0,2){ x=10; y=t; label=5;}; // lato x=10 (Gamma_n)
border ceiling1(t=10,3){ x=t; y=2; label=10;}; // lato y=2 (Gamma_r)
border ceiling2(t=3,1){ x=t; y=2; label=2;}; // lato y=2 (Gamma_d)
border ceiling3(t=1,0){ x=t; y=2; label=10;}; //lato y=2 (Gamma_r)
border left(t=2,0){ x=0; y=t; label=1;}; // lato x=0 (Gamma_d)
mesh Th= buildmesh(floor(100)+right(20)+ceiling1(70)+
          ceiling2(20)+ceiling3(10)+left(20));
// Parameters
real mu=1.e-6;
real sigma=0.5;
real umax = 10;
func ux=umax*(2-y)*y;
func uy=0;
real Cin = 100;
real Cup = 10;
real Cdry = 1;
real alpha = 0.1;
// Variational Problem
fespace Vh(Th,P1);
Vh C,v;
C=0;
problem ADRRiver(C,v) =
    int2d(Th)(  mu*dx(C)*dx(v) + mu*dy(C)*dy(v))
  + int2d(Th)(  ux*dx(C)*v + uy*dy(C)*v)
  + int2d(Th)( sigma*C*v )
  + int1d(Th,10)( alpha*C*v )
  + int1d(Th,10)( -alpha*Cdry*v )
  + on(1,C=Cup)
  + on(2,C=Cin) ;
// Soluzione
ADRRiver;
```

Program 26 - river2 : River pollution problem solved with Finite Elements stabilized with SD and mass lumping

```
problem ADRRiver(C,v) =
    int2d(Th)(  mu*dx(C)*dx(v) + mu*dy(C)*dy(v))
  + int2d(Th)(  umax*hTriangle*(dx(C)*dx(v)))
  + int2d(Th)(  ux*dx(C)*v + uy*dy(C)*v)
  + int2d(Th,qft=qf1pTlump)( sigma*C*v )
  + int1d(Th,10)( alpha*C*v )
  + int1d(Th,10)( -alpha*Cdry*v )
  + on(1,C=Cup)
  + on(2,C=Cin) ;
```

Numerical results

In the first test case, we have a small diffusivity compared with the convection,

$$\mathbb{Pe} = \frac{10 \times 0.1}{2 \times 10^{-6}} = 5 \times 10^{5}.$$

Without stabilization, the numerical solution will oscillate. Fig. 4.32 illustrates these oscillations, the solution being negative in some part of the domain. This is meaningless since by definition the concentration is positive.

In Fig. 4.33 we report the SD stabilized solution. No oscillations nor negative values are present.

In the second test case we assume to have a bacterium reducing significantly the concentration. Since we have $\mathbb{Pe} = 2000 \times 0.01/(6 \times 0.1) \approx 33.3$, we expect stability issues (Fig. 4.34).

Numerical oscillations are eliminated with the mass lumping of the reactive term, as illustrated in Fig. 4.35. Our bacterium strongly reduces the pollutant, which is limited to a small region around Γ_{in}. ◇

Fig. 4.32 Solution for the problem 4.3.2 with $\mu = 10^{-6}$ and $\sigma = 0.5$ with standard Finite Elements. Solution ranges between -30 and 100

Fig. 4.33 Solution to the problem 4.3.2 with $\mu = 10^{-8}$ and $\sigma = 1$ with a SD stabilization method. Solution is everywhere positive. The superimposition of the baseline pollution with that incoming through boundary Γ_{in} is evident

Fig. 4.34 Solution to problem 4.3.2 with $\mu = 0.1$ and $\sigma = 300$ for the standard Galerkin method. The light gray region around Γ_{in} denotes negative values. Solution ranges from -26 to 100

Fig. 4.35 Solution to problem 4.3.2 with $\mu = 0.1$ and $\sigma = 300$ with a mass lumping solver. Solution ranges between 1 and 100. Thanks to the strong bacterial activity, pollutant is confined to a small region in the neighborhood of Γ_{in}

Part III
Time Dependent Problems

5

Equations of parabolic type

In this chapter we consider problems on the numerical approximation of parabolic partial differential equations. We will consider mostly linear problems (with the exclusion of the last exercise in Section 5.1).

Theory of parabolic problems is addressed e.g. in [Sal08] Chapter 2, [Eva10], [QV94], Chapter 11, [Qua09], Chapter 6. In any problem considered here we carry out the well posedness analysis referring to an extension of the Lax-Milgram Lemma and the so called *weak coercivity*. More precisely, we consider problems with the following weak form. Find $u \in V \subseteq H^1(\Omega)$ (where V is a closed Hilbert subspace) s.t. for any $t > 0$

$$\left(\frac{\partial u}{\partial t}, v\right) + a\left(u, v\right) = \mathcal{F}(v) \qquad \forall v \in V, \tag{5.1}$$

with $u(x, 0) = u_0(x)$, initial condition in $L^2(\Omega)$. Sufficient conditions for the well posedness are:

1. $\mathcal{F}(v)$ is a linear continuous functional $\forall v \in V$;
2. $a\left(u, v\right)$ is a bilinear continuous form $\forall u, v \in V$;
3. $a\left(u, v\right)$ is weakly coercive, i.e. for any $v \in V$ there exist two constants $\alpha > 0$ and $\lambda \geq 0$ s.t.

$$a(v, v) + \lambda ||v||_{L^2}^2 \geq \alpha ||v||_V^2. \tag{5.2}$$

Standard coercivity is obtained as a particular case for $\lambda = 0$.

Weak coercivity in parabolic problems is enough because with a change of variables a weakly coercive problem can be reformulated as a (standard) coercive one. As a matter of fact, if (5.2) holds for (5.1), we can set $\omega = e^{-\lambda t}u$. Multiplying (5.1) by $e^{-\lambda t}$, we can verify that ω solves: find $\omega \in V$ s.t.

$$\left(\frac{\partial \omega}{\partial t}, v\right) + a\left(\omega, v\right) + \lambda\left(\omega, v\right) = e^{-\lambda t}\mathcal{F}(v)$$

for any $v \in V$.

Formaggia L., Saleri F., Veneziani A.: Solving Numerical PDEs: Problems, Applications, Exercises. DOI 10.1007/978-88-470-2412-0_5, © Springer-Verlag Italia 2012

If the bilinear form $a\left(\cdot,\cdot\right)$ is weakly coercive, then the bilinear form

$$a^*\left(w,v\right) = a\left(w,v\right) + \lambda\left(w,v\right)$$

is coercive. Well posedness of the problem in u follows from the well posedness of the problem in w [LM54].

For a problem with non homogeneous boundary conditions on $\Gamma_D \subset \partial\Omega$, as done in Chapters 3 and 4, we use the following notation. Set $V \equiv H^1_{\Gamma_D}$. Let $G(x,t)$ (or $G(x,y,t)$ in 2D problems) be an extension of the boundary data (assumed to be regular enough). The problem reads: find $u \in G+V$ s.t. for any $t > 0$ equation (5.1) holds with $u(x,0) = u_0(x)$. This is equivalent to the problem: find $u^* \equiv u - G$ s.t. for any $t > 0$ we have

$$\left(\frac{\partial u^*}{\partial t}, v\right) + a\left(u^*, v\right) = \mathcal{F}(v) - \left(\frac{\partial G}{\partial t}, v\right) - a\left(G, v\right) \qquad \forall v \in V, \qquad (5.3)$$

with $u^*(x,0) = u_0(x) - G$. Conditions 1-3 above (extended to non-homogeneous problems) are sufficient for the well posedness, under the assumption that the data and the domain are regular enough.

We will use in general the abridged notation for non homogeneous Dirichlet problems, even though sometimes we will refer explicitly to (5.3).

In the analysis of parabolic problems the Gronwall's inequality reported in Chapter 2 is an important tool. In fact, when we have weakly parabolic problems this inequality leads to upper bounds for the solution. However, the constant of those bounds in general will increase exponentially with the time, so to become asymptotically meaningless. Uniform estimates with respect to the time dependence can be found when the bilinear form is elliptic in the standard sense.

For the space discretization we will use the finite element method. For the time discretization, in Section 5.1 we consider the finite difference method. In Section 5.2 we consider the finite element discretization of the problem both in space and time.

5.1 Finite difference time discretization

Exercise 5.1.1. Find $u = u(x,t)$ s.t.

$$\begin{cases} \dfrac{\partial u}{\partial t} - \dfrac{\partial}{\partial x}\left(\mu\dfrac{\partial u}{\partial x}\right) + \beta u = 0 & \text{in } Q_T = (0,1) \times (0,\infty), \\[2mm] u = u_0 & \text{for } x \in (0,1), t = 0, \\[2mm] u = \eta & \text{for } x = 0, t > 0, \\[2mm] \mu\dfrac{\partial u}{\partial x} + \gamma u = 0 & \text{for } x = 1, t > 0, \end{cases}$$

where $\mu = \mu(x) \geq \mu_0 > 0$, $u_0 = u_0(x)$ are given functions and $\beta, \gamma, \eta \in \mathbb{R}$.

> 1. Under proper conditions on the coefficients and regularity assumptions on μ and u_0 prove the existence and uniqueness of the solution.
> 2. Write the space discretization of the problem with the finite element method. Prove a stability bound (using the Gronwall's Lemma).
> 3. For $\gamma = 0$ perform the time discretization with the forward Euler method and analyze the stability of the fully discrete problem.

Solution 5.1.1.

Mathematical analysis

Let us write the weak formulation of the problem. Formally we multiply the strong formulation by a test function $v = v(x)$ and integrate on $(0, 1)$. Notice that the test function depends only on x (it is independent of t). Since we have a (non homogeneous) Dirichlet condition at the left end point, we assume $v(0) = 0$. We set $V \equiv H^1_{\Gamma_D}(0, 1) \equiv \{v \in H^1(0, 1), v(0) = 0\}$. After the usual integration by parts for the (space) second derivative, we write the problem for any $t > 0$ with $u(t) \in H^1(0, 1)$ and $u(0, t) = \eta$:

$$\left(\frac{\partial u(t)}{\partial t}, v\right) + \left(\mu\frac{\partial u(t)}{\partial x}, \frac{\partial v}{\partial x}\right) + \beta\left(u(t), v\right) - \left[\mu\frac{\partial u(t)}{\partial x}v\right]_0^1 = 0 \qquad \forall v \in V,$$
(5.4)

with $u(x, 0) = u_0(x)$ for $x \in (0, 1)$.

Taking into account the Robin condition on the right hand side and $v(0) = 0$, equation (5.4) leads to the weak formulation: find $u(t) \in H^1(0, 1)$ and $u(0, t) = \eta$ s.t.

$$\left(\frac{\partial u}{\partial t}, v\right) + \left(\mu\frac{\partial u}{\partial x}, \frac{\partial v}{\partial x}\right) + \beta\left(u, v\right) + \gamma uv|_{x=1} = 0 \qquad \forall v \in V.$$

The second integral is well defined only under some assumptions on μ. Since $\frac{\partial u}{\partial x}$ and $\frac{\partial v}{\partial x}$ belong (at least) to $L^2(0, 1)$, the assumption $\mu \in L^\infty(0, 1)$ guarantees that the integrand function $\mu\frac{\partial u}{\partial x}\frac{\partial v}{\partial x}$ is integrable.

For the non homogeneous Dirichlet condition, we introduce the *lifting* of the data. Let $G \in H^1(0, 1)$ be a function such that $G(0) = \eta$ (and arbitrary elsewhere). Since η is independent of time, the lifting G can be taken independent of t.

The problem reads therefore: find $u \in G + V$ s.t.

$$\left(\frac{\partial u}{\partial t}, v\right) + \left(\mu\frac{\partial u}{\partial x}, \frac{\partial v}{\partial x}\right) + \beta\left(u, v\right) + \gamma uv|_{x=1} = 0$$
(5.5)

for any $v \in V$, and $u(x, 0) = u_0$. The problem is clearly in the form (5.3) setting $a\left(u, v\right) \equiv \left(\mu\frac{\partial u}{\partial x}, \frac{\partial v}{\partial x}\right) + \beta\left(u, v\right) + \gamma uv|_{x=1}$.

From now on, we assume $u_0 \in L^2(0,1)$.

The bilinear form is continuous as a consequence of the Cauchy-Schwarz inequality:

$$\left|\left(\mu \frac{\partial w}{\partial x}, \frac{\partial v}{\partial x}\right)\right| \leq ||\mu||_{L^\infty(0,1)} \left\|\frac{\partial w}{\partial x}\right\|_{L^2(0,1)} \left\|\frac{\partial v}{\partial x}\right\|_{L^2(0,1)}$$

$$\leq ||\mu||_{L^\infty(0,1)} ||w||_V ||v||_V, \tag{5.6}$$

$$|\beta(u,v)| \qquad \leq \beta ||w||_{L^2(0,1)} ||v||_{L^2(0,1)} \leq \beta ||w||_V ||v||_V,$$

$$|\gamma u v|_{x=1}| \qquad \leq |\gamma| C_T^2 ||w||_V ||v||_V.$$

The last inequality is a consequence of the *trace inequality* introduced in Chapter 1.

Let us analyze the (weak) coercivity of $a(\cdot, \cdot)$. It is said that $\mu \geq \mu_0 > 0$ (as it is reasonable in general for physical reasons). It follows that $w \in V$

$$\left(\mu \frac{\partial w}{\partial x}, \frac{\partial w}{\partial x}\right) \geq \mu_0 \left\|\frac{\partial w}{\partial x}\right\|_{L^2(0,1)}^2 \geq \mu_0 C ||w||_V^2,$$

where C depends on the constant of the Poincaré inequality.

If $\gamma \geq 0$ then $\gamma w^2|_{x=1} \geq 0$,

$$a(w,w) \geq \mu_0 C ||w||_V^2 + \beta ||w||_{L^2(0,1)}^2.$$

If $\beta < 0$ the latter inequality leads to the weak coercivity with $\lambda = -\beta$ and coercivity constant $\alpha = \mu_0 C$. If $\beta \geq 0$, then the second term on the right hand side is non-negative, so we have the standard coercivity, i.e. $\lambda = 0$ in (5.2).

If $\gamma < 0$ we can set

$$\gamma w^2|_{x=1} = -|\gamma| w^2|_{x=1} \geq -|\gamma| C_T^2 ||w||_V^2,$$

in view of the trace inequality. Under the assumption

$$\mu_0 - |\gamma| C_T^2 \geq \alpha_0 \geq 0$$

we have

$$a(u,u) + \lambda ||u||_{L^2(0,1)}^2 \geq \alpha_0 ||u||_V^2 \qquad \forall u \in V,$$

where $\lambda = -\beta$ if $\beta < 0$ (weak coercivity) and $\lambda = 0$ if $\beta \geq 0$ (standard coercivity), and the coercivity constant is $\alpha = \alpha_0$.

Under these assumptions, the problem is well posed.

Let us consider the space discretization (or *semi-discretization*) with the Galerkin method (or more precisely Faedo-Galerkin).

Let V_h be a subspace of V with dimension $\dim(V_h) = N_h < \infty$ and with basis function $\{\varphi_i\}$, with $i = 1, 2, \ldots, N_h$.

For any $t > 0$ we look for $u_h \in G_h + V_h$ s.t.

$$\left(\frac{\partial u_h}{\partial t}, v_h\right) + a\left(u_h, v_h\right) = 0 \qquad \forall v_h \in V_h, \tag{5.7}$$

with $u_h(x, 0) = u_{0h}(x)$ (and $x \in (0, 1)$). Here, u_{0h} and G_h are approximations in V_h of u_0 and G.

Let us set $\tilde{u}_h = u_h - G_h$ and rewrite the problem in the form

$$\left(\frac{\partial \tilde{u}_h}{\partial t}, v_h\right) + a\left(\tilde{u}_h, v_h\right) = -\left(\frac{\partial G_h}{\partial t}, v_h\right) - a\left(G_h, v_h\right) \qquad \forall v_h \in V_h, \tag{5.8}$$

with $\tilde{u}_{h,0}(x) \equiv \tilde{u}_h(x, 0) = u_{0h}(x) - G_h(x)$ for $x \in (0, 1)$. Set $F(v_h) \equiv -\left(\frac{\partial G_h}{\partial t}, v_h\right) - a\left(G_h, v_h\right)$, and select the test function $v_h = \tilde{u}_h$:

$$\left(\frac{\partial \tilde{u}_h}{\partial t}, \tilde{u}_h\right) + a\left(\tilde{u}_h, \tilde{u}_h\right) = F(\tilde{u}_h). \tag{5.9}$$

Thanks to the weak coercivity and the identity $\tilde{u}_h \frac{\partial \tilde{u}_h}{\partial t} = \frac{1}{2}\frac{\partial \tilde{u}_h^2}{\partial t}$ we get

$$\frac{1}{2}\frac{d}{dt}||\tilde{u}_h||^2_{L^2(0,1)} + \alpha||\tilde{u}_h||^2_V \leq F(\tilde{u}_h) + \lambda||\tilde{u}_h||^2_{L^2(0,1)} \tag{5.10}$$

where α is the coercivity constant (equal to either μ_0 or α_0 depending on the sign of γ).

Thanks to the continuity of the functional and the Young inequality for an arbitrary $\epsilon > 0$ we have

$$F(\tilde{u}_h) \leq ||F||_{V'}||\tilde{u}_h||_V \leq \frac{1}{4\epsilon}||F||^2_{V'} + \epsilon||\tilde{u}_h||^2_V,$$

where V' is the space dual of V. Choose $\epsilon = \alpha/2$ and after some trivial manipulations (5.10) reads

$$\frac{d}{dt}||\tilde{u}_h||^2_{L^2(0,1)} + \alpha||\tilde{u}_h||^2_V \leq \frac{1}{\alpha}||F||^2_{V'} + 2\lambda||\tilde{u}_h||^2_{L^2(0,1)}.$$

Integrating on the interval $(0, t)$, we get

$$||\tilde{u}_h||^2_{L^2(0,1)}(t) + \alpha \int_0^t \alpha ||\tilde{u}_h||^2_V dt$$

$$\leq ||\tilde{u}_{0h}||^2_{L^2(0,1)} + \frac{1}{\alpha} \int_0^t ||F||^2_{V'} dt + 2\lambda \int_0^t ||\tilde{u}_h||^2_{L^2(0,1)} dt. \qquad (5.11)$$

The last inequality fulfills the assumptions of the Gronwall's Lemma 1.1, with $\phi(t) = ||\tilde{u}_h||^2_{L^2(0,1)}(t)$. We obtain therefore the following estimate

$$||\tilde{u}_h||^2_{L^2(0,1)}(t) \leq g(t) e^{2\lambda t} \qquad (5.12)$$

with $g(t) = \dfrac{1}{\alpha} \displaystyle\int_0^t ||F||^2_{V'} dt + ||\tilde{u}_{0h}||^2_{L^2(0,1)}$.

For each t, the L^2 norm of the space discrete solution is bounded, so the solution belongs to $L^\infty(0, T; L^2(0,1))$ (see Chapter 1). Moreover, plugging (5.12) in (5.11), we have the estimate

$$\int_0^T ||\tilde{u}_h||^2_V dt \leq \frac{1}{\alpha} \left(g(T) + 2\lambda \int_0^T g(t) e^{2\lambda t} dt \right),$$

showing that the discrete solution belongs also to $L^2(0, T; V)$. We write therefore $\tilde{u}_h \in L^\infty(0, T; L^2(0,1)) \cap L^2(0, T; V)$.

Notice that if the standard coercivity holds ($\lambda = 0$) a stability bound can be proved without the Gronwall's Lemma. The boundedness in $L^\infty(0, T; L^2(0,1)) \cap L^2(0, T; V)$ follows from (5.11), with coefficients that do not depend on time.

For the convergence analysis, let us consider first the case of $\lambda = 0$. We remind the following result (see e.g. [Qua09]) with finite elements of order P^r: for any $t > 0$

$$||(u - u_h)||^2_{L^2(0,1)}(t) + 2\alpha \int_0^t ||(u - u_h)(t)||^2_V dt \leq Ch^{2r} N(u) e^t,$$

provided that the exact solution u belongs for any $t > 0$ to $H^{r+1}(0,1)$. Here $N(u)$ is function of u and $\dfrac{\partial u}{\partial t}$.

For a weakly coercive problem, with the usual change of variable we have

$$||(u - u_h)||^2_{L^2(0,1)}(t) + 2\alpha \int_0^t ||(u - u_h)(t)||^2_V dt \leq Ch^{2r} N(u) e^{(1+\lambda)t},$$

that is an extension of the previous inequality [1].

---

[1] For $\gamma = 0$ with some additional assumptions on the regularity of u it is possible to deduce a bound uniform w.r.t. time, see [QV94], Chapter 11 or [EG04].

Finally, we discretize in time the semi-discrete problem with the forward Euler method, for the case $\gamma = 0$. To this aim we introduce the *mass matrix* M, with entries $m_{ij} \equiv \int_0^1 \varphi_i \varphi_j dx$ and the *stiffness matrix* K, with entries $k_{ij} \equiv \int_0^1 \mu \frac{\partial \varphi_i}{\partial x} \frac{\partial \varphi_j}{\partial x} dx$. Let $A \equiv K + \beta M$, then the semi-discrete problem reads

$$M \frac{dU}{dt} + AU = F,$$

where U is the vector of the degrees of freedom of the space discretization. In particular U^0 is the vector of the degrees of freedom associated with the initial data.

Let $[0, \infty)$ be the time interval, we split it into sub-intervals of length $\Delta t > 0$ and denote by $t_n = n\Delta t$ (n positive integer) a generic node of the time discretization. Forward Euler time discretization at t_{n+1} gives

$$\frac{1}{\Delta t} M \left(U^{n+1} - U^n \right) = -AU^n + F^n \qquad (5.13)$$

and consequently

$$U^{n+1} = \left(I - \Delta t M^{-1} A \right) U^n + \Delta t M^{-1} F^n.$$

In order to investigate the absolute stability, we set $F = 0$ to check the evolution of the solution when no forcing terms are present. The previous equation becomes

$$U^{n+1} = \left(I - \Delta t M^{-1} (K + \beta M) \right) U^n = \left((1 - \Delta t \beta) I - \Delta t M^{-1} K \right) U^n.$$

If the spectral radius of the matrix $(1 - \Delta t \beta) I - \Delta t M^{-1} K$ is less than 1, we have absolute stability [QSS00]. Let us denote with ρ_i the eigenvalues of $M^{-1}K$. As we will see in Exercise 5.1.5, $\rho_i > 0$, since both M and K are s.p.d. The eigenvalues of $(1 - \Delta t \beta) I - \Delta t M^{-1} K$ are in the form

$$1 - \Delta t (\beta + \rho_i), \qquad i = 1, \ldots, N_h.$$

If $\rho_i > -\beta$ for any i, then the forward Euler scheme is absolutely stable under the condition

$$\Delta t < \frac{2}{\max\limits_i (\rho_i - \beta)}.$$

This means in particular that, since $\rho_i > 0$, if $\beta > 0$ (absorbing reactive term) the Forward Euler method is conditionally stable. On the contrary, if $\rho_i \leq -\beta$ for a value of i, then the method is *unconditionally unstable*[2].

It is possible to prove (see e.g. [QV94]) that the eigenvalues ρ_i behave as h^{-2} for $h \to 0$. As a consequence, for the stability condition of the forward

---

[2] This is instability of the time discretization, which is clearly another type of stability than the one described in Chapter 4.

Euler method to be verified, the time step needs to be divided by 4 when the mesh size h is divided by 2.   ◇

Exercise 5.1.2. Consider the parabolic problem

$$\begin{cases} \dfrac{\partial u}{\partial t} - \mu \dfrac{\partial^2 u}{\partial x^2} = 0 & \text{in} \quad Q_T \equiv (-a,a) \times (t_0, T], \\[2mm] u(\pm a, t) = -R(t) \equiv \dfrac{1}{\sqrt{4\pi\mu t}} e^{-\frac{a^2}{4\mu t}} & \text{for} \quad t \in (t_0, T], \quad (5.14) \\[2mm] u(x, t_0) = \dfrac{1}{\sqrt{4\pi\mu t_0}} e^{-\frac{x^2}{4\mu t_0}} & \text{for} \equiv u_0(x) \quad x \in (-a,a) \end{cases}$$

where μ is a strictly positive constant and $0 < t_0 < T$.

1. Prove that the problem is well posed and provide an *a priori* estimate for the solution.
2. Verify that the solution is
$$u_{ex}(x,t) = \frac{1}{\sqrt{4\pi\mu t}} e^{-\frac{x^2}{4\mu t}}.$$
3. Discretize in space the problem with piecewise linear finite elements and in time with the Backward Euler scheme. Solve the problem for $a = 3$, $t_0 = 0.5$, $\mu = 0.1$, $T = 1$ with `fem1d` with the following discretization parameters in space and time: $(h = 0.05, \Delta t = 0.05)$, $(h = 0.025, \Delta t = 0.025)$ and $(h = 0.0125, \Delta t = 0.0125)$. Check the theoretical convergence estimates.

Solution 5.1.2.

Mathematical analysis

The differential equation in (5.14) is the well known heat or Fourier equation. A detailed analysis of the equation can be found in [Far93, Sal08] and [Eva10].

For the weak formulation, notice that we have non homogeneous Dirichlet conditions. Let us introduce a lifting of the boundary data. In particular we take $G(x,t) = R(t) = \dfrac{1}{\sqrt{4\pi\mu t}} e^{-a^2/(4\mu t)}$. Set $I \equiv (-a,a)$ and $V \equiv H_0^1(I)$.

The weak formulation of the problem, obtained in a standard way, reads: find $u \in G + V$ such that for any $t \in (t_0, T]$ and for any $v \in V$

$$\int_{-a}^{a} \frac{\partial u}{\partial t} v\, dx + \mu \int_{-a}^{a} \frac{\partial u}{\partial x} \frac{\partial v}{\partial x} dx = 0, \qquad (5.15)$$

with $u(x, t_0) = u_0(x)$.

Since v does not depend on t, we have

$$\frac{\partial}{\partial t}\left(\int\limits_{-a}^{a} uv\,dx\right) + \mu \int\limits_{-a}^{a} \frac{\partial u}{\partial x}\frac{\partial v}{\partial x}\,dx = 0.$$

After time integration in (t_0, T), we have

$$\left(\int\limits_{-a}^{a} uv\,dx\right)(T) + \mu \int\limits_{t_0}^{T}\int\limits_{-a}^{a} \frac{\partial u}{\partial x}\frac{\partial v}{\partial x}\,dx\,dt = \int\limits_{-a}^{a} u_0 v\,dx, \quad \forall v \in V. \tag{5.16}$$

Following an analysis similar to the previous exercise, we look for the solution u of (5.16) in $L^\infty(t_0, T; L^2(I)) \cap L^2(t_0, T; R + V)$. With this choice of the functional spaces, the formulation given above is well defined.

For the well posedness, we show that the bilinear form $a(u,v) \equiv \mu \int\limits_{-a}^{a} \frac{\partial u}{\partial x}\frac{\partial v}{\partial x}\,dx$ is coercive. Since we have Dirichlet boundary conditions, from the Poincaré inequality we have that $\|\partial v/\partial x\|^2_{L^2(I)}$ is equivalent to the norm $\|v\|^2_V$, for any $v \in V$. Therefore there exists a constant α such that

$$a(v,v) = \mu\left\|\frac{\partial v}{\partial x}\right\|^2_{L^2(I)} \geq \alpha\|v\|^2_V.$$

It is easy to verify that the bilinear form is also continuous and this leads to the well posedness. For the *a priori* bound, we write the solution as $u = \tilde{u} + R$, with $\tilde{u} \in V$, set $v = \tilde{u}$ in (5.15) and perform time integration. We obtain

$$\int\limits_{t_0}^{T}\int\limits_{-a}^{a} \frac{\partial \tilde{u}}{\partial t}\tilde{u}\,dx\,dt + \int\limits_{t_0}^{T}\mu\left\|\frac{\partial \tilde{u}}{\partial x}\right\|^2_{L^2(I)}dt = \int\limits_{t_0}^{T} \mathcal{F}(\tilde{u})\,dt,$$

with $\mathcal{F}_t(v) \equiv -\int\limits_{-a}^{a} \frac{dR}{dt}v\,dx$, being the lifting independent of x. Notice that

$$\int\limits_{t_0}^{T}\int\limits_{-a}^{a} \frac{\partial \tilde{u}}{\partial t}\tilde{u}\,dx\,dt = \frac{1}{2}\left(\|\tilde{u}\|^2_{L^2(I)}(T) - \|\tilde{u}_0\|^2_{L^2(I)}\right),$$

where $\tilde{u}_0 \equiv u_0 - R(t_0)$. Therefore we get

$$\left(\|\tilde{u}\|^2_{L^2(I)}(T) - \|\tilde{u}_{t_0}\|^2_{L^2(I)}\right) + 2\mu \int\limits_{t_0}^{T}\left\|\frac{\partial \tilde{u}}{\partial x}\right\|^2_{L^2(I)}dt = 2\int\limits_{t_0}^{T}\mathcal{F}_t(\tilde{u})\,dt.$$

From the definition of $\mathcal{F}$, the Cauchy-Schwarz and Young inequalities we obtain

$$2\left|\int_{t_0}^{T}\mathcal{F}_t(\tilde{u})dt\right| = 2\left|\int_{t_0}^{T}\int_{-a}^{a}\frac{dR}{dt}\tilde{u}dxdt\right| \le 2\int_{t_0}^{T}\left\|\frac{dR}{dt}\right\|_{L^2(I)}||\tilde{u}||_{L^2(I)} \le$$

$$\frac{1}{2\epsilon}\int_{t_0}^{T}\left\|\frac{dR}{dt}\right\|_{L^2(I)}^{2}dt + 2\epsilon\int_{t_0}^{T}||\tilde{u}||_{L^2(I)}^{2}dt \le \frac{1}{2\epsilon}\int_{t_0}^{T}\left\|\frac{dR}{dt}\right\|_{L^2(I)}^{2}dt + \epsilon\int_{t_0}^{T}||\tilde{u}||_{V}^{2}dt,$$

for an arbitrary $\epsilon > 0$. Moreover $||\frac{dR}{dt}||_{L^2} = 2a|\frac{dR}{dt}|$ (since R does not depend on x) and for $\epsilon = \alpha/2$, the coercivity of the bilinear form gives

$$||\tilde{u}||_{L^2(I)}^{2}(T) + \alpha\int_{t_0}^{T}||\tilde{u}||_{V}^{2}dt \le ||\tilde{u}_{t_0}||_{L^2(I)}^{2} + \frac{2a}{\alpha}\int_{t_0}^{T}(\frac{dR}{dt})^2dt. \qquad (5.17)$$

We can readily check that $u_{ex}(x,t) = e^{-x^2/(4\mu t)}/\sqrt{4\pi\mu t}$ is the exact solution, since this function is regular enough to be directly plugged into the strong formulation of the problem. Observe that

$$\frac{\partial u_{ex}}{\partial t} = \left(\frac{x^2}{4\mu t^2} - \frac{1}{2t}\right)u_{ex},$$

$$\frac{\partial^2 u_{ex}}{\partial x^2} = \left(\frac{x^2}{4\mu^2 t^2} - \frac{1}{2\mu t}\right)u_{ex}.$$

Therefore u_{ex} fulfills the equation. Fulfillment of initial and boundary conditions is immediate as well.

Numerical approximation

Discrete formulation of the problem is obtained by introducing a subdivision of the spatial domain. For simplicity, let the subdivision be uniform with step h. The finite dimensional subspace V_h is taken to be X_h^1, formerly introduced in (2.22). This is the space of piecewise linear functions vanishing at the end points $x = \pm a$. Let Δt be the time step. The implicit Euler discretization leads to the following problem: for $n \ge 0$ find the vector $\mathbf{U}^{n+1}$ s.t.

$$\left(\frac{1}{\Delta t}\mathrm{M} + \mathrm{K}\right)\mathbf{U}^{n+1} = \frac{1}{\Delta t}\mathrm{M}\mathbf{U}^{n} + \mathbf{F}^{n+1},$$

where M and K are the mass and stiffness matrices respectively, $\mathbf{U}^n$ is the vector of the nodal values of the solution, at time t^n and $\mathbf{F}^{n+1}$ is the vector with entries $\mathcal{F}_{t^{n+1}}(\varphi_i)$, being φ_i the i-th basis function of V_h.

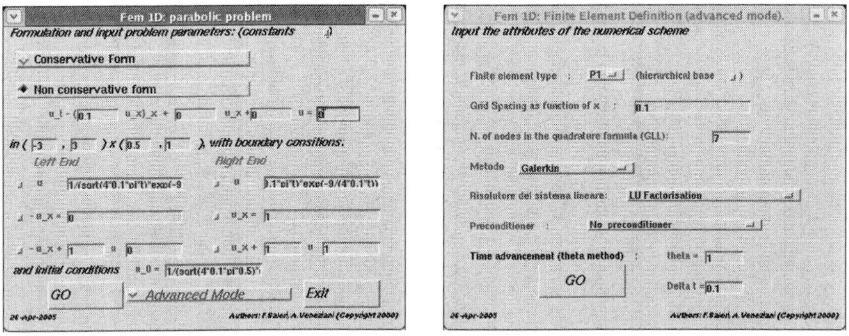

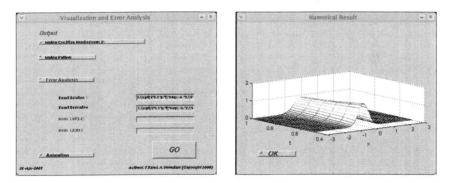

Fig. 5.1 Using `fem1d` for a parabolic problem: screen snapshots of: left, problem definition; right, numerical discretization parameters (in advanced mode)

Fig. 5.2 Using `fem1d` for a parabolic problem: screen snapshots of: left, post-processing specifications, right, solution (space-time representation)

To solve the problem with **fem1d**, we launch the code in MATLAB and select **Parabolic** at the first screenshot. The subsequent screenshot specifies the parabolic problem. We can specify either the conservative form, where the convective term is in the form $(\beta u)'$, or the non-conservative one if it is in the form $\beta u'$. This snapshot is similar to the one we have for elliptic problems. In this case we have to specify the space-time domain in the form

    (a, b) x (t0, T)

where a, b, t0, T are the end points of the domain and the instant and final instants. The initial condition must be specified on the bottom (see Fig. 5.1 on the left).

A button allows to specify whether the problem features constant (in time) coefficients and boundary conditions. When this button is on, the assembling of the matrix and all the terms depending on the boundary conditions is performed once at the beginning of the time loop, with a significant reduction of the computational effort.

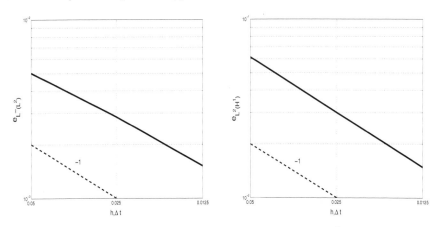

Fig. 5.3 Errors (logarithmic scale) in the norms $L^\infty((t_0,T);L^2(-3,3))$ (left) and $L^2((t_0,T);H^1(-3,3))$ (right) for the errors in Exercise 5.1.2. Dotted curves provide a reference line for a linear behavior of the error

After clicking **Go**, the second screenshot specifies the discretization parameters. If the option **Advanced Mode** is off, we can choose:

1. the degree of finite element (default is 1);
2. the space discretization parameter h (default is 0.1);
3. the parameter θ for the time advancing (θ method, default is 1);
4. the time step Δt.

When the **Advanced Mode** is on, it is possible to specify the stabilization method for possible advection/reaction dominated terms, the option for hyerarchical basis (for finite element with degree greater than 1), the number of nodes for the quadrature rule, the method used for solving the linear system (see Fig. 5.2 on the right). It is also possible to specify a non uniform grid as a function $h = h(x)$.

After clicking **Go**, the screenshot specifies the possible outputs of the computation: pattern of the matrix, condition number of the matrix and possibly - when the exact solution is available - the error. Natural norms for the error computation in a parabolic problem are in the space $L^\infty(t_0,T;L^2(a,b))$ and $L^2(t_0,T;H^1(a,b))$. The last option refers to the visualization mode. The option **Animation** will show the solution as a sequence of time step. When this option is off, solution is displayed as a surface in the plane x,t (see Fig. 5.2).

The results for the error computation are summarized in Fig. 5.3. We have taken $h = \Delta t$. The expected linear behavior is evident. It is possible to prove (see [Qua09]) that for any $n \geq 1$

$$||u_{ex}(t^n) - u_h^n||_{L^2(I)}^2 + 2\alpha\Delta t \sum_{k=1}^{n} ||u_{ex}(t^k) - u_h^k||_V^2 \leq C(\Delta t^2 + h^2).$$

As for elliptic problems, also in this case we point out that the error computation requires quadrature formulas accurate enough. For the time integration in the error computation, we have in this case only in the collocation points of the discretization. We approximate $L^2(0, T)$ norm by its discrete couterpart $||e(t)||_{L^2(0,T)} \approx ||\mathbf{e}||_{\triangle,2} = \left(\sum_{n=0}^{N} (e^n)^2 \Delta t\right)^{1/2}$. $\diamond$

Remark 5.1 Notice that for $t_0 \to 0$ in the last Exercise, we have

$$\lim_{t_0 \to 0} u(x, t_0) = \lim_{t_0 \to 0} \frac{1}{\sqrt{4\pi\mu t_0}} e^{-x^2/(4\mu t_0)} = \delta(0)$$

where δ is the *Dirac Delta* (centered at the origin). The solution is actually the weak solution to the problem: find u such that

$$\begin{cases} \dfrac{\partial u}{\partial t} - \mu\dfrac{\partial^2 u}{\partial x^2} = 0 & x \in (-\infty, \infty), t \in (0, T], \\[2mm] \lim_{a \to \infty} u(\pm a, t) = 0, & u(x, 0) = \delta(0). \end{cases}$$

This is called *fundamental solution* of the heat equation. The fundamental solution is the well known Gaussian function. This is widely used in the theory of probability. In fact, the heat equation is the continuous limit of the probabilistic model of the 1D Brownian motion(see [Sal08]).

Another remark refers to the regularity of the solution with respect to the initial condition. Parabolic problems have a smoothing action. In this case, a C^∞ solution is generated by initial data which are distributions (not even functions).

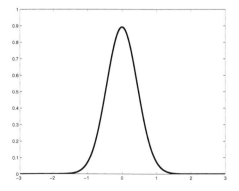

Fig. 5.4 Fundamental solution of the heat equation

Exercise 5.1.3. Consider the parabolic problem: find u such that

$$\begin{cases} \dfrac{\partial u}{\partial t} - \dfrac{\partial^2 u}{\partial x^2} = 0 & \text{in } Q_T \equiv (-3,3) \times (0,1], \\[2mm] u(\pm 3,t) = 0, \quad t \in (0,1], \quad u(x,0) = u_0(x) = \begin{cases} 1 & \text{for } x \in [-1,1], \\ 0 & \text{elsewhere.} \end{cases} \end{cases}$$

1. Give the weak formulation of the problem and prove that it is well posed, providing an *a priori* estimate of the solution.
2. Find the analytical solution.
3. Write the algebraic formulation of the problem discretized with linear finite elements in space and with the θ-method in time. For $\theta = 1$ prove the stability of the discrete solution.
4. Solve the problem with fem1D, for $\theta = 1$ (backward Euler) and $\theta = 0.5$ (Crank-Nicolson), with $\Delta t = 0.1$. Discuss the results.

Solution 5.1.3.

Mathematical analysis

Set $I \equiv (-3,3)$. Following the same procedure of the previous exercises, we have the weak formulation: for any $t > 0$ find $u \in H_0^1(I)$ such that

$$\int_{-3}^{3} \frac{\partial u}{\partial t} v \, dx + \int_{-3}^{3} \frac{\partial u}{\partial x} \frac{\partial v}{\partial x} dx = 0 \tag{5.18}$$

for any $v \in H_0^1(I)$. The solution fulfills the initial condition $u(x,0) = u_0(x)$.

After time integration we have:
find $u \in L^\infty(0,1;L^2(I)) \cap L^2(0,1;H_0^1(I))$ such that

$$\left(\int_{-3}^{3} uv\,dx \right)_{t=1} + \int_0^1 \int_{-3}^{3} \frac{\partial u}{\partial x} \frac{\partial v}{\partial x} dx\,dt = \int_{-3}^{3} u_0 v\,dx$$

for any $v \in H_0^1(I)$. The initial data belong to $L^2(I)$, so that the scalar product on the right hand side is well defined. A stability inequality is obtained by working as in the previous exercises, by taking $v = u$. In particular, notice that thanks to the Poincaré inequality we have

$$||u||^2_{L^2(I)}(1) + \alpha \int_0^1 ||u||^2_{H_0^1(I)}(t)dt \le ||u_0||^2_{L^2(I)} = 2,$$

where α is the coercivity constant of the bilinear form $a(u, v) \equiv \int\limits_{-3}^{3} \dfrac{\partial u}{\partial x} \dfrac{\partial v}{\partial x} dx$.

As noted previously, solution to parabolic problems smooths the initial data. In this case the initial data are $L^2(I)$, and the solution belongs to $L^2(0, T; H_0^1(I))$.

For the computation of the unique analytical solution, we follow a classical approach of *separation of variables*. Let us write the solution $u = TX$ where T depends only on t and X only on x. If we plug this representation of the solution into the equation and notice that $u \neq 0$ (so that $X \neq 0$ and $T \neq 0$), we get

$$\frac{1}{T} \frac{dT}{dt} = \frac{1}{X} \frac{d^2 X}{dx^2}$$

in Q_T. Since the left hand side depends only on t and the right hand one only on x, they are both constant. Let us start by solving the boundary value problem

$$\frac{d^2 X}{dx^2} = kX, \quad X(\pm 3) = 0,$$

with $k \in \mathbb{R}$. We look for non-trivial solution to this problem.

Assume for the moment $k < 0$; we write $k = -\lambda^2$ for the sake of notation. The ordinary differential equation

$$\frac{d^2 X}{dx^2} + \lambda^2 X = 0$$

admits the general solution (see e.g. [BD70])

$$X = C_1 \sin(\lambda x) + C_2 \cos(\lambda x),$$

where C_1 and C_2 are arbitrary constants specified by the boundary conditions. In particular, for $X(\pm 3) = 0$, we have

$$\begin{cases} C_1 \sin(3\lambda) + C_2 \cos(3\lambda) = 0 \\ -C_1 \sin(3\lambda) + C_2 \cos(3\lambda) = 0 \end{cases} \Rightarrow \begin{cases} 2C_1 \sin(3\lambda) = 0 \\ 2C_2 \cos(3\lambda) = 0 \end{cases}.$$

Therefore a possible nontrivial solution to the system is given by

$$C_2 = 0, \lambda = \frac{n\pi}{3} \Rightarrow X(x) = C_1 \sin\left(\frac{n\pi}{3} x\right)$$

with $n > 0$ integer. In this case we represent the space dependent part of the solution with sinusoidal functions[3]. Coefficient C_1 will be determined by the initial condition.

---

[3] Another (equivalent) possibility is a representation with cosine functions, by taking the solution $C_1 = 0$ and $3\lambda = (2n + 1)/2\pi$.

If we assume $k \geq 0$ it is promptly realized that the only particular solution fulfilling the boundary conditions is the trivial one, so we discard this option.

Let us consider the time dependent part, for $k = -\lambda^2 = -\left(\dfrac{n\pi}{3}\right)^2$. We have

$$\frac{dT}{dt} = kT \Rightarrow T(t) = T_0 e^{-\left(\frac{n\pi}{3}\right)^2 t}.$$

By collapsing the two arbitrary constants C_1 and T_0 and putting in evidence the arbitrary index n we have the solution

$$u_n(x,t) = A_n e^{-\left(\frac{n\pi}{3}\right)^2 t} \sin\left(\frac{n\pi}{3}x\right).$$

To fulfill the initial condition of the problem, we advocate the principle of effects superposition (the problem is linear) and write

$$u(x,t) = \sum_{n=1}^{\infty} u_n(x,t) = \sum_{n=1}^{\infty} A_n e^{-\left(\frac{n\pi}{3}\right)^2 t} \sin\left(\frac{n\pi}{3}x\right).$$

The arbitrary constants A_n identify the unique solution of the problem and can be computed by expanding the initial condition

$$u_0(x) = \begin{cases} 1 & \text{for} \quad x \in [-1,1], \\ 0 & \text{elsewhere}, \end{cases}$$

with the Fourier series s.t.

$$u_0(x) = \sum_{n=1}^{\infty} A_n \sin\left(\frac{n\pi}{3}x\right) = u(x,0).$$

Since the sinusoidal functions (see e.g. [Far93]) are an orthogonal set with respect to L^2, we have that for j integer

$$\int_{-3}^{3} u_0 \sin\left(\frac{j\pi}{3}x\right) dx = A_j \int_{-3}^{3} \sin^2\left(\frac{j\pi}{3}x\right) dx \Rightarrow A_j = \frac{\displaystyle\int_{-1}^{1} \sin\left(\frac{j\pi}{3}x\right) dx}{\displaystyle\int_{-3}^{3} \sin^2\left(\frac{j\pi}{3}x\right) dx},$$

so that $A_j = -\dfrac{2}{n\pi} \cos\left(\dfrac{n\pi}{3}\right)$. The analytical solution is completely determined.

Numerical approximation

The semi-discrete formulation (discrete in space, continuous in time) with finite-elements is obtained after introducing a mesh for I with size h (constant). We set $V_h \equiv X_h^1$, space of piecewise linear functions vanishing on the boundary. The semi-discrete solution reads $u_h(x,t) = \sum_{j=1}^{N_h} U_j(t)\varphi_j(x)$. Functions $\{\varphi_j\}$ are a basis set of V_h. Let $\mathbf{U}(t)$ be the vector of components $U_j(t)$, the semi-discrete problem is obtained from (5.18), taking functions φ_i for $i = 1, \ldots, N_h$ as test functions. We have the following system of ordinary differential equations,

$$\mathrm{M}\frac{d\mathbf{U}}{dt} + \mathrm{K}\mathbf{U} = \mathbf{0},$$

where M is the mass matrix, K is the stiffness matrix and $\mathbf{U}(0) = \mathbf{U_0}$ is the vector with entries $U_{0,j} = u_{0h}(x_j)$, being u_{0h} a suitable approximation of the initial data u_0 in V_h. For instance we can choose u_{0h} to be the projection (in L^2) on V_h, solving the problem

$$\int_{-3}^{3} u_{0h}\varphi_j dx = \int_{-3}^{3} u_0\varphi_j dx, \quad j = 1, \ldots N_h.$$

Another option is to take the piecewise linear interpolation of u_0 at the mesh nodes.

For the time discretization we introduce the time step Δt and collocate the semi-discrete problem in the nodes $t^k = k\Delta t$, with $t^N = T = 1$. We discretize the problem with the θ-method which leads to equation

$$\frac{1}{\Delta t}\mathrm{M}\mathbf{U}^{k+1} + \theta\mathrm{K}\mathbf{U}^{k+1} = \frac{1}{\Delta t}\mathrm{M}\mathbf{U}^k - (1-\theta)\mathrm{K}\mathbf{U}^k \qquad (5.19)$$

for $k = 0, 1, \ldots, N-1$. For $\theta = 1$ (backward Euler) at each step we have the system

$$\mathrm{M}\left(\mathbf{U}^{k+1} - \mathbf{U}^k\right) + \Delta t\mathrm{K}\mathbf{U}^{k+1} = \mathbf{0}.$$

To obtain a stability *a priori* estimate for the solution we multiply the latter equation by $\mathbf{U}^{k+1}$, and obtain

$$\left(\mathbf{U}^{k+1}\right)^T \mathrm{M}\left(\mathbf{U}^{k+1} - \mathbf{U}^k\right) + \Delta t\left(\mathbf{U}^{k+1}\right)^T \mathrm{K}\mathbf{U}^{k+1} = 0. \qquad (5.20)$$

The following identity holds[4]

$$\left(\mathbf{U}^{k+1}\right)^T \mathrm{M}\left(\mathbf{U}^{k+1} - \mathbf{U}^k\right) = \qquad (5.21)$$

$$\frac{1}{2}\left(\mathbf{U}^{k+1}\right)^T \mathrm{M}\mathbf{U}^{k+1} + \frac{1}{2}\left(\mathbf{U}^{k+1} - \mathbf{U}^k\right)^T \mathrm{M}\left(\mathbf{U}^{k+1} - \mathbf{U}^k\right) - \frac{1}{2}\left(\mathbf{U}^k\right)^T \mathrm{M}\mathbf{U}^k.$$

Both M and K are s.p.d. matrices so that

$$\left(\mathbf{U}^{k+1} - \mathbf{U}^k\right)^T \mathrm{M}\left(\mathbf{U}^{k+1} - \mathbf{U}^k\right) \geq 0.$$

Equality holds only for $\mathbf{U}^{k+1} = \mathbf{U}^k$, which is the steady state.

From (5.20) and (5.21) we have

$$\left(\mathbf{U}^{k+1}\right)^T \mathrm{M}\mathbf{U}^{k+1} \leq \left(\mathbf{U}^k\right)^T \mathrm{M}\mathbf{U}^k.$$

---

[4] In general, we have $(a - b, a) = 1/2(a - b, a) + 1/2(a - b, a - b) + 1/2(a - b, b) = 1/2(a, a) + 1/2(a - b, a - b) - 1/2(b, b)$.

On the other hand for any k, $\left(\mathbf{U}^k\right)^T \mathbf{M}\mathbf{U}^k = ||u_h^k||^2_{L^2(I)}$ where u_h^k denotes the approximation of u_h at time $t = t^k$.

We obtain therefore

$$||u_h^{k+1}||^2_{L^2(I)} \leq ||u_h^k||^2_{L^2(I)} \quad \forall k = 0, 1, \ldots, N - 1. \tag{5.22}$$

This estimate holds with no restrictions on the time step, so this is an *unconditional stability*.

Norm in L^2 of $\partial u_h / \partial x$ is equivalent to the H^1 norm of u_h, thanks to the Poincaré inequality. Therefore we can set $C||u_h^k||^2_{H^1(I)} \leq ||\dfrac{du_h^k}{dx}||^2_{L^2(I)} = \left(\mathbf{U}^{k+1}\right)^T \mathbf{K}\mathbf{U}^{k+1}$, where C is constant. From (5.20) and (5.21) it follows that

$$C\Delta t ||u_h^{k+1}||^2_{H^1(I)} = \frac{1}{2}||u_h^k||^2_{L^2(I)} - \frac{1}{2}||u_h^{k+1} - u_h^k||^2_{L^2(I)} - \frac{1}{2}||u_h^{k+1}||^2_{L^2(I)}$$

$$\leq \frac{1}{2}||u_h^k||^2_{L^2(I)},$$

that exploiting (5.22) to the initial step gives

$$||u_h^{k+1}||_{H^1(I)} \leq \sqrt{\frac{1}{2C\Delta t}}||u_{0h}||_{L^2(I)}. \tag{5.23}$$

For a fixed Δt, H^1 norm of the discrete solution at the generic instant t^{k+1} with the implicit Euler method is bounded by the L^2 norm of the initial data. It is reasonable to assume that the initial data are approximated in such a way that $||u_{0h}||_{L^2(I)} \leq c||u_0||_{L^2(I)}$ for a constant $c > 0$. In particular, this holds for the L^2 projection advocated before.

In general for $\theta \in [0.5, 1]$ we have unconditional stability. We give a proof based on algebraic considerations. Let $(\lambda_i, \mathbf{w}_i)$ be the solution to the generalized eigenvalue problem

$$\mathbf{K}\mathbf{w}_i = \lambda_i \mathbf{M}\mathbf{w}_i, \tag{5.24}$$

(λ_i are the eigenvalues of $\mathbf{M}^{-1}\mathbf{K}$). Since K and M are s.p.d., λ_i are real and positive for $i = 1, \ldots, N_h$ and the eigenvectors $\mathbf{w}_i$ are linearly independent and M-orthonormal, i.e. $\mathbf{w}_j^T \mathbf{M}\mathbf{w}_i = \delta_{ij}$, where δ_{ij} is the Kronecker symbol.

A generic vector of $\mathbb{R}^{N_h}$ can be written as

$$\mathbf{U}^k = \sum_{i=1}^{N_h} \gamma_i^k \mathbf{w}_i.$$

Multiplying (5.19) by $\mathbf{w}_j^T$, we obtain

$$\sum_{i=1}^{N_h} \mathbf{w}_j^T M \gamma_i^{k+1} \mathbf{w}_i + \Delta t \theta \sum_{i=1}^{N_h} \mathbf{w}_j^T K \gamma_i^{k+1} \mathbf{w}_i =$$

$$\sum_{i=1}^{N_h} \mathbf{w}_j^T M \gamma_i^k \mathbf{w}_i - \Delta t (1 - \theta) \sum_{i=1}^{N_h} \mathbf{w}_j^T K \gamma_i^k \mathbf{w}_i.$$

Thanks to the fact that the eigenvectors are M-orthonormal, we have

$$(1 + \Delta t \theta \lambda_j) \gamma_j^{k+1} = (1 - (1 - \theta) \Delta t \lambda_j) \gamma_j^k,$$

so that

$$\gamma_j^{k+1} = \sigma_j \gamma_j^k, \quad j = 1, \dots N_h, \tag{5.25}$$

with $\sigma_j = \dfrac{1 - (1 - \theta) \Delta t \lambda_j}{1 + \theta \Delta t \lambda_j}$. For $|\sigma_j| < 1$, $\forall j = 1, \dots N_h$ the method is absolutely stable, i.e.

$$-1 - \theta \Delta t \lambda_j < 1 - (1 - \theta) \Delta t \lambda_j < 1 + \theta \Delta t \lambda_j, \quad j = 1, \dots N_h.$$

Inequality on the right is always fulfilled thanks to the fact that $\lambda_j > 0$; inequality on the left gives the condition

$$\theta > \frac{1}{2} - \frac{1}{\lambda_j \Delta t},$$

which is clearly true for any j for $\theta \geq 0.5$. We illustrate the value of $|\sigma|$ for different values of $\Delta t \lambda$ in Fig. 5.5. On the left, we depict $|\sigma|$ as a function of θ. On the right, we illustrate $|\sigma(1)|$ (Backward Euler) and $|\sigma(1/2)|$ (Crank-Nicolson) as a function of $\lambda \Delta t$.

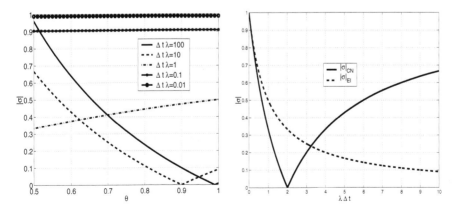

Fig. 5.5 Coefficient $|\sigma|$ function of θ for different values of the time step (left) and function of Δt, for the two cases of Backward Euler ($\theta = 1$) and Crank-Nicolson ($\theta = 0.5$)

In particular, the Crank-Nicolson method ($\theta = 0.5$) is unconditionally stable and the (5.25) implies

$$\left(\mathbf{U}^{k+1}\right)^{T} K \mathbf{U}^{k+1} = \sum_{i,j} \gamma_i^{k+1} \gamma_j^{k+1} \mathbf{w}_i^{T} K \mathbf{w}_j =$$

$$\sum_{i} \left(\gamma_i^{k+1}\right)^2 \lambda_i = \sum_{i} \sigma \left(\gamma_i^{k}\right)^2 \lambda_i \leq \left(\mathbf{U}^{k}\right)^{T} K \mathbf{U}^{k}.$$

From this we have

$$||u_h^{k+1}||_{H^1(I)} \leq ||u_h^k||_{H^1(I)} \Rightarrow ||u_h^k||_{H^1(I)} \leq ||u_{0h}||_{H^1(I)}. \qquad (5.26)$$

If u_0 is continuous in $[-3, 3]$ and $V_h \subset X_h^r$ is given by piecewise polynomial functions, we can take $u_{0h} = \Pi_h^r u_0$, where Π_h^r is the interpolation operator defined in Section 2.1. In this case the (2.4) leads to

$$||u_h^k||_{H^1(I)} \leq ||u_0||_{H^1(I)}$$

and the discrete solution is actually bounded in the H^1 norm by the initial data.

If u_0 is only in $L^2(I)$ (but not in $H^1(I)$) we have to select a stable operator $L^2 \to V_h$, such that

$$||u_{0h}||_{H^1(I)} \leq C||u_0||_{L^2(I)}, \qquad (5.27)$$

with $C > 0$ (for instance, by using a projection operator [EG04]). If this is not possible, inequality (5.26) can become meaningless, as we will see later.

Numerical results

In Figs. 5.6 and 5.7 we report the numerical solutions computed with `fem1D`.

Solution obtained with the backward Euler method shows that the parabolic problem actually features more regular solutions than the initial data

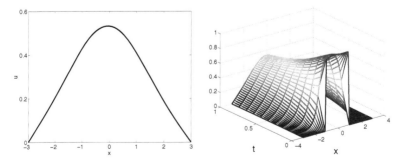

Fig. 5.6 Solution at time $t = 1$ (left) and time evolution (right) for the problem of Exercise 5.1.3 with discontinuous data with the Backward Euler method ($h = 0.1$)

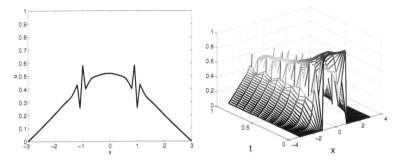

Fig. 5.7 Solution at time $t = 1$ (left) and time evolution (right) for the problem of Exercise 5.1.3 with discontinuous data with the Crank Nicolson method ($h = 0.1$). Discontinuities induce oscillations

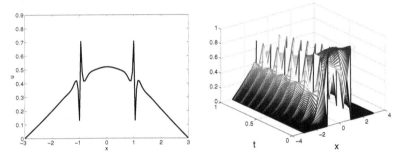

Fig. 5.8 Solution at time $t = 1$ (left) and time evolution (right) for the problem of Exercise 5.1.3 with discontinuous data with the Crank Nicolson method ($h = 0.05$). Oscillations in the initial instants of the simulation are more evident when the step h gets smaller

(discontinuous). However the backward Euler method smooths the solution with the numerical dissipation, with a low order accuracy. The Crank-Nicolson method is more accurate in time (order 2) and less dissipative, so the initial discontinuities produce oscillations that are not dissipated immediately. In this case the initial data belong to $L^2(I)$, and not to $H^1(I)$. Femfd approximates u_0 with

$$u_{0h}(x) = \begin{cases} 0 & \text{per } x \leq -1 - h, \\ \dfrac{x + 1 + h}{h} & \text{per } -1 - h < x \leq -1, \\ 1 & \text{per } -1 < x \leq 1, \\ \dfrac{1 + h - x}{h} & \text{per } 1 < x \leq 1 + h, \\ 0 & \text{per } x > 1 + h. \end{cases}$$

In particular by computing the L^2 norm of u_{0h} we have $||u_{0h}||_{H^1(I)} = \sqrt{2/h}$, that blows up for $h \to 0$, while $||u_0||_{L^2(I)} = \sqrt{2}$. This approximation in V_h does not fulfill (5.27) and the stability estimate of the numerical solution in H^1 is meaningless for $h \to 0$. This results in numerical oscillations in the solution (see Fig. 5.7), which are more evident when the discretization step gets smaller (Fig. 5.8).

In practice when the initial data are not regular and it is not possible nor convenient to find a stable approximation of u_0 in V_h, it is useful to select θ slightly greater than 0.5. This introduces more dissipation dumping the oscillations and maintains an accuracy of almost second order. ◇

Exercise 5.1.4. Consider the initial-boundary value problem

$$\begin{cases} \dfrac{\partial u}{\partial t} - \dfrac{\partial^2 u}{\partial x^2} + u = 0 & x \in (0,1), t \in (0,1), \\ u(0,t) = 0, \quad u(1,t) = e^{-t}, \\ u(x,0) = x. \end{cases}$$

1. Find the analytical solution.
2. Solve the problem with `fem1d`, piecewise linear elements, Backward Euler and Crank-Nicolson time discretizations. Verify the order of convergence. Take $h = 0.1$, $\Delta t = 0.1, 0.05, 0.025$.
3. Discuss the impact of the space discretization error on the global error.

Solution 5.1.4.

Mathematical analysis

By standard arguments on the weak form it is readily proved that the (weak formulation of the) problem is well posed. We find here directly the analytical solution.

First of all, notice that if we introduce the auxiliary variable $w(x,t) = e^t u(x,t)$, we have

$$\frac{\partial w}{\partial t} = e^t \left(\frac{\partial u}{\partial t} + u \right), \qquad \frac{\partial^2 w}{\partial x^2} = e^t \frac{\partial^2 u}{\partial x^2},$$

so that w solves the problem

$$\begin{cases} \dfrac{\partial w}{\partial t} - \dfrac{\partial^2 w}{\partial x^2} = 0 & x \in (0,1), t \in (0,1), \\ w(0,t) = 0, \quad w(1,t) = 1, \\ w(x,0) = x. \end{cases}$$

Let $w^* = w - x$. The problem for w^* reads

$$\begin{cases} \dfrac{\partial w^*}{\partial t} - \dfrac{\partial^2 w^*}{\partial x^2} = 0 & x \in (0,1), t \in (0,1), \\ w(0,t) = 0, \quad w(1,t) = 0, \\ w(x,0) = 0. \end{cases}$$

The unique solution to this problem is clearly the trivial one, so that $w = x$ and consequently the unique solution to the original problem reads

$$u(x,t) = e^{-t}w(x,t) = e^{-t}x.$$

Numerical approximation

Galerkin approximation and time discretization are performed in a way similar to the previous exercises and we do not need specific considerations.

Numerical results

Results with fem1D in the Tables 5.1 and 5.2. Backward Euler is first order accurate in time. This is evident in the Table 5.1 since the error is halved when the time step is halved. The error is reported in the natural norms of parabolic problems defined in (1.13).

Table 5.1 Backward Euler errors

Δt	0.1	0.05	0.025
$L^\infty(0,1;L^2(0,1))$	0.0015	0.00077	0.00039
$L^2(0,1;H^1(0,1))$	0.004	0.0021	0.00107

Table 5.2 Crank-Nicolson errors

Δt	0.1	0.05	0.025
$L^\infty(0,1;L^2(0,1))$	2.77×10^{-5}	6.83×10^{-6}	1.7×10^{-6}
$L^2(0,1;H^1(0,1))$	7.5×10^{-5}	1.8×10^{-5}	4.5×10^{-6}

Crank-Nicolson is second order accurate, as it is evident from Table 5.2.

In general the error is determined by both time and space discretizations. For this reason, a change of one of the discretization parameters (h or Δt) might hide the order of convergence. In the present exercise, h has been kept constant. While in general this can prevent to observe the linear convergence in time, in this exercise this does not occur, because the exact solution, for a fixed $t > 0$ is linear in space, so it belongs to the finite element space.   ◇

Exercise 5.1.5. Find $u(x,t)$ with $0 \le x \le 1$, $t \ge 0$, s.t.

$$
\begin{cases}
\dfrac{\partial u}{\partial t} - \dfrac{\partial}{\partial x}\left(\alpha \dfrac{\partial u}{\partial x}\right) + \dfrac{\partial}{\partial x}(\beta u) + \gamma u = 0, & x \in (0,1), \quad 0 < t < T, \\
u(0,t) = 0 & 0 < t \le T, \\
\alpha \dfrac{\partial u}{\partial x}(1,t) + \delta u(1,t) = 0 & 0 < t \le T, \\
u(x,0) = u_0(x) & 0 < x < 1,
\end{cases}
$$

where α, γ, β, δ are given functions of $x \in (0,1)$.

1. Write the weak formulation.
2. Assuming that:

   a. α_0, α_1, β_1 are positive constants such that $\alpha_0 \leq \alpha(x) \leq \alpha_1$ and $\beta(x) \leq \beta_1$ for any $x \in (0,1)$;

   b. $\dfrac{1}{2}\dfrac{d\beta}{dx}(x) + \gamma(x) \geq 0$ for any $x \in (0,1)$,

   prove that the problem is well posed (under further suitable assumptions) and give an *a priori* estimate of the solution.
3. Repeat the provious point taking $u = g$ on $x = 0$ and $0 < t < T$.
4. Write the finite element space discretization of the problem and prove its stability.
5. Write the time discretization of the semi-discrete problem with the Backward Euler method and prove its unconditional stability.

Solution 5.1.5.

Mathematical analysis

Let V be the space $\{v \in H^1(0,1), v(0) = 0\}$. We multiply the equation by a generic $v \in V$, integrate by parts the term with the second derivative and include the boundary conditions. We obtain the problem: for $t > 0$ find $u \in V$ s.t.

$$\int_0^1 \frac{\partial u}{\partial t} v \, dx + a(u, v) = 0, \quad \forall v \in V, \tag{5.28}$$

with

$$a(u, v) \equiv \int_0^1 \alpha \frac{\partial u}{\partial x}\frac{\partial v}{\partial x} dx + (\delta uv)|_{x=1} + \int_0^1 \frac{\partial \beta}{\partial x} uv \, dx + \int_0^1 \beta \frac{\partial u}{\partial x} v \, dx + \int_0^1 \gamma uv \, dx$$

and initial condition $u(x,0) = u_0(x)$.

By assumption α and β are bounded. Let us assume that $\dfrac{\partial \beta}{\partial x}$ and γ are bounded too[5]. Under these assumptions, the bilinear form of the problem is continuous. As observed in Chapter 4, these assumptions can be relaxed.

---

[5] Notice that we could integrate by parts also the convective term $\displaystyle\int_0^1 \frac{\partial(\beta u)}{\partial x} v \, dx = [\beta uv]_0^1 -$

$\displaystyle\int_0^1 \beta u \frac{\partial v}{\partial x} dx$. This leads to a different weak formulation with different natural boundary conditions and no assumption on $\dfrac{\partial \beta u}{\partial x}$ is required. The analysis of this case is left to the reader.

As a matter of fact, Sobolev Theorem (see Section 1.3) states that a single variable function in $H^1(0,1)$ is continuous and therefore bounded in $[0,1]$. Consequently, the integral $\int_0^1 \frac{\partial \beta}{\partial x} uv dx$ is well defined when $\partial \beta / \partial x \in L^1(0,1)$, since

$$|\int_0^1 \frac{\partial \beta}{\partial x} uv dx| \leq ||\frac{\partial \beta}{\partial x}||_{L^1(0,1)}||u||_{L^\infty(0,1)}||v||_{L^\infty(0,1)} \leq$$

$$C_I ||\frac{\partial \beta}{\partial x}||_{L^1(0,1)}||u||_{H^1(0,1)}||v||_{H^1(0,1)},$$

where C_I is the constant of the embedding $H^1(0,1) \hookrightarrow L^\infty(0,1)$ (see Ch. 2). Likewise, for the continuity of the bilinear form it is enough to assume that $\gamma \in L^1(0,1)$.

Notice that

$$\int_0^1 \beta \frac{\partial u}{\partial x} u dx = \frac{1}{2} \int_0^1 \beta \frac{\partial (u^2)}{\partial x} dx = \frac{1}{2} [\beta u^2]_0^1 - \frac{1}{2} \int_0^1 \frac{\partial \beta}{\partial x} u^2 dx,$$

so that

$$\int_0^1 \frac{\partial \beta}{\partial x} u^2 dx + \int_0^1 \beta \frac{\partial u}{\partial x} u dx = \frac{1}{2} \int_0^1 \frac{\partial \beta}{\partial x} u^2 dx + \frac{1}{2} [\beta u^2]_0^1,$$

and

$$a(u,u) = \int_0^1 \alpha \left(\frac{\partial u}{\partial x}\right)^2 dx + \int_0^1 \left(\frac{1}{2}\frac{\partial \beta}{\partial x} + \gamma\right) u^2 dx + \left(\frac{\beta(1)}{2} + \delta(1)\right) u^2(1,t).$$

Thanks to the Poincaré inequality (remind that on the leftmost end point we have a Dirichlet condition) and the hypothesis $\alpha \geq \alpha_0$, the first term on the right hand side is such that

$$\int_0^1 \alpha \left(\frac{\partial u}{\partial x}\right)^2 dx \geq \alpha_0 C ||u||_V^2.$$

For the assumptions of the problem, also the second term on the right hand side is non-negative. If we assume

$$\frac{\beta(1)}{2} + \delta(1) \geq 0$$

standard stability is proved. For $\delta = 0$, Robin condition reduces to a Neumann condition and the stability is proved if $\beta(1) \geq 0$. The latter assumption means that the Neumann condition is prescribed on the "outflow" boundary.

If $\left(\dfrac{\beta(1)}{2} + \delta(1)\right) < 0$ we may still get a stability result advocating the trace inequality $u(1) \leq C_T ||u||_{H^1(0,1)}$, for a $C_T > 0$ (Chapter 1) if we have that

$$\alpha_0 + \left(\dfrac{\beta(1)}{2} + \delta(1)\right) C_T^2 > 0.$$

In any case, by setting μ equal to the coercivity constant, stability of the solution follows from taking $v = u$, by which

$$\int_0^1 \dfrac{\partial u}{\partial t} u\, dx + a\, (u, u) = \dfrac{1}{2}\dfrac{d}{dt}||u||_{L^2(0,1)}^2 + a\,(u, u).$$

Upon time integration and the application of the coercivity inequality, we obtain

$$||u||_{L^2(0,1)}^2(T) + \dfrac{\alpha_0}{2}\int_0^T ||u||_{H^1(0,1)}^2\, dt \leq ||u_0||_{L^2(0,1)}^2, \qquad (5.29)$$

proving that the solution belongs to $L^\infty(0, T; L^2(0, 1)) \cap L^2(0, T; H^1(0, 1))$ for initial data in $L^2(0, 1)$.

For a non-homogeneous Dirichlet condition, we set $u = \tilde{u} + G$ where $G \in H^1(0, 1)$ is a lifting of the boundary condition, i.e. $G(0) = g$. The problem reads: for any $t > 0$ find $u \in G + V$ s.t.

$$\int_0^1 \dfrac{\partial u}{\partial t} v\, dx + \int_0^1 \alpha \dfrac{\partial u}{\partial x}\dfrac{\partial v}{\partial x}\, dx - \left[\alpha \dfrac{\partial u}{\partial x} v\right]_0^1 + \int_0^1 \dfrac{\partial \beta u}{\partial x} v\, dx + \int_0^1 \gamma u v\, dx = 0,$$

$\forall v \in V$, with $u(x, 0) = u_0(x)$. For $\mathcal{F}(v) = -\displaystyle\int_0^1 \dfrac{\partial G}{\partial t} v\, dx - a\,(G, v)$, inequality

(5.29) modifies in

$$||\tilde{u}||_{L^2(0,1)}^2(T) + \dfrac{\alpha_0}{2}\int_0^T ||\tilde{u}||_{H^1(0,1)}^2\, dt \leq ||\tilde{u}_0||_{L^2(0,1)}^2 + \dfrac{1}{\alpha_0}\int_0^T ||\mathcal{F}||_{V'}^2\, dt.$$

Finally, we may notice that if assumption (b) is removed, well posedness can be proved by showing that the bilinear form $a(u, v)$ is weakly coercive.

Numerical approximation

Given the discretization parameter h, we set $V_h = \{v_h \in X_h^r(0, 1)|\ v_h(0) = 0\}$. Let N_h be the dimension of V_h and $\{\varphi_k,\ k = 1, \ldots, N_h\}$ the associated Lagrangian basis. The semi-discrete form of the problem reads: for any $t \in$

$(0, T)$, find $u_h \in V_h$ s.t.

$$\int_0^1 \frac{\partial u_h}{\partial t} \varphi_k dx + a(u_h, \varphi_k) = 0, \quad k = 1, \dots, N_h,$$

with initial condition $u_h(x, 0) = u_{0h}(x)$, finite-dimensional approximation of the initial data.

Stability of the semi-discrete problem directly inherits from the stability of the continuous one, since continuity and coercivity trivially holds in the subspace V_h,

$$||u_h||^2_{L^2(0,1)}(T) + 2\alpha_0 \int_0^T ||u_h||^2_{H^1(0,1)} dt \leq ||u_{0h}||^2_{L^2(0,1)}. \tag{5.30}$$

Algebraic form of the semi-discrete problem reads

$$M\frac{dU}{dt} + AU = F(t), \tag{5.31}$$

where U is the nodal values vector, M the mass matrix, A the positive (non symmetric) matrix corresponding to the bilinear form $a(\cdot, \cdot)$, F the vector including the non-homogeneous boundary terms.

We split the time interval into sub-intervals with constant step $\Delta t > 0$ and collocate the ordinary differential system (5.31) in $t^n = n\Delta t$. Backward Euler discretization reads, for $n \geq 0$,

$$\frac{1}{\Delta t} M(U^{n+1} - U^n) + AU^{n+1} = F(t^{n+1}).$$

Here U_0 is the nodal values vector of the initial condition u_{0h}.

In the stability analysis we assume that the forcing term is zero. We have

$$U^{n+1} = (I + \Delta t M^{-1}A)^{-1} U^n.$$

Eigenvalues of $A = (I + \Delta t M^{-1}A)^{-1}$ read

$$\mu_i = \frac{1}{1 + \Delta t \lambda_i}, \quad i = 1, \dots, N_h,$$

where λ_i are the eigenvalues of $M^{-1}A$, so they have positive real part since M and A are positive (A is non-symmetric, so in general it has complex eigenvalues). It follows that

$$|\mu_i| = \frac{1}{\sqrt{(1 + \Delta t \operatorname{Re}\lambda_i)^2 + (\Delta t \operatorname{Im}\lambda_i)^2}} < 1, \quad \forall i.$$

Consequently, the spectral radius of A is < 1 for any $\Delta t > 0$, which proves the unconditional stability of the method. $\diamond$

Exercise 5.1.6. Find $u(x,t)$ s.t.

$$
\begin{cases}
\dfrac{\partial u}{\partial t} - \dfrac{1}{\pi^2}\dfrac{\partial^2 u}{\partial x^2} - 3u = 0 & x \in (0,1), t \in (0,5], \\
u(0,t) = 0, \qquad u(1,t) = 0, & t \in (0,5], \\
u(x,0) = u_0(x) = \sin(2\pi x).
\end{cases}
\tag{5.32}
$$

1. Prove that the problem is well posed.
2. Find the exact solution.
3. Solve the problem with `fem1d`. More precisely, with linear finite elements for $h = 0.2$, and for $\Delta t = 0.05$ on the interval $0 < t < 5$, verify that the Forward Euler scheme gives a stable solution. Then solve the problem with $h = 0.1$, $\Delta t = 0.025$. Is the Forward Euler solution stable? Find a value of Δt that guarantees a stable solution and discuss the result.
4. For $h = 0.01$, and $\Delta t = 0.1, 0.05, 0.025$, check out the convergence order of Backward Euler and Crank-Nicolson time discretizations.

Solution 5.1.6.

Mathematical analysis

Well posedness analysis of the problem is standard. Since we have Dirichlet boundary conditions on both the end points, the space for the test functions will be $V \equiv H_0^1(0,1)$.

The weak formulation is obtained in the standard way too: for any $t \in (0,5]$, find $u \in L^2(0,5;V)$ such that for any $v \in V$

$$
\int_0^1 \frac{\partial u}{\partial t}v\,dx + a(u,v) = 0,
\tag{5.33}
$$

with $a(u,v) \equiv \frac{1}{\pi^2}\int_0^1 \frac{\partial u}{\partial x}\frac{\partial v}{\partial x}dx - 3\int_0^1 uv\,dx$ and with $u(x,0) = \sin(2\pi x)$. To prove that the problem is well posed, we show that the bilinear form $a(u,v)$ is continuous and (weakly) coercive. Continuity follows immediately from the continuity of the integrals, since

$$
\left| \frac{1}{\pi^2}\int_0^1 \frac{\partial u}{\partial x}\frac{\partial v}{\partial x}dx - 3\int_0^1 uv\,dx \right| \le 3\|u\|_V\|v\|_V.
$$

Weak coercivity is proved by observing that

$$
a(u,u) = \frac{1}{\pi^2}\int_0^1 \left(\frac{\partial u}{\partial x}\right)^2 dx - 3\|u\|^2_{L^2(0,1)}.
$$

Since the Poincaré inequality holds, there exists a constant $C > 0$ such that

$$a\,(u,u) \geq \frac{C}{\pi^2}||u||_V^2 - 3||u||_{L^2}^2,$$

that proves the weak coercivity (5.2), with $\lambda = 3$ and $\alpha_0 = C/\pi^2$. Problem (5.33) is thus well posed. An *a priori* bound for the solution is obtained by integrating (5.33) over $(0,T)$, with $T \in (0,5]$ and choosing u as test function. We have

$$\int_0^T \int_0^1 \frac{\partial u}{\partial t} u\,dx\,dt + \frac{1}{\pi^2} \int_0^T \int_0^1 \left(\frac{\partial u}{\partial x}\right)^2 dx\,dt = 3 \int_0^T \int_0^1 u^2 dx\,dt. \qquad (5.34)$$

The first term becomes

$$\int_0^T \int_0^1 \frac{\partial u}{\partial t} u\,dx\,dt = \frac{1}{2} \int_0^T \frac{d}{dt} \int_0^1 u^2 dx\,dt = \frac{1}{2} \int_0^T \frac{d}{dt} ||u||_{L^2(0,1)}^2 (t)dt =$$

$$\frac{1}{2}||u||_{L^2(0,1)}^2(T) - \frac{1}{2}||u_0||_{L^2(0,1)}^2.$$

Thanks to the Poincaré inequality, the fact that $||u_0||_{L^2(0,1)}^2 = 1/2$, and the identity $u\frac{\partial u}{\partial x} = \frac{1}{2}\frac{\partial u^2}{\partial x}$, (5.34) becomes

$$||u||_{L^2(0,1)}^2(T) + \frac{2C}{\pi^2} \int_0^T ||u||_V^2 dt \leq \frac{1}{2} + 6 \int_0^T ||u||_{L^2(0,1)}^2 dt. \qquad (5.35)$$

The application of Gronwall's Lemma gives

$$||u||_{L^2(0,1)}^2(t) \leq \frac{1}{2}e^{6t}.$$

From this estimate we have $u \in L^\infty(0,5; L^2(0,1))$. Thanks to this bound in (5.35) we have

$$\frac{2C}{\pi^2} \int_0^T ||u||_V^2 dt \leq \frac{1}{2} + 3 \int_0^T e^{6t} dt,$$

that proves that the solution belongs to $L^2(0,5; V)$.

For the second point, let us denote $w = e^{-3t}u$. We have

$$\frac{\partial w}{\partial t} = e^{-3t}\frac{\partial u}{\partial t} - 3e^{-3t}u, \qquad \frac{\partial^2 w}{\partial x^2} = e^{-3t}\frac{\partial^2 u}{\partial x^2}.$$

Multiply the first of (5.32) by e^{-3t}. We face the problem: find w such that

$$\frac{\partial w}{\partial t} - \frac{1}{\pi^2}\frac{\partial^2 w}{\partial x^2} = 0, \quad w(0,t) = w(1,t) = 0, \quad w(x,0) = \sin(2\pi x).$$

We split the solution as $w(x,t) = X(x)T(t)$. In this way, we reformulate the problem as the couple of ordinary differential equations

$$\begin{cases} \dfrac{d^2 X}{dx^2} = k\pi^2 X \\[2ex] \dfrac{dT}{dt} = kT \end{cases}$$

where k is a constant. The only non-trivial solutions for X compatible with the boundary data $X(0) = X(1) = 0$ are obtained for $k < 0$. We set therefore $k = -\lambda^2$ and solve

$$\frac{d^2 X}{dx^2} = -\lambda^2 \pi^2 X.$$

The general solution reads

$$X(x) = C_1 \sin(\lambda\pi x) + C_2 \cos(\lambda\pi x),$$

where C_1 and C_2 are arbitrary constants. By prescribing the boundary data, we obtain the solution

$$X(x) = C_1 \sin(n\pi x)$$

with C_1 arbitrary and n is integer. The equation for T has the general solution $T(t) = T_0 e^{-n^2 t}$. By collapsing the arbitrary constants C_1 and T_0, and emphasizing the dependence on n, thanks to the principle of effects superposition we have the solution

$$w(x,t) = \sum_n A_n e^{-n^2 t} \sin(\pi n x),$$

where the arbitrary constants are determined by the initial condition. In the present case, it is immediate to conclude that

$$A_2 = 1, \quad A_n = 0 \quad \text{for} \quad n \neq 0.$$

Finally, we obtain the solution

$$u(x,t) = e^{3t} w(x,t) = e^{-t} \sin(2\pi x).$$

Numerical approximation

Discretization of the problem (finite elements in space and finite differences in time) can be carried out in a way similar to the previous exercises.

Numerical results

Stability estimate for the Forward Euler methods (see e.g. [Qua09]) is in the form $\Delta t \leq C h^2$.

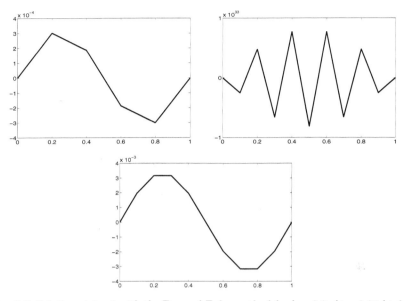

Fig. 5.9 Solution at $t = 5$ with the Forward Euler method for $h = 0.2, \Delta t = 0.05$ (stable) top, left, for $h = 0.1, \Delta t = 0.025$ (unstable) top right, for $h = 0.1, \Delta t = 0.0125$ (stable) bottom

This implies that when the parameter h is divided by 2, time step should be reduced by a factor 4. To check this result, let us take $\Delta t = \Delta t_1$ such that for a given h_1, $\Delta t_1 = Ch_1^2$. Then let us consider a grid with $h = h_1/2$ and solve the problem with $\Delta t_1/2$ and $\Delta t_1/4$ respectively. In particular, for this problem we take $h_1 = 0.2$ and $\Delta t_1 = 0.05$. We verify numerically (see Fig. 5.9) that this choice is stable. For $h = 0.1$ and $\Delta t_1 = 0.025$ on the contrary the Forward Euler method is unstable. Stability holds for $h = 0.1$ and $\Delta t = 0.0125$ (see Fig. 5.9).

In Tables 5.3 and 5.4 we report the results obtained with the methods of Backward Euler and Crank-Nicolson respectively, with linear finite elements and $h = 0.1$. Since the solution is C^∞ with respect to time and space we have the following estimate (see [Qua09])

$$||u_{ex}(t^n) - u_h^n||_{L^2}^2 + 2\alpha\Delta t \sum_{k=1}^{n} ||u_{ex}(t^k) - u_h^k||_V^2 \le C(u_0, u_{ex})(\Delta t^p + h^2)$$

with $p = 2$ for the Forward Euler method and $p = 4$ for Crank-Nicolson.

It is clear that a reduction of the time step implies a significant reduction of the error when the contribution of the time discretization to the total error is dominant. For the implicit Euler method in norm $L^\infty(0, 5; L^2(0, 1))$ time error is the dominating term and the error halves when the time step halves. For the Crank-Nicolson method and the selected steps h and Δt the time discretization error is not dominating and the reduction of the error is less

Table 5.3 Error for the Backward Euler method with $h = 0.1$ and $h = 0.0001$

h		0.1			0.0001	
Δt	0.1	0.05	0.025	0.1	0.05	0.025
$L^\infty(0, 5; L^2(0, 1))$	0.012	0.0059	0.0027	0.012	0.006	0.0032
$L^2(0, 5, V)$	0.08	0.06	0.055	0.06	0.0323	0.01682

Table 5.4 Error for the Crank-Nicolson method with $h = 0.1$ and $h = 0.0001$

h		0.1			0.0001	
Δt	0.1	0.05	0.025	0.1	0.05	0.025
$L^\infty(0, 5; L^2(0, 1))$	0.000653	0.000494	0.000455	0.00022	5.8×10^{-5}	1.7×10^{-5}
$L^2(0, 5, V)$	0.0565	0.057	0.053	0.0057	0.00547	0.0053

evident than expected when the time step (only) halves. Similar arguments for the error in norm $L^2(0, 5; V)$, with both the methods.

When the contribution of the space discretization error is reduced with a remarkable reduction of h, we obtain the results in Tables 5.3 and 5.4.

Crank-Nicolson shows the expected quadratic convergence in norm $L^\infty(0, 5; L^2(0, 1))$. For the norm $L^2(0, 5; V)$ the space discretization error is still dominant. This is somehow expected, since the V-norm (V is a subspace H^1) weights also the error on the first space derivative. We leave to the reader to check that with a further reduction of h the expected behavior of the time discretization error is found.   ◇

Exercise 5.1.7. Find $u(x, t)$ with $0 \leq x \leq 1$, $t \geq 0$, s.t.

$$
\begin{cases}
\dfrac{\partial u}{\partial t} + \dfrac{\partial v}{\partial x} = 0, & \text{in} \quad Q_T \in (0, 1) \times (0, \infty) \\[2mm]
v + \alpha(x)\dfrac{\partial u}{\partial x} - \gamma(x)u = 0, & \text{in} \quad Q_T \\[2mm]
u(0, t) = 0 & v(1, t) = \beta(t), \quad t > 0, \\[2mm]
u(x, 0) = u_0(x), & 0 < x < 1,
\end{cases}
\tag{5.36}
$$

where α, γ and u_0 are given functions of x, β is a given function of t.

1. Get a parabolic problem in its standard form by eliminating v.
2. Write the discretization of the problem with quadratic finite elements and the Backward Euler method for the time discretization.
3. How do you recover the approximation of v once you have computed that of u?

Solution 5.1.7.

Mathematical analysis

Even though the equations are written in an unusual form, this is a parabolic problem. This is promptly realized by eliminating the unknown v. We obtain the problem: for any $t > 0$, find u such that

$$
\begin{cases}
\dfrac{\partial u}{\partial t} - \dfrac{\partial}{\partial x}\left(\alpha \dfrac{\partial u}{\partial x}\right) + \gamma \dfrac{\partial u}{\partial x} + \dfrac{\partial \gamma}{\partial x} u = 0, & \text{in } \ Q_T \\[2mm]
u(0,t) = 0, & t > 0, \\[2mm]
-\alpha(1)\dfrac{\partial u}{\partial x}(1,t) + \gamma(1)u(1,t) = \beta(t), & t > 0, \\[2mm]
u(x,0) = u_0(x) & 0 < x < 1.
\end{cases}
$$

Weak form of the problem is obtained in a standard way. Set

$$
V \equiv \left\{ \varphi : \varphi \in H^1(0,1), \varphi(0) = 0 \right\},
$$

and including the boundary conditions we have to find $u \in V$ s.t.

$$
\int_0^1 \frac{\partial u}{\partial t}\varphi\, dx + \int_0^1 \alpha \frac{\partial u}{\partial x}\frac{\partial \varphi}{\partial x}\, dx + \int_0^1 \gamma \frac{\partial u}{\partial x}\varphi\, dx + \int_0^1 \frac{\partial \gamma}{\partial x}u\varphi\, dx \tag{5.37}
$$
$$
-\gamma(1)u(1,t)\varphi(1) = -\beta(t)\varphi(1), \quad \forall \varphi \in V.
$$

Well posedness analysis can be done in a way similar to Exercises 5.1.1 and 5.1.5, with the appropriate assumptions on the coefficients.

Numerical approximation

For the computation of v, once u has been computed, for instance by Lagrangian finite elements, we have

$$
v = -\alpha \frac{\partial u}{\partial x} + \gamma u.
$$

Since u is known only in the nodal points, the most immediate approach for accomplishing this computation is to perform an approximation of the derivative. For instance, we could use a finite difference approach using the same mesh used for the computation of u, e.g.

$$
v(x_i) \approx v_{h,i} = -\alpha \frac{u_{h,i+1} - u_{h,i-1}}{2h} + \gamma u_i.
$$

The nodal values of u are affected by numerical errors, so that

$$
\frac{\partial u}{\partial x}(x_i) = \frac{u_{h,i+1} + \mathcal{O}(h^p) - u_{h,i-1} + \mathcal{O}(h^p)}{2h} + \mathcal{O}(h^2)
$$

where $\mathcal{O}(h^p)$ denotes the error in the computation of u (in the present case, we know that $p < 1$). This argument points out that this approach can suffer from significant inaccuracies.

Another approach is to assume the discretization v_h of v in a finite dimensional subspace W, spanned by the basis functions $\{\psi_i\}$, with $i = 1, \ldots N_v$. We can represent the solution v_h as

$$v_h = \sum_{j=1}^{N_v} V_j(t)\psi_i = \sum_{k=1}^{N_u} U_k(t)\left(-\alpha\frac{\partial\varphi_k}{\partial x} + \gamma\varphi_k\right)$$

where we have represented the solution $u_h = \sum_{k=1}^{N_u} U_k\varphi_k$ and N_u is the dimension of the space of u_h. At each time step, this leads to solving the linear system

$$M_v \mathbf{V}(t^n) = S\mathbf{V}(t^n),$$

where M_v is the $N_v \times N_v$ mass matrix with entries $\int_0^1 \psi_i\psi_j dx$ and

$$[s_{ij}] = \int_0^1 -\alpha\frac{\partial\varphi_j}{\partial x}\psi_i + \gamma\varphi_j\psi_i dx.$$

At each time step $v_h(t^n)$ is the projection onto W_h of the numerical function $-\alpha\frac{\partial\varphi_j}{\partial x} + \gamma\varphi_j$, computed by solving a s.p.d. system. A drawback of this approach is that it can produce oscillating solutions[6]. This can be avoided by replacing the mass matrix with a lumped one.

A third more sophisticated approach entails the simultaneous computation of u and v. The weak formulation of (5.36) reads as follows. Let $u \in V$ and $v \in W \equiv L^2(0,1)$. For $t > 0$ we look for $u \in V$ and $v \in W$ such that for $t > 0$ and for any $\varphi \in V, v \in W$

$$\int_0^1 \frac{\partial u}{\partial t}\varphi dx - \int_0^1 \frac{\partial\varphi}{\partial x}v dx = -\beta\varphi(1)$$

$$\int_0^1 \left(\alpha\frac{\partial u}{\partial x} - \gamma u\right)\psi dx + \int_0^1 v\psi dx = 0. \tag{5.38}$$

Well posedness analysis of this problem cannot be based on coercivity arguments, but can rely upon the equivalence between (5.37) and (5.38).

---

[6] Oscillations are particularly dangerous when they bring the numerical solutions in meaningless ranges, like, for instance, when they get a negative density.

For the numerical approximation of (5.38) it is important to notice that the non-singularity of the discrete problem is not automatically guaranteed by the non singularity of the continuous one, since we cannot advocate coercivity arguments. We proceed by direct inspection.

Let us consider the finite dimensional subspaces $V_h \subset V$ and $W_h \subset W$, with basis φ_i and ψ_k respectively. Set

$$G = [g_{ij}] \in \mathbb{R}^{N_u \times N_v} \quad \text{with} \quad g_{ij} = \int_0^1 \frac{\partial \varphi_i}{\partial x} \psi_j \, dx,$$

$$G_\alpha = [g_{ij}^\alpha] \in \mathbb{R}^{N_v \times N_u} \quad \text{with} \quad g_{ij}^\alpha = \int_0^1 \alpha \psi_i \frac{\partial \varphi_j}{\partial x} \, dx,$$

$$R = [r_{ij}] \quad \text{with} \quad r_{ij} = \int_0^1 \gamma \psi_j \varphi_i \, dx,$$

and let M_u and M_v be the mass matrices for V_h and W_h respectively. If we discretize in time with the implicit Euler method, at each time step we have

$$\begin{bmatrix} \frac{1}{\Delta t} M_u & -G \\ G_\alpha - R & M_v \end{bmatrix} \begin{bmatrix} \mathbf{U}^{n+1} \\ \mathbf{V}^{n+1} \end{bmatrix} = \begin{bmatrix} \frac{1}{\Delta t} M\mathbf{U}^n - \beta \mathbf{e}_{N_h} \\ 0 \end{bmatrix}$$

where $\mathbf{e}_{Nh} \in \mathbb{R}^{N_h}$ is the vector $[0, 0, 0, \ldots 0, 1]^T$.

Let us assume in particular that $\alpha > 0$ and γ are constants. Moreover, we take $V_h = W_h$, so that $M_u = M_v = M$, $N_u = N_v$ and

$$R = \gamma M, \quad G_\alpha = \alpha G^T.$$

The matrix of the system reads

$$\begin{bmatrix} \frac{1}{\Delta t} M & -G \\ \alpha G^T - \gamma M & M \end{bmatrix}.$$

We proceed by formally reducing the system with a Gaussian elimination,

$$\mathbf{V}^{n+1} = -M^{-1}(\alpha G^T - \gamma M)\mathbf{U}^{n+1}$$

to obtain the system

$$\left(\frac{1}{\Delta t} M + \alpha G^T M^{-1} G - \gamma G \right) \mathbf{U}^{n+1} = \frac{1}{\Delta t} M\mathbf{U}^n - \beta \mathbf{e}_{N_h}.$$

Since $\frac{1}{\Delta t} M + \alpha G^T M^{-1} G$ is s.p.d., for $\gamma = 0$ the system is non-singular. By a continuity argument, we may infer that for γ small enough in comparison with $1/\Delta t$ the system is non-singular.

We finally mention that a more sophisticated approach for recovering the derivatives of a finite element solution in the nodes can be found in [ZZ92]. $\diamond$

Exercise 5.1.8. Consider the problem for $t \in (0,5]$

$$\frac{\partial u}{\partial t} - \nu(t)\triangle u + \sigma(t)u = -4 \qquad (5.39)$$

in the unit square $\Omega \equiv (0,1)^2$ with $\nu(t) = 1/t^2$ and $\sigma(t) = -2/t$, initial condition $u(x,y,0) = 0$ and boundary conditions

$$u = t^2 y^2 \qquad \text{in } x = 0, 0 < y < 1,$$

$$u = t^2(1+y^2) \text{ in } x = 1, 0 < y < 1,$$

$$u = t^2 x^2 \qquad \text{in } y = 0, 0 < x < 1,$$

$$u = t^2(1+x^2) \text{ in } y = 1, 0 < x < 1.$$

1. Write the weak formulation of the problem. Verify that

$$u_{ex} = t^2(x^2 + y^2)$$

is the solution to the problem.
2. Discretize the problem in space with quadratic finite elements. Then discretize the time dependence with Backward Difference Formulas (BDF) of order 1,2 and 3. Discuss the accuracy of the discretization.
3. Write a **Freefem** code solving the problem and verify the expected accuracy.

Solution 5.1.8.

Mathematical analysis

The weak formulation is obtained in a standard way. Let $V \equiv H_0^1(\Omega)$, and $G(x,y,t) \in H^1(\Omega)$ a lifting of the boundary data.

We want to find $u \in L^2(0,5,V+G)$ such that $u(x,y,0) = 0$ and for any $v \in V$

$$\int_\Omega \frac{\partial u}{\partial t} v \, d\omega + \frac{1}{t^2}\int_\Omega \nabla u \cdot \nabla v \, d\omega - \frac{2}{t}\int_\Omega uv = -4\int_\Omega v \, d\omega.$$

Notice that for $t \in (0,5]$

$$\frac{1}{t^2}\int_\Omega \nabla u \cdot \nabla u \, d\omega \geq \frac{C_P}{t^2}\|u\|_V^2 = \frac{C_P}{t^2}\left(\|\nabla u\|_{L^2}^2 + \|u\|_{L^2}^2\right),$$

where $C_P > 0$ is the constant of the Poincaré inequality. Therefore, we have the inequality

$$\frac{1}{t^2}\int_\Omega \nabla u \cdot \nabla u \, d\omega - \frac{2}{t}\int_\Omega u^2 \, d\omega \geq \left(\frac{C_P}{t^2} - \frac{2}{t}\right)\|u\|_{L^2}^2.$$

For $t > 0$, $C_P t^{-2} - 2t^{-1} \geq -C_P^{-1}$, so we conclude that

$$\frac{1}{t^2} \int_\Omega \nabla u \cdot \nabla u \, d\omega - \frac{2}{t^2} \int_\Omega u^2 \geq \frac{C_P}{25} ||\nabla u||_{L^2}^2 - C_P^{-1} ||u||_{L^2}^2.$$

From this weak coercivity inequality we deduce the well posedness of the problem.

The given function u_{ex} clearly fulfills the boundary conditions. Moreover, we may notice that

$$\frac{\partial u_{ex}}{\partial t} - \sigma(t) u_{ex} = 2t(x^2 + y^2) - 2t(x^2 + y^2) = 0; \quad -\nu(t)\triangle u_{ex} = -4$$

that proves that u_{ex} solves the given equation.

Numerical approximation

The discretization in space is obtained by taking V_h as the space of piecewise (continuous) quadratic functions over a conformal mesh $\mathcal{T}_h$ of Ω vanishing on the boundary $\partial\Omega$ and solving for all $v_h \in V_h$

$$\int_\Omega \frac{\partial u_h}{\partial t} v_h \, d\omega + \frac{1}{t^2} \int_\Omega \nabla u_h \cdot \nabla v_h \, d\omega - \frac{2}{t^2} \int_\Omega u_h v_h = -4 \int_\Omega v_h \, d\omega,$$

where $u_h \in L^2(0, 5, V_h + G)$. This leads to the solution of the ordinary differential system for $t \in (0, 5]$

$$M \frac{d\mathbf{U}}{dt} + \frac{1}{t^2} K \mathbf{U} - \frac{2}{t} M \mathbf{U} = \mathbf{F}$$

where $\mathbf{U} = \mathbf{U}(t)$ is the vector of nodal values of the solution and $\mathbf{F}$ is the vector with entries $f_j = -4 \int_\Omega \varphi_j \, d\omega$ where φ_j is the j−th function of the basis of V_h.

For the time discretization, we are required to use BDF schemes. This means that for a generic function w we discretize [QSS00]

$$\frac{\partial w}{\partial t}(t^{n+1}) \approx \frac{\alpha_0}{\Delta t} w(t^{n+1}) - \sum_{k=1}^{p} \frac{\alpha_k}{\Delta t} w(t^{n+1-k})$$

where Δt is the time step (assumed constant), p is the order of the formula and the coefficients α_k for $k = 0, 1, \ldots p$ are properly selected to maximize the accuracy of the formula. We may use the method of undetermined coefficients seen in Section 2.3.

Another possibility is to find the coefficients α_k such that the approximation of the time derivative is exact for polynomial functions of degree $0, 1, \ldots, p$. This leads to the so called Backward Difference Formula (BDF)

Table 5.5 BDF coefficients for order $p = 1, 2, 3$

p	α_0	α_1	α_2	α_3
1	1	1	–	–
2	3/2	2	−1	–
3	11/6	3	−3/2	1/3

schemes [QSS00]. The coefficients of BDF schemes of order 1,2 and 3 computed are reported in Table 5.5. Notice that for variable Δt, like in time adaptive methods, these coefficients should be properly modified. For $p = 1$ we recover the backward Euler scheme.

The linear system to be solved at each time step after BDF time discretization reads

$$\frac{\alpha_0}{\Delta t}\mathbf{M}\mathbf{U}^{n+1} + \left(\frac{1}{t^{(n+1)}}\right)^2 \mathbf{K}\mathbf{U}^{n+1} - \frac{2}{t^{(n+1)}}\mathbf{M}\mathbf{U}^{n+1} = \mathbf{F} + \sum_{k=1}^{p}\frac{\alpha_k}{\Delta t}\mathbf{U}^{n+1-k}.$$

In general, the expected accuracy for a quadratic finite element discretization and a BDF of order p for a regular enough solution is $\mathcal{O}(h^2 + \Delta t^p)$. However, notice that the exact solution is quadratic in both time and space. This means that the expected error is of the order the the machine ε for $p \geq 2$ (the space discretization component of the error being always of the order of the machine ε) if the correct initial conditions are prescribed.

Numerical results

Program 27 - ex618 : Program implementing BDF schemes of order 1,2,3

```
int nn=10;
mesh Th=square(nn,nn);
fespace Xh(Th,P2);
Xh uh,vh,uhrhs,w;
real t=0, Tfin=5, t0=0.;;
real Dt=0.0125;
Xh uh0=0.0;
func f=-4;
func uex=t^2*(x^2+y^2), dxuex=2*x*t^2, dyuex=2*y*t^2;
int bdforder=1,bdfm1;
func sigma=-2./t, nu=1./t^2;
real s,n,tc,errL2sq=0.0,errH1sq=0.0,errLiL2=0.0,errL2H1=0.0,locL2=0.0,
locH1=0.0;
int i;
Xh [int] uhold(3);
real [int,int] bdf(3,4);
// BDF Coefficients
bdf(0,0)=1./Dt;bdf(0,1)=1./Dt;
bdf(0,2)=0.0;bdf(0,3)=0.0;
//
```

```
bdf(1,0)=3./(2*Dt);bdf(1,1)=2./Dt;
bdf(1,2)=-1./(2*Dt);bdf(1,3)=0.0;
//
bdf(2,0)=11./(6*Dt);bdf(2,1)=3./Dt;
bdf(2,2)=-3./(2*Dt);bdf(2,3)=1./(3*Dt);
bdfm1=bdforder-1;

// Vectors storing the previous solutions
t=t0;
 for (i=0;i<bdforder;++i) {uhold[i]=uex;t=t-Dt;}

//Problem definition
problem Problem1(uh,vh) =
    int2d(Th)(tc*uh*vh)+
    int2d(Th)(n*(dx(uh)*dx(vh) + dy(uh)*dy(vh)))
  - int2d(Th)(f*vh)
  -int2d(Th)(uhrhs*vh)
  + on(1,2,3,4,uh=uex);

//Time loop
for (t=Dt+t0;t<=Tfin;t+=Dt)
 {cout << "Time : " << t << endl;
  s = sigma; n=nu;
  tc=s+bdf(bdfm1,0);

 uhrhs=0.;
 for (i=1;i<=bdforder;++i)
   uhrhs=uhrhs+bdf(bdfm1,i)*uhold[i-1];

Problem1;

for (i=bdfm1;i>0;--i) uhold[i]=uhold[i-1]; //(*)

uhold[0]=uh;
 }
```

Program 27 implements in **Freefem** the solution of the problem. Notice that the BDF schemes require to memorize p vectors corresponding to the previous time steps. This is done here with the matrix uhold. When computing uh at time step $n + 1$, the first column of uhold (index 0) corresponds to the solution at time step n, the second one (index 1) corresponds to the solution at time step $n - 1$, and so on. At the end of each time step, for $i \geq 1$ the column $i - 1$ is copied into the column i of uhold and the current solution uh is written in the column uhold[0] (see the line marked by (*)). In this way the columns are "shifted right" of one position.

Results of Table 5.6 show as expected that the error is linear with Δt for $p = 1$, while the solution is numerically exact for $p = 2, 3$. In this case, we have prescribed the supplementary initial conditions required by the numerical schemes of order 2 and 3 with the exact solution. Should these data be not available (as it is the case in general), we have to find a way of prescribing approximate conditions. This can be done for instance by using one-step

Table 5.6 Results of exercise 5.1.8

p	1		2		3	
Δt	$\|e\|_{L^2(H^1)}$	$\|e\|_{L^\infty(L^2)}$	$\|e\|_{L^2(H^1)}$	$\|e\|_{L^\infty(L^2)}$	$\|e\|_{L^2(H^1)}$	$\|e\|_{L^\infty(L^2)}$
0.1	0.320336	0.063622	2.48e-13	4.21e-14	2.87e-13	4.92e-14
0.05	0.157868	0.031734	1.82e-13	2.33e-14	2.13e-13	2.93e-14
0.025	0.077387	0.015689	2.44e-13	4.87e-14	2.37e-13	3.51e-14
0.0125	0.038794	0.007879	2.83e-13	4.81e-14	2.42e-13	3.97e-14

first order methods, which however may affect the overall accuracy of the numerical solution. ◇

Remark 5.2 The extension of the previous exercise to the 3D case is immediate. The exact solution reads $u_{ex,3D} = t^2(x^2 + y^2 + z^2)$ and the right hand side modifies in $f = -6$. This is the test case called `test_bdf` in the LifeV library. The private members of the class Bdf include a matrix for storing the solution at the previous time steps, the coefficients for the approximation of the time derivative (up to the third order) and coefficients for extrapolating a value for the solution given the previous time steps (useful for linearizing non-linear problems like the one presented in Exercise 5.1.10 or in Chapter 7). Member `shift_right` of the Bdf class manages the updating of the time steps stored at the end of each time step (as done in the Freefem code in (*)).

The reader interested in the C++ Object Oriented implementation of the present exercise, is invited to check out the `test_bdf` directory in the `testsuite` of LifeV [lif10].

Exercise 5.1.9. Find u s.t.

$$
\begin{cases}
\dfrac{\partial u}{\partial t} - \nabla \cdot (\mu \nabla u) + \Delta^2 u + \sigma u = 0 & x \in \Omega, \quad t > 0 \\[2mm]
u = u_0 & \text{in} \quad \Omega, t = 0 \\[2mm]
\dfrac{\partial u}{\partial \mathbf{n}} = 0 \quad \text{and} \quad u = 0 & \text{on} \quad \Sigma_T \equiv \partial\Omega, t > 0
\end{cases}
\tag{5.40}
$$

where $\Omega \subset \mathbb{R}^2$ is an open bounded domain with regular boundary $\partial\Omega$ and outward unit vector $\mathbf{n}$, $\Delta^2 = \Delta\Delta$ is the bi-harmonic operator, μ, σ and u_0 are given functions of $\mathbf{x}$ in Ω. Here $\dfrac{\partial u}{\partial \mathbf{n}} = \nabla u \cdot \mathbf{n}$ is the normal derivative of u.
Recall that

$$
H_0^2(\Omega) = \left\{ u \in H^2(\Omega) : u = \frac{\partial u}{\partial \mathbf{n}} = 0 \quad \text{on} \quad \partial\Omega \right\}
$$

and there exist two constants c_1 and c_2 s.t.

$$c_1 \int_\Omega |\Delta u|^2 d\Omega \leq \|u\|_{H^2(\Omega)}^2 \leq c_2 \int_\Omega |\Delta u|^2 d\Omega \qquad \forall u \in H_0^2(\Omega), \qquad (5.41)$$

implying that the L^2 norm of the Laplacian of a function of H_0^2 is equivalent to the norm of H^2.

1. Write the weak formulation of (5.40) and prove that the solution exists and is unique under appropriate conditions on the data.
2. Introduce a finite element semi-discretization of the problem and specify the degree required for a conformal discretization. Remind that if $\mathcal{T}_h$ is a triangulation of Ω and $v_h|_K$ is a polynomial for any $K \in \mathcal{T}_h$, then $v_h \in H^2(\Omega)$ if and only if $v_h \in C^1(\overline{\Omega})$.

Solution 5.1.9.

Mathematical analysis

Multiply the equation by a test function $v \in H_0^2(\Omega)$. We will justify the choice of the space $H_0^2(\Omega)$ (hereafter denoted by V) later on.

$$\int_\Omega \left(\frac{\partial u}{\partial t} - \nabla \cdot (\mu \nabla u) + \Delta^2 u + \sigma u \right) v d\omega = 0 \qquad \forall v \in V.$$

By the Green formula

$$\int_\Omega \Delta^2 u v d\omega = \int_\Omega \nabla \cdot (\nabla \Delta u) v = \int_{\partial\Omega} \nabla(\Delta u) \cdot \mathbf{n} v d\gamma - \int_\Omega \nabla(\Delta u) \cdot \nabla v d\omega$$

and

$$-\int_\Omega \nabla(\Delta u) \cdot \nabla v d\omega = -\int_{\partial\Omega} \Delta u \, (\mathbf{n} \cdot \nabla v) \, d\gamma + \int_\Omega \Delta u \Delta v d\omega.$$

For the selected test function notice that all the boundary terms vanish.

We assume that $u_0 \in L^2(\Omega)$. The weak formulation reads: for any $t > 0$ find $u \in L^2(0, T; V) \cap L^\infty 0, T; L^2(\Omega)$ s.t.

$$\int_\Omega \frac{\partial u}{\partial t} v d\omega + a(u, v) = 0 \quad \forall v \in V,$$

with $u(x, 0) = u_0(x)$, where $a(u, v) \equiv \int_\Omega (\mu \nabla u \cdot \nabla v + \Delta u \Delta v + \sigma uv) \, d\omega$. This bilinear form is continuous in V, provided that μ and σ are regular enough, for instance $\mu \in L^\infty(\Omega)$ and $\sigma \in L^\infty(\Omega)$. We will assume this regularity hereafter. However this hypothesis can be relaxed. For the Sobolev embedding

Sobolev (see Ch. 1) in two dimensions $H^2(\Omega) \hookrightarrow C^0(\bar{\Omega})$ and thus $\int_\Omega \sigma uv d\omega$ is well defined if $\sigma \in L^1(\Omega)$. Similarly, $\nabla w \in H^1(\Omega) \hookrightarrow L^6(\Omega)$ if $w \in H^2(\Omega)$. Therefore $\nabla u \cdot \nabla v$ belongs $L^3(\Omega)$ and it is enough to assume $\mu \in L^{3/2}(\Omega)$ for the continuity of $a(\cdot, \cdot)$.

The bilinear form is weakly coercive provided that $\mu(\mathbf{x}) \geq \mu_0 > 0$. Let σ^- the negative part of σ (i.e. the function equal to σ when $\sigma < 0$ and 0 elsewhere). Let $\sigma^* = ||\sigma^-||_{L^\infty(\Omega)}$. Thanks to (5.41)

$$a(u,u) \geq \alpha ||u||_V^2 - \sigma^* ||u||_{L^2(\Omega)}^2,$$

where α depends on μ_0 and c_2 in (5.41). The bilinear form is then weakly coercive if $\sigma < 0$ on a subset of Ω with non-null measure and coercive for $\sigma^* = 0$.

Formally we have a problem similar to those considered in the previous exercises. The difference is in the space V, more regular compared to the spaces in the other examples. We conclude that the problem is well posed. Notice that also in this case, the initial data need "just" to belong to the space $L^2(\Omega)$.

Numerical approximation

As suggested in the text of the exercise, the finite dimensional space V_h is a subspace of V if it is given by functions of $C^1(\overline{\Omega})$. In general, functions of $H^2(\Omega)$ with $\Omega \subset \mathbb{R}^2$ are not in $C^1(\Omega)$. Therefore we force the piecewise polynomial functions of V_h to belong to $C^1(\overline{\Omega})$ explicitly.

For a triangular grid, the number of degrees of freedom for finite elements of degree r is (see Exercise 2.2.3)

$$n_l = \frac{(r+1)(r+2)}{2}.$$

To identify the proper degree r, we need to compute the number of degrees of freedom that guarantee $C^1(\overline{\Omega})$ continuity. Continuity of a polynomial function of degree r on the edges can be forced by the continuity in $r+1$ points. Two of these points are the vertexes of the triangle so we prescribe $r-1$ internal points on each edge (see Exercise 2.2.4). Continuity of the functions implies also continuity of the tangential derivatives on the internal points of the edge. Therefore we have to impose continuity to the normal derivative to each edge and to the derivatives at the vertexes. The latter corresponds to continuity of derivatives with respect to x and y in the three vertexes, corresponding to $3 \times 2 = 6$ constraints. Normal derivatives along each edge are polynomials of degree $r-1$. Continuity requires r conditions, two of them have already been prescribed in the vertexes, so we need $r-2$ additional conditions.

Overall, we need:

1. 3 constraints for the continuity in the vertexes;
2. $3(r-1)$ constraints for the continuity on the edges;
3. 6 constraints for the continuity of the derivatives on the vertexes;
4. $3(r-2)$ constraints for the continuity of the normal derivatives.

Selection of r must accommodate all the required constraints. Hence the number of degrees of freedom on each element n_l needs to be

$$n_l = \frac{(r+1)(r+2)}{2} \geq 3 + 6 + 3r - 3 + 3r - 6 = 6r, \qquad (5.42)$$

with r positive integer. The smallest integer r by which (5.42) is satisfied is $r = 9$, corresponding to $n_l = 55$ local degrees of freedom. Adoption of this high degree elements is troublesome both for the stability of the underlying interpolation and the practical implementation.

A possibility is to relax the constaints partially. A well known choice is the Argyris triangle which is the finite element of minimal degree, with $r = 5$ and $n_l = 21$, forcing constraints on the function and its first and second derivatives in the vertexes, and the normal derivatives in the mid points of the edges, see e.g. [BS02, EG04]. $\diamond$

Remark 5.3 Let us consider the following variant to the previous problem (see also Remark 3.4): find u s.t.

$$\begin{cases} \dfrac{\partial u}{\partial t} - \nabla \cdot (\mu \nabla u) + \Delta^2 u + \sigma u = 0 & x \in \Omega, \quad t > 0 \\[2mm] u = u_0 & \text{in} \quad \Omega, \, t = 0 \qquad (5.43) \\[2mm] \Delta u = 0 \quad \text{and} \quad u = 0 & \text{on} \quad \Sigma_T \equiv \partial\Omega, \, t > 0 \end{cases}$$

where in this case the boundary conditions refer to the value of u and its Laplacian. A possible way for solving this problem is the introduction of the auxiliary variable

$$w = -\Delta u,$$

which is supposed to vanish on the boundary according to the boundary conditions. Set $\mathbf{v} = [u, w]^T$. We write the equation in the vector form

$$\begin{bmatrix} \mathcal{I} & 0 \\ 0 & 0 \end{bmatrix} \frac{\partial \mathbf{v}}{\partial t} - \nabla \cdot \left(\begin{bmatrix} \mu & \mathcal{I} \\ \mathcal{I} & 0 \end{bmatrix} \nabla \mathbf{v} \right) + \begin{bmatrix} \sigma \mathcal{I} & O \\ 0 & -\mathcal{I} \end{bmatrix} \mathbf{v} = \mathbf{0},$$

with $\mathbf{v}(x, y, 0) = [u_0, -\Delta u_0] \equiv \mathbf{v}_0$. The problem can be therefore cast in the from of a mixed parabolic-elliptic problem. We can write the weak form: find $\mathbf{v} \in H_0^1(\Omega) \times H_0^1(\Omega)$ such that for all $\varphi, \psi \in H_0^1(\Omega)$

$$\int_\Omega \left(\frac{\partial u}{\partial t} \varphi + \mu \nabla u \cdot \nabla \varphi + \nabla w \cdot \nabla \varphi + \sigma u \varphi \right) d\omega + \int_\Omega (\nabla u \cdot \nabla \psi - w \psi) \, d\omega = 0,$$

completed by the initial conditions.

For the numerical solution of the problem we discretize the space dependence with finite elements of the same order for u and w. For the time advancing we use the Backward Euler scheme. Then at each time step we solve a linear system in the matrix (let us assume μ constant for simplicity)

$$\left[\begin{matrix} \left(\dfrac{1}{\Delta t} + \sigma \right) M + \mu K & K \\ K & -M \end{matrix} \right].$$

The system is promptly recognized to be non-singular, by observing that Gaussian elimination of the unknowns associated with w leads to a system with the matrix

$$\left(\frac{1}{\Delta t} + \sigma \right) M + \mu K + KM^{-1}K$$

which is clearly s.p.d. For boundary conditions on the Laplacian of u this "augmented" formulation of the problem can be therefore a convenient approach.

Exercise 5.1.10. (*) Let us consider the evolution of temperature u in a brake disc for a car, like the one represented in Fig.5.10. The disc is made of cast iron with a *thermal diffusivity* function of the temperature, according to the constitutive law

$$k(u) = K_2 u^2 + K_1 u + K_0 \ge \mu_0 > 0,$$

where K_2, K_1 and K_0 are coefficients measured by experimental benchmarks.

We want to study the temperature propagation on the transverse section of the disc, represented in Fig. 5.10, assumimg that the brake is actioned periodically with period t_s seconds.

When the brake shoes touch the disc, a heat flux Φ is generated by friction entering the brake through surface $\Gamma_g \subset \partial\Omega$. This flux is zero when the brake shoes are not touching the disc. In these time intervals, and over the entire process on the part of the surface of the disc never touched by the shoes, we assume a convective thermal exchange with the surrounding air.

1. Find a mathematical model for the temperature evolution in the brake, assuming to start from a uniform constant distribution of temperature u_0; density and specific heat of the cast iron are constant.
2. Write the time discretized problem by using the implicit Euler method.
3. Discretize in space the problem with linear finite elements.
4. Propose a method for solving the nonlinear algebraic system obtained after the complete discretization of the problem and discuss its convergence.

5. Perform the numerical simulation of the problem with the geometrical data given in Fig. 5.11 and the coefficients listed hereafter. Density $\rho = 7.2 \times 10^{-2}\ Kg/mm^3$, specific heat $c_p = 500\ W/mK$, convective heat exchange coefficient $\alpha = 80\ W/mmK$, conductivities coefficients of $k(u)$ as a function of the temperature $k(200) = 50\ W/mK$, $k(300) = 47\ W/mK$ e $k(700) = 37\ W/mK$. Environment temperature $u_{amb} = 20°C$ (68 °F) e $u_0 = 60°C$ (140 °F). Brakes are performed every 5 seconds and they last 5 seconds. Simulate a sequence of 12 brakes, on an interval of $T = 120$. Denoting by t_0 the starting time of a brake action, $\Phi(t) = \Phi_{MAX} \sin(\pi(t - t_0)/5)$ for $t \le t_0 + 5$ and $\varphi(t) = 0$, with $\Phi_{MAX} = 50\ W/mm^2$.

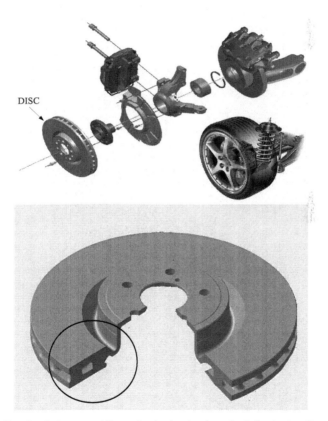

DISC

Fig. 5.10 Sketch of the assembling of a brake (top) and of the brake disc, the circle identify the section of interest (bottom) (courtesy of Brembo ©, Italy)

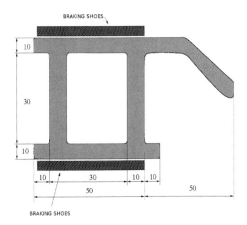

Fig. 5.11 Simplified geometry of Exercise 5.1.10

Solution 5.1.10.

Mathematical modeling

The problem is described by the heat equation with the thermal conductivity function of the temperature. If we denote by u the temperature in the domain in Fig. 5.11, as a function of x, y and t, we have to solve the equation

$$\rho c_p \frac{\partial u}{\partial t} - \nabla \cdot (k(u)\nabla u) = 0 \quad \mathbf{x} \in \Omega, \quad t > 0,$$

with the boundary condition

$$\begin{cases} k(u)\nabla u \cdot \mathbf{n} = \alpha_g(t)(u_{amb} - u) - \Phi(t) & \mathbf{x} \in \Gamma_g, t > 0 \\ k(u)\nabla u \cdot \mathbf{n} = \alpha(u_{amb} - u) & \mathbf{x} \in \Gamma_r \equiv \partial\Omega \setminus \Gamma_g, t > 0 \end{cases}$$

and the initial condition $u = u_0$ for $t = t_0 = 0$. The coefficient of advective thermal exchange α for the cast iron is assumed to be constant and $\alpha_g(t) = 0$ during the braking action and α in the other instants. The negative sign for Φ is due to the fact that the heat flux is incoming.

Mathematical analysis

Let V be $H^1(\Omega)$. We formulate the problem as follows: for any $t > 0$ find $u \in L^2(0, T, V)$ s.t.

$$\left(\frac{\partial u}{\partial t}, v\right) + a(u; u, v) + \alpha_g \int_{\Gamma_g} uv d\gamma + \alpha \int_{\Gamma_r} uv d\gamma =$$

$$\alpha_g \int_{\Gamma_g} u_{amb} v d\gamma + \alpha \int_{\Gamma_r} u_{amb} v d\gamma - \int_{\Gamma_g} \Phi v d\gamma \tag{5.44}$$

with the initial condition $u(\mathbf{x}, 0) = u_0(\mathbf{x})$. In (5.44) we set

$$a(w; u, v) = \int_{\Omega} k(w) \nabla u \cdot \nabla v d\omega.$$

Let us assume that u is regular enough so that the bilinear form $a(u; u, v)$ is well defined. In particular, if we assume that $u \in H^2(\Omega)$, being Ω regular enough, thanks to the Sobolev embedding Theorem (see Chapter 1), then u is continuous in $\overline{\Omega}$ and $a(u; u, v)$ is well defined.

Since the problem is nonlinear, the well posedness analysis of this problem cannot be accomplished with the analytical tools we have used so far. Possible methods for this task can be found in [Sal08, Eva10]. Here we just consider the bound for the (possible) solution as a function of the data. Let us assume that $u \in L^2(0, T; H^2(\Omega))$, $\Phi \in L^2(\Gamma_g)$, $u_{amb} \in L^2(\partial\Omega)$. Notice that α_g is piecewise constant, so it belongs to $L^2(0, T)$. Finally it is suggested to assume that

$$k(u) \geq \mu_0 > 0, \tag{5.45}$$

with $\mu_0 > 0$. Set $v = u$ in (5.44), and notice that for any $t > 0$ we have $0 \leq \alpha_g(t) \leq \alpha$, so for an arbitrary $\epsilon > 0$ we have

$$\frac{1}{2}\frac{d}{dt}||u||_{L^2}^2 + \mu_0||\nabla u||_{L^2}^2 + \leq \left(\alpha||u_{amb}||_{L^2(\partial\Omega)} + ||\varphi||_{L^2(\Gamma_g)} \right)||u||_V \leq$$

$$\frac{1}{4\epsilon} \left(\alpha||u_{amb}||_{L^2(\partial\Omega)}^2 + ||\varphi||_{L^2(\Gamma_g)}^2 \right) + \epsilon \left(||u||_{L^2}^2 + ||\nabla u||_{L^2}^2 \right).$$

Set $\epsilon = \mu_0/2$ and integrate in time between 0 and T. We obtain

$$||u||_{L^2}^2(T) + \mu_0 \int_0^T ||\nabla u||_{L^2}^2 dt + \leq$$

$$||u_0||_{L^2}^2 + \frac{1}{\mu_0} \left(\alpha||u_{amb}||_{L^2(\partial\Omega)}^2 + ||\varphi||_{L^2(\Gamma_g)}^2 \right) + \mu_0 \int_0^T ||u||_{L^2}^2 dt. \tag{5.46}$$

By the Gronwall Lemma, we have the following bound

$$\begin{array}{c} ||u||_{L^\infty(L^2)}^2 \leq C_1 \\ \\ ||\nabla u||_{L^2(L^2)}^2 \leq C_2 \end{array} \Rightarrow ||u||_{L^2(H^1)}^2 \leq C_3, \tag{5.47}$$

where C_1, C_2, C_3 depend on the data and on T.

| Numerical approximation |

Time discretization with the backward Euler method is obtained by splitting the time domain into intervals with a constant (for simplicity) time step Δt, such that $T = N\Delta t$. From (5.44) we get the problem: find for $n = 0, \ldots, N-1$, $u^{n+1} \in V$ s.t.

$$
\frac{1}{\Delta t}\left(u^{n+1}, v\right) + a(u^{n+1}; u^{n+1}, v) + \alpha \int_{\Gamma_r} u^{n+1} v d\gamma + \alpha_g^{n+1} \int_{\Gamma_g} u^{n+1} v d\gamma =
$$
$$
\frac{1}{\Delta t}\left(u^n, v\right) + \alpha \int_{\Gamma_r} u_{amb} v d\gamma + \alpha_g^{n+1} \int_{\Gamma_g} u_{amb} v d\gamma + \int_{\Gamma_g} \phi^{n+1} v d\gamma
$$

(5.48)

with $u^0 = u_0$. For the space discretization, we introduce a finite dimensional subspace $V_h \subset V$ which the numerical solution belongs to. The problem reads: for all $n = 0, 1, \ldots, N-1$, find $u_h^{n+1} \in V_h$ s.t.

$$
\frac{1}{\Delta t}\left(u_h^{n+1}, v_h\right) + a(u_h^{n+1}; u_h^{n+1}, v_h) + \alpha \int_{\Gamma_r} u^{n+1} v d\gamma + \alpha_g^{n+1} \int_{\Gamma_g} u^{n+1} v d\gamma =
$$
$$
\frac{1}{\Delta t}\left(u_h^n, v_h\right) + \alpha \int_{\Gamma_r} u_{amb} v d\gamma + \alpha_g^{n+1} \int_{\Gamma_g} u_{amb} v d\gamma + \int_{\Gamma_g} \Phi^{n+1} v d\gamma
$$

(5.49)

with $u_h^0 = u_{0h}$, where u_{0h} is an approximation of u_0 in V_h. Algebraic formulation of the problem (5.49) is

$$
\frac{1}{\Delta t}\mathrm{M}\mathbf{U}^{n+1} + \mathrm{K}(\mathbf{U}^{n+1})\mathbf{U}^{n+1} + \mathrm{R}^{n+1}\mathbf{U}^{n+1} = \frac{1}{\Delta t}\mathrm{M}\mathbf{U}^n + \mathbf{F}^{n+1},
$$

where $u_h^{n+1} = \sum_{j=1}^{N_h} U_j^{n+1}\Phi_j$, $\mathbf{U}^{n+1} = [U_j^{n+1}]$, R corresponds to the discretization of the boundary terms, $\mathbf{F} = \mathbf{F}(u_{amb}, \Phi, \alpha, \alpha_g)$ comes from the discretization of the forcing term and K is the matrix with entries

$$
K_{ij}(\mathbf{U}^{n+1}) = a\left(u_h^{n+1}; \varphi_j, \varphi_i\right).
$$

Thanks to (5.45) K(**U**) is s.p.d. for any value of **U**. In a standard way we therefore obtain the stability estimate

$$
\mathbf{U}^n \leq C_4, \quad n = 0, \ldots, N, \tag{5.50}
$$

where C_4 depends on the initial condition and the forcing term and it is independent of Δt and h.

The fully discrete problem results in a nonlinear algebraic system in the form

$$
\mathcal{A}(\mathbf{U}^{n+1}) = 0.
$$

We can solve this system with a *fixed point method* (see e.g. [QSS00]). Let us consider for instance the auxiliary linear problem at the time level t^{n+1}: given $u_h^{n+1,k}$, find $u_h^{n+1,k+1} \in V_h$ s.t.

$$\frac{1}{\Delta t}\left(u_h^{n+1,k+1}, v_h\right) + a\left(\boxed{u_h^{n+1,k}}; u_h^{n+1;k+1}, v_h\right) + \alpha \int_{\Gamma_r} u_h^{n+1,k+1} v d\gamma +$$

$$\alpha_g^{n+1} \int_{\Gamma_g} u_h^{n+1,k+1} v d\gamma = \frac{1}{\Delta t}\left(u_h^n, v_h\right) + \alpha \int_{\Gamma_r} u_{amb} v d\gamma + \qquad (5.51)$$

$$\alpha_g^{n+1} \int_{\Gamma_g} u_{amb} v d\gamma + \int_{\Gamma_g} \Phi^{n+1} v d\gamma$$

with $u_h^0 = u_{h0}$, leading to the linear system

$$\left[\frac{1}{\Delta t}M + K(U^{n+1,k}) + R\right] U^{n+1,k+1} = \frac{1}{\Delta t}MU^n + F^{n+1}.$$

If the sequence of solutions of this system converges, i.e.

$$\lim_{k\to\infty} U^{n+1,k} = U_{fin}, \qquad (5.52)$$

then U_{fin} is solution to the nonlinear system (5.49). The nonlinear problem is solved by a sequence of linear problems. The issue is to guarantee that the convergence (5.52) holds. As a matter of fact, we have

$$U^{n+1,k+1} = \left(\frac{1}{\Delta t}M + K(U^{n+1,k}) + R\right)^{-1}\left(\frac{1}{\Delta t}MU^n + F^{n+1}\right),$$

$$U^{n+1} = \left(\frac{1}{\Delta t}M + K(U^{n+1}) + R\right)^{-1}\left(\frac{1}{\Delta t}MU^n + F^{n+1}\right). \qquad (5.53)$$

For the sake of notation from now on we omit the time index $n+1$ and we will write $A(U) = K(U) + R$ and $\mathcal{F} = \frac{1}{\Delta t}MU^n + F^{n+1}$. With this notation, from (5.53) we get

$$U^{k+1} - U = $$
$$\Delta t \left[(I + \Delta t M^{-1}A(U^k))^{-1} - (I + \Delta t M^{-1}A(U))^{-1}\right] M^{-1}\mathcal{F}. \qquad (5.54)$$

For Δt small enough, the spectral radius of $\Delta t M^{-1}A(U)$ is < 1, so we can exploit the Neumann expansion (1.20)

$$\left(I + \Delta t M^{-1}A(U)\right)^{-1} = \sum_{j=0}^{\infty}(-\Delta t)^j M^{-j}A^j(U),$$

leading to

$$\mathbf{U}^{k+1} - \mathbf{U} =$$

$$\Delta t \sum_{j=0}^{\infty} (-\Delta t)^j \left[\mathrm{M}^{-j} \mathrm{A}^j(\mathbf{U}^k) - \mathrm{M}^{-j} \mathrm{A}^j(\mathbf{U}) \right] \mathrm{M}^{-1}\mathcal{F} =$$

$$-\Delta t^2 \mathrm{M}^{-1} \left(\mathrm{A}(\mathbf{U}^k) - \mathrm{A}(\mathbf{U}) \right) \mathrm{M}^{-1}\mathcal{F} + \mathcal{O}(\Delta t^3) =$$

$$-\Delta t^2 \mathrm{M}^{-1} \left(\mathrm{K}(\mathbf{U}^k) - \mathrm{K}(\mathbf{U}) \right) \mathrm{M}^{-1}\mathcal{F} + \mathcal{O}(\Delta t^3).$$

For $\mathbf{U}$ and $\mathbf{U}^k$ bounded, the non-linearity of the problem is such that there exists $C > 0$ with

$$\|\mathrm{K}(\mathbf{U}^k) - \mathrm{K}(\mathbf{U})\| \leq C(\mathbf{U}) \|\mathbf{U}^k - \mathbf{U}\|. \tag{5.55}$$

For Δt small enough we conclude that

$$\|\mathbf{U}^{k+1} - \mathbf{U}\| < \|\mathbf{U}^k - \mathbf{U}\|.$$

This implies that the single fixed-point iteration is a *contraction* provided that Δt is small enough, i.e. $\Delta t \leq \Delta t_{cr}$. The Banach Theorem (see [QSS00]) states that, under this condition, the single fixed point iteration is convergent. The limitation on Δt depends on $C(\mathbf{U})$, so in general the critical time step Δt_{cr} is different at each time step. To complete the analysis, we want to guarantee that

$$\min_k \Delta t_{cr}^{(k)} > 0.$$

The bound on Δt is proportional to $1/\sqrt{C(\mathbf{U})}$ and the constant C depends on dk/du. In particular $C(\mathbf{U}) \leq |K_2| \|\mathbf{U}\| + |K_1|$. Under inequality (5.50), we can give a uniform bound to $C(\mathbf{U})$ and conclude that there exists Δt_{max} independent of time that guarantees the convergence. In practice it is difficult to have a precise estimate for Δt_{max}, however the theoretical result suggests that this method can be used even for long time intervals.

An alternative fixed point method with a quadratic convergence rate (provided that the initial guess is close enough to the solution) is the Newton method (see [QSS00]). Let J be the Jacobian matrix for the system, with entries

$$\mathrm{J}_{ij}(\mathbf{U}) = \frac{\partial \mathcal{A}_i}{\partial U_j}(\mathbf{U}),$$

then, the Newton method resorts to solving the sequence of linear systems

$$\mathrm{J}(\mathbf{U}^{n+1,k}) \left(\mathbf{U}^{n+1,k+1} - \mathbf{U}^{n+1,k} \right) = -\mathcal{A}(\mathbf{U}^{n+1,k}), \tag{5.56}$$

for $k = 0, 1, \ldots$, with an initial guess for $\mathbf{U}^{n+1}$ that we denote $\mathbf{U}^{n+1,0}$.

The convergence of the Newton method requires that $\mathbf{U}^{n+1,0}$ is close enough to the solution. It is difficult to estimate the neighborhood of the exact solution that guarantees the success of the method. A typical practical

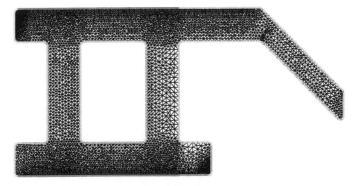

Fig. 5.12 Mesh for the problem of the brake disc

choice is to take $\mathbf{U}^{n+1,0} = \mathbf{U}^n$, the solution at the previous time step. If Δt is small enough, this guess is close to the solution so to ensure the convergence[7].

Any fixed point method requires a stopping criterion. In practice, iterations are performed up to the fulfillment of the *convergence test*

$$||\mathbf{U}^{n+1,k+1} - \mathbf{U}^{n+1,k}|| \leq \epsilon,$$

being $\epsilon > 0$ a prescribed tolerance.

The Newton method has been applied to the algebraic system, according to a workflow that we could call *Discretize then Linearize*. Another possibility is to follow a *Linearize then Discretize* procedure. In the latter case, the differential problem is first linearized then approximated with a linear system. An example of the second approach will be presented in Chapter 7 for the Navier-Stokes equations.

In general a drawback of using fixed point methods (and Newton's in particular) in a time dependent problem is that the iterations for solving the nonlinear problem are nested in the time advancing loop. The computational cost for assembling the matrix and solving the linear system is multiplied by the number of fixed point iterations and eventually of the time steps. This entails high computational costs.

Notice that also the linear system is often solved with an iterative solver. In this case, the nested loops are three (time advancing (non-linear iterations (linear iterations))). Several approaches can be pursued for reducing the CPU cost. Working with implicit schemes, it has been noticed that an inaccurate solution of the linear system (e.g. performing just a few iterations of an iterative method, typically GMRes or Multigrid solvers) does not inhibit the convergence of the nonlinear iterations.

---

[7] More sophisticated approaches can be pursued as well, such as the *continuation method*, see e.g. [QV94], Chapter 10.

On the other hand, explicit time advancing methods are a possible alternative. For instance the forward explicit Euler at each time step leads to the linear system

$$\frac{1}{\Delta t} \mathbf{M U}^{n+1} = \frac{1}{\Delta t} \mathbf{M U}^n - \mathbf{K}(\mathbf{U}^n)\mathbf{U}^n - \mathbf{R U}^n + \mathbf{F}^{n+1}. \qquad (5.57)$$

Should the mass matrix be lumped as we do for reaction dominated problems (see Chapter 4), this system will be actually diagonal, with a great computational saving. However the explicit time advancing introduces a stability condition on the time step. In particular (see [Qua09]) the bound is of the type $\Delta t \leq Ch^2$, a fine mesh requires a very small time step.

In conclusion, with an explicit time advancing method we reduce the cost of each time step, but the number of necessary time steps could be large because of the stability constraint. Implicit time advancing schemes may be unconditionally stable, but they require to solve an involved non-linear problem at each time step.

It is not evident which is the best strategy and the most convenient strategy depends on the specific problem and on the basis of different arguments, concerning for instance the desired accuracy for the solution.

There is a third possibility. In a time dependent problem often is the time discretization error that eventually dominates the accuracy. For this reason, the nonlinear system needs not to be solved with an accuracy greater than that of the time discretization error. The computational cost can be therefore reduced by solving the nonlinear system in an approximate way with an error of the same order of the time discretization. For instance, we can linearize the implicit Euler method as follows. For $n = 0, 1, 2, \ldots, N-1$, find $u^{n+1} \in V$ s.t.

$$\frac{1}{\Delta t} \left(u^{n+1}, v \right) + a(\tilde{u}; u^{n+1}, v) + \alpha \int_{\partial \Omega} u^{n+1} v d\gamma = \frac{1}{\Delta t} \left(u^n, v \right) +$$

$$\alpha \int_{\partial \Omega} u_0 v d\gamma + \int_{\Gamma_{gan}} \Phi^{n+1} v d\gamma, \qquad (5.58)$$

with $u^0 = u_0$. In (5.58) $\tilde{u}$ is an approximation of u^{n+1}. Since we have a first order method, the time accuracy is retained with an approximation of order $\mathcal{O}(\Delta t)$. From the Taylor expansion we easily realize that $\tilde{u} = u^n$ is an approximation of order 1. We call this method *semi-implicit*, since the nonlinear term of the implicit Euler is linearized with an explicit approximation. This approach clearly reduces the computational cost and can be regarded as the result of a fixed point method stopped at the first iteration. This cost reduction is attained at the price of a stability bound on Δt, since the method is partially explicit.

However, notice that:

1. The bound on Δt is in general less restrictive than in a completely explicit method; in particular, since the Laplacian term is partially treated in implicit, the stability bound on Δt will not depend in general on h^2 but on a lower power of the mesh size.
2. The fixed point method (5.51) introduces a bound on the time step, as well. The bound is not for the stability, but for the convergence of the iterations. Therefore, also with this method the use of arbitrarily large time step is prevented.

Upon this discussion, the semi-implicit method is a good trade-off for the problem at hand.

Numerical results

The first step for solving the problem is to build the curve of the thermal conductivity for the cast iron as a function of the temperature, by using the experimental data. This can be done in Matlab with the given data, for instance with the following instructions

```
x=[200;300;700];y=[50;47;37];
polyfit(x,y,2)
```

giving for the function $k(u) = c_2 u^2 + c_1 u + c_0$ the coefficients

$$c_2 = 0.00001, \quad c_1 = -0.035, \quad c_0 = 56.6.$$

Notice that even with more than 3 measures we could have used the same instructions, generating again a quadratic function approximating the data in the least-squares sense. The latter is more convenient when the data are noisy (as it is almost invariably the case).

Fig. 5.13 shows that the thermal conductivity obtained by interpolation is positive in the range of temperature of interest.

Program 28 gives the FreeFem code solving the problem with the implicit Euler method linearized with the fixed point method (5.51). The stopping criterion is based on the quantity $E = |\Omega|^{-1} \int_\Omega \left(u_h^{k+1} - u_h^k\right)^2 d\omega$, where $|\Omega|$ is the area of the section of the disc. Iterations are stopped when $E \leq 10^{-2}$. Fig. 5.14 illustrates the temperature in the disc with this method after 1 and 2 minutes respectively. Each time step requires 2 or 3 iterations for the convergence of the fixed point. Time step $\Delta t = 0.1$ s is enough for the fixed point iterations to converge. The complete code is available on the web site of the book.

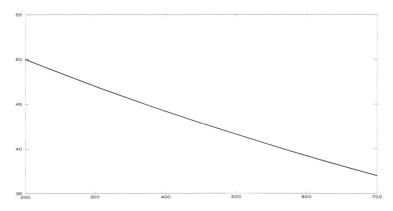

Fig. 5.13 Coefficient $k(u)$ interpolated from the available data

Program 28 - brake-im : Brake disc problem with implicit time advancing method with fixed point iterations

```
// ... functions

func real flux(real tsuT, real T)
{
 real fmax=10;
 real res;
 if (tsuT <= T/2.)
  {res=fmax*sin(2.*pi*tsuT/T);}
 else
   res=0;
 return res;
}

func real alphag(real tsuT, real T, real cstc)
{
 real res;
 if (tsuT <= T/2.)
  {res=0;}
 else
   res=cstc;
 return res;
}

// ... problem

problem Brake(u,v) =
    int2d(Th)((c1*ulast*ulast+c2*ulast+c3)*(dx(u)*dx(v)+dy(u)*dy(v)))
  + int2d(Th)(rhocp*dtm1*u*v)
  + int1d(Th,20)(cstc*u*v)
  - int1d(Th,20)(cstc*uamb*v)
  + int1d(Th,10)(alphag((i%iT)*dt,T,cstc)*u*v)
```

```
    - int1d(Th,10)(alphag((i%iT)*dt,T,cstc)*uamb*v)
    - int1d(Th,10)(flux((i%iT)*dt,T)*v)
    - int2d(Th)(rhocp*dtm1*uold*v);

// Time loop
for (i=1;i<=N;i++)
{
 k=0; resL2=1.;
 while (k<=nmax & resL2>=toll) // fixed point iteration
 {
  Brake;
  w[]=u[]-ulast[];
  resL2 = int2d(Th)(w*w)/abrake; //abrake=area of the disc
  ulast[]=u[];
  k++;
 }
 uold[]=u[];
}
```

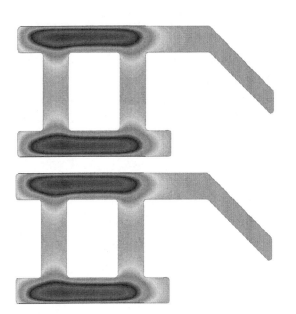

Fig. 5.14 Temperature in the brake section with the implicit time advancing + fixed point iterations. Solution after 1 minute (top) and 2 minutes (bottom). Dark colors are associated with high temperature. Temperature ranges between 22.05 °C (71.69 °F) and 141.21 °C (286.17 °F) in the top part and 22.07 °C (71.72 °F) and 142.21 °C (287.97 °F) at the bottom

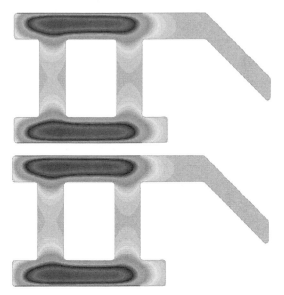

Fig. 5.15 Temperature in the brake section with the implicit time advancing + fixed point iterations. Solution after 1 minute (top) and 2 minutes (bottom). Dark colors are associated with the high temperature. Temperature ranges between 22.05 °C (71.69 °F) and 141.16 °C (286.09 °F) in the top part and 22.07 °C (71.73 °F) and 142.1 °C (287.78 °F) in the bottom

Program 29 considers the semi-implicit approach (5.58). Results at $t = 1$ minute and $t = 2$ minutes are in Fig. 5.15. We note that the difference with the implicit case are below 1°C (33.8 °F), which is perfectly acceptable for this application.

For the CPU time

Implicit (+ fixed point):   1631.15 s;
Semi-implicit: 781.71 s.

These indications have to be intended in a comparative sense[8].

Program 29 - Brake-si : Brake disc problem with semi-implicit time advancing method

```
problem BrakeSI(u,v) =
    int2d(Th)((c1*uold*uold+c2*uold+c3)*(dx(u)*dx(v)+dy(u)*dy(v)))
  + int2d(Th)(rhocp*dtm1*u*v)   + int1d(Th,20)(cstc*u*v)
  - int1d(Th,20)(cstc*uamb*v)
  + int1d(Th,10)(alphag((i%iT)*dt,T,cstc)*u*v)
  - int1d(Th,10)(alphag((i%iT)*dt,T,cstc)*uamb*v)
```

---

[8] It is worth reminding that the basic versions of FreeFem and of Matlab produce parsed code, which is not efficient in general. The indications of CPU time is purely indicative.

```
  - int1d(Th,10)(flusso((i%iT)*dt,T)*v)
  - int2d(Th)(rhocp*dtm1*uold*v);
for (i=1;i<=N;i++) // Time loop
{
  BrakeSI;
  uold=u;
}
```

Program 30 - Brake-ex : Brake disc problem with explicit (Euler) time advancing method

```
problem BrakeExp(u,v) =
int2d(Th)((c1*uold*uold+c2*uold+c3)*(dx(uold)*dx(v)+dy(uold)*dy(v)))
  + int2d(Th)(rhocp*dtm1*u*v)   + int1d(Th,20)(cstc*uold*v)
  - int1d(Th,20)(cstc*uamb*v)
  + int1d(Th,10)(alphag((i%iT)*dt,T,cstc)*uold*v)
  - int1d(Th,10)(alphag((i%iT)*dt,T,cstc)*uamb*v)
  - int1d(Th,10)(flusso((i%iT)*dt,T)*v)
  - int2d(Th)(rhocp*dtm1*uold*v);
for (i=1;i<=N;i++)
{
  BrakeExp;
  uold=u;
}
```

The time step $\Delta t = 0.1$ s for the semi-implicit method results to be stable. On the contrary, the explicit Euler scheme with this step is unstable as it can be verified with Program 30 in Fig. 5.16.

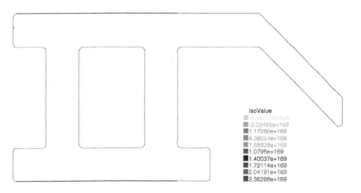

Fig. 5.16 Explicit Euler method for the brake problem: numerical solution is unstable (it ranges around 10^{169}, which is clearly meaningless)

These results corroborate the expectation that the semi-implicit approach is a good trade-off between stability requirements and computational costs. $\Diamond$

Remark 5.4 Numerical data given in this exercise are clearly for an idealized situation, not corresponding to a real brake. Also, the assumption that the iron cast specific heat is constant is unrealistic. It is actually function of the temperature u.

Selection of geometry and material for a correct cooling of the brake is an important step in brakes design (in view also of industrial constraints). The brake disc can develop fractures induced by the thermo-mechanical stress. Numerical simulations remarkably support the design, giving quantitative indications on the thickness of the disc and its optimal shape.

Remark 5.5 Among the emerging applications where partial differential equations are progressively used we mention the dynamics of population in biology. The evolution of the density of individuals of a certain population in a geographic region can be described by a parabolic problem. This is obtained by the balance between incoming individuals (new born or immigrated) and the outgoing (death or emigrated). The diffusive term in this case represents the natural spreading of individuals in the region. The first order term accounts the effect of an ordered directional dynamics as the one induced by the tide for a population of fishes. The zero-th order term includes the reproductive dynamics and the interaction with other population within the same territory. In Fig. 5.1 we report for instance the result of a simulation of the spreading of rabies in Raccoons in the State of New York (USA). In the model, three population of raccoons have been modeled, the Susceptible (S), the Exposed (E) and the Infectious (I) individuals, corresponding to different stages of the epidemic. In more complex models a Recovered population (see [Kel11]) could be considered. This SEI model leads to a system of parabolic equations with nonlinear reactive terms. The coefficients of the different terms of the equations include possible geographical features. For instance, the presence of the Husdon river has been modeled by a local modification of the diffusivity tensors. Simulations have been carried out moving from a satellite map of the state, meshed with the free code `NetGen` [Sch]. A complete discussion of this problem is beyond the scope of the book. The interested reader is referred to [Kel11, KGV11]. On the web site associated to this book it is possible to download the mesh we have used for the simulations, for the sake of reproducibility of the computations.

Fig. 5.17 Left: Nodes of the reticulation of the New York State (USA) adopted for the simulation of the spread of rabies in Raccoons. Center: Snapshot of Susceptible individuals of the SEI model at a given instant of the simulation period. Right: Snapshot of Infectious individuals of the SEI model at a given instant of the simulation period

5.2 Finite element time discretization

Time discretization can be carried out with a Galerkin approach and finite element in particular. In this case, the time dependence is treated as the space dependence. For a problem with d space dimensions, we will resort to a $d+1$ dimensional problem. Notice however that for the time variable we have to solve an initial value problem. Moreover, the heat equation is of first order in time and second order in space. The different features of the dependence on space and time and the computational costs suggest in any case a different treatment of the two variables (even though both with finite element). For more details, see e.g. [QV94], Chapter 11, [QSS00], Chapter 13, [Joh87], Chapter 6.

Exercise 5.2.1. Find u such that

$$\begin{cases} \dfrac{\partial u}{\partial t} - \mu \dfrac{\partial^2 u}{\partial x^2} + \beta \dfrac{\partial u}{\partial x} + \sigma u = f & 0 < x < 1, \quad 0 < t < T, \\[2mm] u(0,t) = u(1,t) = 0, & 0 \le t \le T, \\[2mm] u(x,0) = u_0(x), & 0 < x < 1. \end{cases}$$

Assume that μ and σ are real positive constants, $\beta \in \mathbb{R}$, $f \in L^2(0,T;L^2(0,1))$ and $u_0 \in L^2(0,1)$.

Write the weak formulation, analyze the well posedness and give an *a priori* bound for the solution. Perform the finite element discretization for both space and time variables. Give a stability estimate for the discrete solution in the case $f = 0$.

Solution 5.2.1.

Mathematical analysis

The weak formulation of the problem is obtained as usual by multiplying the differential equation by a function $v \in H_0^1(0,1)$. We obtain the problem: for $t > 0$ find $u \in H_0^1(0,1)$ such that for any $v \in H_0^1(0,1)$

$$\int_0^1 \frac{\partial u}{\partial t} v\,dx + a(u,v) = \int_0^1 f v\,dx, \tag{5.59}$$

where the bilinear form

$$a(u,v) \equiv \mu \int_0^1 \frac{\partial u}{\partial x}\frac{\partial v}{\partial x}\,dx + \beta \int_0^1 \frac{\partial u}{\partial x} v\,dx + \sigma \int_0^1 uv\,dx$$

is continuous and coercive. Coercivity in particular follows from

$$\beta \int_0^1 \frac{\partial u}{\partial x} u dx = \beta \frac{1}{2} \int_0^1 \frac{\partial u^2}{\partial x} dx = \frac{\beta}{2} \left[u^2 \right]_0^1 = 0$$

thanks to the boundary conditions on u. Therefore,

$$a(u,u) \geq \min(\mu,\sigma) \left(\left\| \frac{\partial u}{\partial x} \right\|_{L^2(\Omega)}^2 + \|u\|_{L^2(\Omega)}^2 \right) = \min(\mu,\sigma) \|u\|_{H_0^1(0,1)}^2,$$

with $\min(\mu,\sigma) > 0$ for the assumptions on the data. Moreover, the assumptions on the forcing term and the initial data guarantee that each term in (5.59) is well defined.

Well posedness follows from standard arguments. An *a priori* estimate for the solution is obtained by taking $v = u(t)$ in (5.59) and integrating by parts. We obtain

$$\int_0^T \frac{\partial}{\partial t} \|u\|_{L^2(\Omega)}^2 dt + 2 \min(\mu,\sigma) \int_0^T \|u\|_{H^1(0,1)}^2 dt \leq 2 \int_0^T \int_0^1 f u dx dt.$$

The Young inequality on the right hand side yields

$$\left| 2 \int_0^T \int_0^1 f u dx dt \right| \leq \frac{1}{\epsilon} \int_0^T \|f\|_{L^2(0,1)}^2 dt + \epsilon \int_0^T \|u\|_{L^2(0,1)}^2 dt,$$

for any $\epsilon > 0$. For $\epsilon = \min(\mu,\sigma)/2$, we obtain

$$\|u\|_{L^2(0,1)}^2(T) + \min(\mu,\sigma) \int_0^T \|u\|_{H^1(0,1)}^2 dt \leq$$

$$\|u_0\|_{L^2(0,1)}^2 + \frac{2}{\min(\mu,\sigma)} \int_0^T \|f\|_{L^2(0,1)}^2 dt.$$

Consequently, $u \in L^\infty(0,T;L^2(0,1)) \cap L^2(0,T;H_0^1(0,1))$.

There is another possible way for formulating weakly the problem.

For a test function $\psi : (0,T) \to \mathbb{R}$ and noticing that $\left(\frac{\partial u}{\partial t}, v \right) = \frac{d}{dt}(u,v)$, we have the following space-time weak formulation: find u such that

$$\int_0^T \left(\frac{d}{dt} \int_0^1 u(t) v dx \right) \psi(t) dt + \int_0^T a(u,v) \psi(t) dt = \int_0^T \left(\int_0^1 f v dx \right) \psi(t) dt, \quad (5.60)$$

with $u = u_0$ for $t = 0$. The latter equation holds for all $v \in H_0^1(0,1)$ and for any ψ with a compact support in $(0,T)$ and regular enough. This is the formulation used for the space-time finite elements.

Numerical approximation

Let us introduce a subdivision of the time interval $(0,T)$ with uniform (for simplicity) step Δt. For the time being we assume also a uniform (in space and time) spatial mesh with step h. The space-time domain is represented in Fig. 5.18. We actually consider a 2D problem with a quadrilateral reticulation. In view of the different nature of the problem in space and time, we assume to work with discontinuous functions in time. This allows the solution of a "local" system, referring to each time interval. This allows also to take different space grids at each time interval (see Fig. 5.19).

Let I_n be the interval (t^{n-1}, t^n). The region

$$S_n = [0,1] \times \overline{I}_n,$$

is called *space-time slab*. In each *slab* S_n, space mesh introduces a set of rectangles R_{jn} (see Fig. 5.18). Let W_{rn}^k be the space of functions defined on $(0,1) \times I_n$, polynomials of order k in time and piecewise polynomials of order r in space,

$$W_{rn}^k \equiv$$

$$\left\{ w : (0,1) \times I_n \to \mathbb{R} \,\middle|\, w(x,t) = \sum_{i=0}^{k} \psi_i(t) v_{ih}(x), x \in (0,1), t \in (0,T) \right\}.$$

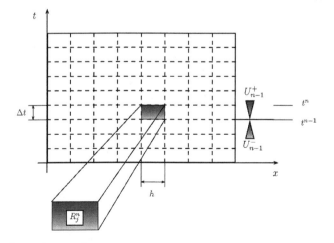

Fig. 5.18 Space-time mesh

Here:

1. $\psi_i(t)$ are the basis functions of $\mathbb{P}^k(I_n)$. A possible (non-Lagrangian) choice is

$$\psi_j(t) = t^j, \quad j = 0, \ldots, k.$$

   A Lagrangian basis requires the introduction of intermediate instants in each sub-interval.
2. $v_{ih}(x)$ is a piecewise polynomial function with degree r in each interval of the mesh in the slab and vanishes at the end points of $(0, 1)$. We can write

$$v_{ih}(x) = \sum_{j=1}^{N_h-1} V_{ij}\varphi_j(x),$$

   where N_h is the number of space sub-intervals of $(0, 1)$ in the slab S_n.

The global definition of the discrete space, with discontinuous functions in time over $(0, T)$ reads

$$W_r^k \equiv \left\{ w : (0, T) \times (0, 1) \to \mathbb{R} \mid w|_{I_n} \in W_{rn}^k, n = 1, \ldots, N \right\}.$$

In the sequel, a function $w \in W_{rn}^k$ belongs to W_r^k by the trivial (zero) extension to the slabs different from S_n. More precisely, for $w \in W_{rn}^k$, we consider the function $w_{ext} \in W_r^k$

$$w_{ext} = \begin{cases} w \text{ for } t \in I_n, \\ 0 \text{ for } t \notin I_n. \end{cases}$$

With a little abuse of notation, we indicate w_{ext} with w.

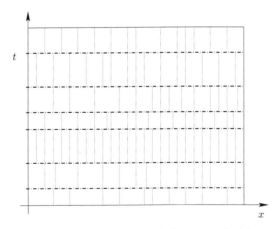

Fig. 5.19 Space-time mesh with different reticulations on each slab

To obtain the weak form discretized in space and time we multiply the differential equation by a function v of W_r^k and integrate both in x and t,

$$\int_0^T \left(\int_0^1 \frac{\partial u}{\partial t} v dx + a(u,v) \right) dt = \int_0^T \int_0^1 f v dx dt, \qquad (5.61)$$

for all $v \in W_r^k$. Notice that v now depends also on t.

Equation (5.61) is only formal. Actually, the term $\int_0^T \int_0^1 \frac{\partial u}{\partial t} v dx \, dt$ is not well defined, since u and v are discontinuous across the slabs. Moreover, in this formulation space and time variables are treated in the same way. However, the time dependence features an initial value condition and moves in the "direction" specified by the time advancing. We use this circumstance for splitting the problem (5.61) into a sequence of problems in each slab S_n. Since we have time-discontinuous functions, we will actually choose functions vanishing in all the slabs but one. As we will see, the drawback is that the number of degrees of freedom is larger than the one of the continuous case.

If we just reduce the time integrals in (5.61) to integrals on I_n, for $n = 1, 2, \ldots, N$, we would get a sequence of decoupled and independent problems. In the first slab we can prescribe the initial data and however the subsequent slabs would result independent of the initial conditions. We need to formulate a method such that the data are transmitted from a slab to the subsequent one. We could *penalize* the discontinuity in time of the numerical solution. This corresponds to adding in each slab a proper term proportional to the jump of the solution at the interface between two slabs. This term should vanish for a continuous solution (in time).

More precisely, we set

$$[u_l] \equiv u_l^+ - u_l^-, \quad \text{con } u_l^\pm = \lim_{s \to \pm 0} u(t^l \pm s), \quad \forall l = 1, 2, \ldots$$

Moreover, let $[u_0] \equiv u_0^+ - u_0$, where u_0 is the initial data. We write the formulation

$$\int_{I_n} \left(\int_0^1 \frac{\partial u}{\partial t} v dx + a(u,v) \right) dt + \boxed{\sigma \int_0^1 [u_{n-1}] v_{n-1}^+ dx} = \int_{I_n} \int_0^1 f v dx dt, \quad (5.62)$$

for any $v \in W_{rn}^k$, with σ suitable coefficient.

The boxed term is weakly forcing the time continuity of the numerical solution by penalizing the jump. In this way we actually introduce the propagation of the solution from one slab to the subsequent. However we have to define a reasonable value for σ. Let us start formally from (5.61). In general $v \in W_{rn}^k$ is not (classically) differentiable in time, because it is discontinuous. For instance, if v equals α on I_n and 0 elsewhere, we write

$$v(t) = \alpha \left(H(t - t_{n-1}) - H(t - t_n) \right),$$

where $H(t - \bar{t})$ is the well known Heaviside function

$$H(t - \bar{t}) = \begin{cases} 0 \text{ for } t < \bar{t} \\ 1 \text{ for } t \geq \bar{t} \end{cases}.$$

The characteristic function of the interval (t_{n-1}, t_n) reads $\chi_n \equiv H(t - t_{n-1}) - H(t - t_n)$. Distributional derivative reads

$$\frac{dv}{dt} = \alpha\delta(t_{n-1}) - \alpha\delta(t_n),$$

where $\delta(\bar{t})$ is the Dirac δ distribution, centered in $\bar{t}$, so that

$$\int_0^T \frac{dv}{dt}\psi = \alpha\psi(t_{n-1}) - \alpha\psi(t_n)$$

for any $\psi \in H^1(0, T)^9$.

For a linear $v(t)$ in I_n (and zero in the other slabs) we could write

$$v(t) = (\alpha + \gamma(t - T_{n-1}))\chi_n,$$

for $\alpha, \gamma \in \mathbb{R}$. In particular, $v_{n-1}^+ = \alpha$ and $v_n^- = \alpha + \Delta t\gamma$. Time derivative in this case is

$$\frac{dv}{dt} = \alpha\delta(t_{n-1}) - (\alpha + \Delta t\gamma)\delta(t_n) + \gamma\chi_n.$$

In general, for a time dependent function $v(t)$,

$$\frac{\partial v}{\partial t} = v_{n-1}^+\delta(t_{n-1}) - v_n^-\delta(t_n) + \dot{v}\chi_{I_n}$$

where $\dot{v}$ is the time derivative of v inside the slab (where the function v is regular).

In the first term in (5.61), we revert the order of space and time integration. After an integration by part for the time variable, we formally obtain

$$\int_0^1 \int_0^T \frac{\partial u}{\partial t}v\,dt\,dx = \int_0^1 [uv]_0^T\,dx - \int_0^1 \int_0^T u\frac{\partial v}{\partial t}\,dt\,dx. \tag{5.63}$$

Again, we say "formally" because the integral on the right hand side is not well defined. In fact, the Dirac δ in the derivative of v are applied to u, that in general is not continuous at the interface between two slabs. Let us define

---

[9] The integral $\int_0^T \frac{dv}{dt}\psi$ has to be understood in a generalized sense. More precisely, we should write $< \frac{dv}{dt}, \psi >$ to point out that dv/dt belongs to $H^{-1}(0, 1)$, the dual space of $H_0^1(0, T)$.

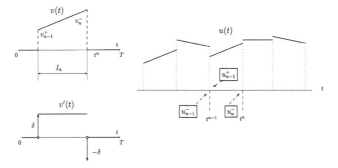

Fig. 5.20 Linear function on a slab I_n (top, left) and its derivative (bottom, left). On the right a piecewise linear discontinuous function on the time domain

arbitrarily (but reasonably) this term. With the given notation, we set

$$-\int_0^T u\frac{\partial v}{\partial t}dtdx \equiv -v_{n-1}^+ u_{n-1}^- + v_n^- u_n^- - \int_{t_{n-1}}^{t_n} u\dot{v}dtdx.$$

In practice, we assume that the Dirac δ applied to u returns the value in the opposite direction to the time. The rationale behind this choice is similar to the one of the upwind correction of a centered derivative, namely *significant data are assumed to be in the upstream direction*. Here "upstream" means backward in time. With this assumption in (5.63) we get

$$\int_0^1\int_0^T \frac{\partial u}{\partial t}vdtdx = \int_0^1\left([uv]_0^T - v_{n-1}^+ u_{n-1}^- + v_n^- u_n^- - \int_{t_{n-1}}^{t_n} u\dot{v}dt\right)dx. \quad (5.64)$$

Assume for the moment being that S_n is not either the first or the last slab. We have $v(0) = v(T) = 0$ and consequently

$$\int_0^1\int_0^T \frac{\partial u}{\partial t}vdtdx = \int_0^1\left(-v_{n-1}^+ u_{n-1}^- + v_n^- u_n^- - \int_{t_{n-1}}^{t_n} u\dot{v}dt\right)dx.$$

If we back integrate by parts on S_n, where all the functions are infinitely regular, we get

$$\int_0^1\int_0^T \frac{\partial u}{\partial t}vdtdx = \int_0^1\Big(- v_{n-1}^+ u_{n-1}^- + v_n^- u_n^- -$$

$$v_n^- u_n^- + v_{n-1}^+ u_{n-1}^+\Big)dx = \int_0^1\left(v_{n-1}^+\left[u_{n-1}^+\right] + \int_{t_{n-1}}^{t_n} \frac{\partial u}{\partial t}vdt\right)dx. \quad (5.65)$$

Reverting again the order of space and time integration we obtain the weak form: find $u \in W_r^k$ s.t.

$$\int_{I_n} \left(\int_0^1 \frac{\partial u}{\partial t} v dx + a(u,v) \right) dt + \int_0^1 [u_{n-1}] v_{n-1}^+ dx = \int_{I_n} \int_0^1 f v dx dt, \qquad (5.66)$$

for all $v \in W_r^k$ (with the exception of the first and the last slab).

On the first slab, the time derivative of V does not give an impulse (a Dirac δ), since the end point of the slab and of the entire domain coincide. However, in this case $v(0) \neq 0$ and we have

$$\int_0^1 \int_0^T \frac{\partial u}{\partial t} v dt dx = \int_0^1 \left(-u_0^- v_0^+ \right) dx + \int_0^1 \left(v_1^- u_1^- - \int_{t_0}^{t_1} u v dt \right) dx,$$

where, for the sake of a uniform notation, we denote the initial condition with u_0^-. In practice, in the first time slab the *upwind* data are given by the initial condition.

Similarly, in the last time slab we do not have an impulse in the time derivative of v at the last instant. However in general $v_N^- \neq 0$ so that we obtain

$$\int_0^1 \int_0^T \frac{\partial u}{\partial t} v dt dx = \int_0^1 u_N^- v_N^- dx + \int_0^1 \left(-v_{N-1}^+ u_{N-1}^- - \int_{t_{N-1}}^T u v dt \right) dx.$$

We conclude that (5.65) and (5.66) hold also for $n = 1$ and $n = N$.

The complete the weak formulation on the time interval $(0,T)$ is obtained by summing up the (5.66) on any slab: find $u \in W_r^k$ s.t.

$$\sum_{n=1}^N \left[\int_{I_n} \left(\int_0^1 \frac{\partial u}{\partial t} v dx + a(u,v) \right) dt + \int_0^1 [u_{n-1}] v_{n-1}^+ dx \right] = \\ \sum_{n=1}^N \int_{I_n} \int_0^1 f v dx dt, \qquad (5.67)$$

for all $v \in W_r^k$. Equation (5.66) is in turn obtained by this equation by taking null test functions on all the intervals apart I_n, for any $n = 1, \ldots, N$. If we compare this equation with (5.62) we realize that the penalization term occurs with $\sigma = 1$.

The stability estimate with $f = 0$ is obtained by setting $v = u$ in (5.66).
We have

$$\int_0^1 \left(u_{n-1}^+ - u_{n-1}^- \right) u_{n-1}^+ dx =$$

$$\frac{1}{2} \left(||u_{n-1}^+||_{L^2(0,1)}^2 + ||[u_{n-1}]||_{L^2(0,1)}^2 - ||u_{n-1}^-||_{L^2(0,1)}^2 \right),$$

so that thanks to the coercivity of the bilinear form $a(\cdot,\cdot)$,

$$\frac{d}{dt} \int_{I_n} ||u||_{L^2(0,1)}^2 + 2\alpha \int_{I_n} ||u||_{H^1(0,1)}^2 dt + ||u_{n-1}^+||_{L^2(0,1)}^2 +$$
$$||[u_{n-1}]||_{L^2(0,1)}^2 \leq ||u_{n-1}^-||_{L^2(0,1)}^2.$$

It follows that

$$||u_n^-||_{L^2(0,1)}^2 + 2\alpha \int_{I_n} ||u||_{H^1(0,1)}^2 dt + ||[u_{n-1}]||_{L^2(0,1)}^2 \leq ||u_{n-1}^-||_{L^2(0,1)}^2.$$

By summing on all the slabs, we get the stability estimate

$$||u_N^-||_{L^2(0,1)}^2 + 2\alpha \sum_{n=1}^N \int_{I_n} ||u||_{H^1(0,1)}^2 dt + \sum_{n=1}^N ||[u_{n-1}]||_{L^2(0,1)}^2 \leq ||u_0^+||_{L^2(0,1)}^2. \quad \Diamond$$

Exercise 5.2.2. Find u such that

$$\begin{cases} \dfrac{\partial u}{\partial t} - \mu \dfrac{\partial^2 u}{\partial x^2} = f, & (x,t) \in (0,1) \times (0,T), \\ u(0,t) = u(1,t) = 0, & 0 < t \leq T, \quad u(x,0) = u_0(x), 0 \leq x \leq 1. \end{cases}$$

Assume μ to be a positive constant, $f \in L^2(0,T; L^2(0,1))$, and $u_0 \in L^2(0,1)$.

Discretize the problem with space-time finite elements. Assume a constant space grid in time and use P^0 finite elements for the time variable (piecewise constant in time). Write the discrete problem and verify that it coincides with the backward Euler method apart from the numerical treatment of the forcing term f.

Solution 5.2.2.

Numerical approximation

Well posedness of the heat equation has been already discussed in the previous exercises. For the space-time finite element discretization we first introduce a splitting of the time interval as done in the previous exercise. We assume again a constant time step Δt and that the solution is constant in time over each time slab. Let S be the space of piecewise constant functions and $H_0^1(0,1)$ in space. The basis for piecewise constant functions in time is given by the characteristic function of I_n

$$\psi_n(t) = \begin{cases} 1 & t \in I_n, \\ 0 & \text{elsewhere.} \end{cases}$$

Multiply the given equation by these functions and integrate on the interval $(0, T)$. We obtain the formulation: find $u_{\Delta t} \in S$ s.t.

$$\int_0^T \left(\frac{\partial u_{\Delta t}}{\partial t} - \mu \frac{\partial^2 u_{\Delta t}}{\partial x^2} \right) \psi_n dt = \int_0^T f \psi_n dt$$

for any $n = 1, \ldots, N$.

Proceeding as in the previous exercise, we get

$$\int_{I_n} \left(\frac{\partial u_{\Delta t}}{\partial t} - \mu \frac{\partial^2 u_{\Delta t}}{\partial x^2} \right) dt + u_{\Delta t}(t_{n-1}^+) - u_{\Delta t}(t_n^-) = \mathcal{F}^n(f),$$

where

$$\mathcal{F}^n(f) = \int_0^T f \psi_n dt = \int_{I_n} f dt.$$

Since the solution is constant on each slab, we set for each n

$$u_{\Delta t}(t_{n-1}^+) = u_{\Delta t}(t_n^-) = u_{\Delta t}^n.$$

Moreover, $\dfrac{\partial u_{\Delta t}}{\partial t} = 0$ on each slab, so that the semi-discrete (discrete in space, continuous in time) problem reads

$$-\Delta t \mu \frac{\partial^2 u_{\Delta t}^n}{\partial x^2} + u_{\Delta t}^n - u_{\Delta t}^{n-1} = \mathcal{F}^n(f).$$

Space discretization is performed in a standard way, leading to the discrete problem

$$\left(u_{h,\Delta t}^n, \varphi_j \right) - \left(u_{h,\Delta t}^{n-1}, \varphi_j \right) + \Delta t \mu \left(\frac{\partial u_{h,\Delta t}^n}{\partial x}, \frac{\partial \varphi_j}{\partial x} \right) = \left(\mathcal{F}^n(f), \varphi_j \right),$$

for $j = 1, \ldots, N_h$, where $(\cdot, \cdot)$ is the scalar product in $L^2(0,1)$ and the φ_j (with $j = 1, 2, \ldots, N_h$) are the basis function of the finite dimensional space for the spatial discretization. Setting $u_{h,\Delta t}^n = \sum_{i=1}^{N_h} u_i^n \varphi_i$, the fully discrete problem reads

$$\mathbf{M} \mathbf{U}^n + \Delta t \mu \mathbf{K} \mathbf{U}^n = \mathbf{M} \mathbf{U}^{n-1} + \mathbf{F}^n, \tag{5.68}$$

with M and K mass and stiffness matrices respectively and $\mathbf{U}^n$ the vector with the nodal values U_i^n. The first term on the left hand side coincides with the term obtained with a backward Euler scheme (and a space discretization with finite element of order r). However, the scheme is different from a backward Euler method for the term on the right hand side. We have in fact

$$F_j^n = \int_{I_n} \int_0^1 f \varphi_j \, dx \, dt,$$

while with a backward Euler method we would have,

$$F_{DFj}^n = \Delta t \int_0^1 f(t^n) \varphi_j \, dx.$$

This can be actually regarded as an approximation of F_j^n when the integral is approximated with the rectangle formula $\int_a^b f(t)dt \approx (b-a)f(b)$. $\diamond$

Remark 5.6 If we allow the space grid to change in time, we need to introduce the basis function set depending on n, $\varphi_j^n(x)$ for each slab. In this case, (5.68) modifies in

$$\mathrm{M}^{n,n}\mathbf{U}^n + \Delta t \mathrm{K}^{n,n}\mathbf{U}^n - \mathrm{M}^{n-1,n}\mathbf{U}^{n-1},$$

where

$$m_{ij}^{n,n} = \int_0^1 \varphi_i^n \varphi_j^n \, dx, \quad k_{ij}^{n,n} = \int_0^1 \frac{\partial \varphi_i^n}{\partial x} \frac{\partial \varphi_j^n}{\partial x} \, dx, \quad m_{ij}^{n-1,n} = \int_0^1 \varphi_i^{n-1} \varphi_j^n \, dx.$$

Notice that $\mathrm{M}^{n-1,n}$ is in general a rectangular matrix since the number of degrees of freedom in the two slabs can be different.

Exercise 5.2.3. Repeat the Exercise 5.2.2 with P^1 finite elements for the time discretization (discontinuous piecewise linear functions). Write the full discretization of the problem.

Solution 5.2.3.

Numerical approximation

Let us perform first the time discretization with a constant time step Δt, then the space discretization. Let W be the space of piecewise discontinuous linear functions

$$W \equiv \left\{ w : w|_{I_n} = a^n(x)t + b^n(x) \quad a^n, b^n \in H_0^1(0,1) \right\},$$

where $I_n = (t_{n-1}, t_n)$. We introduce a Lagrange basis functions set in time in each slab,

$$\psi_0^n(t) = \begin{cases} \dfrac{t^n - t}{\Delta t} & t \in I_n, \\[2mm] 0 & \text{elsewhere,} \end{cases} \quad \text{and} \quad \psi_1^n(t) = \begin{cases} \dfrac{t - t^{n-1}}{\Delta t} & t \in I_n, \\[2mm] 0 & \text{elsewhere.} \end{cases}$$

Consequently

$$W \equiv \left\{ w : w|_{I_n} = W_{n-1}^+ \psi_0^n + W_n^- \psi_1^n, \ W_{n-1}^+, W_n^- \in H_0^1(0,1) \right\}.$$

Multiply the equation by ψ_0^n and by ψ_1^n and integrate on $(0, T)$. The integral clearly reduces on the slab I_n and the discontinuity is treated as in the previous exercise. We obtain the 2×2 system of differential equation in x

$$\begin{cases} \beta_{0,0}^n u_{n-1}^+(x) - \sigma_{0,0}^n \mu \dfrac{\partial^2 u_{n-1}^+}{\partial x^2} + \beta_{1,0}^n u_n^-(x) + \\[4mm] \qquad -\sigma_{1,0}^n \mu \dfrac{\partial^2 u_n^-}{\partial x^2} + \psi_0^n(t^{n-1})\left(u_{n-1}^+(x) - u_{n-1}^-(x)\right) = \mathcal{F}_0^n, \\[5mm] \beta_{0,1}^n u_{n-1}^+(x) - \sigma_{0,1}^n \mu \dfrac{\partial^2 u_{n-1}^+}{\partial x^2} + \beta_{1,1}^n u_n^-(x) + \\[4mm] \qquad -\sigma_{1,1}^n \mu \dfrac{\partial^2 u_n^-}{\partial x^2} + \psi_1^n(t^{n-1})\left(u_{n-1}^+(x) - u_{n-1}^-(x)\right) = \mathcal{F}_1^n, \end{cases} \tag{5.69}$$

where, for $i, j = 0, 1$,

$$\beta_{i,j}^n = \int_{I_n} \frac{d\psi_i^n}{dt} \psi_j^n \, dt, \quad \sigma_{i,j}^n = \int_{I_n} \psi_i^n \psi_j^n \, dt \quad \text{and} \quad \mathcal{F}_i^n = \int_{I_n} f \psi_i^n \, dt.$$

We obtain therefore

$$\beta^n = \begin{bmatrix} -\dfrac{1}{2} & -\dfrac{1}{2} \\[3mm] \dfrac{1}{2} & \dfrac{1}{2} \end{bmatrix} \quad \text{and} \quad \sigma^n = \begin{bmatrix} \dfrac{1}{3} & \dfrac{1}{6} \\[3mm] \dfrac{1}{6} & \dfrac{1}{3} \end{bmatrix} \Delta t.$$

Since $\psi_0^n(t^{n-1}) = 1$ and $\psi_1^n(t^{n-1}) = 0$ system (5.69) reads

$$\begin{cases} \left(\dfrac{1}{2} - \dfrac{\mu \Delta t}{3} \dfrac{\partial^2}{\partial x^2}\right) u_{n-1}^+ + \left(\dfrac{1}{2} - \dfrac{\mu \Delta t}{6} \dfrac{\partial^2}{\partial x^2}\right) u_n^- = u_{n-1}^- + \displaystyle\int_{I_n} f\psi_0^n \, dt, \\[5mm] \left(-\dfrac{1}{2} + \dfrac{\mu \Delta t}{6} \dfrac{\partial^2}{\partial x^2}\right) u_{n-1}^+ + \left(\dfrac{1}{2} - \dfrac{\mu \Delta t}{3} \dfrac{\partial^2}{\partial x^2}\right) u_n^- = \displaystyle\int_{I_n} f\psi_1^n \, dt. \end{cases} \tag{5.70}$$

Discretization in space of this system is carried out by introducing the finite dimensional space $V_h \subset H_0^1(0,1)$ with basis φ_i $(i = 1, \ldots, N_h)$. Set

$$u_{n-1}^+(x) = \sum_{j=1}^{N_h} u_j^{n-1,+} \varphi_j(x), \quad u_{n-1}^-(x) = \sum_{j=1}^{N_h} u_j^{n-1,-} \varphi_j(x),$$

$$u_n^-(x) = \sum_{j=1}^{N_h} u_j^{n,-} \varphi_j(x), \tag{5.71}$$

multiply (5.70) by φ_i, integrate over $(0,1)$ and perform the "usual" integration by parts. We obtain the algebraic system

$$
\begin{cases}
\left(\dfrac{M}{2} + \dfrac{\mu \Delta t}{3} K \right) \mathbf{U}_{n-1}^+ + \left(\dfrac{M}{2} + \dfrac{\mu \Delta t}{6} K \right) \mathbf{U}_n^- = M \mathbf{U}_{n-1}^- \\[2mm]
\quad + \displaystyle\int_0^1 \int_{I_n} f \psi_0^n \varphi_i \, dt\, dx, \\[4mm]
\left(-\dfrac{M}{2} + \dfrac{\mu \Delta t}{6} K \right) \mathbf{U}_{n-1}^+ + \left(\dfrac{M}{2} + \dfrac{\mu \Delta t}{3} K \right) \mathbf{U}_n^- = \displaystyle\int_0^1 \int_{I_n} f \psi_1^n \varphi_i \, dt\, dx,
\end{cases}
\tag{5.72}
$$

where M and K are the mass and stiffness matrices respectively and vectors $\mathbf{U}_n^\pm$ contain the nodal values of the solution in t^n from the left (with $-$) and from the right $(+)$, $\mathbf{U}_{n,i}^\pm = \lim_{s \to 0^\pm} u_h(x_i, t^n + s)$.

This system represents the discretization of the heat equation with continuous finite elements in space and linear discontinuous finite elements in time. Discontinuity allows to solve the problem as a sequence of problems in each slab. With continuous in time finite elements, we solve a problem over the entire time interval, with consequent high computational costs (in 2D and 3D in particular). ◇

Remark 5.7 We notice again that if the space grid changes in each time slab, the first term on the right hand side of (5.72)$_1$ becomes $M^{n-1,n} \mathbf{U}_n^-$, where $M^{n-1,n} = \int_0^1 \varphi_j^{n-1} \varphi_i^n \, dx$ where φ_j^{n-1} denotes the basis functions in the slab $[t^{n-2}, t^{n-1}]$.

Remark 5.8 Concerning the accuracy, it is possible to prove that (see [Joh87]) for $W = L^\infty(0,T, L^2(\Omega))$ if the exact solution u is regular enough ($u_{\Delta t, h}$ denotes the numerical one):

1. for *constant in time finite elements*

$$||u - u_{\Delta t, h}||_W \le C \left(\Delta t || \frac{\partial u}{\partial t} ||_{L^\infty(L^\infty)} + h^2 ||u||_{L^\infty(H^2)} \right) ;$$

2. for *linear in time finite elements*

$$||u - u_{\Delta t, h}||_W \le C \left(\Delta t || \frac{\partial^2 u}{\partial t^2} ||_{L^\infty(L^\infty)} + h^2 ||u||_{L^\infty(H^2)} \right) .$$

Here C is independent of the discretization parameters Δt and h (assumed constant).

6

Equations of hyperbolic type

In this chapter we deal with a typical class of problems arising from *conservation laws*, that is *evolution problems of hyperbolic type* involving either pure-advection phenomena or advection-reaction ones. We will focus mainly on linear problems in one spatial dimension. The non-linear counterparts of these problems have a great interest in several contexts, but are characterized by specific aspects that go beyond the scope of this text. An excellent introduction to this subject is the book of LeVeque [LeV90], with the accompanying software library `ClawPack` (see also [LeV02]).

The method of finite differences is still very much in use for the numerical study of these problems. That is why many of the exercises refer to space discretizations by the techniques seen in Section 2.3. The code `fem1d`, albeit structurally based on finite elements, can be employed in these exercises, too, by activating suitable options.

The chapter is divided in two parts: scalar equations and systems. The latter is extremely interesting from the applicative viewpoint, as we will see in the last exercise.

We will consider only first-order hyperbolic problems, where only first order derivatives appear in the differential equations. The reason being that second-order hyperbolic problems may be converted to systems of first order by a suitable change of variables. Among these, we mention the well-known *wave equation*, or *D'Alembert equation* (see e.g. [Qua09]).

6.1 Scalar advection-reaction problems

This section deals with the general linear problem of advection-reaction type. Precisely, we will consider two situations: (i) in one the problem is defined on the entire $\mathbb{R}$, (ii) in the other on a bounded interval. Hence the former reads

Formaggia L., Saleri F., Veneziani A.: Solving Numerical PDEs: Problems, Applications, Exercises. DOI 10.1007/978-88-470-2412-0_6, © Springer-Verlag Italia 2012

as follows: find $u(x,t)$ such that

$$\begin{cases} \dfrac{\partial u}{\partial t} + a\dfrac{\partial u}{\partial x} + a_0 u = 0, & t > 0, \ x \in \mathbb{R}, \\ u(x,0) = u^0(x), & x \in \mathbb{R}, \end{cases} \tag{6.1}$$

where $a, a_0 \in R$, and $u^0 : \mathbb{R} \to \mathbb{R}$ is a sufficiently regular given function. The latter problem reads: find $u(x,t)$ such that

$$\begin{cases} \dfrac{\partial u}{\partial t} + a\dfrac{\partial u}{\partial x} + a_0 u = f, & t > 0, \ x \in (\alpha, \beta), \\ u(x,0) = u^0(x), & x \in (\alpha, \beta), \\ u(\alpha, t) = \varphi(t), & t > 0. \end{cases} \tag{6.2}$$

Here and in the sequel, when not otherwise stated we assume $a > 0$, so that the point $x = \alpha$ denotes the *inflow*. As we know (see for instance [Qua09], Ch. 12), the boundary condition (Dirichlet in the present case, given by the function φ) must be assigned at the inflow only. We will also suppose $a_0 \geq 0$.

The problem on $\mathbb{R}$ is interesting for the applications because the study of stability of a numerical scheme by the use of *Von Neumann*'s technique is based on it. Therefore we will use problem (6.1) to assess the general stability properties of numerical schemes for hyperbolic equations. The results thus found provide useful information also in the case of bounded intervals.

We recall that the differential equation stemming from a conservation may be written in *conservation form*

$$\frac{\partial u}{\partial t} + \frac{\partial F(u)}{\partial x} + S(u) = f, \tag{6.3}$$

where $F(u)$ is called flux, S source term, and they both depend on u, while the forcing term f may be a function of x and t but is independent form u. We recover this form for problems (6.1) and (6.2) by setting $F = au$ and $S = a_0 u$. Conservation form is very important when dealing with non linear problems and discontinuous solutions [LeV90], issues that are not treated in this text.

Von Neumann's analysis takes a Fourier-series decomposition of the solution of problem (6.1), assuming a periodicity of period 2π. This choice does not affect the generality of the stability result, which will be independent of it.

If we consider the expansion of the initial datum given by

$$u^0(x) = \sum_{k=-\infty}^{\infty} \alpha_k e^{ikx},$$

i being the imaginary unit and $\alpha_k \in \mathbb{C}$ the kth Fourier coefficient[1], one can show [Qua09] that the numerical approximation given by a finite-difference

---

[1] Clearly, one considers the series' real part only.

scheme for problem (6.1) satisfies

$$u_j^n = \sum_{k=-\infty}^{\infty} \gamma_k^n \alpha_k e^{ikx_j}, \quad j = 0, \pm 1, \pm 2, \ldots, \quad n = 1, 2, \ldots,$$

where γ_k is the so-called *amplification coefficient* of the kth harmonic, and is specific to the scheme considered. It depends in general on the kind of space discretization, on the grid's spacing h and on the time step Δt. If the grid for problem (6.1) has $N + 1$ (equidistant) nodes in each period (N even), then on any interval of length 2π we have

$$\|\mathbf{u}^n\|_{\triangle,2}^2 = 2\pi \sum_{k=-\frac{N}{2}}^{\frac{N}{2}-1} |\alpha_k|^2 |\gamma_k|^2, \tag{6.4}$$

where $\mathbf{u}^n$ is the numerical solution at time t^n, for all $n \geq 0$, while the discrete norm $\| \cdot \|_{\triangle,2}$ is defined in (1.14).

If we consider a pure advection problem (when $a_0 = 0$), the general solution of problem (6.1) at the nodes x_j and time t^n can be written as

$$u_j^n = \sum_{k=-\infty}^{\infty} g_k^n \alpha_k e^{ikx_j}, \quad j = 0, \pm 1, \pm 2, \ldots, \quad n = 1, 2, \ldots,$$

where g_k is a complex coefficient of unitary modulus (this derives from the fact that the exact solution of problem (6.1) satisfies $u(x, t) = u^0(x - at)$).

One way to highlight the effect of numerical errors is therefore to compare γ_k with g_k. In particular we define

$$\epsilon_{a,k} \equiv \frac{|\gamma_k|}{|g_k|}, \quad \epsilon_{d,k} \equiv \frac{\sphericalangle(\gamma_k)}{\sphericalangle(g_k)},$$

the ratios between the coefficients' absolute values and phase angles. The first ratio will account for the effects of discretization on the amplitude of the kth harmonic, and goes under the name of *dissipation* or *amplification error*, whereas the second weights the effects on the phase of the kth harmonic, i.e. on its *velocity*, and is called *dispersion error*.

A scheme is said *strongly stable* in norm $\| \cdot \|_{\triangle,p}$, for some integer $p \geq 1$, if it satisfies for any $n \geq 0$ the equation

$$\|\mathbf{u}^{n+1}\|_{\triangle,p} \leq \|\mathbf{u}^n\|_{\triangle,p} \tag{6.5}$$

when applied to problem (6.1) or (6.2) with homogeneous boundary condition.

Another definition of strong stability is present in the literature, namely

$$\lim_{n \to \infty} \|\mathbf{u}^n\|_{\triangle,p} = 0, \tag{6.6}$$

for any integer $p \geq 1$. This is somewhat similar to the absolute stability of numerical schemes for ODEs [QSS00].

The techniques illustrated in this chapter lead to a numerical resolution of the differential problems considered here by schemes with the following general form

$$\mathbf{u}^{n+1} = B\mathbf{u}^n + \mathbf{c}^n, \tag{6.7}$$

where B is the so-called *iteration matrix* and the vector $\mathbf{c}^n$ depends only on the boundary data and the source term.

Strong stability in norm $\| \cdot \|_{\triangle,p}$ corresponds to asking $\|B\|_p \leq 1$, where $\|B\|_p$ is the matrix p-norm, see (1.18). If the spectral radius ρ of B satisfies $\rho(B) < 1$, then there is a p for which (6.5) holds; in particular if B is normal then $\rho(B) \leq 1$ implies strong stability for $p = 2$ [QSS00]. The condition $\rho(B) < 1$ is sufficient to have (6.6) for some norm.

In general, the stability properties of numerical schemes of explicit type depend on the value of the dimensionless quantity $|a|\Delta t/h$, aka *CFL number* (or simply CFL) from the names of Courant, Friedrichs and Lewy who introduced it. A necessary condition for the strong stability of an explicit time-advancing scheme is the following inequality, called *CFL condition*,

$$\mathrm{CFL} \leq \sigma_c,$$

where the critical value σ_c depends on the particular scheme. We recall that the CFL condition is only necessary, not sufficient, for strong stability.

This inequality prescribes that the speed $h/\Delta t$ of the numerical solution must be related to the speed $|a|$ at which the exact solution travels.

Table 6.1 summarizes the stability features of the main methods when $a_0 = 0$ and for $p = 2$.

Table 6.1 Summary of the stability features of some time-advancing methods

Method	Finite Differences	Finite Elements
Implicit Euler/Centered	Uncond. Stable	Uncond. Stable
Explicit Euler/Centered	Unstable	Unstable
Upwind	$\sigma_c = 1$	$\sigma_c = \dfrac{1}{3}$
Lax-Friedrichs	$\sigma_c = 1$	Unstable
Lax-Wendroff	$\sigma_c = 1$	$\sigma_c = \dfrac{1}{\sqrt{3}}$

Even if a weaker definition of stability can be given (cf. [Qua09], Ch. 12), in practice we will require that a scheme for hyperbolic equations be strongly stable.

Exercise 6.1.1. Check that $|\gamma_k| \leq 1$ is necessary and sufficient to have strong stability in norm $\| \cdot \|_{\triangle,2}$ in a numerical scheme for problem (6.1).

Solution 6.1.1. Sufficiency follows directly from (6.4), given that

$$\|\mathbf{u}^{n+1}\|_{\triangle,2} = |\gamma_k|\,\|\mathbf{u}^n\|_{\triangle,2}.$$

For the opposite implication we need to show that if a numerical scheme satisfies $\|\mathbf{u}^n\|_{\triangle,2} \le \|\mathbf{u}^{n-1}\|_{\triangle,2}$ for any $n > 0$ and any chosen initial datum, then the absolute value of the amplification coefficient of the kth Fourier mode satisfies $|\gamma_k| \le 1$. From (6.4) we have

$$\sum_k |\alpha_k|^2 |\gamma_k|^{2n} \le \sum_k |\alpha_k|^2 |\gamma_k|^{2n-2},$$

and hence

$$0 \ge \sum_k |\alpha_k|^2 (|\gamma_k|^{2n} - |\gamma_k|^{2n-2}) =$$

$$\sum_k |\alpha_k|^2 \left[(|\gamma_k|^n + |\gamma_k|^{n-1})(|\gamma_k|^n - |\gamma_k|^{n-1}) \right].$$

This inequality must hold for any initial datum, so also for any α_k. Thus it holds only if $|\gamma_k|^n \le |\gamma_k|^{n-1}$ for any k and $n > 0$, i.e. only if $|\gamma_k| \le 1$.   ◇

Exercise 6.1.2. Consider the problem

$$\frac{\partial u}{\partial t} + \frac{\partial u}{\partial x} = 0, \quad -3 < x < 3, \quad 0 < t \le 1, \tag{6.8}$$

with initial condition

$$u^0(x) = \begin{cases} \sin(\nu\pi x) & \text{for} \quad -1 \le x \le 1, \\ 0 & \text{otherwise}, \end{cases}$$

and boundary condition

$$u(-3, t) = 0.$$

Compute the exact solution and give an *a priori* estimate for the L^2 norm.

Solve the problem numerically with femld using the Implicit Euler, Upwind and Lax-Wendroff methods with $\nu = 1$, $\nu = 2$, both by finite elements and finite differences. Discuss the solutions thus found as the CFL number varies.

Solution 6.1.2.

Mathematical analysis

Let us first observe that since $a = 1 > 0$ the inflow point is $x = -3$ and indeed the boundary condition is (correctly) assigned at this point. Multiplying equation (6.8) by u and integrating on the space-time domain for any $\bar{t} > 0$, we get

$$\frac{1}{2} \int_0^{\bar{t}} \int_{-3}^{3} \left(\frac{\partial u^2}{\partial t} + \frac{\partial u^2}{\partial x} \right) dx dt = 0.$$

Recalling the definition of L^2 norm in x, the fundamental theorem of calculus transforms the left-hand side into

$$\frac{1}{2} \left(||u(\bar{t})||^2_{L^2(-3,3)} - ||u(0)||^2_{L^2(-3,3)} \right),$$

while the right side gives

$$\frac{1}{2} \left(\int_0^{\bar{t}} u^2 |_{x=3} dt - \int_0^1 u^2 |_{x=-3} dt \right).$$

Bearing in mind the initial and boundary conditions we obtain the estimate

$$||u(\bar{t})||^2_{L^2(-3,3)} + \int_0^{\bar{t}} u^2 |_{x=3} dt = ||u^0||^2_{L^2(-3,3)}, \quad \forall \bar{t} \in (0,1) \qquad (6.9)$$

that bounds the $L^2(-3,3)$ norm of the solution for almost all $t \in (0,1)$, i.e. $u \in L^\infty(0,1; L^2(-3,3))$, since the second term on the left is certainly positive. Note that the positive sign depends on the fact that the boundary condition was correctly given at the inflow. Had it be given at the outflow point we would have been unable to bound the L^2 norm of the solution.

We know by the theory of characteristics ([Qua09], Ch. 12) that for (6.2) the exact solution is constant along the characteristic lines $dx/dt = a$. Therefore, the exact solution is

$$u(x,t) = \begin{cases} \sin(\nu\pi(x - t)) & \text{on} \quad -1 \le x - t \le 1, \\ 0 & \text{elsewhere.} \end{cases}$$

If the characteristic line passing for the generic point P with coordinates (x,t) in the space-time plane intersects the boundary $x = -3$ for $t > 0$, the solution is taken from the boundary data.

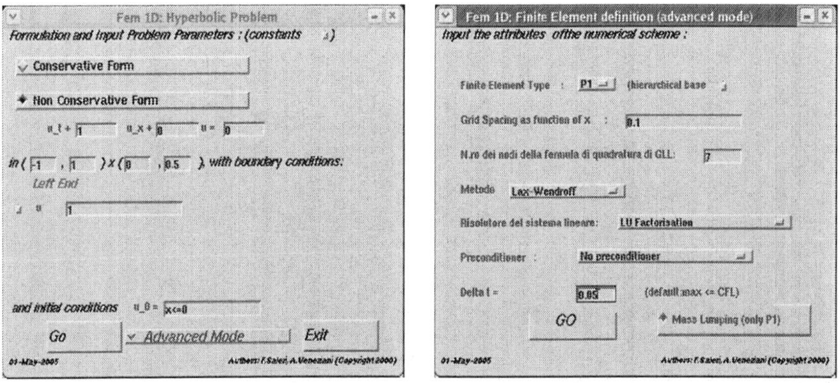

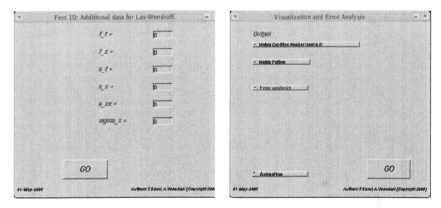

Fig. 6.1 Using `fem1d` on hyperbolic problems: interface window (left) and methods employed (right)

Fig. 6.2 Using `fem1d` on hyperbolic problems: additional information for the Lax-Wendroff method (left) and visualization options (right)

Numerical approximation

Let us see how to use `fem1d` for problems of hyperbolic type, and in particular for solving the exercise. Once `fem1d` is launched, choose the option `hyperbolic`. As for elliptic and parabolic problems, the first window (Fig. 6.1 left) refers to the problem's definition, which can be stated both in conservation form and in non-conservation form. Either form will work. The program lets us indicate whether the problem parameters are independent of time: this allows to save computing time, because in this case the discrete problem is assembled only once at the beginning of the computation.

As in the parabolic case, the domain has the form (-3, 3) x (0, 1), where the first interval is the spatial range, the second one the time domain.

The code `fem1d` allows only for Dirichlet conditions in hyperbolic problems: it "suggests" where the boundary conditions should be imposed, by opening the window for the inflow boundaries, determined according to the sign of the spatial derivative. In our case this coefficient is 1, positive, and the inflow boundary is on the left.

Then we assign the initial data. If this has support entirely contained in the domain, as in the present case where u_0 is $\sin(\nu\pi x)$ for $-1 \leq x \leq 1$, it will suffice to write

```
u_0 = sin(x).*(x>=-1 & x<=1)
```

In MATLAB in fact, 1 is returned when the condition in brackets is true, 0 when it is false, which is precisely what we need.

Like with elliptic and parabolic problems, the code can operate either in a simple mode or an advanced one with more options. We proceed here with the latter.

The second window is dedicated to prescribing the numerical methods to be used and assigning the discretization parameters (Fig. 6.1 right). We are asked to choose the degree of the finite element space (available options: P1, P2, P3) and the space-discretization step, which in general may be space dependent.

Besides specifying the precision of the numerical quadrature and the kind of solution for linear systems (also present for elliptic and parabolic problems), we are then asked to decide the method of advancement in time. The possibilities are implicit Euler, explicit Euler, Upwind and Lax-Wendroff. Once one of these is selected the step Δt suggested by the code adapts to the maximum step allowed by the CFL condition. Naturally, the user can choose a different Δt.

In the problem of concern we wish to use finite-difference methods as well: this is possible with `fem1d`, provided we impose lumping on all mass matrices. Indeed, for 1D problems choosing mass lumping (the general technique being illustrated in Ch. 4) the finite element scheme implemented in `fem1d` reduces to its finite difference equivalent. As mentioned in the introduction, the stability condition changes if one uses finite differences rather than finite elements. The code then suggests a different time step if one selects the option **Mass lumping**.

Note that when we use Lax-Wendroff we must specify further data, namely the space- and time-derivatives of the coefficients. This is the content of an ad-hoc window that is activated only if we select this method (Fig. 6.2 left). In our case all coefficients are constant, i.e. all derivatives zero, so we can immediately press Go.

The last window (Fig. 6.2 right) concerns post-processing and does not differ substantially from that of parabolic problems: we just mention the visualization option by space-time diagram or animated solution.

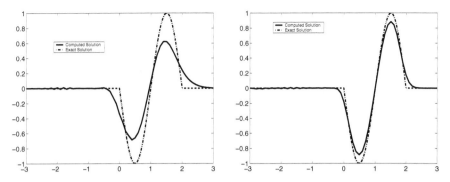

Fig. 6.3 Implicit Euler method by finite differences for Exercise 6.1.2 with $\nu = 1$. On the left, the case CFL=1, on the right CFL=0.25

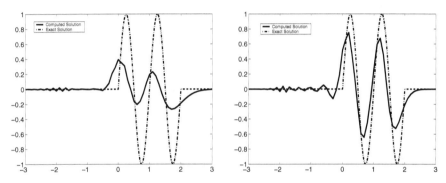

Fig. 6.4 Explicit Euler method by finite differences for Exercise 6.1.2 with $\nu = 2$. On the left we have CFL=1, on the right CFL=0.25. The comparison with the previous figure highlights how the dissipation and dispersion errors depend on the frequencies of the solution's harmonics

Numerical results

Here are some numerical results obtained by various choices of scheme, frequency (the value of ν) and CFL number. All results refer to the final time (T=1).

Fig. 6.3 refers to the implicit Euler method by finite differences with $\nu = 1$, CFL=1 (left) and CFL=0.25 (right). The effects of dissipation and dispersion induced by the numerical discretization are evident: in particular, the smaller amplitude of the sine wave is related to dissipation, while its delay with respect to the exact solution depends on the dispersion. Both dissipation and dispersion errors in the implicit Euler method decrease as the CFL number gets smaller, as we can see by comparing the pictures. The same phenomena appear in a clearer way if the wave frequency is doubled, i.e. when we take $\nu = 2$, as Fig. 6.4 shows. With implicit Euler finite elements the results are very similar.

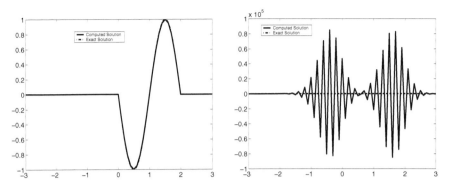

Fig. 6.5 Upwind method for Exercise 6.1.2 with $\nu = 1$. On the left the finite differences, on the right the finite elements, both with CFL=1. Finite differences are stable, in contrast to finite elements, for this value of the CFL

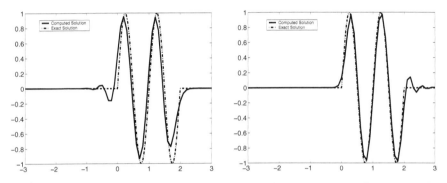

Fig. 6.6 Lax-Wendroff method for Exercise 6.1.2 with $\nu = 2$. Finite differences on the left, finite elements on the right, both for CFL=0.5

Fig. 6.5 shows the solution coming from Upwind with CFL=1 and $\nu = 1$, by finite differences (left) or finite elements (right). For the case of finite differences one can appreciate how the numerical solution for CFL=1 is little affected by dissipation, much less than the solution generated through the implicit Euler method. For the same value of CFL finite elements are unstable: the theory shows in fact that (strong) stability is reached for CFL$\leq 1/3$.

In Fig. 6.6 we have the results of the Lax-Wendroff method with $\nu = 2$ by finite differences (left) and finite elements (right), for CFL=0.5. Here both methods are stable, as the stability condition for finite differences is CFL ≤ 1, for finite elements CFL $\leq 1/\sqrt{3} \approx 0.577$. Note how the discretization by finite elements is more accurate: the error in norm $L^{\infty}(0, 1; L^2(-3, 3))$ equals 0.15809 for finite elements, 0.3411 for finite differences. The finite-element case is characterized by a dispersion error that moves the numerical solution forward with respect the exact one, whereas finite differences delay it.   ◇

Exercise 6.1.3. Show that the Lax-Wendroff scheme discretized by finite differences applied to (6.1) with $a_0 = 0$ is strongly stable in norm $\|\cdot\|_{\Delta,2}$ if $|a|\lambda \leq 1$, where $\lambda = \Delta t/h$.
Consider the problem (6.1) for $a_0 = 0$ and initial datum

$$u^0(x) = \begin{cases} 1 & x \leq 0, \\ 0 & x > 0, \end{cases}$$

and its discretization with a Lax-Wendroff finite difference scheme on a uniform grid $x_j = jh$ for $-\infty \leq j \leq \infty$ and $h > 0$.
Take the value $\lambda a = 0.5$ and show that $u_0^1 = 9/8$, independently of h. From this result, what can we say about the strong stability of the Lax-Wendroff scheme in norm $\|\cdot\|_{\Delta,\infty}$?
Using femld do the computations on the domain $(-1,1)$, for $a = 1$ and imposing $u = 1$ on the left boundary. Consider the numerical solution at time $T = 0.5$ on grids with 21 and 41 nodes, keeping the CFL number constant and equal to 0.5. Comment on the solutions found.

Solution 6.1.3.

Numerical approximation

Let us recall the Lax-Wendroff scheme has this expression

$$u_j^{n+1} = u_j^n - \frac{\lambda a}{2}(u_{j+1}^n - u_{j-1}^n) + \frac{\lambda^2 a^2}{2}(u_{j-1}^n - 2u_j^n + u_{j+1}^n). \qquad (6.10)$$

Its stability in norm $\|\cdot\|_{\Delta,2}$ can be studied using Von Neumann's analysis. For this we need to find the amplification coefficient γ_k of the kth harmonic. We look for a solution of the form

$$u_j^n = \sum_{k=-\infty}^{\infty} u_j^{n,k} = \sum_{k=-\infty}^{\infty} \alpha_k e^{ikjh}\gamma_k^n,$$

where α_k depends only on the initial datum. Then it is enough to consider a single harmonic, so we substitute $u_j^{n,k} = \alpha_k e^{ikjh}\gamma_k^n$ in the Lax-Wendroff scheme (6.10), getting

$$u_j^{n+1,k} = u_j^{n,k} - \frac{\lambda a}{2}(u_{j+1}^{n,k} - u_{j-1}^{n,k}) + \frac{\lambda^2 a^2}{2}(u_{j-1}^{n,k} - 2u_j^{n,k} + u_{j+1}^{n,k})$$

$$= \alpha_k e^{ikjh}\left[1 - \frac{\lambda a}{2}(e^{ikh} - e^{-ikh}) + \frac{\lambda^2 a^2}{2}(e^{-ikh} - 2 + e^{ikh})\right].$$

Using the identities $i\sin(x) = \dfrac{e^{ix} - e^{-ix}}{2}$ and $\cos(x) = \dfrac{e^{ix} + e^{-ix}}{2}$ we infer

$$u_j^{n+1,k} = u_j^{n,k}\left[1 - i\lambda a \sin(kh) + \lambda^2 a^2(\cos(kh) - 1)\right].$$

By definition, the amplification coefficient for the kth harmonic is thus

$$\gamma_k = \frac{u_j^{n+1,k}}{u_j^{n,k}} = 1 - i\lambda a \sin(kh) + \lambda^2 a^2 (\cos(kh) - 1).$$

Strong stability in norm $\|\cdot\|_{\triangle,2}$ requires $|\gamma_k| \leq 1$, for all k. To simplify calculations we observe first of all that $|\gamma_k| \leq 1$ implies $\gamma_k^2 \leq 1$, so we can use the latter inequality to avoid annoying square roots. We also set $\sigma = \lambda a$ and $\phi_k = kh$. This gives

$$\gamma_k^2 = [1 + \sigma^2(\cos(\phi_k) - 1)]^2 + \sigma^2 \sin^2(\phi_k)$$
$$= (1 + \sigma^2 \cos(\phi_k))^2 - 2\sigma^4 \cos(\phi_k) - \sigma^2 + \sigma^4 - \sigma^2 \cos^2(\phi_k)$$
$$= (\sigma^4 - \sigma^2)(1 - \cos(\phi_k))^2 + 1.$$

Therefore $\gamma_k^2 \leq 1$ if $(\sigma^4 - \sigma^2)(1 - \cos(\phi_k))^2 \leq 0$. This condition holds if and only if $\sigma^4 \leq \sigma^2$. But remembering σ is positive by definition, that is equivalent to $\sigma = \lambda|a| \leq 1$, which is nothing but the CFL condition.

Numerical results

Consider now the scheme for the given problem. The initial datum will be approximated by the nodal values

$$u_j^0 = u^0(x_j) = \begin{cases} 1 & j \leq 0, \\ 0 & j > 0. \end{cases}$$

Hence if $\lambda a = 0.5$ we obtain, by direct substitution in scheme (6.10),

$$u_0^1 = 1 - \frac{1}{4}(0 - 1) + \frac{1}{8}(0 - 2 + 1) = 1 + \frac{1}{4} - \frac{1}{8} = \frac{9}{8}.$$

Clearly $\|u^0\|_{\triangle,\infty} = 1$, while the result found implies $\|u^1\|_{\triangle,\infty} \geq 9/8$ (actually, one can verify that it is in fact an equality since at all other nodes the solution is $\leq 9/8$). The scheme is therefore not strongly stable in this norm. Fig. 6.7 illustrates the solution at the first temporal step. Thus, we have verified that a scheme strongly stable in a given norm is not necessarily so in a different norm.

In Fig. 6.8 we report the solution at the final time obtained by **fem1d** with CFL=0.5 for $h = 0.1$ (left) and $h = 0.05$ (right). The initial discontinuity has moved rightwards, as expected, by $aT = 1 \times 0.5 = 0.5$. As hand computations predicted, the solution is not strongly stable in norm $\|\cdot\|_{\triangle,\infty}$ and shows values larger than 1 around the initial discontinuity: this phenomenon, by which the numerical solution does not respect the initial datum's monotonicity, is known as *overshooting* (if the solution is bigger that the datum) or *undershooting* (if the solution can be smaller). Lack of strong stability in the infinity norm is much more relevant in non-linear problems, where excessive over/undershooting may lead the numerical solution outside the range of validity of the equations. For instance, they may generate a negative value

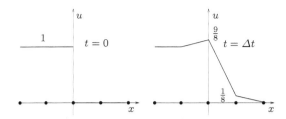

Fig. 6.7 Solution to the problem of Exercise 6.1.3. On the left the initial solution, on the right the result after the first temporal step

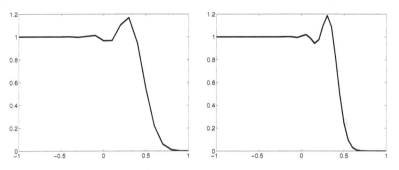

Fig. 6.8 Numerical evidence for Exercise 6.1.3: solution with CFL=0.5 and grid spacing $h = 0.1$, left, and $h = 0.05$, right. The typical overshooting of a non-monotone method like Lax-Wendroff is evident

for the density in a gas-dynamics problem modeled by Euler equations (an example of non-linear hyperbolic system). ◇

Exercise 6.1.4. Verify that the solution to problem (6.1) produced by the *upwind* scheme with finite differences satisfies $\|\mathbf{u}^n\|_{\triangle,\infty} \leq \|\mathbf{u}^0\|_{\triangle,\infty}$, provided $\lambda a < 1$. Repeat the stability analysis in norm $\|\cdot\|_{\triangle,2}$ using the Von Neumann method.
As done in Exercise 6.1.3, take $\lambda a = 0.5$ and check that in this case $u_0^1 = 1$, independently of λa.
Using a grid with $h = 0.01$, find the solution with `fem1d` for CFL=1, CFL=0.5 and CFL=0.25, verifying the answer and discussing the results.

Solution 6.1.4.

Numerical approximation

The upwind scheme by finite differences can be written, for $a > 0$, as

$$u_j^{n+1} = (1 - \lambda a)u_j^n + \lambda a u_{j-1}^n, \qquad n \geq 0, \tag{6.11}$$

so for CFL≤ 1

$$\|\mathbf{u}^{n+1}\|_{\triangle,\infty} = \max_{j\in\mathbb{Z}}|(1-\lambda a)u_j^n + \lambda a u_{j-1}^n| \leq$$

$$\max(1-\lambda a, \lambda a)\|\mathbf{u}^n\|_{\triangle,\infty} \leq \|\mathbf{u}^n\|_{\triangle,\infty}.$$

This scheme, in contrast to Lax-Wendroff, is thus strongly stable in norm $\|\cdot\|_{\triangle,\infty}$.

For the Von Neumann analysis we take, as in Exercise 6.1.3, the kth Fourier mode, written as $u_j^{n,k} = \alpha_k e^{ikjh}\gamma_k^n$, and we substitute it in scheme (6.11). This gives

$$u_j^{n+1,k} = \alpha_k e^{ikjh}\gamma_k^{n+1} = \alpha_k e^{ikjh}\gamma_k^n\left[(1-\lambda a) + \lambda a e^{-ikh}\right].$$

Then we get the following equation for the amplification coefficient

$$\gamma_k = (1-\lambda a) + \lambda a e^{-ikh} = 1 - \lambda a[1 - \cos(kh)] - i\lambda a\sin(kh).$$

Setting $\phi_k = kh$ and $\sigma = \lambda a$ we find

$$|\gamma_k|^2 = (1 - \sigma + \sigma\cos\phi_k)^2 + \sigma^2\sin^2\phi_k =$$
$$(1-\sigma)^2 + 2\sigma(1-\sigma)\cos\phi_k + \sigma^2(\cos^2\phi_k + \sin^2\phi_k) =$$
$$(1-\sigma)[(1-\sigma) + 2\sigma\cos\phi_k] + \sigma^2 =$$
$$(1-\sigma)[(1-\sigma) + 2\sigma\cos\phi_k - \sigma - 1] + 1 = 2(1-\sigma)\sigma(\cos\phi_k - 1) + 1.$$

Hence $|\gamma_k| \leq 1$ if $(1-\sigma)(\cos\phi_k - 1) \leq 0$ (by assumption, $\sigma > 0$). Since $\cos\phi_k - 1 \leq 0$ the inequality holds for any ϕ_k (and so for any harmonic k) if and only if $\sigma \leq 1$.

Reproducing the argument of Exercise 6.1.3, the first step of upwind furnishes, at the node $x_0 = 0$, the value

$$u_0^1 = (1-\lambda a)u_0^0 - \lambda a u_{-1}^0 = (1-\lambda a) - \lambda a = 1.$$

Beware: This partial result confirms the unconditional stability in sup norm but cannot be used as a proof!

Numerical results

Let us compute with `fem1d`. Figs. 6.9, 6.10 and 6.11 give the results at the first (left) and final (right) temporal step in finite-difference upwind, for CFL=1, 0.5 and 0.25 respectively. The left pictures confirm the result we have found for the numerical solution at the origin after the first step. The right pictures highlight how finite-difference upwind with CFL=1 gives a very small dissipation error (see [Qua09], Ch. 12). For smaller values of CFL, the rounder shape that the initial discontinuity assumes in moving rightwards makes the error more evident. The speed of the traveling wave is nonetheless calculated in a satisfactory way, indicating a small dispersion.

It is worth observing how the solution found by upwind is always "monotone" and does not manifest the over- or undershooting phenomena of the Lax-Wendroff method. This is an intrinsic feature of the upwind scheme,

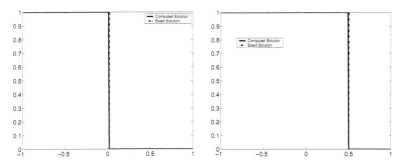

Fig. 6.9 Numerical results for Exercise 6.1.4 with $h = 0.01$ and CFL $= 1$. Left, the solution at time $t = 0.01$, and, right, at $t = 0.5$

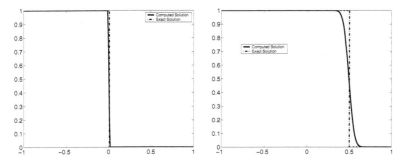

Fig. 6.10 Numerical results for Exercise 6.1.4 with $h = 0.01$ and CFL $= 0.5$. Left, the solution at time $t = 0.005$ and, right, at $t = 0.5$

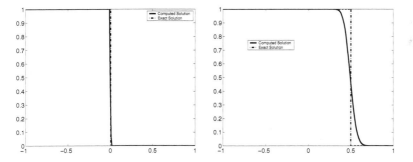

Fig. 6.11 Numerical results for Exercise 6.1.4 with $h = 0.01$ and CFL $= 0.25$. Left, the solution at time $t = 0.0025$ and, right, at $t = 0.5$

not related to the specific example, and is called *monotonicity*. It is important precisely in correspondence of discontinuities (*shocks*) of the solution, for it allows computations without unpleasant oscillations. However, it is obtained only at the expense of accuracy, as upwind is of order one. Many modern methods that deal with accurate calculation of discontinuous solutions (*shock-capturing schemes*) combine methods of high order, where the solution is regular, with monotone methods around shock regions. For more details we suggest consulting, for instance, [LeV90].   ◇

Exercise 6.1.5. Discuss the accuracy of the Lax-Friedrichs scheme for problem 6.1, using strong stability in norm $\|\cdot\|_{\triangle,1}$ and the fact that its local truncation error satisfies $|\tau_j^n| = \mathcal{O}(\dfrac{h^2}{\Delta t} + h^2 + \Delta t)$. (See [Qua09].) Consider the problem with periodic boundary conditions

$$\begin{cases} \dfrac{\partial u}{\partial t} + 2\dfrac{\partial u}{\partial x} = 0, & 0 < x < 2\pi, \, t > 0, \\ u(x,0) = \sin(x), & u(0,t) = u(2\pi,t), \, t > 0. \end{cases}$$

Discretize it in MATLAB and check the convergence estimate found above. Hint: vary Δt and h while maintaining the CFL equal 0.5; check the error's behavior at $t = \pi$, doubling several times the number N of grid elements, starting from $N = 20$.

Solution 6.1.5.

Numerical approximation

For arbitrary sign of a the Lax-Friedrich's scheme reads

$$\begin{aligned} u_j^{n+1} &= \frac{1}{2}(u_{j+1}^n + u_{j-1}^n) - \frac{\lambda|a|}{2}(u_{j+1}^n - u_{j-1}^n) = \\ &\quad \frac{1}{2}(1 - \lambda|a|)u_{j+1}^n + \frac{1}{2}(1 + \lambda|a|)u_{j-1}^n. \end{aligned} \quad (6.12)$$

Calling $y_j^n = u(x_j, t^n)$ the exact solution to (6.1) at the discretization nodes at time $t = t^n$, by definition of local truncation error τ_j^n we have

$$y_j^{n+1} = \frac{1}{2}(1 - \lambda|a|)y_{j+1}^n + \frac{1}{2}(1 + \lambda|a|)y_{j-1}^n + \Delta t \tau_j^n.$$

Subtracting the previous equations one from the other tells that the error $e_j^n = y_j^n - u_j^n$ fulfills

$$e_j^{n+1} = \frac{1}{2}(1 - \lambda|a|)e_{j+1}^n + \frac{1}{2}(1 + \lambda|a|)e_{j-1}^n + \Delta t \tau_j^n.$$

Under the assumption $\lambda|a| \leq 1$ the scheme is stable and both $1 - \lambda|a|$ and $1 + \lambda|a|$ are ≥ 0. On the other hand the truncation error of Lax-Friedrich's scheme satisfies $|\tau_j^n| \leq C(h^2/\Delta t + h^2 + \Delta t)$ for some constant $C > 0$. For a given CFL number σ we have $h^2/\Delta t = \mathcal{O}(h)$. Indeed,

$$\frac{h^2}{\Delta t} = \frac{|a|h^2}{|a|\Delta t} = \frac{|a|h}{\sigma}.$$

So, we may conclude that for a fixed CFL number and h small enough the truncation error satisfies $|\tau_j^n| \leq C^*(h + \Delta t)$, by taking $C^* = C(1 + \frac{|a|}{\sigma})$. Thus we can write

$$\|e^{n+1}\|_{\triangle,1} \le \frac{1}{2}(1 - \lambda|a|)\|e^n\|_{\triangle,1} + \frac{1}{2}(1 + \lambda|a|)\|e^n\|_{\triangle,1} + \Delta t \max_{j \in \mathbb{Z}} |\tau_j^n| \le$$
$$\|e^n\|_{\triangle,1} + \Delta t C^*(h + \Delta t).$$

Suppose the initial error e^0 is zero (having interpolated $\mathbf{u}^0$ from the initial datum). Then, operating recursively and calling T the final time, we get $\|e^{n+1}\|_{\triangle,1} \le \|e^0\|_{\triangle,1} + n\Delta t C^*(\Delta t + h) = C^* t^n (\Delta t + h) \le C^* T(\Delta t + h)$. The scheme convergence is therefore $\mathcal{O}(\Delta t + h)$, i.e. of order one both in time and space. Note the role played by the truncation error (consistency) and by the CFL condition (stability) in yielding the result.

Numerical results

The Lax-Friedrichs scheme is not contemplated by femld. But since the programming is in this case extremely simple, we have developed here a MATLAB script, given in the code hypLF.m, where the reader can find the details on how the scheme has been programmed, specialized to the problem in exam. If we numerate the nodes 1 to n, with $x_1 = 0$ and $x_n = 2\pi$, the periodicity can be implemented by setting, for the two boundary nodes,

$$u_N^{n+1} = \frac{1}{2}(u_2^n + u_{N-1}^n) - \frac{\lambda a}{2}(u_2^n - u_{N-1}^n) \quad \text{and} \quad u_1^{n+1} = u_N^n$$

respectively. Computing the ratio $\dfrac{\|e^M\|_{\triangle,1}}{h}$, for $t^M = \pi$, $h = \dfrac{2\pi}{N}$ and $\lambda a = \dfrac{a\Delta t}{h} = 0.5$, we get

```
N=20,   error=3.363574e+00, error/h = 1.070659e+01
N=40,   error=2.245515e+00, error/h = 1.429539e+01
N=80,   error=1.309590e+00, error/h = 1.667422e+01
N=160,  error=7.059657e-01, error/h = 1.797727e+01
N=320,  error=3.659651e-01, error/h = 1.863845e+01
N=640,  error=1.860935e-01, error/h = 1.895533e+01
```

showing how the error is indeed proportional to h (at fixed CFL), the proportionality factor being, here, about 1.8.   ◇

Exercise 6.1.6. Consider the advection-reaction problem

$$\begin{cases} \dfrac{\partial u}{\partial t} + a\dfrac{\partial u}{\partial x} + a_0 u = f, & x \in (0, 2\pi), \, t > 0 \\ u(x,0) = u^0(x), & x \in (0, 2\pi), \end{cases}$$

with suitable inhomogeneous boundary conditions $\varphi(t)$. Here $a = a(x)$, $a_0 = a_0(x)$, $f = f(x,t)$ and $u_0 = u^0(x)$ are sufficiently regular given functions.

1. Provide suitable boundary conditions in terms of the problem data. Assuming $a(x) > 0$, $a_0(x) \geq 0$, give a stability estimate, under suitable hypotheses.
2. Find the exact solution for $T = 2\pi$, $f = 0$, $a = 1$, $u_0 = \sin(x)$, $\varphi(t) = \sin(-t)$ in two cases, $a_0 = 0$ and $a_0 = 1$.
3. Discretize the problem in case a and a_0 are constant (with $a > 0$), using a backward-centered Euler scheme by finite differences. Discuss how to deal with boundary nodes, verify numerically the strong stability for $a = 1$, $a_0 = 0$, $h = 2\pi/20$ and $\Delta t = \pi/20$, and check the properties of the iteration matrix.
4. Using `fem1d` find the numerical solution for the two cases at point 2.
5. In the case $a_0 = 1$, by varying Δt and h compare the error $\|u - u_h\|_{\Delta,2}$ at the final time T with the expected accuracy $\mathcal{O}(\Delta t + h^2)$.

Solution 6.1.6.

Mathematical analysis

The boundary conditions are related to the sign of the advection coefficient a at the boundary.

If $a(0) > 0$ the boundary $x = 0$ is an "inflow" point, so one needs to assign u. Conversely, if $a(0) \leq 0$ the boundary is an "outflow" point and the solution is there determined by the differential equation itself, so it cannot be prescribed. Likewise, at $x = 2\pi$ we will have to force a boundary condition only if $a(2\pi) < 0$. Therefore the problem may admit 2, 1 or no boundary constraints at all! In the last case the solution is entirely determined by the initial datum. In the sequel we will suppose $a(x) > 0$ for all x, so that the boundary condition will be of the form $u(0, t) = \varphi(t)$, for $t > 0$.

Let us seek a stability estimate. For this, we multiply the equation by u and integrate between 0 and 2π to obtain

$$\frac{1}{2}\frac{d}{dt}\int_0^{2\pi} u^2 dx + \frac{1}{2}\int_0^{2\pi} a\frac{d}{dx}(u^2)dx + \int_0^{2\pi} a_0 u^2 dx = \int_0^{2\pi} f u dx.$$

The second integral can be rewritten by using the identity $a\dfrac{d(u^2)}{dx} = \dfrac{d}{dx}(au^2) - u^2\dfrac{da}{dx}$. Multiplying by 2 both sides gives

$$\frac{d}{dt}\|u(t)\|^2_{L^2(0,2\pi)} + a(2\pi)u^2(2\pi, t) - a(0)\psi^2(t) +$$

$$\int_0^{2\pi}\left(2a_0(x) - \frac{d}{dx}a(x)\right)u^2(x, t)\,dx = 2\int_0^{2\pi} f(x, t)u(x, t)dx. \quad (6.13)$$

So far we have proceeded formally, neglecting the issues of existence and boundedness of the integrals involved. But now we can verify that they exist and are finite if, for any t, $u(\cdot, t) \in L^2(0, 2\pi)$ and $f(\cdot, t) \in L^2(0, 2\pi)$ plus $a_0 \in L^\infty(0, 2\pi)$ and $\dfrac{da}{dx} \in L^\infty(0, 2\pi)$. Let us therefore assume the data belong to those spaces, and we will later check that these assumptions imply the solution is in $L^2(0, 2\pi)$ for all $t > 0$.

At this point we have two options for attaining an estimate. The first possibility works if we can find a $\mu_0 > 0$ such that $a_0(x) - \dfrac{1}{2}\dfrac{d}{dx}a(x) \geq \mu_0$ for any $x \in (0, 2\pi)$, that is if a is decreasing or grows sufficiently slowly. Then Young's inequality

$$\int_0^{2\pi} fu\,dx \leq \frac{\epsilon}{2}\|u\|_{L^2(0,2\pi)}^2 + \frac{1}{2\epsilon}\|f\|_{L^2(0,2\pi)}^2,$$

for $\epsilon = \mu_0$ gives

$$\frac{d}{dt}\|u(t)\|_{L^2(0,2\pi)}^2 + a(2\pi)u^2(2\pi, t) + \mu_0\|u(t)\|_{L^2(0,2\pi)}^2 \leq$$
$$\frac{1}{\mu_0}\|f(t)\|_{L^2(0,2\pi)}^2 + a(0)\psi^2(t).$$

Integrating between time 0 and T we obtain the following stability result:

$$\|u(T)\|_{L^2(0,2\pi)}^2 + a(2\pi)\int_0^T u^2(2\pi, t)dt \leq$$
$$\|u^0\|_{L^2(0,2\pi)}^2 + \frac{1}{\mu_0}\int_0^T \|f(t)\|_{L^2(0,2\pi)}^2 dt + a(0)\int_0^T \psi^2(t)dt.$$

The other possibility is that $2a_0(x) - \dfrac{d}{dx}a(x) \geq 2\mu_0$ fails for any $\mu_0 > 0$. We can still reach a conclusion by means of Grönwall's Lemma 1.1. Using (1.7) with $\epsilon = 1$ in (6.13) and moving the advection-reaction term to the right we arrive at

$$\frac{d}{dt}\|u(t)\|_{L^2(0,2\pi)}^2 + a(2\pi)u^2(2\pi, t)$$
$$\leq \|f(t)\|_{L^2(0,2\pi)}^2 + a(0)\psi^2(t) + \int_0^{2\pi}\left(-2a_0 + \frac{d}{dx}a + 1\right)u^2\,dx$$
$$\leq \|f(t)\|_{L^2(0,2\pi)}^2 + a(0)\psi^2(t) + \left[1 + \left\|-2a_0 + \frac{da}{dx}\right\|_{L^\infty(0,2\pi)}\right]\|u(t)\|_{L^2(0,2\pi)}^2.$$

Integrating from 0 to t and recalling $a(2\pi) > 0$, we can write the inequality

$$\|u(t)\|^2_{L^2(0,2\pi)} + a(2\pi)\int_0^t u^2(2\pi, s)ds \leq \int_0^t A\varphi(s)ds + g(t),$$

where $A = 1 + \left\|-2a_0 + \dfrac{da}{dx}\right\|_{L^\infty(0,2\pi)}$, $\varphi(t) = \|u(t)\|^2_{L^2(0,2\pi)} + a(2\pi)u^2(2\pi, t)$
and

$$g(t) = \|u_0\|_{L^2(0,2\pi)} + \int_0^t \left[\|f(s)\|^2_{L^2(0,2\pi)} + a(0)\psi^2(s)\right]ds.$$

Now we are in the position of applying the Lemma, with the result that $\forall T > 0$

$$\|u(T)\|^2_{L^2(0,2\pi)} + a(2\pi)u^2(2\pi, T) \leq$$
$$e^{AT}\left[\|u_0\|_{L^\infty(0,2\pi)} + \int_0^T \left[\|f(\tau)\|^2_{L^2(0,2\pi)} + a(0)\psi^2(\tau)\right]d\tau\right].$$

In contrast to the previous situation, because of the presence of the exponential e^{AT} the estimate is not significant in the long term, i.e. for large values of T.

To compute the exact solution in the case $f = 0$, $a = 1$, $T = 2\pi$, we follow the characteristic approach. We split the domain into two regions, Ω_1 for $x \geq at$ and Ω_2 for $x < at$. In Ω_1 (see Fig. 6.12) the solution is determined by the initial condition $u_0(x)$, entering the domain with velocity $a = 1$. In Ω_2 the solution is determined by the boundary condition $\varphi(t)$ entering the region with velocity $a = 1$. Let us consider first the case $a_0 = 0$. A generic point P_1 with coordinates (x,t) in Ω_1 lies on the characteristic line intersecting the x axis at $P_{1,0}$ with coordinates $(x - at, 0)$. The solution in Ω_1 reads

$$u(x, at) = u_0(x - at).$$

In Ω_2 the characteristic line of a generic point P_2 with coordinates (x,t) intersects the boundary $x = 0$ in $P_{2,0}$ with coordinates $(0, t - x/a)$. The solution here reads

$$u(x, t) = \varphi\left(t - \frac{x}{a}\right).$$

If the compatibility condition $u_0(0) = \varphi(0)$ holds the solution is continuous along $x = t$.

Let us consider now the case $a_0 \neq 0$. In Ω_1 we introduce the auxiliary variable $w_1(x, t) = u(x, t)e^{a_0 x/a}$. It is readily verified that for $f = 0$ variable w solves the equation

$$\frac{\partial w_1}{\partial t} + a\frac{\partial w_1}{\partial x} = 0, \qquad w_{1,0}(x) = u_0(x)e^{a_0 x/a},$$

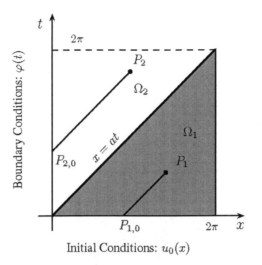

Fig. 6.12 The solution at a point P_1 depends only on the initial data, since the characteristic through that point intersects the x axis $t = 0$ before the boundary $x = 0$. The solution at point P_2 depends only on the boundary data

so the equation for w_1 is purely convective. Moving from the previous analysis, we have that in Ω_1

$$w_1(x,t) = w_0(x - t) = u_0(x - at)e^{a_0(x-at)/a} \Rightarrow u(x,t) = u_0(x - at)e^{-a_0t}.$$

In Ω_2 we work in a similar way with the auxiliary variable $w_2(x,t) = u(x,t)e^{a_0t}$. The new variable satisfies the equation

$$\frac{\partial w_2}{\partial t} + a\frac{\partial w_2}{\partial x} = 0, \qquad w_2(0,t) = \varphi(t)e^{a_0t}.$$

We conclude

$$w_2(x,t) = \varphi\left(t - \frac{x}{a}\right)e^{a_0(t-x/a)} \Rightarrow u(x,t)\varphi\left(t\frac{x}{a}\right)e^{-a_0x/a}.$$

In the specific case of our problem, exact solution reads

$$u(x,t) = \begin{cases} \sin(x - t)e^{-t} & \text{in } \Omega_1 \\[2mm] \sin(x - t)e^{-x} & \text{in } \Omega_2. \end{cases}$$

Numerical approximation

Time-discretization by centered- and backward-Euler finite differences requires the domain to be divided in N sub-intervals. The nodes at which we need to compute the approximate solution are $x_j = jh$, $j = 1, \ldots, N$, with $h = 2\pi/N$, since at $x_0 = 0$ we will set $u_0^n = \psi(t^n)$, $t^n = n\Delta t$ and ψ being the boundary data; moreover, we set $u_j^0 = \sin(x_j)$, $j = 1, \ldots, N$. Centered finite differences cannot be used at x_N, for this would require the knowledge of the solution at $x_{N+1} = 2\pi + h$, outside the domain. Therefore at that point we will use a decentered discretization of *upwind* type for the first derivative. As $a > 0$, the latter will involve the unknown only at the nodes x_N, x_{N-1}. Denoting, as customary, $\lambda = \Delta t/h$, we will have

$$
\begin{cases}
u_1^{n+1} + \dfrac{a\lambda}{2}u_2^{n+1} + a_0 u_1^{n+1}\Delta t = u_1^n + \dfrac{a\lambda}{2}\psi(t^n), \\
u_j^{n+1} + \dfrac{a\lambda}{2}(u_{j+1}^{n+1} - u_{j-1}^{n+1}) + a_0 u_j^{n+1}\Delta t = u_j^n, \quad j = 2,\dots,N-1 \\
u_N^{n+1} + a\lambda(u_N^{n+1} - u_{N-1}^{n+1}) + a_0 u_N^{n+1}\Delta t = u_N^n,
\end{cases}
$$

for $n = 0,1,\dots$. Now setting $\mathbf{u}^n = [u_1^n,\dots,u_N^n]^T$ and $\mathbf{b}^n = [a\lambda\psi(t^n)/2, 0, \dots, 0]^T$, we can write

$$
A\mathbf{u}^{n+1} = \mathbf{u}^n + \mathbf{b}^n,
$$

where A is the tridiagonal matrix

$$
A = \begin{bmatrix}
1 + a_0\Delta t & a\dfrac{\lambda}{2} & \cdots & & 0 \\
-a\dfrac{\lambda}{2} & 1 + a_0\Delta t & a\dfrac{\lambda}{2} & \cdots & 0 \\
& \vdots & \vdots & & \\
0 & & \cdots & 0 \;\; -a\lambda & 1 + a\lambda + a_0\Delta t
\end{bmatrix}.
$$

A MATLAB script that implements the proposed problem can be found in `hyp1DBEuler.m`. Since upwind scheme is only first order accurate, we may anticipate some local degradation of the accuracy at the outflow $x = 2\pi$.

We can write the iterative scheme in the canonical form

$$
\mathbf{u}^{n+1} = B\mathbf{u}^n + \mathbf{c}^n,
$$

with $B = A^{-1}$, while $\mathbf{c}^n = B\mathbf{b}^n$ depends on the boundary datum solely. For the stability in norm $\|\cdot\|_{\triangle,2}$, we compute the 2-norm of matrix B. Matrix A is easy to be inverted, being tridiagonal[2]. It will suffice to type `B=inv(full(A))` and `norm2=norm(B,2)`, which returns `norm2=0.864`, less than one, as expected. The spectral radius can be found by with `max(abs(eig(B)))` and gives 0.864. Had we computed the ∞-norm we would have got 1.207: the proposed scheme is not strongly stable in norm $\|\cdot\|_{\triangle,\infty}$. Please note that here we could not use the Von Neumann analysis directly because of the particular treatment of the boundary node. The script `hyp1DBEuler.m` relies on Program 31 which solves a generic linear advection-reaction problem with backward Euler.

An important remark: to compute B we have used the statement `full(A)`, which converts the *sparse matrix* A (tridiagonal) to a full matrix, in order to use the MATLAB statements `norm` and `eig`. This is in general not advisable, particular if A is large. Indeed MATLAB uses a special storage for sparse matrices and we have used this type of storage in Program `advreact1`. Yet, in

---

[2] Thomas algorithm [QSS00] can be used for this kind of matrices.

general the inverse of a sparse matrix is not sparse and trying to do `inv(A)` could easily use up all the available memory. If you want to compute the spectral radius (or the norm) of the inverse of a sparse matrix without the need of building the inverse explicitly you should use an iterative method like one of those described, for instance, in [Saa92].

Program 31 - advreact1 : Solution of 1D advection-reaction problem with centered finite difference and backward Euler

```
%  [mesh,uh,time,rho]=advtrasp1(uOfun,phi,a,a0,T, alpha,beta,n,dt,iflag)
%
%  Solution of   u_t+a u_x + a0 u =0 in (alpha,beta) 0<t<T
%  with centered finite difference and temporal discretization by backward
%  Euler.
%
%  uOfun      Initial data as function of  x (or string)
%  phi          Boundary data as function of t (or string). It is imposed on the
%                   correct boundary node, according to the sign of a
%  n                Number of nodes
%  dt         Time step
%  mesh     Mesh nodes
%  uh           Of dimension (n,:), this matrix contains in column i the
%                   solution at the ith time step
%  time     Vector with time steps
%  rho        Spectral radius or p-norm of the iteration matrix
%  iflag    0    rho returns spectral radius (default)
%                >0 rho returns p norm, with p=iflag.
%
```

Numerical results

We obtain the graphs of Fig. 6.13. Note how the scheme's dissipation error makes the amplitude of the approximate solution considerably smaller than the exact one.

To answer the last question we can compute the error in 2-norm as Δt and h vary. For instance, we may choose $h = 2\pi/N$ and $\Delta t = \pi/(2M)$, for $N, M = 5, 10, ..., 640$. The MATLAB command **surf** generates Fig. 6.14, telling us how the error decreases linearly with respect to Δt, but more than linearly with respect to h. The results have been obtained using the script `hyp1DBEuler.m`. In fact, the use of upwind at the right boundary node does not degrade the convergence w.r.t. h, since the truncation error introduced by the upwinding acts only on one interval. Alternatively, one could use a second order upwind scheme (see Section 2.3) similar to the $(2.31)_1$ in Exercise 2.3.1.

To verify experimentally the convergence in h we need to be careful. In fact, for any given Δt the numerical solution does not converge to the exact solution as $h \to 0$, but rather to the solution of the semi-discretized problem. For the order of convergence in h we must get rid of the effects of the time-discretization error. For that, we have chosen a specific temporal step, $\Delta t =$

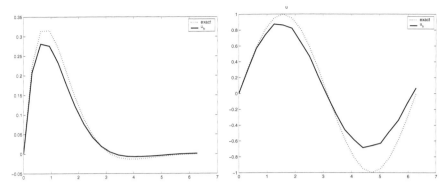

Fig. 6.13 Solution to the problem of Exercise 6.1.6. Left, the case $a_0 = 1$, right $a_0 = 0$. Both pictures show the exact and numerical solutions at time $t = 2\pi$

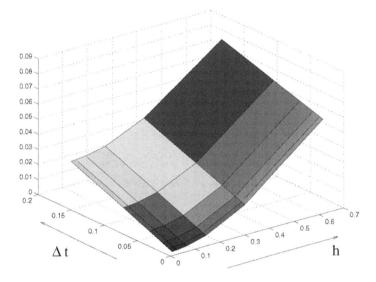

Fig. 6.14 Error behavior in 2-norm for the problem of Exercise 6.1.6 (for $a = a_0 = 1$), as Δt and h vary. From a qualitative point of view the convergence is patently linear in Δt, and more than linear in h

$\pi/80$, and computed a reference solution u_r using a very fine grid (3200 elements). This solution will be taken as "reference solution" of the problem discretized only in time. Hence, with fixed Δt, for $h = 2\pi/N$ with $N = 5, 10, \ldots, 640$ we computed $\|u_h - \Pi_h u_r\|_{\triangle,2}$, where $\Pi_h u_r$ denotes the vector of values of u_r at the nodes where u_h was computed. The result is shown in Fig. 6.15, and confirms quadratic convergence. For this result we have used script `hyp1DBE_DtFixed.m`.   ◇

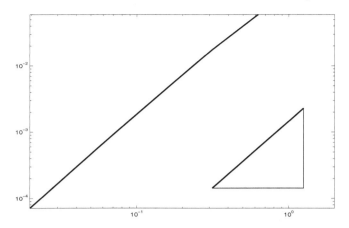

Fig. 6.15 Error behavior in 2-norm as h varies, for the problem of Exercise 6.1.6 (for $a = a_0 = 1$) with $\Delta t = \pi/80$. The reference slope corresponds to quadratic convergence

Exercise 6.1.7. Take problem (6.2) with $a > 0$ and $a_0 \geq 0$ constant, and its discretization by *discontinuous finite elements* on a grid of constant spacing h.

1. Determine a stability estimate for the semi-discretized problem (i.e. discretized with respect to the spatial variable x only).
2. Time-discretize the problem using the implicit Euler scheme, highlighting its algebraic structure. Find a stability estimate for the discrete solution in the case $a_0 > 0$.
3. Write a program that implements the scheme. Consider the case $a_0 = 0$, $u^0 = 1$, $T = \pi/2$, $\varphi(t) = 0$ using 10, and then 100, intervals and compare the results with the exact solution (to be found).

Solution 6.1.7.

Mathematical analysis

We recall [Qua09, Ch. 8] that the core of the method of discontinuous finite elements lies in detecting an approximation belonging at each time t to the vector space

$$W_h = Y_h^r = \{v_h \in L^2(\alpha, \beta): \quad v_h|_{K_i} \in \mathbb{P}^r(K_i), \, i = 1, \ldots, m\},$$

where $r \geq 0$ is the degree of the finite element, $K_i = [x_i, x_{i+1}]$ and x_i is the generic point of the grid $\mathcal{T}_h$ that partitions the interval $[\alpha, \beta]$. The grid's vertexes are numbered from 1 to $m+1$. The space Y_h^r is just L^2-conforming:

this means the numerical solution admits discontinuities in correspondence to the vertexes (whereas the grid is continuous in each element K_i, being a polynomial). A similar situation was seen in Ch. 5, whose Section 5.2 is dedicated to temporal discretizations by finite elements. In the present case, though, the discontinuities are relative to the spatial variable.

We will denote by $u_h(t) \in W_h$ the approximation of u at time t, and set $u_h^-|_{x_i} = \lim_{x \to x_i^-} u_h(x, t)$, $u_h^+|_{x_i} = \lim_{x \to x_i^+} u_h(x, t)$, thus omitting, for simplicity reasons, to indicate the time-dependency whenever obvious. To avoid misunderstandings let us stress that the symbols $+$ and $-$ refer to the limits in the spatial, rather than temporal, variable (unlike in Section 5.2).

The space-discretized problem then reads: for any $t > 0$ find $u_h = u_h(t) \in W_h$ such that, $\forall v_h \in W_h$,

$$
\int_\alpha^\beta \frac{\partial u_h}{\partial t} v_h dx + \sum_{i=1}^m \left[a \int_{x_i}^{x_{i+1}} v_h \frac{\partial u_h}{\partial x} dx + a(u_h^+|_{x_i} - U_h^-|_{x_i}) v_h^+|_{x_i} \right] +
$$
$$
a_0 \int_\alpha^\beta u_h v_h dx = \int_\alpha^\beta f v_h dx. \quad (6.14)
$$

Above, $U_h^-|_{x_i} = u_h^-|_{x_i}$, for $i = 1, \ldots, m-1$, while $U_h^-|_\alpha(t) = \varphi(t)$. In analogy to Section 5.2 we do not demand the continuity between adjacent elements to be strong, but rather have a *jump term* $a(u_h^+|_{x_i} - U_h^-|_{x_i}) v_h^+|_{x_i}$, which is also how one imposes the boundary condition (weak imposition). Such term looks so because we assumed $a > 0$. If $a < 0$ that term would be replaced by $a(U_h^+|_{x_{i+1}} - u_h^-|_{x_{i+1}}) v_h^-|_{x_{i+1}}$ (where now $U_h^+|_{x_m} = \varphi$, to impose weakly a boundary condition on the right).

In practice, if one thinks of the differential equation at the single element K_i, the jump term appears always at the "inflow" node, coherently with our problem's hyperbolic nature. (Exactly as for the time-discretization by discontinuous finite elements, described in Section 5.2, the jump term appeared at the "initial time" of each time slab to impose weakly the initial datum.)

Setting $v_h = u_h$ and reminding that $u_h \dfrac{\partial u_h}{\partial t} = \dfrac{1}{2} \dfrac{\partial u_h^2}{\partial t}$ and $u_h \dfrac{\partial u_h}{\partial x} = \dfrac{1}{2} \dfrac{\partial u_h^2}{\partial x}$, we get

$$
\frac{1}{2} \frac{d}{dt} \|u_h\|_{L^2(\alpha,\beta)}^2 + \sum_{i=1}^m \left[\frac{a}{2} \int_{x_i}^{x_{i+1}} \frac{\partial u_h^2}{\partial x} dx + a(u_h^+|_{x_i})^2 - aU_h^-|_{x_i} u_h^+|_{x_i} \right] +
$$
$$
a_0 \|u_h\|_{L^2(\alpha,\beta)}^2 = (f, u_h). \quad (6.15)
$$

where (u, v) indicates the scalar product in $L^2(\alpha, \beta)$. As $\displaystyle\int_{x_i}^{x_{i+1}} \frac{\partial u_h^2}{\partial x} dx =$ $(u_h^-|_{x_{i+1}})^2 - (u_h^+|_{x_i})^2$, the second term in (6.15) becomes

$$\frac{a}{2} \sum_{i=1}^{m} \left[(u_h^-|_{x_{i+1}})^2 - (u_h^+|_{x_i})^2 + 2(u_h^+|_{x_i})^2 - 2U_h^-|_{x_i} u_h^+|_{x_i} \right] =$$

$$\frac{a}{2} \sum_{i=1}^{m} \left[(u_h^-|_{x_{i+1}})^2 + (u_h^+|_{x_i})^2 - 2U_h^-|_{x_i} u_h^+|_{x_i} \right].$$

On the other hand $\displaystyle\sum_{i=1}^{m}(u_h^-|_{x_{i+1}})^2 = \sum_{i=2}^{m+1}(u_h^-|_{x_i})^2 = \sum_{i=2}^{m}(u_h^-|_{x_i})^2 + (u_h^-|_{\beta})^2.$
Rearranging terms, and since at internal nodes we have $U_h^- = u_h^-$, we find

$$\frac{a}{2} \sum_{i=1}^{m} \left[(u_h^-|_{x_{i+1}})^2 + (u_h^+|_{x_i})^2 - 2U_h^-|_{x_i} u_h^+|_{x_i} \right] =$$

$$\frac{a}{2} \sum_{i=2}^{m} \left[(u_h^-|_{x_i})^2 + (u_h^+|_{x_i})^2 - 2u_h^-|_{x_i} u_h^+|_{x_i} \right] + \frac{a}{2}[(u_h^-|_{\beta})^2 + (u_h^+|_{\alpha})^2] - a\varphi_h u_h^+|_{\alpha} =$$

$$\frac{a}{2} \sum_{i=2}^{m} \left(u_h^-|_{x_i} - u_h^+|_{x_i} \right)^2 + \frac{a}{2}[(u_h^-|_{\beta})^2 + (u_h^+|_{\alpha})^2] - a\varphi u_h^+|_{\alpha}.$$

So, indicating with $[u_h]_i = u_h^+|_{x_i} - u_h^-|_{x_i}$ the *jump* of u_h at x_i, equality (6.15) transforms into

$$\frac{1}{2} \frac{d}{dt} \|u_h\|_{L^2(\alpha,\beta)}^2 + \frac{a}{2} \sum_{i=2}^{m} [u_h]_i^2 + \frac{a}{2}[(u_h^-|_{\beta})^2 + (u_h^+|_{\alpha})^2] + a_0\|u_h\|_{L^2(\alpha,\beta)}^2 =$$

$$(f, u_h) + a\varphi u_h^+|_{\alpha}.$$

Let us use Young's inequality (1.7) on the right-hand side. For the first term we do not specify the constant ϵ yet, while for the second we take $\epsilon = 1$, so to cancel out the factors containing $u_h^+|_{\alpha}$.
Then

$$\frac{1}{2} \frac{d}{dt} \|u_h\|_{L^2(\alpha,\beta)}^2 + \frac{a}{2} \sum_{i=2}^{m} [u_h]_i^2 + \frac{a}{2}[(u_h^-|_{\beta})^2 + (u_h^+|_{\alpha})^2] + a_0\|u_h\|_{L^2(\alpha,\beta)}^2 \leq$$

$$\frac{1}{2\epsilon} \|f\|_{L^2(\alpha,\beta)}^2 + \frac{\epsilon}{2} \|u_h\|_{L^2(\alpha,\beta)}^2 + \frac{a}{2}\varphi^2 + \frac{a}{2}(u_h^+|_{\alpha})^2.$$

Now set $\epsilon = a_0$, simplify and multiply by 2 to get, eventually,

$$\frac{d}{dt}\|u_h\|^2_{L^2(\alpha,\beta)} + a\sum_{i=2}^{m}[u_h]^2_i + a(u_h^-|_\beta)^2 + a_0\|u_h\|^2_{L^2(\alpha,\beta)} \leq$$

$$\frac{1}{a_0}\|f\|^2_{L^2(\alpha,\beta)} + a\varphi^2.$$

Integrate from 0 to T and obtain the following estimate,

$$\|u_h(T)\|^2_{L^2(\alpha,\beta)} + a\int_0^T \sum_{i=2}^{m}[u_h(\tau)]^2_i + (u_h^-(\tau)|_\beta)^2 + a_0\|u_h(\tau)\|^2_{L^2(\alpha,\beta)}\,d\tau \leq$$

$$\|u_h^0\|^2_{L^2(\alpha,\beta)} + \int_0^T \left[\frac{1}{a_0}\|f(\tau)\|^2_{L^2(\alpha,\beta)} + a\varphi^2(\tau)\right]d\tau,$$

u_h^0 being the approximation of the initial datum.

Notice that to arrive at this we had to assume $a_0 > 0$. In case $a_0 = 0$ and $f \neq 0$, we may still obtain a stability estimate via Grönwall's Lemma, but then the result would no longer be (in general) uniform with respect the final time T, rendering the estimate meaningless for $T \to \infty$. It is still possible to retain a uniform estimate in T in the case $f = 0$.

Numerical approximation

The backward Euler scheme applied to (6.14) arises from approximating the time derivative by a backward finite difference. Equation (6.14) is discretized as

$$\frac{1}{\Delta t}\int_\alpha^\beta u_h^{n+1}v_h dx + \sum_{i=1}^{m}\left[a\int_{x_i}^{x_{i+1}} v_h \frac{\partial u_h^{n+1}}{\partial x}dx+\right.$$

$$\left. a(u_h^{+,n+1}|_{x_i} - U_h^{-,n+1}|_{x_i})v_h^+|_{x_i}\right] = \frac{1}{\Delta t}\int_\alpha^\beta u_h^n v_h dx + \int_\alpha^\beta f^n v_h dx, \quad (6.16)$$

for $n = 0, 1, \ldots$

In order to define the algebraic structure we must take into account that the spatial discretization is discontinuous between the elements. First, let us denote by $U^n = [\mathbf{u}_1^n, \mathbf{u}_2^n, \ldots, \mathbf{u}_m^n]^T$ the vector of unknowns at time t^n, where $\mathbf{u}_i^n = [u_{i,0}^n, \ldots, u_{i,r}^n]^T$ are the degrees of freedom of the ith element. Since $u_h^n|_{K_i}$ is a degree-r polynomial, we can represent it by Lagrange interpolation on a set of nodes $\{z_s^i \in K_i, s = 0, \ldots, r\}$, chosen evenly spaced. Secondly, we set $z_0^i = x_{i-1}$, $z_r^i = x_i$, so that (here we drop the time index for the sake of

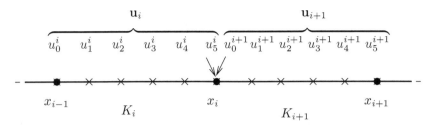

Fig. 6.16 Distribution of nodes and degrees of freedom for $r = 4$

notation) $u_s^i = u_h(z_s^i)$, for $s = 1, \ldots, r-1$, while $u_r^i = u_h^-|_{x_i}$ and $u_0^r = u_h^+|_{x_{i-1}}$. The last element in $\mathbf{u}_i$ and the first in $\mathbf{u}_{i+1}$ refer to the same vertex x_{i+1}, albeit containing the values from the left and from the right of u_h respectively, i.e. $u_h^-(x_i)$ and $u_h^+(x_i)$. Fig. 6.16 illustrates the node distribution for $r = 4$. To finish the construction of the linear system note that in numbering the vector with the degrees of freedom, $\mathbf{U} = [U_1, \ldots, U_{N_h}]^T$, there is the correspondence $u_s^i = U_{(r+1)(i-1)+s+1}$, for $i = 1, \ldots, m$ and $s = 0, \ldots, r$ (see Table 6.2). The overall number of unknowns is $N_h = m(r+1)$.

Table 6.2 Correspondence between the indexes of the local numbering (i, s) for the degree of freedom u_s^i and the index in the global vector U

(i, s)	$(1, 0)$	$(1, 1)$	$\ldots$	$(1, r)$	$(2, 0)$	$\ldots$	$(m, 0)$	$\ldots$	(m, r)
gl. numb.	1	2	$\ldots$	$r+1$	$r+2$	$\ldots$	$r(m-1) + m$	$\ldots$	$m(r+1)$

We point out an important difference with the continuous case: the value of u_h at α (precisely at α^+) is an unknown, even though at α a boundary condition is assigned. In fact, the constraint is imposed only *weakly*, meaning that in general U_0^n will be other than $\varphi(t^n)$, although $\lim_{h \to 0} |U_0^n - \varphi(t^n)| = 0$ still holds (if the scheme is convergent).

We indicate by M_i the *local mass matrix* (of size $(r+1) \times (r+1)$) relative to the element K_i, defined by $[\mathrm{M}_i]_{kl} = \int_{x_{i-1}}^{x_i} \psi_k^i \psi_l^i \, dx$, while C_i is the matrix, coming from the advective term, with entries $[\mathrm{C}_i]_{kl} = a \int_{x_{i-1}}^{x_i} \psi_k^i \dfrac{d\psi_l^i}{dx} \, dx$. With ψ_s^i we denote the local basis functions for the element K_i, i.e. Lagrange's characteristic polynomials of degree r defined on K_i and associated to the nodes z_s^i, for $s = 0, \ldots, r$, while k and l go from 1 to $r + 1$.

For instance, for linear finite elements we will have

$$\mathrm{C}_i = \frac{a}{2} \begin{bmatrix} -1 & 1 \\ -1 & 1 \end{bmatrix}, \quad \mathrm{M}_i = \frac{h_i}{6} \begin{bmatrix} 2 & 1 \\ 1 & 2 \end{bmatrix}.$$

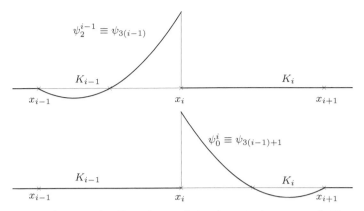

Fig. 6.17 Basis functions for discontinuous finite elements, in case $r = 2$. The two basis functions are associated to the same vertex x_i in the grid. Note that they are discontinuous, and non-zero only at one grid element

Let us define the *block-diagonal matrices* M, C of size $N_h \times N_h$, whose m diagonal blocks are M_i and C_i, $i = 1, \dots, m$:

$$
M = \begin{bmatrix} M_1 & 0 & \cdots & 0 \\ 0 & M_2 & & \vdots \\ \vdots & & \ddots & 0 \\ 0 & \cdots & 0 & M_m \end{bmatrix}, \quad
C = \begin{bmatrix} C_1 & 0 & \cdots & 0 \\ 0 & C_2 & & \vdots \\ \vdots & & \ddots & 0 \\ 0 & \cdots & 0 & C_m \end{bmatrix}.
$$

Now (6.16) can be written in a concise form, choosing as usual v_h in the set of basis functions $\{\psi_s^i, \quad i = 1, \dots, m, \ s = 0, \dots r\}$. Observe that to define correctly the basis for the discontinuous finite elements it is extremely important to remind that the functions ψ_s^i *vanish* outside the interval K_i: $\psi_s^i(x) = 0$ if $x \notin K_i$. Consequently, we have $\lim_{x \to x_i^-} \psi_0^i(x) = 0$ but $\lim_{x \to x_i^+} \psi_0^i(x) = 1$, and similarly, $\lim_{x \to x_{i+1}^-} \psi_r^i(x) = 1$, yet $\lim_{x \to x_{i+1}^+} \psi_r^i(x) = 0$ (Fig. 6.17). Besides, it is convenient to order basis functions in a unique way, consistent with the indexing of the degrees of freedom U^n. Therefore we set $\psi_{(r+1)(i-1)+s+1} = \psi_s^i$, for all $i = 1, \dots, m, \ s = 0, \dots, r$. The set of basis functions now is $\{\psi_j, \quad j = 1, \dots, N_h\}$. With this choice u_h^n can be expanded canonically as

$$
u_h^n(x) = \sum_{j=1}^{N_h} U_j^n \psi_j(x), \quad x \in [\alpha, \beta].
$$

Unlike the continuous case, to the nodes corresponding to internal vertexes are associated two degrees of freedom (so two basis functions), relative to the left and right limits respectively. Fig. 6.17 shows the situation for two basis functions associated to the vertex x_i, for $r = 2$. Choosing $v_h = \psi_i$, therefore, the ith row of the linear system associated to (6.16) can be written as follows, where M_{ij}, C_{ij} are the entries of M, C: for $a > 0$,

$$\sum_{j=1}^{N_h} \left[(\frac{1}{\Delta t} + a_0) M_{ij} + C_{ij} \right] U_j^{n+1} + a U_0^{n+1} \psi_i^+(\alpha) +$$

$$\sum_{k=1}^{m-1} a \left(U_{(r+1)k+1}^{n+1} - U_{(r+1)k}^{n+1} \right) \psi_i^+(x_k) = a\varphi(t^n)\psi_i^+(\alpha) \frac{1}{\Delta t} \sum_{j=1}^{N_h} M_{ij} U_j^n +$$

$$\int_\alpha^\beta f^n \psi_i(x) dx, \quad i = 1, \ldots, N_h. \quad (6.17)$$

If $a < 0$, instead,

$$\sum_{j=1}^{N_h} \left[(\frac{1}{\Delta t} + a_0) M_{ij} + C_{ij} \right] U_j^{n+1} - a U_{N_h}^{n+1} \psi_i^-(\beta) +$$

$$\sum_{k=1}^{m-1} a \left(U_{(r+1)k+1}^{n+1} - U_{(r+1)k}^{n+1} \right) \psi_i^-(x_k) = -a\varphi(t^n)\psi_i^-(\beta) \frac{1}{\Delta t} \sum_{j=1}^{N_h} M_{ij} U_j^n +$$

$$\int_\alpha^\beta f^n \psi_i(x) dx, \quad i = 1, \ldots, N_h.$$

Let us consider expression (6.17). Due to the properties of Lagrange's characteristic polynomial and the definition of basis functions, $\psi_i^+(x_k) = \lim_{x \to x_k^+} \psi_i(x) \neq 0$ only if $i = (r+1)k+1$, and in that case $\psi_{(r+1)k+1}^+(x_k) = 1$, if x_k is the left vertex of K_k (see Fig. 6.17, for $r = 2$). Hence $\psi_i^+(\alpha) = 1$ if $i = 1$, zero otherwise.

In conclusion, the linear system is of the form $AU^{n+1} = \mathbf{b}$, where

$$A = \left(\frac{1}{\Delta t} + a_0 \right) M + C + L$$

and L has all zero entries except

if $a > 0$:
  $L_{1,1} = a$ and $L_{i,(i-1)} = -a$,
  $L_{ii} = a$, for $i = (r+1)s+1, s = 1, \ldots, m-1$;

if $a < 0$:
  $L_{N_h,N_h} = -a$ and $L_{i,i} = -a$,
  $L_{i,i+1} = a$, for $i = (r+1)s, s = 1, \ldots, m-1$.

The matrix L "binds" diagonal blocks together. Without it, in fact, the system would be block-diagonal and the solution at each element would evolve completely independently from the others (actually, one may verify that the

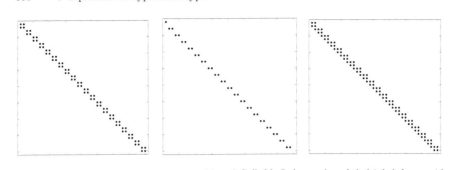

Fig. 6.18 The pattern of the matrices M and C (left), L (center) and A (right) for a grid of 20 linear finite elements, with $a > 0$

resulting linear system would be singular). We can observe analogies with Section 5.2 for time-discontinuous finite elements. Fig. 6.18 shows the pattern of M, C, L and A when $a > 0$.

The source-term vector **b** is given by

$$\mathbf{b} = \frac{1}{\Delta t}MU^n + [a\varphi(t^n), 0, \ldots, 0]^T + [f_1^n, \ldots, f_{N_h}^n]^T,$$

where $f_i^n = \int_\alpha^\beta f^n \psi_i \, dx$.

The "almost" diagonal block-structure of A gives way to special solving techniques (static elimination of internal degrees of freedom, also known as Schur complement technique), whose study goes beyond the scope of this exercise.

Numerical results

A possible MATLAB implementation of the scheme is given in Program 32.

Program 32 - femtraspdisc : It solves a one dimensional transport problems with linear discontinuos finite elements

```
%  [X,u,time,rho,ut]=femtraspdisc(u0fun,phi,a,a0,T, alpha,beta,m,dt)
%
%  Solution of u_t+a u_x + a0 u =0 in (alpha,beta) 0<=t<=T
%  with linear discontinuous finite elements and backward Euler temporal
%  discretization
%
%  NAME      Content
%  u0fun     Initial data, as function of x (or string)
%  phi       boundary data as a function of  t (or string)
%  m         number of elements
%  T         Desired final time. The actual final time is given by
```

```
%                round(T/dt)*dt
%   dt           time step
%   alpha        left boundary of the domain
%   beta         right boundary of the domain
%   X            Vector of length 2*m containing mesh vertexes. Nodes at the end
%                of each element are repeated. This is the mesh data to be used
%                with the program plotdisc to plot the result
%   ut           matrix of dimension (2*m,:) containing in each column the solution
%                at the corresponding time step
%   u            column vector of length 2*m containing the solution at the
%                final time
%   time         vector containing the timesteps. length(time) returns the
%                number of time steps employed by the program
%   rho          Spectral radius of the iteration matrix
%
```

Repeating the last part of Exercise 6.1.6 we obtain a graph similar to the left one in Fig. 6.19, and for $a_0 = 0$ the right one. Note how the solution is discontinuous even if the exact solution is continuous (initial and boundary data are continuous). To finish the exercise there remains to settle the third question.

Following the procedure of the previous exercise we find that the exact solution is $u(x,t) = u^0(x-t)$ for $x - t > 0$ and equals 0 for $x < t$, so for $t = \pi/2$ it has a discontinuity at $x = \pi/2$ (around 0.78). However, the numerical solution, shown in Fig. 6.20, smooths out the discontinuity, despite the space of discontinuous finite elements can represent it accurately, even exactly if the discontinuity occurs at a vertex. This is empirical evidence that the temporal backward-Euler scheme is strongly dissipative, even with discontinuous finite elements. We can also observe that the numerical solution at the left boundary node is just an approximation, because of the weak

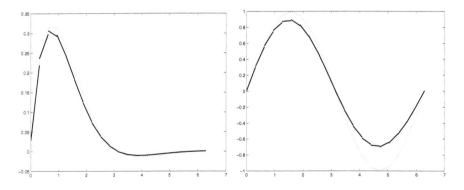

Fig. 6.19 Solution to the problem of Exercise 6.1.6 by linear discontinuous finite elements and time-discretization by backward Euler. On the left the case $a_0 = 1$, on the right $a_0 = 0$. Both pictures show the numerical solution at time $t = 2\pi$ and the corresponding exact one (dotted curve)

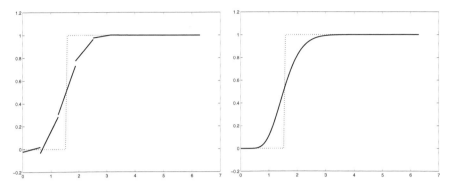

Fig. 6.20 Solution to the advection problem with $u_0(x) = 1$ and $\varphi(t) = 0$, computed by linear discontinuous finite elements and backward Euler time-discretization. The numerical solution with $m = 10$ finite elements is on the left, the case $m = 100$ on the right. Both pictures show the initial datum $\mathbf{u}^0$ and the numerical solution at time $t = \pi/4$

constraint on the datum. This approximation is all the more evident for $m = 10$.   ◇

6.2 Systems of linear hyperbolic equations of order one

In this section we will consider differential systems of the general type

$$\frac{\partial \mathbf{u}}{\partial t} + \mathrm{H}\frac{\partial \mathbf{u}}{\partial x} + \mathrm{K}\mathbf{u} = \mathbf{b} \quad x \in (a, b), \quad t > 0 \tag{6.18}$$

where $\mathbf{u} = \mathbf{u}(x, t) \in \mathbb{R}^d$, H is a diagonalizable square matrix of dimension d with real eigenvalues, K is a matrix of (constant) reaction coefficients and $\mathbf{b} = \mathbf{b}(x, t)$ a source term. The two last terms may be zero, in which case we have a *pure transport* problem. The system of equations is supplemented by an initial condition $\mathbf{u}(x, 0) = \mathbf{u}_0(x)$, $\mathbf{u}_0$ being a given function, and appropriate boundary conditions. We remark that also hyperbolic systems of second order, like the wave equation

$$\frac{\partial^2 w}{\partial t^2} - \gamma^2 \frac{\partial^2 w}{\partial x^2} = f,$$

may be recast as a system of two first order equations of the type (6.18) by the change of variables $u_1 = \dfrac{\partial w}{\partial t}$ and $u_2 = \dfrac{\partial w}{\partial x}$, and setting

$$\mathrm{H} = \begin{bmatrix} 0 & \gamma^2 \\ -1 & 0 \end{bmatrix}, \quad \mathrm{K} = 0, \quad \mathbf{b} = [f, 0]^T.$$

For this type of problems we may extend the classical numerical methods based on one step in time and three points in space we have already used in the previous sections. We will partition the domain into a grid of spacing h

and $N + 1$ nodes, and advance explicitly in time with a step Δt to obtain, for each internal node j and time step $t^n \to t^{n+1}$ a scheme of the form

$$\mathbf{U}_j^{n+1} = \mathbf{U}_j^n - \frac{\Delta t}{h}\left(\mathcal{F}_{j+1/2}^n - \mathcal{F}_{j-1/2}^n\right) - \Delta t \mathcal{B}_j^n, \quad j = 1, \ldots N - 1, \quad (6.19)$$

where $\mathcal{F}_{j+1/2}^n$ is the *numerical flux* associated to the interface between nodes j and $j + 1$, $\mathcal{B}_j^n$ contains the contribution of the reaction and source terms and it is zero in pure transport problems, where K and b are both null. Here, $\mathbf{U}_j^n \in \mathbb{R}^d$ is a vector containing the approximation of $\mathbf{u}(x_j, t^n)$ we are seeking. The numerical flux $\mathcal{F}_{j+1/2}^n$ depends on $\mathbf{U}_j^n$ and $\mathbf{U}_{j+1}^n$, and possibly on other quantities as well. In particular, for the Lax-Wendroff scheme the numerical flux may depend also on the reaction term matrix K. The term $\mathcal{B}_j^n$ in a finite difference scheme depends on $\mathbf{U}_j^n$ (as well as on K and b). The affix n indicates that quantities are computed at time step t^n (we are the considering here only explicit schemes).

Since H is diagonalizable there exists a non-singular matrix R such that $H = R\Lambda R^{-1}$, where $\Lambda = \operatorname{diag}(a_1, \ldots, a_d)$ is the diagonal matrix of the eigenvalues of H indicated by a_i, $i = 1, \ldots, d$ with the convention $a_1 \geq a_2 \geq \ldots \geq a_d$ (see Section 1.6). The columns of R are (right) eigenvectors for H and we will indicate them by $\mathbf{r}_i$, for $i = 1, \ldots d$.

We indicate by $\mathbf{l}_i^T$ the i-th row of R^{-1}. The vector $\mathbf{l}_i^T$ is a left eigenvector of H. We will set $L = R^{-1}$ so that $\Lambda = LHR$ and, conversely, $H = R\Lambda L$.

Left-multiplying by L equations (6.18) we obtain the following system of equations for the variables $\mathbf{w} = [w_1, \ldots, w_d]$ defined by $\mathbf{w} = L\mathbf{u}$ and called *characteristic variables*,

$$\frac{\partial \mathbf{w}}{\partial t} + \Lambda \frac{\partial \mathbf{w}}{\partial x} + LKR\mathbf{w} = L\mathbf{b}. \quad (6.20)$$

Note that for pure transport problems (or whenever LKR is a diagonal matrix) this system reduces to d uncoupled equations for each characteristic variable w_i of the type illustrated in the previous section (however the characteristic variables may still be coupled by the boundary conditions, as we will see).

If for a given $x \in (a, b)$ and $1 \leq i \leq d$ we consider the equation of the *characteristic line* $t \to y_i(t; x)$,

$$\frac{dy_i}{dt}(t; x) = a_i \quad t > 0, \quad i = 1, \ldots, d, \quad (6.21)$$

with $y_i(0; x) = x$, we can verify that the characteristic variables w_i satisfies the system of ordinary differential equation

$$\frac{d}{dt}w_i(y(t; x), t) + \mathbf{l}_i^T KR\mathbf{w}(y(t; x), t) = \mathbf{l}_i^T \mathbf{b}(y(t; x), t), \quad i = 1, \ldots, i \quad (6.22)$$

for $t > 0$, with $w_i(y(0; x), 0) = \mathbf{l}_i^T \mathbf{u}_0(x)$. The system is in fact uncoupled whenever LKR is a diagonal matrix.

The general solution of our differential problem then reads $\mathbf{u} = \sum_{i=1}^d w_i \mathbf{r}_i$. In a pure transport problem $K = 0$ and $\mathbf{b} = \mathbf{0}$, then $\frac{d}{dt}w_i(y, t) = 0$ and the characteristic variable is constant along the corresponding characteristic line [Qua09].

In any case, the slope of the characteristic line depends on the sign of the eigenvalues a_i. Thus, system (6.18) admits a number of boundary conditions at each end of the interval (a, b) which depends on it. More precisely, the number of boundary conditions equals the number of *incoming characteristic variables*. At the left end, $x = a$, it is equal to the number of positive eigenvalues, at the right end, $x = b$, to that of the negative ones.

The problem of imposing boundary conditions in hyperbolic systems is quite crucial. To fix the ideas let us assume that the first m eigenvalues, $a_1, \dots, a_m$, be (strictly) positive (we allow also the case $m = 0$), and consider the left end of the domain, $x = a$, (the case $x = b$ being analogous, we just need to take the negative eigenvalues instead). We consider a wide class of boundary conditions of the form

$$\mathbf{c}_i^T \mathbf{u}(a, t) = \psi(t), \quad t > 0, \quad i = 1, \dots, m$$

where $\mathbf{c}_i$ and ψ_i are given vectors and functions, respectively. Those boundary conditions are admissible if the $d \times d$ matrix

$$Z_a = \begin{bmatrix} \mathbf{c}_1 & \cdots & \mathbf{c}_m & \mathbf{l}_{m+1} & \cdots & \mathbf{l}_d \end{bmatrix}^T \quad (6.23)$$

is invertible. In particular, if $\mathbf{c}_i = \mathbf{l}_i$ for $i = 1, \dots, m$ then $Z_a = L$ and the boundary conditions are called *non reflective*. They correspond to imposing the conditions directly on the incoming characteristic variables. In this case, if K is null the equations for the characteristic variables are effectively decoupled. For a different choice of $\mathbf{c}_i$'s, the boundary conditions couple incoming and outgoing characteristic variables, and thus "waves" transported by the outgoing characteristics may be reflected back into the domain.

From the numerical point of view things are complicated by the fact that we need to compute at each time step a value for the $2d$ unknowns contained in $\mathbf{U}_0^{n+1}$ and $\mathbf{U}_N^{n+1}$, while the boundary conditions provide at most d relations[3]. Two strategies are here illustrated for the so-called *numerical boundary conditions*, the first is based on the *extrapolation of the outgoing characteristic variables*, the second on the *compatibility relations*.

The extrapolation of the characteristic variables relies on the solution of equations (6.21) and (6.22). To fix the ideas, let us consider again the left end node $x_0 = a$ and time step $t^n \to t^{n+1}$. The indexes $m+1, \dots, d$ are associated to the non positive eigenvalues of H and thus to the outgoing characteristics. We solve (6.21) to find the foot x_i^* of the corresponding characteristic line. If we employ a forward Euler scheme we have

$$x_i^* = x_0 - a_i \Delta t, \quad i = m + 1, \dots, d.$$

Note that if $\sigma_c \leq 1$, the CFL condition implies that all x_i^* are located in the interval $[x_0, x_1]$, see Fig. 6.21.

---

[3] Remind that no boundary condition is associated to a null eigenvalue.

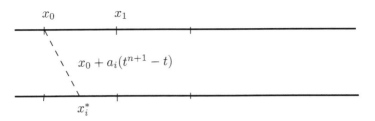

Fig. 6.21 The foot x_i^* of the ith characteristics passing from (x_0, t^{n+1}) falls necessarily in the interval $[x_0, x_1]$ if the CFL number is not greater than one. We are here assuming $a_i < 0$

We now approximate $\mathbf{u}(x^*, t^n)$ with $\mathbf{U}^*$ by interpolation (usually linear interpolation, i.e. $\mathbf{U}^* = (x^* - x_0)\mathbf{U}_0^n/h + (x_1 - x^*)\mathbf{U}_1^n/h)$ from the known approximation at time t^n, and set $w_i^* = \mathbf{l}_i^T \mathbf{U}^*$. Using again forward Euler[4] on equation (6.22) we have

$$w_i^{n+1} = w_i^* + \Delta t \left[-\mathbf{l}_i^T \mathbf{K} \mathbf{U}_0^n + \mathbf{l}_i^T \mathbf{b}(x_0, t^{n+1}) \right], \quad i = m+1, \dots, d.$$

Finally, the numerical boundary conditions on u are obtained by solving the linear system

$$\mathbf{Z}_a \mathbf{U}_0^{n+1} = \left[\psi_1^{n+1} \ \dots \ \psi_m^{n+1} \ w_{m+1}^{n+1} \ \dots \ w_d^{n+1} \right]^T, \tag{6.24}$$

with $\mathbf{Z}_a$ given by (6.23).

The second technique avoids the need of computing the characteristic variables (it is useful in non linear problems, where their calculation may be difficult, if not impossible), and it employs the numerical scheme suitably extended to the boundary nodes. We start by noting that the equations for the outgoing characteristics at $x = a$ may be "extended" to the boundary, thus formally we may write that the solution of our problem satisfies

$$\mathbf{l}_i^T \left(\frac{\partial \mathbf{u}}{\partial t} + \mathbf{H}\frac{\partial \mathbf{u}}{\partial x} + \mathbf{K}\mathbf{u} - \mathbf{b} \right) = \mathbf{0} \quad \text{for } x = a, t > 0, i = m+1, \dots, d.$$

The equations of this set are sometimes called *compatibility relations*[QV99, Chapter 7] and provide additional $d - m$ relations. We now need to find a way of using our numerical scheme also at the boundary nodes.

To this purpose, we add two additional nodes to each end of the interval, called *ghost nodes* and here indicated by x_{-1} and x_{N+1}, respectively and we extend the numerical solution at time t^n to those nodes by

$$\begin{aligned} \mathbf{U}_{-1}^n &= \mathbf{U}_0^n & \mathbf{U}_{N+1}^n &= \mathbf{U}_N & \text{constant extrapolation,} \\ \mathbf{U}_{-1}^n &= 2\mathbf{U}_0^n - \mathbf{U}_1^n & \mathbf{U}_{N+1}^n &= 2\mathbf{U}_N - \mathbf{U}_{N-1} & \text{linear extrapolation.} \end{aligned}$$

---

[4] Using the Euler scheme the convergence in time is reduced to first order, other schemes are possible to recover second order accuracy, if needed.

Linear extrapolation is required if we do not want to degrade the convergence to first order. We can now extend the numerical scheme to the first node, obtaining

$$\mathbf{U}_0^* = \mathbf{U}_0^n - \frac{\Delta t}{h} \left(\boldsymbol{\mathcal{F}}_{-1/2}^n - \boldsymbol{\mathcal{F}}_{1/2}^n \right) - \Delta t \boldsymbol{\mathcal{B}}_0^n. \tag{6.25}$$

We set $w_i^{n+1} = \mathbf{l}_i \mathbf{U}_0^*$, for $i = m+1, \ldots, d$, and solve linear system (6.24) to compute the desired approximation at the boundary. Node x_N is treated analogously.

Exercise 6.2.1. Consider the system of differential equations

$$\begin{cases} \dfrac{\partial u_1}{\partial t} + 17\dfrac{\partial u_1}{\partial x} + 6\dfrac{\partial u_2}{\partial x} = 0, \\[2mm] \dfrac{\partial u_2}{\partial t} - 45\dfrac{\partial u_1}{\partial x} - 16\dfrac{\partial u_2}{\partial x} = 0, \end{cases} \quad x \in (0,1), \quad t > 0,$$

where $u_1(x,0) = u_1^0(x)$, $u_2(x,0) = u_2^0(x)$ for $x \in (0,1)$ and appropriate Dirichlet non-homogeneous boundary conditions are given.

1. Check the hyperbolic nature of the problem and determine how many boundary conditions should be given at the two end points of the domain.
2. Diagonalize the problem so to arrive at the form

$$\frac{\partial w_i}{\partial t} + a_i \frac{\partial w_i}{\partial x} = 0, \quad i = 1, 2.$$

3. Find the domain of dependence of the solution at the generic point $(\bar{x}, \bar{t})$.
4. Determine the conditions needed for the centered/forward Euler scheme, the centered/backward Euler one and the Lax-Wendroff scheme to be strongly stable.
5. Starting from the diagonalized system, write the Lax-Wendroff scheme in the primitive variables u_1 and u_2, identifying the compatibility conditions. Use general boundary conditions written as linear combinations of u_1 and u_2.
6. Taking suitable initial data, study the numerical properties of the system in case the boundary conditions are given via the characteristic and primitive variables. Check numerically the consistency of the stability condition thus found.

Solution 6.2.1.

Mathematical analysis

First of all, by introducing the vector variable $\mathbf{u} = [u_1, u_2]^T$ the differential system can be written in the matrix form

$$\frac{\partial \mathbf{u}}{\partial t} + \mathrm{H}\frac{\partial \mathbf{u}}{\partial x} = \mathbf{0},$$

where the matrix

$$\mathrm{H} = \begin{bmatrix} 17 & 6 \\ -45 & -16 \end{bmatrix}$$

has $a_1 = 2$ and $a_2 = -1$ as eigenvalues. Indeed, the characteristic polynomial is $|\mathrm{H}-aI| = (17-a)(-16-a)+45\times 6 = a^2-a-2$. Since H has real eigenvalues, the system is hyperbolic and we will need to impose one boundary condition at each end point of the domain, since the eigenvalue signs are opposite. As H has distinct eigenvalues, a well known result of linear algebra assures that it is diagonalizable (cf. Section 1.6). The diagonalized system in this case reads

$$\frac{\partial w_i}{\partial t} + a_i\frac{\partial w_i}{\partial x} = 0, \quad i = 1, 2,$$

with characteristic variables given by $w_i = \mathbf{l}_i^T\mathbf{u}$.

The command MATLAB [R,D]=eig(H) returning in D the matrix Λ and in R precisely R, can be used to define the diagonalized system explicitly. The commands L=inv(R); l1=L(1,:); l2=L(2,:) return in the variables l1 and l2 the left eigenvectors: $\mathbf{l}_1 = [16.1555, 5.3852]^T$ and $\mathbf{l}_2 = [15.8114, 6.3246]^T$ (approximated to the fourth digit). By definition of characteristic variable we obtain $w_1 = \mathbf{l}_1^T\mathbf{u} = 16.1555u_1 + 5.3852u_2$ and $w_2 = \mathbf{l}_2^T\mathbf{u} = 15.8114u_1 + 6.3246u_2$.

The diagonalized problem now reads

$$\begin{cases} \dfrac{\partial w_1}{\partial t} + 2\dfrac{\partial w_1}{\partial x} = 0, \\ \dfrac{\partial w_2}{\partial t} - \dfrac{\partial w_2}{\partial x} = 0, \end{cases} \quad x \in (0, 1), \quad t > 0,$$

with $w_1(x, 0) = w_{1,0}(x) = 16.1555u_1^0(x) + 5.3852u_2^0(x)$ and $w_2(x, 0) = w_{2,0}(x) = 15.8114u_1^0(x) + 6.3246u_2^0(x)$, for $x \in (0, 1)$. The two differential equations are now independent. However, the values of w_1 and w_2 can be coupled by the boundary conditions. For instance, if we impose $u_{1,0}(0, t) = \psi(t)$ at the left boundary (which corresponds to taking $\mathbf{c}_1^T = [1, 0]$), the two characteristic variables are related by

$$w_1(0, t) = l_{11}\psi(t) + l_{12}u_2(0, t), \; w_2(0, t) = l_{21}\psi(t) + l_{22}u_2(0, t),$$

where l_{ij} is the jth component of $\mathbf{l}_i$. Eliminating $u_2(0, t)$ we obtain

$$w_1(0, t) - \frac{l_{12}}{l_{22}}w_2(0, t) = (l_{11} - \frac{l_{21}}{l_{22}})\psi(t).$$

The only way to have a complete decoupling between w_1 and w_2 is to impose directly w_1 at the left boundary and w_2 at the right.

Note finally that the characteristic variables w_1, w_2 are not uniquely defined. The same diagonalized problem could have arisen from taking, instead of l_i, the vectors $\tilde{l}_i = \alpha l_i$ $(i = 1, 2)$ for any $\alpha \neq 0$: *the characteristic variables are known up to a multiplicative constant*, which is normally fixed by the boundary conditions.

The domain of dependence D associated to the point $(\bar{x}, \bar{t})$, with $\bar{x} \in (0, 1)$ and $\bar{t} > 0$, consists by definition of the feet of the two characteristic lines in the (x, t)-plane through the point with slope $\tilde{a}_1 = 1/2$ and $\tilde{a}_2 = -1$. But the domain is bounded, so $D = \{x_l, x_r\}$ with $x_l = \max(\bar{x} - 2\bar{t}, 0)$ and $x_r = \min(\bar{x} + \bar{t}, 1)$. In Fig. 6.22 we can see the domain of dependence for three points. Note that for (x_1, t_1) the left characteristic's end lies on $x = 0$, so the solution at that point will depend both on the initial condition (via the right characteristic) and the value of w_1 at the left boundary (since w_1 is the characteristic variable associated to the wave coming in from that boundary). For (x_2, t_2), instead, the dependence domain is associated exclusively to the initial values, therefore the solution there will not be affected by boundary data. Conversely, the solution at (x_3, t_3) depends on the boundary values, as the ends of both characteristics belong to the lines delimiting the spatial domain. More exactly, it depends on $w_1(0, A)$ and $w_2(1, B)$.

Remark 6.1 If the boundary conditions are both non-reflective, $w_1(0, t) = \psi_1(t)$ and $w_2(1, t) = \psi_2(t)$, the solution at (x_3, t_3) depends only on the boundary data $\psi_1(A)$ and $\psi_2(B)$. In such case, in fact, starting from time $t = 1$ the solution values at any point of the domain depend only on the boundary data: "waves" produced by the initial condition have left the domain for good. If instead the conditions are reflective, we saw that the value of $w_1(0, A)$ will be a function of $w_2(0, A)$, which is equal to $w_{2,0}(x_A)$. Similarly $w_2(1, B)$ depends on $w_1(1, B)$, equal to $w_{1,0}(x_B)$. Therefore the solution at (x_3, t_3) depends not only on $\psi_1(A)$ and $\psi_2(B)$, but also on the initial conditions at x_A and x_B. In general, when the boundary conditions are reflective the initial condition is (partially) reflected, so it continues to influence the solution of the problem at any $t > 0$.

Numerical approximation

The centered/forward Euler scheme is unconditionally (strongly) unstable, and this is clearly true also at present, exactly as the unconditional stability of the backward Euler method. The Lax-Wendroff method undergoes a CFL-type stability condition. The analysis of the diagonalized problem tells us that both conditions $\dfrac{2\Delta t}{h} \leq 1$ and $\dfrac{\Delta t}{h} \leq 1$ must hold; it is thus necessary that $\Delta t \leq \dfrac{h}{2}$. The ratio $\Delta t \dfrac{\max(|a_1|, |a_2|)}{h} = \dfrac{2\Delta t}{h}$ is the CFL number of our problem.

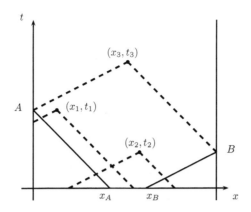

Fig. 6.22 The domain of dependence, marked by ×, of three points on the plane (x, t), relative to the hyperbolic problem of Exercise 6.2.1

Numerical results

Suppose dividing the domain in a grid of N nodes x_j, $j = 1, \ldots, N$, uniform by $h = x_{j+1} - x_j$ for all j, and set $t^n = n\Delta t$ for some $\Delta t > 0$ and $n = 1, \ldots, M$, where $M = T/\Delta t$. Let us also denote by $\mathbf{W}_j^n = [w_{1,j}^n, w_{2,j}^n]^T$ the approximation of the characteristic variables at x_j and at time t^n, while $\mathbf{U}_j^n = [u_{1,j}^n, u_{2,j}^n]^T$ will be the analogous quantity for the primitive variables.

The numerical scheme for the two characteristic variables can be written, at the internal nodes, in the usual form

$$w_{i,j}^{n+1} = w_{i,j}^n - \frac{\Delta t}{h} \left[f_{i,j+1/2}^n - f_{i,j-1/2}^n \right], \quad j = 1, \ldots N - 1$$

where $f_{i,j+1/2}$ is the *numerical flux* that characterizes the scheme, and in particular for the Lax-Wendroff scheme we have

$$f_{i,j\pm1/2}^n = \frac{a_i}{2} (w_{i,j\pm1}^n + w_{i,j}^n) \mp \frac{\Delta t a_i^2}{h} (w_{i,j\pm1}^n - w_{i,j}^n). \tag{6.26}$$

In vectorial form the scheme reads

$$\mathbf{W}_j^{n+1} = \mathbf{W}_j^n - \frac{\Delta t}{h} \left[\mathbf{F}_{j+1/2}^n - \mathbf{F}_{j-1/2}^n \right], \quad j = 1, \ldots N - 1,$$

with $\mathbf{F}_{j\pm1/2}^n = [f_{1,j\pm1/2}^n, f_{2,j\pm1/2}^n]^T$.

The boundary nodes must be dealt with apart. We wish to impose

$$\mathbf{c}_1^T \mathbf{u}(0, t) = \psi_1(t) \quad \text{and} \quad \mathbf{c}_2^T \mathbf{u}(1, t) = \psi_2(t), \quad \text{for } t > 0, \tag{6.27}$$

where $\mathbf{c}_i \in \mathbb{R}^2$ are two vectors of given coefficients. For example, setting $\mathbf{c}_1 = [1, 0]^T$ we obtain the boundary condition $u_1(0, t) = \psi_1(t)$ seen earlier, while $\mathbf{c}_1 = \mathbf{l}_1$ corresponds to imposing w_1 at the left boundary (non-reflective condition). In terms of the characteristic variables we have

$$\mathbf{c}_1^T \mathbf{R} \mathbf{W}(0,t) = \psi_1(t), \quad \mathbf{c}_2^T \mathbf{R} \mathbf{W}(1,t) = \psi_2(t),$$

which relate the value of the two characteristic variables at the boundary through the vectors $\mathbf{c}_i^T \mathbf{R}$, unless $\mathbf{c}_i = \mathbf{l}_i$, in which case $\mathbf{l}_1^T \mathbf{R} = [1,0]^T$ and $\mathbf{l}_2^T \mathbf{R} = [0,1]^T$ (recall that the matrix with rows $\mathbf{l}_i^T$ is $\mathbf{L} = \mathbf{R}^{-1}$). Observe further that the choice $\mathbf{c}_1 = \mathbf{l}_2$ (just like $\mathbf{c}_2 = \mathbf{l}_1$) is *not admissible* because it would correspond to having w_2 at the left boundary (and w_1 at the right one), which is incompatible with the sign of the advection term in the respective equations. In general, *we must exclude those $\mathbf{c}_i$ that would correspond to impose the value of just the characteristic variables outgoing from the boundary in question.* Indeed, one may verify that in this situation matrix $\mathbf{Z}_a$ is singular.

Equations (6.27) are not sufficient to close up the algebraic problem at the boundary. We choose here to adopt the compatibility relations to close the system, following the technique indicated in the introductory part of this section.

Now we may return to the primitive variables. Reminding that $\mathbf{u} = \mathbf{R}\mathbf{w}$, $\mathbf{H} = \mathbf{R}\Lambda\mathbf{R}^{-1}$ and the expression for the Lax-Wendroff numerical fluxes, we have

$$\boldsymbol{\mathcal{F}}_{j\pm1/2}^n = \mathbf{R}\mathbf{F}_{j\pm1/2}^n = \frac{1}{2}\mathbf{H}(\mathbf{U}_{j\pm1}^n + \mathbf{U}_j^n) - \frac{\Delta t}{2h}\mathbf{H}^2(\mathbf{U}_{j\pm1}^n - \mathbf{U}_j^n) \quad (6.28)$$

and, for $n = 0, 1, \ldots, M$,

$$\mathbf{U}_j^{n+1} = \mathbf{U}_j^n - \frac{\Delta t}{h}(\boldsymbol{\mathcal{F}}_{j+1/2}^n - \boldsymbol{\mathcal{F}}_{j-1/2}^n), \quad j = 1, \ldots, N-1,$$

whereas at the boundary we will have to impose

$$\begin{cases} \mathbf{c}_1^T \mathbf{U}_0^{n+1} = \psi_1(t^{n+1}), \\ \mathbf{l}_2^T \mathbf{U}_0^{n+1} = \mathbf{l}_2^T \left[\mathbf{U}_0^n - \frac{\Delta t}{h}(\boldsymbol{\mathcal{F}}_{1/2}^n - \boldsymbol{\mathcal{F}}_{-1/2}^n) \right], \end{cases}$$

and

$$\begin{cases} \mathbf{c}_2^T \mathbf{U}_N^{n+1} = \psi_2(t^{n+1}), \\ \mathbf{l}_1^T \mathbf{U}_N^{n+1} = \mathbf{l}_1^T \left[\mathbf{U}_N^n - \frac{\Delta t}{h}(\boldsymbol{\mathcal{F}}_{N+1/2}^n - \boldsymbol{\mathcal{F}}_{N-1/2}^n) \right], \end{cases}$$

where the numerical fluxes may be computed by extending the approximation at the ghost nodes x_{-1} and x_{N+1} using linear extrapolation, to avoid loosing accuracy.

Note that if we use constant extrapolation instead, the numerical fluxes at the boundary become

$$\boldsymbol{\mathcal{F}}_{-1/2}^n = \mathbf{H}\mathbf{U}_1^n, \quad \text{and} \quad \boldsymbol{\mathcal{F}}_{N+1/2}^n = \mathbf{H}\mathbf{U}_N^n, \quad (6.29)$$

which corresponds to using an *upwind* scheme for the outgoing characteristic variable at the boundary nodes. Indeed, left-multiplying by the vector $\mathbf{l}_i$ automatically selects the correct component: for instance, $\mathbf{l}_2^T \mathbf{H} \mathbf{U}_1^n = a_2 \mathbf{l}_2^T \mathbf{U}_1^n = a_2 w_{2,1}^n$.

Numerical results

To check the properties of the reflective boundary conditions we have set up a numerical simulation (MATLAB details can be found in the script `syshyp.m`). The initial datum was found using $w_1(x,0) = \sin(2\pi x)$ and $w_2(x,0) = 0$ (see Fig 6.23). We have divided the domain into 80 intervals of width $h = 1/80$, and the time step was chosen to be $\Delta t = 0.95 \frac{h}{2}$, so to stay within the stability range with CFL number 0.95. The numerical solution was computed at $T = 0.25$ using two distinct boundary conditions. The first case involves non-reflective boundary conditions, $\mathbf{c}_1 = \mathbf{l}_1$, $\mathbf{c}_2 = \mathbf{l}_2$, with $\psi_1 = \psi_2 = 0$. Then we looked at the case $\mathbf{c}_1 = \mathbf{l}_1$, $\mathbf{c}_2 = [0,1]^T$, corresponding to a non-reflective condition at the left and to setting $u_2 = 0$ at the right. The computation relies on the script `hyp_example.m`. The script uses several MATLAB functions which are provided on the web site of this book. For the sake of space we only give a list in Table 6.3. Their detailed description may be found inside each file.

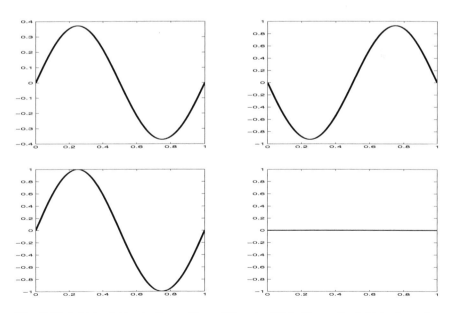

Fig. 6.23 Initial solution to the problem of Exercise 6.2.1. Above, the initial values of u_1 and u_2, below the corresponding characteristic variables w_1, w_2

Table 6.3 List of main MATLAB function provided with the book for the solution of linear systems of hyperbolic equations. The detailed description may be found in each file

hyp_setproblem	Prepares the data of the hyperbolic system to be solved
hyp_getspectrum	Computes eigenvector/values in the right order
hyp_setdiscretization	Defines the parameters for the discretization
hyp_initialsolution	Computes the initial solution
hyp_advance	Advances one step with the chosen scheme

The result, in terms of primitive and characteristic variables, is shown in Fig. 6.24. The solution in the second case is drawn with dashes. We may notice the following:

1. The wave associated to w_1 has moved right by 0.5, coherently with the value of a_1.
2. In case of non-reflective boundary conditions, w_1 leaves the right boundary without reflections, in fact w_2 stays constant, and equal to the initial datum. In the proximity of the left boundary the solution is null, in agreement with the constraint $w_1 = w_2 = 0$ valid there.
3. In case of a reflective condition on the right, part of the wave associated to w_1 is reflected and re-enters the domain as a wave associated to w_2.
4. The solution near the left boundary is unchanged in either case. The reflected wave does not have the time to reach the left boundary: in fact, $a_2 = -1$ forces it to affect only the portion $0.75 \le x \le 1$ of the domain.

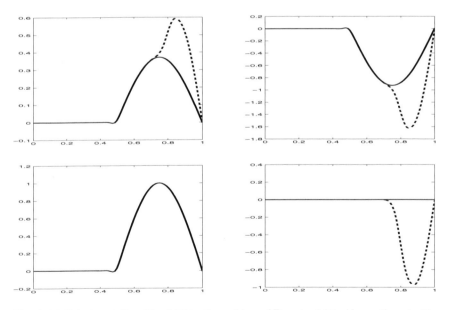

Fig. 6.24 Solution at time $T = 0.25$ to the problem of Exercise 6.2.1. Above, the primitive variables u_1 and u_2, below the respective characteristic variables w_1, w_2. The solid curve gives the solution with non-reflective boundary conditions, the dashed curve that with $u_2 = 0$ on the right boundary. Note the wave reflection associated to the latter case

5. In case of reflective condition, the effect of superposing the two components (associated to w_1 and w_2) on u_1, u_2 is self-evident.

The same calculation was then performed with $\Delta t = 1.1\frac{h}{2}$, just beyond the stability range. Fig. 6.25 shows the result at $T = 0.25$ for the primitive variables as well as for the characteristic ones. Notice that w_2 is not touched by the instability. In fact $|a_2| = 2|a_1|$, therefore the chosen CFL (equal 1.1) satisfies the stability condition for w_2. On the other hand even in the case of non-reflective boundary conditions the oscillations of w_1 (already grown several times wider than the initial data!) still have not reached the boundary and so cannot affect w_2. In Fig. 6.26 we have the result at $T = 0.6$. Now, in case of reflective boundary conditions the instability has propagated to w_2, as well. Clearly u_1 and u_2 will be both affected by the instability: thus, the more restrictive choice for Δt is necessary for the stability of the system.

As final remark, observe how in the last case the instability of w_1 arises from the wave "tail". This can be observed by launching the program for a short interval, say $T = 0.1$, as in Fig. 6.27. This is due to the fact that the first space derivative of the solution is discontinuous at the point where the wave meets the line $w_1 = 0$ emanating from the left boundary. The derivative's discontinuities are natural points where the numerical instability is "turned on" if Δt is larger than the critical value. Notice, however, that the instability

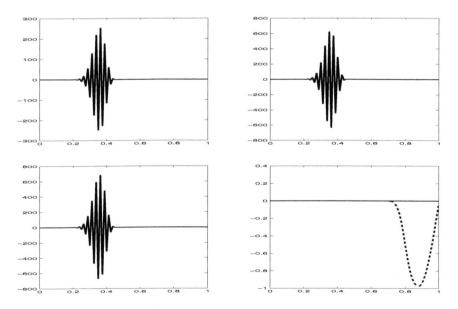

Fig. 6.25 Solution at $T = 0.25$ for the problem of Exercise 6.2.1, in case $\Delta t = 1.1\frac{h}{2}$. Above, the primitive variables u_1 and u_2, below the characteristic variables w_1 and w_2. The solid curve is the solution obtained with reflective boundary conditions, the dashed curve the one from imposing $u_2 = 0$ on the right. The solution for w_1 is entirely dominated by spurious oscillations

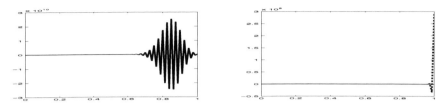

Fig. 6.26 Solution at $T = 0.6$ to the problem of Exercise 6.2.1, for $\Delta t = 1.1\frac{h}{2}$. We show only the characteristic variables w_1, w_2. Note that for reflective boundary conditions (dashed), the reflection of the spurious oscillations of w_1 on the right does affect w_2

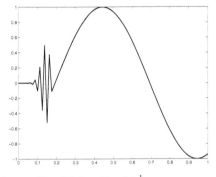

Fig. 6.27 Behavior of w_1 at $T = 0.1$ for $\Delta t = 1.1\frac{h}{2}$

would have appeared anyway (more slowly) even if the initial data had been more regular.   ◇

Exercise 6.2.2. (*) The flow of an incompressible fluid of density ρ in a pipeline system with elastic walls (see Fig. 6.28) reveals an interesting propagation phenomenon.

Consider a cylindrical pipe on a downhill slope, with cross-section $A_0 = A_0(x)$ where x denotes the coordinate of the cylinder's axis. The volumetric flow rate inside the duct (technically, the *discharge*) is $Q = Q(x,t)$. Let us suppose the velocity u is constant at each axial cross-section, so that $Q = uA$ throughout the fluid. Due to the elastic properties the cross-section deforms in accordance with the flow, so its area $A = A(x,t)$ is a function of both time and position. The difference $P = P(x,t) = p - p_{ref}$ between the internal pressure and the reference pressure p_{ref}, equal to the pressure at steady state and design conditions, is called *overpressure*, and is responsible for the pipe's deformation with respect to the design conditions. We further assume the fluid is not very viscous, so that the effects of viscosity can be appreciated only near the walls, and can be assimilated to a force against motion (friction).

Next, suppose the wall is elastic and the deformations small. Poisson's law for pressurized pipes leads to the following relationship between P and the diameter d

$$P = sE\frac{d - d_0}{d_0}, \tag{6.30}$$

E being the Young module of the material of which the wall is made, s its thickness and d_0 the diameter of the pipe at design conditions.

1. Determine a system of PDEs, in A and Q only, describing the system, assuming one can linearize the problem around the design conditions $A = A_0$, $Q = Q_0 = qA_0 = \text{constant}$, where q is the fluid's speed at design conditions. For the linearization purposes one may also assume that A_0 (which is not necessarily constant in space at design conditions) is in fact a small perturbation of a constant state, so also A_0 may be taken constant.

2. Write the linearized system in the *quasi-linear* form

$$\frac{\partial \mathbf{u}}{\partial t} + \mathrm{H}\frac{\partial \mathbf{u}}{\partial x} + \mathrm{K}\mathbf{u} = \mathbf{b},$$

where $\mathbf{u}$ is the vector of the unknowns $[A, Q]^T$ while H, K are suitable matrices, and $\mathbf{b}$ a forcing term. Characterize the problem and then provide suitable boundary conditions. Discuss its diagonalizability, highlighting differences and analogies with the previous exercise.

3. Study the stability under non-reflective boundary conditions and regular data, in case $\mathbf{b} = \mathbf{0}$.

4. Using the functions listed in Table 6.1 solve the problem relative to a pipeline 100 meters long such that $c = \sqrt{sE/2\rho} = 300$ m/s. The design conditions prescribe the discharge of $Q = 8$ m^3/s, with average speed $q = 8$ m/s corresponding to a mean cross-section 1 m^2. Consider the closure at the outlet valve in two cases: slow closure, occurring in 10 seconds, and instant closure, taking 0.5 seconds. In either case consider a linear variation in time for the outflow Q_u. Upon linearization of relation (6.30) with respect to the area, estimate the excess pressure P exerted on the vessel walls. Take $K_r = 0.1$ s^{-1} as friction coefficient and suppose there is a reservoir at the upstream end that keeps P null. Compare the solution obtained after 0.1 and 10 seconds.

5. Using a suitable linearization, find how to implement the (more realistic) condition $P = -R(Q - Q_0)$ (where R is a resistance parameter) at the inlet. Write the condition in the usual form $\mathbf{c}_1^T \mathbf{U}_0 = \psi$, for suitable $\mathbf{c}_1$ and ψ.

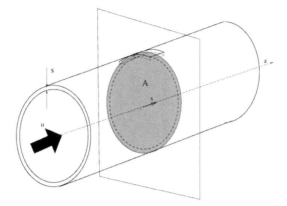

Fig. 6.28 A pipe with circular cross-section

Solution 6.2.2.

Mathematical modeling

Under the given assumptions the differential equations governing the motion of the fluid inside the pipe (see for example [FQV09], Chapter 10, for an explanation on their derivation) are

$$\frac{\partial A}{\partial t} + \frac{\partial Q}{\partial x} = 0, \tag{6.31}$$

$$\frac{\partial Q}{\partial t} + \frac{\partial}{\partial x}\left(\frac{Q^2}{A}\right) + \frac{A}{\rho}\frac{\partial P}{\partial x} + K_r Q = -\frac{dp_{ref}}{dx},$$

where K_r is a coefficient accounting for the viscosity (constant, for simplicity) and ρ the density (also constant). The two equations express the conservation of mass and of linear momentum, respectively.

We recall that here P represents the excess pressure, that's why we have the contribution of the reference pressure in the right hand side of the momentum equation. Since we know the design condition, by imposing that (A_0, Q_0) be a constant steady state solution (i.e. the solution of the equations after setting $\partial_t A = \partial_t Q = 0$ in the equations), we obtain after a few computations

$$\frac{dp_{ref}}{dx} = -K_r Q_0 = -K_r A_0 q, \tag{6.32}$$

an expression which will be used later on and that tells us that in design conditions the pressure gradient in the pipe has to overcome friction. Clearly the reference pressure will be linked to the slope of the pipe and gravity force, yet we will see that we do not need this information to obtain the wanted result. To obtain (6.32) we have assumed that A_0 is constant in space, by which (6.30) leads that correspondingly $\partial_x p = 0$.

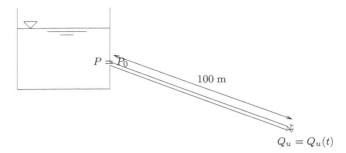

Fig. 6.29 The pipeline system of Exercise 6.2.2

The phenomenon we are looking at is that of a *hydraulic shock*, also known as *water hammer*, typical of pressurized pipelines (e.g., between a basin and a turbine) when the downstream valve is shut suddenly. The inertia of the fluid mass causes a surge in the pressure that starts at the outlet valve and travels the length of the pipe backwards to the top; there this pulse gets partially reflected as a depression wave that moves downhill, to be reflected again at the bottom valve, and so forth. The reflection phenomenon actually involves the characteristic variables of the problem, so that not just the pressure but also the flow Q are subject to oscillations. For the analysis we need to express P as function of A.

Let us rewrite (6.30) in terms of the cross-section's area, with $A = \pi d^2/4$, so that

$$P = \beta \frac{\sqrt{A} - \sqrt{A_0}}{\sqrt{A_0}} = \beta \left(\sqrt{\frac{A}{A_0}} - 1 \right),$$

where $\beta = Es$. Therefore,

$$\frac{\partial P}{\partial x} = \frac{\beta}{2\sqrt{A_0 A}} \frac{\partial A}{\partial x} - \frac{\beta}{2A_0} \sqrt{\frac{A}{A_0}} \frac{dA_0}{dx} = \frac{\beta}{2\sqrt{A_0 A}} \frac{\partial A}{\partial x},$$

since we have taken $A_0 = $ constant. We have also assumed small deformations, so we may linearize the above expression by taking $A/A_0 \simeq 1$. Hence, if we define $c = \sqrt{\frac{A_0}{\rho} \frac{dP}{dA}} = \sqrt{\frac{\beta}{2\rho}}$, we have

$$\frac{A}{\rho} \frac{\partial P}{\partial x} = c^2 \frac{\partial A}{\partial x}. \tag{6.33}$$

The positive parameter c depends on the mechanical properties of the wall and of the fluid's density, and is called *speed of sound*, since it represents the speed of propagation of small perturbations of pressure.

The non-linear term

$$\frac{\partial}{\partial x}(Q^2/A) = 2\frac{Q}{A} \frac{\partial Q}{\partial x} - \frac{Q^2}{A^2} \frac{\partial A}{\partial x}$$

is approximated as follows

$$\frac{\partial}{\partial x}(Q^2/A) = 2u\frac{\partial Q}{\partial x} - u^2\frac{\partial A}{\partial x} \simeq 2q\frac{\partial Q}{\partial x} - q^2\frac{\partial A}{\partial x}.$$

This expression is only partially correct. Indeed, even when $A - A_0$ is small, the difference $u - q$ is not necessarily so, and it is not allowed to approximate u by q. In the case of flow in pipelines, nonetheless, the error made is generally small compared to the other terms involved, and it is normally accepted that the linearization of the inertial term is a reasonable approximation.

Substituting in the original equations and using (6.32) we obtain

$$\frac{\partial A}{\partial t} + \frac{\partial Q}{\partial x} = 0, \tag{6.34}$$

$$\frac{\partial Q}{\partial t} + 2q\frac{\partial Q}{\partial x} + (c^2 - q^2)\frac{\partial A}{\partial x} + K_r Q = K_r Q_0,$$

with $0 < x < L$, $t > 0$. We may verify that $(A_0, Q_0 = qA_0)$ is a steady state solution of this system of differential equations.

Mathematical analysis

Equations (6.34) are a first-order system in the variables A and Q. Its non-conservative quasi-linear form is easy to write, for we have

$$\mathrm{H} = \begin{bmatrix} 0 & 1 \\ c^2 - q^2 & 2q \end{bmatrix}, \quad \mathrm{K} = \begin{bmatrix} 0 & 0 \\ 0 & K_r \end{bmatrix}, \quad \mathbf{b} = \begin{bmatrix} 0 \\ K_r Q_0 \end{bmatrix}.$$

The eigenvalues $a_{1,2}$ of H satisfy

$$|\mathrm{H} - a\mathbf{I}| = \begin{vmatrix} -a & 1 \\ c^2 - q^2 & 2q - a \end{vmatrix} = a^2 - 2aq + q^2 - c^2 = 0,$$

solved by $a_{1,2} = q \pm c$. The system is thus hyperbolic and the boundary conditions are determined by the signs of $q + c$ and $q - c$. In particular if $\frac{|q|}{c} < 1$ the eigenvalues have different sign and we need a condition at $x = 0$ and one at $x = L$. When $\frac{q}{c} > 1$ instead, we will need two conditions at $x = 0$, or two at $x = L$ in case $\frac{q}{c} < -1$. In case $\frac{|q|}{c} = 1$ the system admits only one boundary condition: at $x = 0$ if $q > 0$, at $x = L$ otherwise. We must find the left eigenvectors $\mathbf{l}_1$ and $\mathbf{l}_2$ of H.

Proceeding as indicated in the introductory part of this section, we can write the differential system for the characteristic variables as

$$\frac{\partial \mathbf{w}}{\partial t} + \Lambda\frac{\partial \mathbf{w}}{\partial x} + \tilde{\mathrm{K}}\mathbf{w} = \mathrm{L}\mathbf{b}, \tag{6.35}$$

with $\tilde{K} = LKL^{-1}$. The boundary conditions are given by $w_{i,0} = \mathbf{l}_i^T \begin{bmatrix} A_0 \\ Q_0 \end{bmatrix}$, $i = 1, 2$, w_1 and w_2 being the *characteristic variables* of our problem. They may be promptly computed,

$$w_1 = \frac{c+q}{2c}\left[(c-q)A + Q\right], \quad w_2 = \frac{c+q}{2c}\left[-(c-q)A + Q\right],$$

and we may note that $w_1 + w_2 = Q$.

If $K_r = 0$ and $\mathbf{b} = \mathbf{0}$ then K, and hence $\tilde{K}$, is zero and system (6.35) in the unknowns $\mathbf{W} = [w_1, w_2]^T$ is diagonal. This takes us back to the situation examined in Exercise 6.2.1.

Now suppose $a_1 > 0$ and $a_2 < 0$, and consider non-reflective boundary conditions,

$$w_1(0, t) = \psi_1(t), \quad w_1(L, t) = \psi_2(t) \qquad t > 0, \tag{6.36}$$

with Ψ_1 and Ψ_2 given functions. To test stability we multiply the first of (6.35) by w_1, the second by w_2, integrate from 0 to L and use the boundary conditions, to get

$$\begin{cases} \dfrac{d}{dt}\|w_1(t)\|_{L^2(0,L)}^2(t) + |a_1|w_1^2(L, t) = |a_1|\psi_1^2(t), \\[2mm] \dfrac{d}{dt}\|w_2(t)\|_{L^2(0,L)}^2(t) + |a_2|w_2^2(0, t) = |a_2|\psi_2^2(t), \end{cases} \tag{6.37}$$

where $\|w_i(t)\|_{L^2(0,L)}^2 = \displaystyle\int_0^L w_i^2(x, t)\, dx.$

For a shorter expression we use the norm

$$\|\mathbf{W}\|_{L^2(0,L)} = \sqrt{\|w_1\|_{L^2(0,L)}^2 + \|w_2\|_{L^2(0,L)}^2}.$$

Adding the equations (6.37) gives

$$\frac{d}{dt}\|\mathbf{W}(t)\|_{L^2(0,L)}^2 + |\Lambda| \begin{bmatrix} w_1^2(L, t) \\ w_2^2(0, t) \end{bmatrix} = |\Lambda|\Psi(t), \tag{6.38}$$

where $\Psi = [\psi_1^2, \psi_2^2]^T$ and $|\Lambda| = \mathrm{diag}(|a_1|, |a_2|)$. Now we integrate in time between 0 and t to obtain

$$\|\mathbf{W}(t)\|_{L^2(0,L)}^2 + \int_0^t |a_1|w_1^2(L, \tau) + |a_2|w_2^2(0, \tau)\, d\tau$$

$$\leq \|\mathbf{W}_0\|_{L^2(0,L)}^2 + |\Lambda| \int_0^t \Psi(\tau)\, d\tau. \tag{6.39}$$

This means that for any $t > 0$ the solution $\mathbf{W}$ is controlled in L^2-norm by the data. For the inequality to make sense the initial datum must belong in $L^2(0, L)$ and the boundary datum in $L^2(0, T)$, if T denotes the final time.

The previous expression gives an a-priori estimate in the primitive variables $\mathbf{u}$, since the L^2-norms of $\mathbf{u}$ and W are equivalent, for

$$\|\mathbf{R}\|_2^{-1}\|\mathbf{u}(t)\|_{L^2(0,L)}^2 \leq \|\mathbf{W}(t)\|_{L^2(0,L)}^2 \leq \|\mathbf{L}\|_2\|\mathbf{u}(t)\|_{L^2(0,L)}^2,$$

where $\|\mathbf{R}\|_2$ is the matrix 2-norm (see Section 1.6).

If $K_r \neq 0$, similar calculations replace (6.39) with

$$\|\mathbf{W}(t)\|_{L^2(0,L)}^2 + \int_0^t |a_1|w_1^2(L,\tau) + |a_2|w_2^2(0,\tau)d\tau$$

$$+ \int_0^t \mathbf{W}^T(\tau)\widetilde{\mathbf{K}}\mathbf{W}(\tau)\,d\tau = |\Lambda|\int_0^t \Psi(\tau)d\tau.$$

An equivalent a-priori estimate would arise in case $\mathbf{W}^T\widetilde{\mathbf{K}}\mathbf{W} \geq 0$ for any $\mathbf{W}$, in other words for $\widetilde{\mathbf{K}}$ *non-negative*. By simple computations, maybe with the help of MATLAB's symbolic toolbox, one finds that $\mathbf{x}^T\widetilde{\mathbf{K}}\mathbf{x}$, for $\mathbf{x} = [x_1, x_2]^T$, equals $\dfrac{K_r}{2c}(x_2 - x_1)[(q-c)x_1 + (q+c)x_2]$. Unfortunately this has not a definite sign. The analysis of the general case is beyond the scope of the exercise.

> ## Numerical approximation

The numerical approximation[5] goes along the lines of the previous exercise, with the sole difference that now we have both a zero-order (reaction) term and a forcing (source) term. We will use, here as well, a finite difference on a grid of uniform step h over the interval $[0, L]$ formed by $N + 1$ nodes $x_0, \ldots, x_N$. Let $\mathbf{U}_i^n = [A_i^n, Q_i^n]^T$ be the approximation of $\mathbf{U}$ at the node x_i at time $t^n = n\Delta t$, where Δt is the temporal step, and let us call $\lambda = \Delta t/h$. The presence of reaction and source terms makes the computation of the Lax-Wendroff terms much more complex. The simple approximation given by $\mathcal{B}_j^n = \mathbf{K}\mathbf{U}_j^n - \mathbf{b}_j^n$ in (6.19) would reduce the order of convergence of the scheme to one. A possibility is to derive the scheme with the usual technique accounting also for those term. This leads, however, to rather complex expressions involving also the time derivatives of the source term $\mathbf{b}$. We here follow another approach, by using a second order centered discretization (in time) for the source and reaction terms in the Taylor expansion normally used to derive the scheme.

---

[5] A simplified analytical solution of the equations can be obtained with the Allievi theory, see [All25].

For the sake of space we omit the details of the derivation and we still write the final scheme in the form (6.19), but now the numerical fluxes are given by

$$\mathcal{F}^n_{j\pm1/2} = (I + \frac{\Delta t}{2}K)^{-1}\left[\frac{1}{2}(H - \frac{\Delta t}{2}HK)(U^n_{j\pm1} + U^n_j) - \frac{\Delta t}{2h}H^2(U^n_{j\pm1} - U^n_j)\right.$$
$$\left. + \frac{\Delta t}{2}H(b_{j\pm1} + b_j)\right]$$

and

$$\mathcal{B}^n_j = -(I + \frac{\Delta t}{2}K)^{-1}\left(b^n_j + b^{n+1}_j\right).$$

while the term U^n in the scheme must be replaced by

$$\widehat{U}^n = (I + \frac{\Delta t}{2}K)^{-1}(I - \frac{\Delta t}{2}K)U^n.$$

Here, I is the $d \times d$ identity matrix, so if $K = 0$ the numerical fluxes reduce to the usual ones for a Lax-Wendroff scheme, given in (6.28), and $\widehat{U}^n = U^n$. Note that for this particular problem b is constant.

At the boundary we can again impose a suitable linear combination of A and Q, completed by the compatibility equations. Assuming $a_1 > 0$ and $a_2 < 0$,

$$\begin{cases} c^T_1 U^{n+1}_0 = \psi_1(t^{n+1}), \\ 1^T_2 U^{n+1}_0 = 1^T_2\left[\widehat{U}^n_0 - \frac{\Delta t}{h}(\mathcal{F}^n_{0+1/2} - \mathcal{F}^n_{0-1/2}) - \Delta t\mathcal{B}^n_0\right] \end{cases}$$

and

$$\begin{cases} c^T_2 U^{n+1}_N = \psi_2(t^{n+1}), \\ 1^T_1 U^{n+1}_N = 1^T_1\left[\widehat{U}^n_N - \frac{\Delta t}{h}(\mathcal{F}^n_{N+1/2} - \mathcal{F}^n_{N-1/2}) - \Delta t\mathcal{B}^n_N\right]. \end{cases}$$

Clearly, $\mathcal{F}^n_{0-1/2}$ and $\mathcal{F}^n_{N+1/2}$ must be computed with the help of the ghost node technique and a suitable extrapolation. In this case linear extrapolation should be preferred if we want to maintain the accuracy of the Lax-Wendroff scheme. We point out that the loss of accuracy caused by constant extrapolation may be appreciated only for a Δt much smaller than the one normally used in practice. That's why constant extrapolation is often adopted, for its simplicity. To end the study we have to consider the other sign possibilities for a_1 and a_2. Let us discuss only $a_1 > 0$, $a_2 > 0$. We must impose two conditions at $x = 0$, and we may give both components of U, i.e. $U^{n+1}_0 = [A^*(t^{n+1}), Q^*(t^{n+1})]^T$ for given A^* and Q^*. At the right we cannot impose anything: indeed, the signs of a_1 and a_2 at the node x_N ensure that we may compute

$$\mathbf{U}_N^{n+1} = \widehat{\mathbf{U}}_N^n - \frac{\Delta t}{h}(\boldsymbol{\mathcal{F}}_{N+1/2}^n - \boldsymbol{\mathcal{F}}_{N-1/2}^n) - \Delta t \boldsymbol{\mathcal{B}}_N^n,$$

directly.

The scheme stability requires to satisfy a CFL condition, i.e.

$$\Delta t \le \frac{h}{\max(|a_1|, |a_2|)}.$$

Numerical results

For the numerical solution we rely on the program listed in Table 6.1, and in the function `hyp_pipe.m` we describe a possible implementation. We need to specify the boundary conditions. We will impose at the left node $A = A_0$, so $\mathbf{c}_1 = [1, 0]^T$ and $\psi_1 = A_0$, while at the right $Q_N^n = \max(0, Q_0(1 - \alpha t))$, where $Q^0 = 8$ m^3/s, and α is 0.1 (slow closure) and 2 (instantaneous closure). The domain was divided into 100 nodes, Δt chosen so that the CFL number is 0.98. The Lax-Wendroff scheme was used.

Fig. 6.30 shows the results 0.1 and 10 seconds after shutting the valve. We have both the slow (dashed line) and the instant closure (solid line). Note how variation of pressure and fluxes are much more marked in the latter case even after 10 seconds (long after the valve closure). In particular, a strong reverse flux is present in the fast closure case. At $t = 0.1$ seconds the upstream wave associated to w_2 has just reached one third of the length of the pipe. Indeed, the upper part of the pipe has not yet been affected and the flow is still at its design condition.

As for the proposed resistance condition, by exploiting (6.33) we may use the approximation $P(A) \simeq \frac{dP}{dA}(A - A_0)$ and the definition of c to obtain

$$\frac{\rho c^2}{A_0} A + RQ = \rho c^2 + Rq,$$

already in the desired form. $\diamond$

Remark 6.2 The code `femld` features among the "model problems" the study of blood flowing in an artery, where the propagation phenomena are generated by the interaction of a viscous fluid with the elastic arterial wall. The numerical model provided with `femld`, though, refers to the non-linearized problem, whose treatment goes beyond the scope of this book.

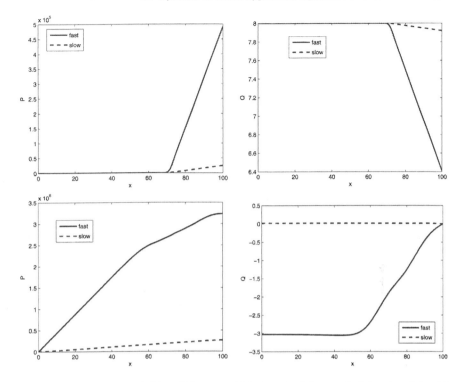

Fig. 6.30 Overpressure (left) and flow rate (right) at $t = 0.1$ seconds (above) and $t = 10$ seconds (below). The dashes represent slow closure, the solid curve fast closure

7

Navier-Stokes equations for incompressible fluids

Incompressible fluid dynamics is a challenging topic, basically for the "saddle point" nature of the problem. Under certain assumptions, the velocity field that solves this problem is the minimum of an energy constrained by incompressibility. This makes this problem significantly different from elliptic problems (corresponding to free minimization). Pressure is the *Lagrange multiplier* associated with the incompressibility constraint. For this reason, the numerical solution may be in general expensive to compute and large efforts have been devoted to develop efficient solution schemes.

The present chapter is split into two sections. In the first one we consider steady Stokes and Navier-Stokes problems, with specific attention to the space discretization. Selection of finite dimensional spaces for the velocity and pressure is not arbitrary, being constrained by the fulfillment of the so called *inf-sup* or LBB condition (Ladyzhenskaja-Babuska-Brezzi - see e.g. [QV94], Chapter 7). Most of the exercises in this section are oriented to investigate the role of the LBB condition.

In the second section, we consider time-dependent problems. In particular, we consider time advancing methods for the separate (or segregated) computation of velocity and pressure, aiming at reducing the computational costs. We do not consider some advanced topics like turbulence. We just want to cover some basic topics, as a first step in a complex and fascinating field.

The interested reader can refer for more details to [ESW05], [Qua09], Chapter 10, [QV94], Chapters 9, 10 and 13, [CHQZ88, DFM02] (for Spectral Methods), [Qua93] (Finite Elements), [FP99] (Finite Volumes), [Pro97] (Projection methods), [Tur99]. For the algebraic aspects of the problem, we mention also [BGL05].

Regarding the notation, velocity is indicated with $\mathbf{u}$, scaled pressure (i.e. divided by the density) with p (unit of measure: m^2/s^2), Ω is a domain with boundary Γ. In large part of the chapter $\Omega \subset \mathbb{R}^2$. The space of vector functions (in two or three dimensions) with components in $H^1(\Omega)$ is denoted by $\mathbf{H}^1(\Omega)$. Similar notation for $\mathbf{L}^2(\Omega)$.

Formaggia L., Saleri F., Veneziani A.: Solving Numerical PDEs: Problems, Applications, Exercises. DOI 10.1007/978-88-470-2412-0_7, © Springer-Verlag Italia 2012

Finite dimensional spaces for the numerical approximation of the problem will be denoted by:

1. V_h for the velocity; the dimension of V_h is denoted by $d_\mathbf{u}$. Most often $d_\mathbf{u} = d \times d_c$ where d is the space dimension and d_c is the dimension of the finite dimensional space of each velocity component;
2. Q_h for the pressure, with dimension d_p.

Space coordinates will be denoted with the indexed notation $\mathbf{x} = (x_1, x_2, x_3)$.

7.1 Steady problems

Exercise 7.1.1. Consider the following steady Stokes problem

$$\begin{cases} -\nu\triangle\mathbf{u} + \nabla p = \mathbf{f} & \text{in } \Omega \subset \mathbb{R}^2, \\ \nabla \cdot \mathbf{u} = 0 & \text{in } \Omega, \end{cases} \tag{7.1}$$

with condition $\mathbf{u} = \mathbf{g}$ on Γ, where $\mathbf{g}$ is a given function in $\mathbf{H}^{1/2}(\Gamma)$.

1. Show that the problem admits a solution only under the compatibility condition $\int_\Gamma \mathbf{g} \cdot \mathbf{n} \, d\gamma = 0$ and write the weak formulation identifying proper functional spaces eventually introducing suitable assumptions on the data. Give an a priori estimate for the velocity.
2. Let Ω be the unit square $[0, 1] \times [0, 1]$ in the plane (x_1, x_2). Let $\mathbf{f} = \mathbf{0}$ and $\mathbf{g} = [g_1, g_2]^T$ with

$$g_1 = \begin{cases} 0 & \text{for } x_2 = 0, x_2 = 1, 0 < x_1 < 1, \\ x_2 - x_2^2 & \text{for } x_1 = 0, x_1 = 1, 0 < x_2 < 1, \end{cases} \qquad g_2 = 0.$$

Verify that $\int_\Gamma \mathbf{g} \cdot \mathbf{n} d\gamma = 0$. Find a solution in the form $\mathbf{u} = [u_1, u_2]^T$ with $u_2 = 0$.
3. Write the numerical problem with finite elements P2 for the velocity and P1 for the pressure. Comment on this pair of spaces.
4. Solve the problem with **Freefem++** for $h = 0.05, 0.025, 0.0125$, $\nu = 0.1$. Comment on the computed numerical solution.

Solution 7.1.1.

Mathematical analysis

The condition on $\mathbf{g}$ is a compatibility condition with the divergence free constraint. Divergence Theorem states that if $\mathbf{u}$ is a solenoidal vector

$$0 = \int_\Omega \nabla \cdot \mathbf{u} \, d\Omega = \int_\Gamma \mathbf{u} \cdot \mathbf{n} d\gamma = \int_\Gamma \mathbf{g} \cdot \mathbf{n} d\gamma.$$

If this condition on $\mathbf{g}$ is not fulfilled, the problem does not have a solution. Hereafter, we assume this condition to hold. Note that this condition is required because the Dirichlet data $\mathbf{g}$ is applied to the whole boundary Γ. In the cases where the Dirichlet data is applied only on a portion of $\partial\Omega$ it is not required anymore.

A consequence of the Fredholm alternative (see, for instance, [Sal08]) is that when the compatibility on $\mathbf{g}$ is enforced the problem has more than one solution. Indeed, one may note that pressure appears in our problem only under the gradient. Therefore, it is known up to a constant. To fix this constant, and recover a unique solution for the problem at hand, a common strategy is to look for pressures with zero average, i.e. such that $\int_\Omega p\,d\Omega = 0$. We will however see that other strategies may be more convenient at numerical level.

To derive the weak formulation we set:

1. $V \equiv \mathbf{H}_0^1(\Omega)$;
2. $Q \equiv L_0^2(\Omega) = \{q \in L^2(\Omega) : \int_\Omega q\,d\Omega = 0\}$. Given a pressure field $\widetilde{p}$, pressure $p = \widetilde{p} - \int_\Omega \widetilde{p}\,d\Omega$ belongs to Q.

We will also make use of the spaces

$$V_{\mathrm{div}} = \{\mathbf{v} \in \left[\mathrm{H}^1(\Omega)\right]^d : \nabla \cdot \mathbf{v} = 0\}, \quad V_{\mathrm{div}}^0 = \{\mathbf{v} \in V_{\mathrm{div}} : \mathbf{v} = \mathbf{0} \text{ on } \Gamma\}. \quad (7.2)$$

Let $\mathbf{G}(\mathbf{x})$ (with $\mathbf{x} \in \Omega$) be a divergence free *lifting* (or extension) of $\mathbf{g}$, i.e. $\mathbf{G} \in V_{\mathrm{div}}$ with $\mathbf{G} = \mathbf{g}$ on Γ. To prove the existence of this extension can be nontrivial: see [Lad63] Chapter 1, Section 2 for a proof. However, even if divergence free extensions are useful for well posedness analysis, they are not required for the numerical treatment.

Clearly if $\mathbf{u}$ is the solution of our problem and $\tilde{\mathbf{u}} = \mathbf{u} - \mathbf{G}$ we have that $\tilde{\mathbf{u}} \in V$, and more precisely $\tilde{\mathbf{u}} \in V_{\mathrm{div}}$.

Finally, we introduce the following notation:

1. $\displaystyle\int_\Omega \nabla\mathbf{u} : \nabla\mathbf{v}\,d\Omega \equiv \sum_{i,j=1}^2 \int_\Omega \frac{\partial u_i}{\partial x_j}\frac{\partial v_i}{\partial x_j}\,d\omega$;

2. $\displaystyle a(\mathbf{u},\mathbf{v}) \equiv \nu\,(\nabla\mathbf{u}, \nabla\mathbf{v}) = \nu\int_\Omega \nabla\mathbf{u} : \nabla\mathbf{v}\,d\Omega$ for any $\mathbf{u},\mathbf{v} \in \mathrm{H}^1(\Omega)$;

3. $\displaystyle b(\mathbf{v},q) \equiv -\int_\Omega \nabla\cdot\mathbf{v}q\,d\Omega$ for any $\mathbf{v} \in \mathrm{H}^1(\Omega)$ and $q \in Q$.

These functional spaces guarantee that the bilinear forms are well defined and continuous (see e.g. [Qua09], Chapter 10).

For the weak form of the problem, let us multiply the first of (7.1) by a function $\mathbf{v} \in V$, the second one by $q \in Q$, and integrate on Ω. Since $\mathbf{v}$ has a null trace on the boundary, by the Green formula we have

$$-\nu\int_\Omega \Delta\mathbf{u}\cdot\mathbf{v}\,d\Omega = -\nu\int_\Gamma (\nabla\mathbf{u}\cdot\mathbf{n})\cdot\mathbf{v}d\gamma + \nu\int_\Omega \nabla\mathbf{u} : \nabla\mathbf{v}\,d\Omega.$$

With the notation of the previous chapters we get the weak form:
find $\mathbf{u} \in \mathbf{G} + V$ and $p \in Q$ such that for any $\mathbf{v} \in V$ and $q \in Q$

$$
\begin{cases}
a(\mathbf{u}, \mathbf{v}) + b(\mathbf{v}, p) = (\mathbf{f}, \mathbf{v}), \\
b(\mathbf{u}, q) = 0,
\end{cases}
$$

corresponding to:
find $\tilde{\mathbf{u}} \in V$ and $p \in Q$ such that for any $\mathbf{v} \in V$ and $q \in Q$

$$
\begin{cases}
a(\tilde{\mathbf{u}}, \mathbf{v}) + b(\mathbf{v}, p) = \mathcal{F}(\mathbf{v}), \\
b(\tilde{\mathbf{u}}, q) = -b(\mathbf{G}, q) = 0,
\end{cases}
\tag{7.3}
$$

where $\mathcal{F}(\mathbf{v}) \equiv (\mathbf{f}, \mathbf{v}) - a(\mathbf{G}, \mathbf{v})$. Functional $\mathcal{F}$ is linear on V, being the sum of linear functionals. Moreover, if for instance[1] $\mathbf{f} \in \mathbf{L}^2(\Omega)$ it is bounded for the selected functional spaces and therefore continuous, i.e. $\mathcal{F} \in V'$.

The required stability estimate can be obtained by selecting in (7.3) $\mathbf{v} = \tilde{\mathbf{u}}$ and $q = p$. If we subtract member-wise the two equations obtained, we get

$$
a(\tilde{\mathbf{u}}, \tilde{\mathbf{u}}) = \mathcal{F}(\tilde{\mathbf{u}}).
$$

Bilinear form $a(\cdot, \cdot)$ is coercive thanks to the Poincaré inequality. If α denotes the coercivity constant, we get the stability estimate

$$
\alpha \|\tilde{\mathbf{u}}\|_V^2 \leq a(\tilde{\mathbf{u}}, \tilde{\mathbf{u}}) \leq \|\mathcal{F}\|_{V'} \|\tilde{\mathbf{u}}\|_V \Rightarrow \alpha \|\tilde{\mathbf{u}}\|_V \leq \|\mathcal{F}\|_{V'}.
$$

The third point of the problem is solved as follows. On the horizontal sides of the square $\mathbf{g} \cdot \mathbf{n} = g_2 = 0$. Notice that g_1 is the same on the two vertical sides, so that

$$
\int_\Gamma \mathbf{g} \cdot \mathbf{n} = -\int_0^1 g_1(0, x_2) dx_2 + \int_0^1 g_1(1, x_2) dx_2 = 0.
$$

Therefore, boundary data fulfill the compatibility condition. Moreover, they are compatible with a velocity filed where $u_2 = 0$. We can then look for a possible solution of the form $\mathbf{u} = [u_1, 0]^T$. Mass conservation equation yields

$$
\nabla \cdot \mathbf{u} = \frac{\partial u_1}{\partial x_1} + \frac{\partial u_2}{\partial x_2} = \frac{\partial u_1}{\partial x_1} = 0,
$$

which implies that u_1 is independent of x_1. In the second component of the momentum equation, $u_2 = 0$ yields

$$
-\nu \left(\frac{\partial^2 u_2}{\partial x_1^2} + \frac{\partial^2 u_2}{\partial x_2^2} \right) + \frac{\partial p}{\partial x_2} = \frac{\partial p}{\partial x_2} = 0,
$$

which means that the pressure p depends only on x_1 and not on x_2.

---

[1] It is sufficient that $\mathbf{f}$ be a member of V'.

Finally, the first component of the momentum equation reads

$$-\nu\left(\frac{\partial^2 u_1}{\partial x_1^2} + \nu\frac{\partial^2 u_1}{\partial x_2^2}\right) + \frac{\partial p}{\partial x_1} = -\nu\frac{\partial^2 u_1}{\partial x_2^2} + \frac{\partial p}{\partial x_1} = 0,$$

leading to

$$\nu\frac{\partial^2 u_1}{\partial x_2^2} = \frac{\partial p}{\partial x_1}.$$

On the left we have a function only of the variable x_2, on the right there is a function of x_1. The equation holds therefore if both sides are equal to a constant c_1. Thus

$$p = c_1 x_1 + c_2, \qquad u_1 = \frac{1}{2\nu}c_1 x_2^2 + c_3 x_2 + c_4.$$

From the boundary conditions we infer

$$c_1 = -2\nu, \quad c_3 = 1, \quad c_4 = 0,$$

yielding

$$u_1 = -x_2^2 + x_2, \quad u_2 = 0, \qquad p = -2\nu x_1 + c_2.$$

Constant c_2 is determined by forcing a null average pressure,

$$\int_\Omega p\,d\Omega = -\nu + c_2 = 0,$$

giving $c_2 = \nu$ and $p = \nu(1 - 2x_1)$. This is the well known 2D *Poiseuille solution*.

Numerical approximation

Galerkin formulation of the problem requires to introduce the finite-dimensional functional spaces V_h and Q_h. Approximate solutions $\tilde{\mathbf{u}}_h$ and p_h belong to these spaces and are such that for any $\mathbf{v}_h \in V_h$ and $q_h \in Q_h$ we have

$$\begin{cases} a(\tilde{\mathbf{u}}_h, \mathbf{v}_h) + b(\mathbf{v}_h, p_h) = \mathcal{F}(\mathbf{v}_h), \\[2mm] b(\tilde{\mathbf{u}}_h, q_h) = -b(\mathbf{G}_h, q_h). \end{cases} \qquad (7.4)$$

Here, we do not assume that that the lifting of the boundary data $\mathbf{G}_h \in V_h$ is divergence free.

Let us introduce a mesh $\mathcal{T}_h$ with size $h > 0$ and let $V_h = \{\mathbf{v}_h \in X_h^2 \times X_h^2 : \mathbf{v}_h|_\Gamma = \mathbf{0}\}$. Approximate pressure is assumed to be in $Q_h = X_h^1$. We need however to enforce a condition the have uniqueness for the pressure. While in the analysis this was done by considering functions with zero average, in finite element discretizations the simplest way to prescribe a constraint on the pressure is to fix its value in a given point of the domain. Typically the set value is 0. This choice is in fact inconsistent with the pressure belonging to $L^2(\Omega)$ since a function in $L^2(\Omega)$ may be discontinuous and its "value in a point" has no meaning. We may thus expect problems in the numerical solution when $h \to 0$. However, in many practical situations the pressure is indeed continuous, in which case fixing its value in a point is acceptable. After this discussion one may ask why not imposing the zero average constraint directly, which is well posed for elements of $L^2(\Omega)$. Of course this is possible, but not practical. Imposing $\int_\Omega p_h = 0$ implies to link all the nodal values of pressure, creating a single full row in the matrix that eventually governs the computation of p_h. This contrasts with the usual sparsity pattern of finite element matrices and requires a special treatment. Imposing the value at a point, instead, can be handled with the usual techniques for the imposition of Dirichlet data (see Appendix A for details). With some other discretization methods, for instance spectral method, the null-average constraint may be forced directly.

We will the seek the discrete pressure in the subspace of X_h^1 of functions vanishing in a point of the domain. From the practical viewpoint, a simple choice is to take this point as the last one in the list of the pressure degrees of freedom. This can be realized from the algebraic formulation of (7.4). Yet, for the time being, let us take $Q_h = X_h^1$, without any constraint. Let φ_i and ψ_j be the basis functions of V_h and Q_h respectively. Let us introduce the matrices $K \in \mathbb{R}^{d_u \times d_u}$ e $D \in \mathbb{R}^{d_p \times d_u}$ whose entries are

$$k_{ij} \equiv \nu \int_\Omega \nabla \varphi_j : \nabla \varphi_i \, d\Omega, \qquad d_{kj} \equiv \int_\Omega \nabla \psi_k \cdot \nabla \varphi_j \, d\Omega,$$

$$1 \le i, j \le d_\mathbf{u}, \, 1 \le k \le d_p.$$

The algebraic form of the discrete problem reads

$$\begin{bmatrix} K & D^T \\ D & \mathbf{0} \end{bmatrix} \begin{bmatrix} \mathbf{U} \\ \mathbf{P} \end{bmatrix} = \begin{bmatrix} \mathbf{F} \\ -\mathbf{DG} \end{bmatrix}, \tag{7.5}$$

where $\mathbf{U}$ and $\mathbf{P}$ are vectors with the nodal values of velocity and pressure, respectively, $\mathbf{G}$ is the vector with the nodal values of the extension, $\mathbf{F}$ is the result of the discretization of $\mathcal{F}$ ($F_i = \mathcal{F}(\varphi_i)$). In order to fix the pressure $p = 0$ in the last pressure node, a possible technique is to cancel the last row (and column) of the matrix of the linear system (7.5), and correspondingly the last entry of the right-hand side. The choice of the last node makes this operation handy even for matrices stored with a sparse format. Or we might use any of the other techniques presented in Section A.2.1.

An alternative approach, which does not require any modification on the matrix, resorts to using an iterative method for solving the linear system. Actually the matrix obtained before fixing the pressure is singular (otherwise we would have a unique pressure field), however many common iterative methods (like Conjugate Gradient or GMRES) are not affected by this circumstance and are able to converge to a solution anyway. The numerical

solution found will be different according to the initial guess used to start
the iterative procedure. However all the solutions differ by a constant. It is
therefore possible to fix the pressure *a posteriori* by shifting the computed
solution in such a way it assumes the required value. For instance, if p_{old} is
the solution provided by the iterative solver and we wish to correct it so that
the new pressure has value $\bar{p}$ in $\mathbf{x}$, we just set $p_{new}(\mathbf{x}) = p_{old}(\mathbf{x}) - \bar{p}$. We
follow this approach in Program 33[2]. Note that if we use post-processing also
the imposition of zero average pressure is not complicated.

Even after forcing the uniqueness of the pressure, system (7.5) is non
singular only under specific conditions on the spaces V_h and Q_h, namely the
fulfillment of the LBB condition mentioned in the introductory part of this
chapter. This is a remarkable difference with respect to elliptic problems,
where well posedness of the discrete problem is an immediate consequence of
the well posedness of the continuous one. We remind that the chosen spaces
V_h and Q_h (usually denoted with $P^2 - P^1$) fulfill the LBB condition.

Program 33 - stokes1 : Steady Stokes problem

```
mesh Th=square(40,40);
fespace Vh2(Th,P2), Qh(Th,P1);
Vh2 u2,v2,u1ex = y-y^2,u1,v1; Qh p=0,q;
func g=y-y^2; real  nu=0.1;
solve Stokes ([u1,u2,p],[v1,v2,q],solver=GMRES,eps=1.e-9) =
    int2d(Th)(nu * ( dx(u1)*dx(v1) + dy(u1)*dy(v1)
             + dx(u2)*dx(v2) + dy(u2)*dy(v2) )
             - p*dx(v1) - p*dy(v2) - dx(u1)*q - dy(u2)*q)
 + on(1,3,u1=0,u2=0) + on(2,4,u1=g,u2=0);
real pref = p[][Qh.ndof-1];
p = p - pref;
```

Numerical results

When using second order polynomials for the velocity and first order for
the pressure, not only we satisfy the LBB condition, but also we build a finite
dimensional space that contains the exact solution of the problem at hand.
Up to round-off errors (and approximation error if we solve the system by an
iterative procedure), we expect to find the exact solution. After computation
with the step sizes suggested in the exercise, we get the following table,

h	0.05	0.025	0.0125
$\|\mathbf{e}_h\|_{L^2}$	$9.45041e - 07$	$2.01579e - 07$	$3.46727e - 07$

---

[2] `Freefem++` denotes the space coordinates with x and y.

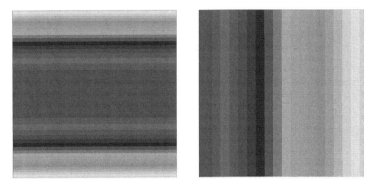

Fig. 7.1 Speed isolines (on the left) and pressure isolines (on the right) for Exercise 7.1.1

where the errors $\mathbf{e}_h = \mathbf{u} - \mathbf{u}_h$ are very small. They are mostly linked to the tolerance adopted in the iterative method (we have used GMRES to solve the linear system). In Fig. 7.1 we report the map of the speed (left) and pressure (right) computed by Program 33.   ◇

Exercise 7.1.2. Let us consider the following steady non-homogeneous Navier-Stokes problem

$$\begin{cases} (\mathbf{u} \cdot \nabla)\,\mathbf{u} - \nu \triangle \mathbf{u} + \nabla p = \mathbf{f} \text{ in } & \Omega \subset \mathbb{R}^2, \\ \nabla \cdot \mathbf{u} = 0 & \text{in } \Omega, \end{cases} \qquad (7.6)$$

with the condition $\mathbf{u} = \mathbf{g}$ on Γ, where $\mathbf{g} \in H^{1/2}(\Gamma)$ is a given function.

1. Find the weak formulation of the problem and introduce suitable assumptions on the boundary data in order to give a stability estimate for the (possible) solution.
2. Let Ω be the unit square $[0,1] \times [0,1]$ in the plane (x_1, x_2). Let $\mathbf{f} = \mathbf{0}$ and $\mathbf{g} = [g_1, g_2]^T$ given as in Exercise 7.1.1. Check that solution found in the previous Exercise is solution to the present problem too.
3. Write the discretized formulation of the problem and the generic Newton iteration for the solution of the nonlinear algebraic system resulting after finite element discretization.
4. Solve the problem with **Freefem++**, using the same data as in the previous exercise.

Solution 7.1.2.

Mathematical analysis

In order to manage the nonlinear term, we introduce the trilinear form

$$c(\mathbf{u}, \mathbf{w}, \mathbf{v}) \equiv \int_\Omega (\mathbf{u} \cdot \nabla) \mathbf{w} \cdot \mathbf{v} \, d\Omega$$

defined for any $\mathbf{w}, \mathbf{u}, \mathbf{v} \in \mathbf{H}^1(\Omega)$. We show first that this definition is sound, i.e. that this integral is bounded. We have to prove that the integrand function belongs to $\mathbf{L}^1(\Omega)$. Let us express the different components

$$(\mathbf{u} \cdot \nabla) \mathbf{v} \cdot \mathbf{w} = \sum_{i,j=1}^{2} u_j \frac{\partial v_i}{\partial x_j} w_i.$$

The integrand is therefore the sum of terms that are the product of three functions, w_j, $\frac{\partial u_i}{\partial x_j}$ and v_i. If $\mathbf{u} \in \mathbf{H}^1(\Omega)$, it follows that $\partial u_i / \partial x_j \in L^2(\Omega)$. Moreover, for the Sobolev embedding theorem (see Section 1.3) we have that in two and three dimensions $H^1(\Omega) \hookrightarrow L^4(\Omega)$, which means that

$$f \in H^1(\Omega) \Rightarrow f \in L^4(\Omega), \quad \text{with} \quad \|f\|_{L^4(\Omega)} \leq C\|f\|_{H^1(\Omega)},$$

for a $C > 0$. Thus $w_j v_i \in L^2(\Omega)$, being the product of two L^4 functions, and consequently by Hölder inequality $w_j \frac{\partial u_i}{\partial x_j} v_i \in L^1(\Omega)$, which eventually proves that trilinear form $c(\cdot, \cdot, \cdot)$ is well defined.

We prove an important property of the trilinear form, holding for functions $\mathbf{u} \in V_{\text{div}}^0$. It is the *skew-symmetry* property:

$$c(\mathbf{u}, \mathbf{v}, \mathbf{w}) = -c(\mathbf{u}, \mathbf{w}, \mathbf{v}).$$

To prove it, we use the Green formula

$$c(\mathbf{u}, \mathbf{v}, \mathbf{w}) = \int_\Omega (\mathbf{u} \cdot \nabla) \mathbf{w} \cdot \mathbf{v} \, d\Omega =$$

$$\int_\Gamma \mathbf{u} \cdot \mathbf{n} \, (\mathbf{w} \cdot \mathbf{v}) \, d\gamma - \int_\Omega (\nabla \cdot \mathbf{u}) \, (\mathbf{w} \cdot \mathbf{v}) \, d\Omega - \int_\Omega (\mathbf{u} \cdot \nabla) \mathbf{v} \cdot \mathbf{w} \, d\Omega.$$

The first term on the right-hand side vanishes since $\mathbf{w}$ and $\mathbf{v}$ have null trace on Γ. Second integral vanishes since $\mathbf{u}$ is divergence free. This concludes the proof.

As an immediate consequence we have

$$c(\mathbf{u}, \mathbf{v}, \mathbf{v}) = 0. \tag{7.7}$$

If we exploit the lifting $\mathbf{G} \in V_{\text{div}}$ introduced in the previous exercise, the weak form of the problem reads:

find $\mathbf{u} \in \mathbf{G} + V$, $p \in Q$ s.t. for all $\mathbf{v} \in V$ and $q \in Q$

$$
\begin{cases}
a(\mathbf{u}, \mathbf{v}) + c(\mathbf{u}, \mathbf{u}, \mathbf{v}) + b(\mathbf{v}, p) = (\mathbf{f}, \mathbf{v}), \\
\\
b(\mathbf{u}, q) = 0.
\end{cases}
\tag{7.8}
$$

To deduce a stability estimate, we proceed like in the previous exercise. We denote $\tilde{\mathbf{u}} \equiv \mathbf{u} - \mathbf{G}$ and select $\mathbf{v} = \tilde{\mathbf{u}}$ and $q = p$. Equation (7.7) and coercivity of the bilinear form $a(\cdot, \cdot)$ yield

$$
\alpha \|\tilde{\mathbf{u}}\|_V^2 + c(\tilde{\mathbf{u}}, \mathbf{G}, \tilde{\mathbf{u}}) \leq (\mathbf{f}, \tilde{\mathbf{u}}) - a(\mathbf{G}, \tilde{\mathbf{u}}) - c(\mathbf{G}, \mathbf{G}, \tilde{\mathbf{u}}).
\tag{7.9}
$$

Functional

$$
\mathcal{F}(\mathbf{v}) \equiv (\mathbf{f}, \mathbf{v}) - a(\mathbf{G}, \mathbf{v}) - c(\mathbf{G}, \mathbf{G}, \mathbf{v})
$$

is continuous as a consequence of the continuity of each term of the sum.

As for the non-linear term in the left-hand side of (7.9), we have, applying Hölder inequality and the Sobolev embedding theorem,

$$
|c(\tilde{\mathbf{u}}, \mathbf{G}, \tilde{\mathbf{u}})| \leq \|\tilde{\mathbf{u}}\|_{L^4(\Omega)}^2 \|\mathbf{G}\|_{L^2(\Omega)} \leq C_1 \|\tilde{\mathbf{u}}\|_{H^1(\Omega)}^2 \|\mathbf{G}\|_{H^1(\Omega)}.
$$

The lifting operator is a continuous operator from $H^{1/2}(\Gamma)$ to V_{div} (see e.g. [Qua09]), in the sense that there exists a constant C_2 such that for any $\mathbf{g} \in H^{1/2}(\Gamma)$ satisfying the compatibility condition there exist a $\mathbf{G} \in V_{\mathrm{div}}$ such that

$$
\|\mathbf{G}\|_{H^1(\Omega)} \leq C_2 \|\mathbf{g}\|_{H^{1/2}(\Gamma)}.
$$

Finally, we assume that the boundary data are small enough, more precisely we assume that

$$
\|\mathbf{g}\|_{H^{1/2}(\Gamma)} < \frac{\alpha}{C_2}
\tag{7.10}
$$

so that there exists a real number α_0 s.t. $\alpha - C_2 \|\mathbf{g}\|_{H^{1/2}} = \alpha_0 > 0$.

With that assumption, from (7.9) we finally get the estimate

$$
\|\tilde{\mathbf{u}}\|_{H^1} \leq \frac{1}{\alpha_0} \|\mathcal{F}\|_{V'},
\tag{7.11}
$$

where the right-hand side depends on the forcing term $\mathbf{f}$ and the boundary data $\mathbf{g}$ only.

In order to check that the solution to the problem of the previous exercise is solution of the Navier-Stokes problem at hand we just verify that the nonlinear term vanishes for this solution. Since $u_2 = 0$, the nonlinear term reads

$$
(\mathbf{u} \cdot \nabla) \mathbf{u} = \begin{bmatrix} u_1 \dfrac{\partial u_1}{\partial x_1} + u_2 \dfrac{\partial u_1}{\partial x_2} \\ u_1 \dfrac{\partial u_2}{\partial x_1} + u_2 \dfrac{\partial u_2}{\partial x_2} \end{bmatrix} = \begin{bmatrix} u_1 \dfrac{\partial u_1}{\partial x_1} \\ 0 \end{bmatrix}.
$$

Since u_1 does not depend on x_1, also the first component vanishes.

Numerical approximation

We discretize the problem as done in the previous exercise, using the same notations. Let us introduce, in particular, spaces V_h and Q_h, matrices K and D and vectors $\mathbf{U}$, $\mathbf{P}$ and $\mathbf{F}$. Let $C(\mathbf{U})$ be the matrix with entries

$$C_{ij}(\mathbf{U}) = \sum_{l=1}^{d_{\mathbf{u}}} \int_{\Omega} (U_l \varphi_l \cdot \nabla)\, \varphi_j \cdot \varphi_i \, d\Omega$$

corresponding to the discretization of the nonlinear term. For reasons that we will explain later on, let us introduce also the matrix $C^*(\mathbf{U})$ with entries

$$C_{ij}^*(\mathbf{U}) = \sum_{l=1}^{d_{\mathbf{u}}} \int_{\Omega} (\varphi_j \cdot \nabla)\, U_l \varphi_l \cdot \varphi_i \, d\Omega.$$

The discretized problem corresponds to the nonlinear algebraic system with size $d_{\mathbf{u}} + d_p$

$$\begin{bmatrix} K + C(\mathbf{U}) & D^T \\ D & 0 \end{bmatrix} \begin{bmatrix} \mathbf{U} \\ \mathbf{P} \end{bmatrix} = \begin{bmatrix} \mathbf{F} \\ -DG \end{bmatrix}.$$

A possible method for solving this non-linear system is the *Newton method*, see Chapter 5. Let us recall briefly the idea.

For a given nonlinear system $\mathbf{f}(\mathbf{u}) = 0$ and a guess $\mathbf{u}_{(k)}$ of the solution, the new guess $\mathbf{u}_{(k+1)}$ is obtained by solving the linear system

$$J(\mathbf{u}_{(k)})(\mathbf{u}_{(k+1)} - \mathbf{u}_{(k)}) = -\mathbf{f}(\mathbf{u}_{(k)}), \tag{7.12}$$

where $J(\mathbf{u}_{(k)})$ is the *Jacobian matrix* with entries

$$J_{ij}(\mathbf{x}_{(k)}) = \frac{\partial f_i}{\partial x_j}(\mathbf{u}_{(k)}).$$

This correspond to a linearization of the non linear problem around $\mathbf{u}_{(k)}$ using a Taylor's expansion truncated at the first order.

In our case, non-linearity stems only from the convective term $\mathbf{f}_{NL} \equiv C(\mathbf{U})\mathbf{U}$ in the momentum equation. By definition

$$\mathbf{f}_{NL,i}(\mathbf{U}) = \sum_{j,k=1}^{d_{\mathbf{u}}} \int_{\Omega} U_j (\varphi_j \cdot \nabla)\, U_k \varphi_k \cdot \varphi_i \, d\Omega, \quad \text{for } i = 1, \dots, d_{\mathbf{u}},$$

so that

$$\frac{\partial \mathbf{f}_{NL,i}}{\partial U_l}(\mathbf{U}) = \sum_{j}^{d_{\mathbf{u}}} \int_{\Omega} U_j (\varphi_j \cdot \nabla)\, U_l \varphi_l \cdot \varphi_i \, d\Omega +$$

$$\sum_{k}^{d_{\mathbf{u}}} \int_{\Omega} U_l (\varphi_l \cdot \nabla)\, U_k \Rightarrow J_{NL}(\mathbf{U}) = \varphi_k \cdot \varphi_i \, d\Omega = C(\mathbf{U}) + C^*(\mathbf{U}).$$

All other terms in the discrete problem are linear, so their contribution to the Jacobian is trivial. At each iteration, Newton method (7.12) reads

$$
\begin{bmatrix} \mathrm{K} + \mathrm{C}(\mathbf{U}^{(k)}) + \mathrm{C}^*(\mathbf{U}^{(k)}) & \mathrm{D}^T \\ \\ \mathrm{D} & 0 \end{bmatrix} \begin{bmatrix} \mathbf{U}^{(k+1)} - \mathbf{U}^{(k)} \\ \\ \mathbf{P}^{(k+1)} - \mathbf{P}^{(k)} \end{bmatrix} = \begin{bmatrix} \mathbf{F} \\ \\ -\mathrm{D}\mathbf{G} \end{bmatrix}
$$
$$
- \begin{bmatrix} \mathrm{K} + \mathrm{C}(\mathbf{U}^{(k)}) & \mathrm{D}^T \\ \\ \mathrm{D} & 0 \end{bmatrix} \begin{bmatrix} \mathbf{U}^{(k)} \\ \\ \mathbf{P}^{(k)} \end{bmatrix}.
$$

Noting that $\mathrm{C}(\mathbf{U}^{(k)})\mathbf{U}^{(k)} = \mathrm{C}^*(\mathbf{U}^{(k)})\mathbf{U}^{(k)}$, we have the following simplified form

$$
\begin{bmatrix} \mathrm{K} + \mathrm{C}(\mathbf{U}^{(k)}) + \mathrm{C}^*(\mathbf{U}^{(k)}) & \mathrm{D}^T \\ \\ \mathrm{D} & \mathbf{0} \end{bmatrix} \begin{bmatrix} \mathbf{U}^{(k+1)} \\ \\ \mathbf{P}^{(k+1)} \end{bmatrix} = \begin{bmatrix} \mathbf{F} + \mathrm{C}(\mathbf{U}^{(k)})\mathbf{U}^{(k)} \\ \\ -\mathrm{D}\mathbf{G} \end{bmatrix}. \quad (7.13)
$$

Newton method converges with a second order convergence rate if the initial guess $\mathbf{U}^0$ is accurate enough. Its computational cost is high, since at each step we have to recompute the Jacobian matrix. "Quasi-Newton" methods are a trade-off, trying to reduce or eliminate the number of Jacobian recalculations. For more details, see e.g. [QSS00].

This is based on the sequence *Discretize then Linearize*. We could follow also the other way, i.e. *Linearize then Discretize*. Again we discuss only the non linear term. Newton linearization of $(\mathbf{u} \cdot \nabla)\,\mathbf{u}$ in the neighborhood of a guess of the solution $\mathbf{u}_{(k)}$ is performed as follows. We need to compute the derivative (more precisely the *Fréchet derivative*), of the differential operator. It is simpler to compute the *Gateaux derivative* (also called directional derivative) at a point $\mathbf{u}$ and then verify that it is linear and bounded on for any $\mathbf{u} \in V$ (in which case the linear operator is the Fréchet derivative). The *Gateaux derivative* of our non-linear operator, applied to a vector $\mathbf{v}$ is defined as

$$
[(\mathbf{u} \cdot \nabla)\,\mathbf{u}]'\,|_{\mathbf{u}=\mathbf{w}}(\mathbf{v}) = \lim_{\epsilon \to 0} \frac{((\mathbf{w} + \epsilon\mathbf{v}) \cdot \nabla)\,(\mathbf{w} + \epsilon\mathbf{v}) - (\mathbf{w} \cdot \nabla)\,\mathbf{w}}{\epsilon}.
$$

The limit is easily computed and the resulting operator is

$$
[(\mathbf{u} \cdot \nabla)\,\mathbf{u}]'\,|_{\mathbf{u}=\mathbf{w}}(\bullet) = (\bullet \cdot \nabla)\,\mathbf{w} + (\mathbf{w} \cdot \nabla)\,\bullet,
$$

where $\bullet$ is the *argument* where the operator is applied. We have already deduced in the previous discussion about the trilinear form that this operator is linear and bounded from V to V'. The derivative of the other operators is trivial since they are all linear, and for a linear differential operator L we have that $L'(\bullet) = L(\bullet)$.

The Newton iteration then reads: given $\mathbf{u}_{(0)}$, for $k = 0, 1, \ldots$ find the new guess by solving

$$\begin{cases} -\nu \triangle \mathbf{u}_{(k+1)} + \left(\mathbf{u}_{(k+1)} \cdot \nabla\right) \mathbf{u}_{(k)} + \left(\mathbf{u}_{(k)} \cdot \nabla\right) \mathbf{u}_{(k+1)} + \nabla p_{(k+1)} = \\ \qquad\qquad \mathbf{f} + \left(\mathbf{u}_{(k)} \cdot \nabla\right) \mathbf{u}_{(k)}, \\ \nabla \cdot \mathbf{u}_{(k+1)} = 0, \end{cases}$$

up to the convergence.

A possible convergence criterion is to require that the difference between $\mathbf{u}_{(k)}$ and $\mathbf{u}_{(k+1)}$ is small, i.e.

$$\|\mathbf{u}_{(k+1)} - \mathbf{u}_{(k)}\| \leq \epsilon.$$

where $\|\cdot\|$ is a suitable norm (typically H^1 or L^2 norm) and ϵ is a fixed tolerance. A more restrictive constraint is to add a control also to two consecutive pressure solutions. Galerkin finite element discretization of this linearized problem is pretty standard.

Numerical results

Program 34 is a **Freefem++** code implementing Newton method for Navier-Stokes. Numerical results correspond to the exact solution up to the rounding errors (plots of the solution correspond to the ones of the previous exercises). Notice that, for $\nu = 0.1$, Newton method converges in two iterations at most for all mesh sizes tested.

Program 34 - NewtonNavierStokes : Newton method for the steady Navier-Stokes problem

```
problem NewtonNavierStokes([u1,u2,p],[v1,v2,q],
                    solver=GMRES,eps=1.e-15) =
    int2d(Th)(nu * ( dx(u1)*dx(v1) + dy(u1)*dy(v1)
            +        dx(u2)*dx(v2) + dy(u2)*dy(v2) ))
  + int2d(Th)(u1*dx(u1last)*v1 + u1*dx(u2last)*v2+
            u2*dy(u1last)*v1 + u2*dy(u2last)*v2
          + u1last*dx(u1)*v1 + u1last*dx(u2)*v2+
          u2last*dy(u1)*v1 + u2last*dy(u2)*v2)
  - int2d(Th)(u1last*dx(u1last)*v1+ u1last*dx(u2last)*v2+
            u2last*dy(u1last)*v1 + u2last*dy(u2last)*v2)
  - int2d(Th)(p*dx(v1) + p*dy(v2))
  - int2d(Th)(dx(u1)*q + dy(u2)*q)
  + on(1,3,u1=0,u2=0)
  + on(2,4,u1=g,u2=0);

while (i<=nmax & resL2>tol)
{
  NewtonNavierStokes;
    w1[]=u1[]-u1last[];
```

```
    w2[]=u2[]-u2last[];
    resL2 = int2d(Th)(w1*w1) + int2d(Th)(w2*w2);
    resL2 = sqrt(resL2);
    u1last[]=u1[];
    u2last[]=u2[];
    i++;
}
```

$\diamondsuit$

Remark 7.1 Skew-symmetry of the trilinear form is fundamental in proving stability of the solution. This property follows from two features:

1. $\mathbf{u}$ has null trace at the boundary;
2. $\mathbf{u}$ is divergence free.

A key to prove the stability estimate at continuous level was the use of (7.7). Unfortunately this equation does not extend, in general, to the discrete setting. Indeed, in general

$$\int_\Omega (\nabla \cdot \mathbf{u}_h)(\mathbf{v}_h \cdot \mathbf{w}_h)\, d\Omega \neq 0$$

even if $\int_\Omega \nabla \cdot \mathbf{u}_h q_h\, d\Omega = 0$ for all $q_h \in Q_h$, since $\mathbf{v}_h \cdot \mathbf{w}_h \notin Q_h$ in general. The stability estimate can therefore be not valid at discrete level. To recover skew-symmetry a *consistent modification* of the trilinear form has been proposed (see [Tem84]),

$$\widehat{c}(\mathbf{u}, \mathbf{v}, \mathbf{w}) = c(\mathbf{u}, \mathbf{v}, \mathbf{w}) + \frac{1}{2} \int_\Omega (\nabla \cdot \mathbf{u})\, \mathbf{w} \cdot \mathbf{v}\, d\Omega,$$

which at continuous level is equivalent to the original one, since the additional term is zero as $\nabla \cdot \mathbf{u} = 0$. However, this form is skew-symmetric also at discrete level. Indeed, since $\mathbf{v}_h$ and $\mathbf{w}_h$ have null trace on the boundary we have

$$\frac{1}{2}c(\mathbf{u}_h, \mathbf{v}_h, \mathbf{w}_h) = -\frac{1}{2} \int_\Omega (\nabla \cdot \mathbf{u}_h)\, \mathbf{w}_h \cdot \mathbf{v}_h\, d\Omega - \frac{1}{2}c(\mathbf{u}_h, \mathbf{w}_h, \mathbf{v}_h).$$

It follows that

$$\widehat{c}(\mathbf{u}_h, \mathbf{v}_h, \mathbf{w}_h) = \frac{1}{2}\left(c(\mathbf{u}_h, \mathbf{v}_h, \mathbf{w}_h) - c(\mathbf{u}_h, \mathbf{w}_h, \mathbf{v}_h)\right),$$

by which $c(\mathbf{u}_h, \mathbf{v}_h, \mathbf{v}_h) = 0$, as desired. We can then reproduce at discrete level the stability estimate valid at continuous level. For this reason, this trilinear form is often used in numerical codes, being consistent with the original one and giving rise to more stable numerical schemes.

Remark 7.2 Assumption on smallness of boundary data for the stability estimate can be relaxed. Actually, we have the following result. For any $\gamma > 0$, there exists a lifting $\mathbf{G} \in V_{\text{div}}$ such that for any $\mathbf{v} \in V$

$$\int_\Omega (\mathbf{v} \cdot \nabla)\, \mathbf{G} \cdot \mathbf{v}\, d\Omega \leq \gamma \|\nabla \mathbf{v}\|_{L^2}^2$$

and $\|\mathbf{G}\|_{H^1} \leq C_1 \|\mathbf{g}\|_{H^{1/2}}$, for a constant $C_1 > 0$. The (non trivial) proof can be found in [Tem95]. Without assumptions on smallness of the data, we can choose $\gamma = \dfrac{\alpha}{2}$, and we get the following inequality

$$\frac{\alpha}{2}\|\tilde{\mathbf{u}}\|_V^2 \leq (\mathbf{f}, \tilde{\mathbf{u}}) + |a(\mathbf{G}, \tilde{\mathbf{u}})| + |c(\mathbf{G}, \mathbf{G}, \tilde{\mathbf{u}})|.$$

From the Cauchy-Schwarz and Young inequalities we get

$$\|\tilde{\mathbf{u}}\|_V^2 \le \frac{4}{\alpha^2}\left(\|\mathbf{f}\|_{V'}^2 + \left(C_2 + C_3\|\mathbf{g}\|_{H^{1/2}}^2\right)\right)\|\mathbf{g}\|_{H^{1/2}}^2. \tag{7.14}$$

Note that α is in fact proportional to ν (it comes from the application of Poincaré inequality to the form $a(\mathbf{u}, \mathbf{v})$). As a consequence if $\nu \ll 1$ this estimate, as well as (7.11), becomes meaningless (even if still valid).

Exercise 7.1.3. Let us consider the steady Stokes problem

$$\begin{cases} -\nu\triangle\mathbf{u} + \nabla p = \mathbf{f} & \text{in } \Omega, \\ \nabla\cdot\mathbf{u} = 0 & \text{in } \Omega, \\ \mathbf{u} = \mathbf{g} & \text{on } \Gamma, \end{cases} \tag{7.15}$$

where $\Omega \subset \mathbb{R}^d$ $(d = 2, 3)$. Discuss the significance of the *inf-sup* condition on the algebraic setting of problem (7.15) discretized with finite elements.

Show that the discretization of the following *quasi-compressibility problem*

$$\begin{cases} -\nu\triangle\mathbf{u}_\epsilon + \nabla p_\epsilon = \mathbf{f} & \text{in } \Omega, \\ \nabla\cdot\mathbf{u}_\epsilon = -\epsilon p_\epsilon & \text{in } \Omega, \\ \mathbf{u}_\epsilon = \mathbf{g} & \text{on } \Gamma, \end{cases} \tag{7.16}$$

with $\epsilon > 0$ yields a stable problem for any finite elements pair. Discuss the approximation of (7.15) with (7.16) with ϵ small enough.

Check your answer with `Freefem++` for $\nu = 0.1$ and Ω given by the unit square, with homogeneous Dirichlet conditions on the bottom side and the two lateral sides and velocity $\mathbf{u} = (1,0)^T$ on the upper side (*lid-driven cavity problem*).

Solution 7.1.3.

Numerical approximation

From previous exercises, weak formulation of Stokes problem reads: find $\tilde{\mathbf{u}} \in V$, $p \in Q$ such that for any $\mathbf{v} \in V \equiv \mathbf{H}_0^1(\Omega)$ and $q \in Q \equiv L_0^2(\Omega)$

$$\begin{cases} a(\tilde{\mathbf{u}}, \mathbf{v}) + b(\mathbf{v}, p) = (\mathbf{f}, \mathbf{v}) - a(\mathbf{G}, \mathbf{v}), \\ b(\tilde{\mathbf{u}}, q) = -b(\mathbf{G}, q), \end{cases} \tag{7.17}$$

where $\mathbf{G}$ is a lifting of the boundary data (not necessarily divergence free). We introduce the finite-dimensional spaces $V_h \subset V$ and $Q_h \subset Q$, with basis

$\{\varphi_i\}$ $(i = 1, 2, \ldots, d_{\mathbf{u}})$ and $\{\psi_j\}$ $(j = 1, 2, \ldots, d_p)$ respectively. As seen in the previous exercises we finally get the algebraic system

$$\begin{bmatrix} K & D^T \\ D & O \end{bmatrix} \begin{bmatrix} \mathbf{U} \\ \mathbf{P} \end{bmatrix} = \begin{bmatrix} \mathbf{F}_1 \\ \mathbf{F}_2 \end{bmatrix}.$$

If the *inf-sup* condition is fulfilled we expect that the system is non singular. Let's look at the issue in more detail. System (7.18) may be equivalently written as

$$\begin{aligned} K\mathbf{U} + D^T\mathbf{P} &= \mathbf{F}_1 \\ D\mathbf{U} &= \mathbf{F}_2. \end{aligned} \qquad (7.18)$$

Since K is symmetric positive definite, from the symmetry and coercivity of the associated bilinear form, we may formally compute the velocity in function of the pressure,

$$\mathbf{U} = K^{-1}\mathbf{F}_1 - K^{-1}D^T\mathbf{P}. \qquad (7.19)$$

The latter plugged in the second set of equation of (7.18) yields

$$DK^{-1}D^T\mathbf{P} = DK^{-1}\mathbf{F}_1 - \mathbf{F}_2, \qquad (7.20)$$

that is a $d_p \times d_p$ system for the pressure governed by the *pressure matrix*[3] $DK^{-1}D^T$.

If the pressure matrix is invertible we can compute a unique pressure and consequently velocity. Thus, the entire system is invertible if and only if the pressure matrix is invertible. From a result illustrated in Section 1.6 we know that this condition is equivalent to ask that the null-space of D^T is trivial, that is

$$D^T\mathbf{Q} = \mathbf{0} \quad \text{if and only if } \mathbf{Q} = \mathbf{0}. \qquad (7.21)$$

We now show that this condition is the algebraic counterpart of the *inf-sup* condition. If condition (7.21) is not satisfied there exist a $\mathbf{Q}$ with $\mathbf{Q} \neq \mathbf{0}$ such that $D^T\mathbf{Q} = \mathbf{0}$. Let consider the associated discrete pressure $p_h^* = \sum_{k=1}^{d_p} Q_k\psi_k$, by the definition of D we have

$$D^T\mathbf{Q} = \mathbf{0} \quad \Rightarrow b(\varphi_j, p_h^*) = 0, \text{ for } j = 1, \ldots, d_{\mathbf{u}}.$$

But this imply that $b(\mathbf{v}_h, p_h^*) = 0$ for all $\mathbf{v}_h \in V_h$, in contradiction with the *inf-sup* condition. So the latter is sufficient to have an invertible algebraic system. Pressure p_h^* (and its algebraic counterpart $\mathbf{Q}$) is called *spurious pressure* and it is immediate to verify that in this situation if p_h is a solution of our problem, so is $p_h + p_h^*$.

---

[3] In general, for a system like (7.18), matrix $-DK^{-1}D^T$ is said *pressure Schur complement* of the system.

Let's look at the converse statement. If the null-space is trivial $\mathbf{Z} = D^T\mathbf{Q} \neq \mathbf{0}$ for all $\mathbf{Q} \neq \mathbf{0}$. It is sufficient to take $\mathbf{V} = D^T\mathbf{Q}$ to have

$$\mathbf{V}^T D^T \mathbf{Q} = \|\mathbf{V}\|^2 \geq 0.$$

Moreover, there exist a $\gamma > 0$ such that

$$D^T\mathbf{Q} \geq \gamma\|\mathbf{Q}\| \quad \forall\mathbf{Q}, \tag{7.22}$$

indeed it is sufficient to take $\gamma = \min\|D^T\mathbf{X}\|$ over all vectors $\mathbf{X}$ of unit norm. Let v_h and q_h be the finite elements function associated with $\mathbf{V}$ and $\mathbf{Q}$, respectively, i.e. $v_h = \sum_i V_i\varphi_i$ and $q_h = \sum_i Q_i\psi_i$. With S we indicate the pressure mass matrix

$$S_{ij} = \int_\Omega \psi_i\psi_j \, d\Omega, \tag{7.23}$$

which is s.p.d. Since also K is s.p.d., we can write

$$\|q_h\|^2_{L^2(\Omega)} = \mathbf{Q}^T S \mathbf{Q} \leq \lambda^S_{max}\|\mathbf{Q}\|^2,$$
$$\alpha\|v_h\|_V \leq a(v_h, v_h) = \mathbf{V}^T K \mathbf{V} \leq \lambda^K_{max}\|\mathbf{V}\|^2 \tag{7.24}$$

where λ^S_{max} and λ^K_{max} are the maximal eigenvalues of the respective matrices. Exploiting the coercivity of the bilinear form a, equations (7.24) and (7.22) and the special choice of $\mathbf{V}$ we obtain

$$b(v_h, q_h) = \mathbf{V}D^T\mathbf{Q} = \|\mathbf{V}\|\|D^T\mathbf{Q}\| \geq \gamma\|\mathbf{V}\|\|\mathbf{Q}\|$$
$$\geq \frac{\gamma\sqrt{\alpha}}{\sqrt{\lambda^S_{max}\lambda^K_{max}}}\|v_h\|_V \|q_h\|_{L^2(\Omega)}.$$

Since the procedure may be repeated for any $\mathbf{Q}$ (and thus any q_h) we conclude that we satisfy an *inf-sup* condition with $\beta_h = \frac{\gamma\sqrt{\alpha}}{\sqrt{\lambda^S_{max}\lambda^K_{max}}}$. This is a somehow weaker condition than the standard one because it may happen that $\lim_{h\to 0} \beta_h = 0$. To satisfy the *inf-sup* condition in the standard form the finite element spaces should be such that β_h can be bounded from below independently of h. If this is not the case, the algebraic system is still invertible for all $h > 0$, but we may expect a degeneration of the conditioning when h is small. In conclusion, the condition on the null-space of D^T is "almost" equivalent to the *inf-sup* condition in the sense that if the *inf-sup* is satisfied by the discrete spaces then $\dim \text{Ker}(D^T) = 0$. However just the satisfaction of the latter condition implies only a weaker form of the *inf-sup*, with a constant that may go to zero as $h \to 0$, and this, in general, is unwanted.

Let us consider now the *quasi-compressibility problem*. The name is justified, since the incompressibility condition is violated, so ϵ should be selected suitably small. The weak form may be derived in the standard way and it reads:

find $\mathbf{u}_\epsilon \in V$, $p_\epsilon \in L^2(\Omega)$ such that $\mathbf{v} \in V$ and $q \in L^2(\Omega)$:

$$\begin{cases} a(\mathbf{u}_\epsilon, \mathbf{v}) + b(\mathbf{v}, p_\epsilon) = (\mathbf{f}, \mathbf{v}) - a(\mathbf{G}, \mathbf{v}), \\ b(\mathbf{u}_\epsilon, q) - \epsilon(p_\epsilon, q) = b(\mathbf{G}, q), \end{cases} \tag{7.25}$$

where $(p_\epsilon, q) = \int_\Omega p_\epsilon q \, d\Omega$ is the $L^2(\Omega)$ scalar product in $L^2(\Omega)$. Here $\mathbf{u}_\epsilon$ is the lifted solution, the final velocity is obtained by summing $\mathbf{G}$.

Notice that in this case the pressure occurs in the formulation of the problem not only under differentiation, so we do not need to force null average for having uniqueness. The term $\mathbf{G}$ is, a usual, the lifting of the Dirichlet datum $\mathbf{g}$. Negative sign in front of the term (p_ϵ, q) is to the fact that $b(\mathbf{u}_\epsilon, \mathbf{v})$ corresponds to the weak form of $-\nabla \cdot \mathbf{u}$.

Stability estimate is found for the homogeneous problem, so for this purpose we set $\mathbf{G} = \mathbf{0}$. If we select in (7.25) as test functions $\mathbf{v} = \mathbf{u}_\epsilon$ and $q = p_\epsilon$, by subtracting the two equations we get

$$a(\mathbf{u}_\epsilon, \mathbf{u}_\epsilon) + \epsilon \|p_\epsilon\|_{L^2(\Omega)}^2 = (\mathbf{f}, \mathbf{u}_\epsilon),$$

so thanks to the coercivity of a,

$$\alpha\|\mathbf{u}_\epsilon\|_V^2 + \epsilon\|p_\epsilon\|_{L^2(\Omega)}^2 \leq (\mathbf{f}, \mathbf{u}_\epsilon) \quad \Rightarrow \quad \|\mathbf{u}_\epsilon\|_V \leq \frac{1}{\alpha}\|\mathbf{f}\|_{L^2(\Omega)}. \tag{7.26}$$

By which, assuming $\mathbf{f} \in \mathbf{L}^2(\Omega)$ and using the Cauchy-Schwarz inequality

$$\|\mathbf{u}_\epsilon\|_V \leq \frac{1}{\alpha}\|\mathbf{f}\|_{L^2(\Omega)}.$$

Using again (7.26), Cauchy-Schwarz inequality and the previous relation, we have an *a priori estimate* on the pressure as well,

$$\|p_\epsilon\|_{L^2(\Omega)} \leq \frac{1}{\sqrt{\epsilon\alpha}}\|\mathbf{f}\|_{L^2(\Omega)}.$$

The discretization of (7.25), carried out with the usual means, leads to the algebraic system

$$\begin{bmatrix} \mathrm{K} & \mathrm{D}^T \\ \mathrm{D} & -\epsilon\mathrm{S} \end{bmatrix} \begin{bmatrix} \mathbf{U}_\epsilon \\ \mathbf{P}_\epsilon \end{bmatrix} = \begin{bmatrix} \mathbf{F}_1 \\ \mathbf{F}_2 \end{bmatrix},$$

where S is the pressure mass matrix defined in (7.23). Working as for the incompressible Stokes problem, from (7.19) we have that $\mathbf{P}_\epsilon$ satisfies system

$$\left(\epsilon\mathrm{S} + \mathrm{D}\mathrm{K}^{-1}\mathrm{D}^T\right)\mathbf{P}_\epsilon = \mathrm{D}\mathrm{K}^{-1}\mathbf{U}_\epsilon. \tag{7.27}$$

In this case pressure matrix $\mathrm{H} = \epsilon\mathrm{S} + \mathrm{D}\mathrm{K}^{-1}\mathrm{D}^T$ is non singular independently of the satisfaction of the *inf-sup* condition by spaces V_h and Q_h. This because it is a sum of the positive definite matrix $\epsilon\mathrm{S}$ and the non-negative matrix $\mathrm{D}\mathrm{K}^{-1}\mathrm{D}^T$ (the fact that $\mathrm{D}\mathrm{K}^{-1}\mathrm{D}^T$ is not negative derives from the positive definiteness of K, and thus of K^{-1}). Indeed H is symmetric and positive definite. If $\mathbf{Y} \neq \mathbf{0}$

$$\mathbf{Y}^T \left(\epsilon S + DK^{-1}D^T \right) \mathbf{Y} = \epsilon \mathbf{Y}^T S \mathbf{T} + \mathbf{Y}^T DK^{-1}D^T \mathbf{Y} > 0,$$

since the first term in the right-hand side is positive and the second is not negative. However if the kernel of D^T is nontrivial, pressure matrix tends to become singular when $\epsilon \to 0$. Coefficient ϵ should be selected as a trade-off between the need of a non singular matrix and of a small perturbation on the original problem.

Remark 7.3 Quasi-compressibility problem (7.16) can be generalized by replacing ϵp_ϵ by $\epsilon \mathcal{L} p_\epsilon$, where $\mathcal{L}$ is a *self-adjoint positive operator*.

Among the other authors who have analyzed quasi-compressibility methods, we mention A. Prohl [Pro97], who treated (together with R. Rannacher) the quasi-compressibility method as a step for the analysis of the Chorin-Temam projection scheme (see Exercise 7.2.4).

Beside the choice $\mathcal{L} p_\epsilon = p_\epsilon$, another possibility is (see [BP84])

$$\mathcal{L} p_\epsilon = -\triangle p_\epsilon$$

with homogeneous Neumann boundary conditions ($\partial p/\partial \mathbf{n} = 0$ on Γ). With this choice at the continuous level we assume $p \in H^1(\Omega)$ with $\int_\Omega p \, d\Omega = 0$. With this condition, the Laplacian operator $-\triangle$ is self-adjoint and positive thanks to the *Poincaré-Wittinger inequality* (see Ch. 1). In this case mass matrix S is replaced by the matrix stiffness matrix with entries $\int_\Omega \nabla \psi_j \cdot \nabla \psi_i \, d\Omega$. Neumann homogeneous boundary condition on pressure is required by the modified problem because of the presence of the Laplace operator. Even if Neumann conditions are in general "less perturbing" on the solution than Dirichlet ones (and easier to treat numerically), the introduction of such an unphysical condition induces an error, particularly near the boundary.

Another choice, valid for unsteady problems is $\mathcal{L} p_\epsilon = dp_\epsilon/dt$. At the discrete level, this still yields a pressure mass matrix, even if the right hand side of the continuity equation will now include the pressure at the previous time steps.

Program 35 - Stokes-pen1 : Stokes quasi-compressibility problem with pressure mass matrix

```
solve Stokes ([u1,u2,p],[v1,v2,q],solver=Crout) =
    int2d(Th)(nu * ( dx(u1)*dx(v1) + dy(u1)*dy(v1)
    +            dx(u2)*dx(v2) + dy(u2)*dy(v2) )
        - p*dx(v1) - p*dy(v2)
        - p*q*epsilon
//        - (dx(p)*dx(q)+dy(p)*dy(q))*epsilon
        - dx(u1)*q - dy(u2)*q)
    + on(1,2,4,u1=0,u2=0)  + on(3,u1=1.,u2=0);
um = sqrt(u1^2+u2^2);
```

Numerical results

In Fig. 7.2 we report the solution of the velocity field (on the left) and pressure isolines (on the right) on a grid with $h = 1/30$, P^2 elements for the velocity and P^1 for the pressure. In Fig. 7.3 we report the P^1-P^1 solution (piecewise linear elements for both velocity and pressure) with a quasi-compressibility method using the pressure mass matrix and[4] $\epsilon = 10^{-5}$. We can note that the computed velocity is acceptable, while pressure solution still oscillates, since the matrix is ill-conditioned because of a too small ϵ. For $\epsilon = 1$ (Fig. 7.4) oscillations are reduced, even if in this way the perturbation on the solution induced by the quasi-compressibility is more evident, as can be realized observing the position of the vortex and the range of variability of the pressure compared with the solution in (7.2). The code generating these solutions is in Program 35.

In Fig. 7.5 we report the results obtained with a quasi-compressibility method using the Laplace operator, with $\epsilon = 10^{-4}$. Pressure values are reasonable. A snapshot of the code corresponding to this case is in Program 36.

Identification of a good trade-off value for ϵ is nontrivial. The advantage is in the resolution of smaller linear systems because one can use P^1-P^1 elements, with a reduction of CPU time. Even if for an interpreted code like `Freefem++` indication of CPU times is questionable, P^2-P^1 computation in Fig. 7.2 required more than twice the time of solution of Fig. 7.5.   ◇

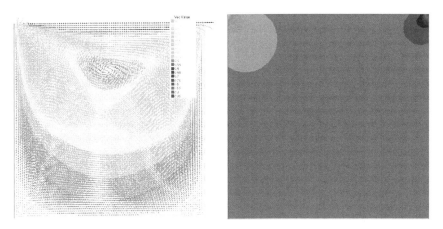

Fig. 7.2 P^2-P^1 computation of the lid-driven cavity solution (velocity on the left, pressure on the right). *Inf-sup* condition is fulfilled. Range of the pressure is [-15,14] (from the top left corner, to the top right one). CPU time: 2.48 s

---

[4] We solve this system with a direct method. For $\epsilon = 0$ `Freefem++` clearly would give an error message, since the matrix is singular.

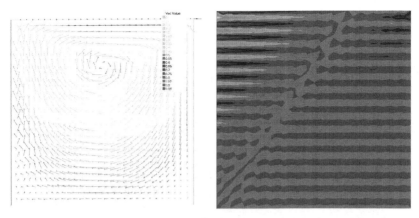

Fig. 7.3 Lid-driven cavity solution with P^1-P^1 elements: velocity on the left, pressure on the right. Quasi-compressibility introduced with a pressure mass matrix with $\epsilon = 10^{-5}$. Pressure ranges from -44 to +35 with oscillations induced by the ill-conditioning of the matrix

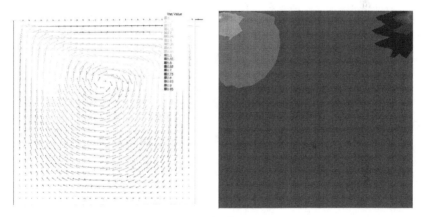

Fig. 7.4 Lid-driven cavity solution with P^1-P^1 elements: velocity on the left, pressure on the right. Quasi-compressibility introduced with a pressure mass matrix with $\epsilon = 1$. Pressure ranges from -6 to 4, oscillation are strongly reduced

Program 36 - Stokes-pen2 : Stokes quasi-compressibility problem with pressure stiffness matrix

```
solve Stokes ([u1,u2,p],[v1,v2,q],solver=Crout) =
  int2d(Th)(nu*( dx(u1)*dx(v1)+dy(u1)*dy(v1)+dx(u2)*dx(v2)+dy(u2)*dy(v2) )
    - p*dx(v1) - p*dy(v2) - dx(u1)*q - dy(u2)*q
    - (dx(p)*dx(q)+dy(p)*dy(q))*epsilon)
    + on(1,2,4,u1=0,u2=0) + on(3,u1=1.,u2=0);
```

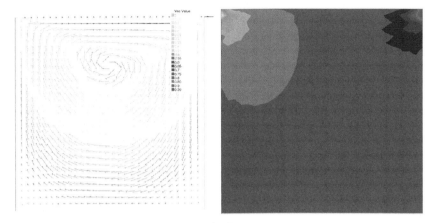

Fig. 7.5 Lid-driven cavity solution with P^1-P^1 elements: velocity on the left, pressure on the right. Quasi-compressibility introduced with a pressure stiffness matrix with $\epsilon = 10^{-4}$. Pressure ranges from -8 and 5. CPU Time: 1.07 s

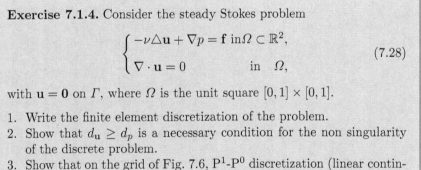

Exercise 7.1.4. Consider the steady Stokes problem

$$\begin{cases} -\nu \triangle \mathbf{u} + \nabla p = \mathbf{f} \ \text{in} \varOmega \subset \mathbb{R}^2, \\ \nabla \cdot \mathbf{u} = 0 \qquad \quad \text{in} \quad \varOmega, \end{cases} \tag{7.28}$$

with $\mathbf{u} = \mathbf{0}$ on $\varGamma$, where $\varOmega$ is the unit square $[0,1] \times [0,1]$.

1. Write the finite element discretization of the problem.
2. Show that $d_{\mathbf{u}} \geq d_p$ is a necessary condition for the non singularity of the discrete problem.
3. Show that on the grid of Fig. 7.6, P^1-P^0 discretization (linear continuous elements for the velocity and piecewise constant discontinuous for the pressure) does not fulfill the *inf-sup* condition. Use a simple counting of the degrees of freedom.
4. Show that Q^1-Q^0 discretization (bilinear elements for the velocity and piecewise constant for the pressure) on the quadrangular grid of Fig. 7.7 does not fulfill the *inf-sup* condition by finding a spurious mode for the pressure.

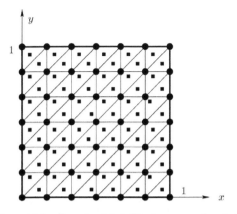

Fig. 7.6 Triangular grid for Exercise 7.1.4. Black squares denote the degrees of freedom for the pressure, circles are the degrees of freedom of velocity

Solution 7.1.4.

Numerical approximation

Discretization of the problem is obtained following the same guidelines of previous exercises. We do not need to perform any lifting, since the Dirichlet data are homogeneous. With the same notation of previous exercises, the algebraic formulation reads

$$
\begin{bmatrix} K & D^T \\ D & O \end{bmatrix} \begin{bmatrix} U \\ P \end{bmatrix} = \begin{bmatrix} F_1 \\ 0 \end{bmatrix}.
$$

Following Exercise 7.1.3, we may conclude that the algebraic system is invertible if and only if the null space of D^T is trivial.

The number of columns of a matrix is equal to the sum of its rank and the dimension of its null space. Moreover, the rank of a matrix coincides with that of its transpose. Therefore, if $\dim(\mathrm{Ker}(D^T)) = 0$, we have $d_{\mathbf{u}} = \mathrm{rank}(D) + \dim(\mathrm{Ker}(D))$ and $d_p = \mathrm{rank}(D^T) + \dim(\mathrm{Ker}(D^T)) = \mathrm{rank}(D^T) = \mathrm{rank}(D)$ yielding $d_{\mathbf{u}} - d_p = \dim(\mathrm{Ker}(D)) \geq 0$. Condition

$$
d_{\mathbf{u}} \geq d_p \tag{7.29}
$$

is therefore necessary (even if not sufficient!) to have $\dim(\mathrm{Ker}(D^T)) = 0$.

Consider now the grid of Fig. 7.6 and the adoption of P^1 elements for the velocity and P^0 for the pressure (discontinuous). Let us compute the number of degrees of freedom of velocity and pressure:

1. the number of degrees of freedom for the velocity is twice the number of internal vertexes;

2. the number of degrees of freedom for the pressure is the number of elements minus one (the one lost for forcing the null average).

If n is the number of mesh nodes on each side of the square domain (including the corner points), we have:

1. total number of nodes (internal and on the boundary): n^2;
2. number of boundary nodes: $4n - 4$;
3. number of elements: $2(n-1)^2$. Since the mesh has been made by splitting quadrilateral (of number $(n-1)^2$) into 2 triangles.

Since the velocity is prescribed at the boundary the corresponding nodes do not enter the count of degrees of freedom. So we have $d_{\mathbf{u}} = 2\left(n^2 - 4n + 4\right)$ degrees of freedom for the velocity and $d_p = 2(n-1)^2 - 1$ degrees of freedom for pressure. Inequality $d_{\mathbf{u}} \geq d_p$ with these data has solution $n \leq 7/4$, so that for the grid at hand (and any reasonable grid of that type!) condition (7.29) is not satisfied.

For the last point of the exercise, we refer to Fig. 7.7, where the element have been drawn in a checkerboard fashion. We may check if the necessary condition (7.29) is verified when using Q^1 elements for the velocity and Q^0 elements for pressure. If n denotes the number of nodes on each side of the boundary of the square domain, a direct computation yields:

1. $(n-2)^2$ internal nodes, i.e. $d_{\mathbf{u}} = 2(n-2)^2$ velocity degrees of freedom;
2. $(n-1)^2$ elements, i.e. $d_p = (n-1)^2 - 1$ pressure degrees of freedom.

By a direct solution of inequality $d_{\mathbf{u}} \geq d_p$, we find that this is fulfilled for $n \geq 4$. For n large enough, (7.29) is therefore verified. However, this is only a necessary condition. We are indeed able to find a pressure spurious mode, i.e. a non-null vector $\mathbf{Y}$ such that $D^T\mathbf{Y} = \mathbf{0}$. Let $\mathcal{Q}_{ij}$ be the quadrilateral with vertexes (counterclockwise sense) $V_0 = (x_i, y_j)$, $V_1 = (x_{i+1}, y_j)$, $V_2 = (x_{i+1}, y_{j+1})$, $V_3 = (x_i, y_{j+1})$, as in Fig. 7.7. Let q_h be a generic function of Q^0. Its constant value in $\mathcal{Q}_{ij}$ is indicated as $q_{i+\frac{1}{2}, j+\frac{1}{2}}$.

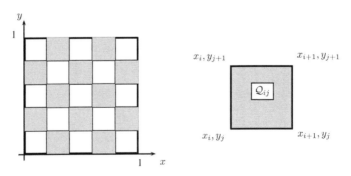

Fig. 7.7 Quadrilaterals grid for the Exercise 7.1.4. On the right a detail for the notation used for the generic quadrilateral $\mathcal{Q}_{ij}$

Basis functions φ_k for each velocity components in Q^1 on $\mathcal{Q}_{ij}$ (for $x_i \leq x \leq x_{i+1}$ and $y_i \leq y \leq y_{i+1}$) read

$$\varphi_0 = \frac{1}{h^2}\,(x_{i+1} - x)\,(y_{j+1} - y)\,,\ \varphi_1 = \frac{1}{h^2}\,(x - x_i)\,(y_{j+1} - y)\,,$$

$$\varphi_2 = \frac{1}{h^2}\,(x - x_i)\,(y - y_j)\,,\qquad \varphi_3 = \frac{1}{h^2}\,(x_{i+1} - x)\,(y - y_j)\,,$$

and consequently

$$\frac{\partial \varphi_0}{\partial x} = -\frac{1}{h}\,(y_{j+1} - y)\,,\ \frac{\partial \varphi_0}{\partial y} = -\frac{1}{h}\,(x_{i+1} - x)\,,$$

$$\frac{\partial \varphi_1}{\partial x} = \frac{1}{h}\,(y_{j+1} - y)\,,\ \frac{\partial \varphi_1}{\partial y} = -\frac{1}{h}\,(x - x_i)\,,$$

$$\frac{\partial \varphi_2}{\partial x} = \frac{1}{h}\,(y - y_j)\,,\ \frac{\partial \varphi_2}{\partial y} = \frac{1}{h}\,(x - x_i)\,,$$

$$\frac{\partial \varphi_3}{\partial x} = -\frac{1}{h}\,(y - y_j)\,,\ \frac{\partial \varphi_3}{\partial y} = \frac{1}{h}\,(x_{i+1} - x)\,.$$

By a direct computation we have

$$\int_{\mathcal{Q}_{ij}} q_h \frac{\partial u_h^1}{\partial x}\,dxdy = q_{i+\frac{1}{2},j+\frac{1}{2}} \sum_{k=0}^{3} h u_h^1(V_k) \int_{y_j}^{y_{j+1}} \frac{\partial \varphi_k}{\partial x}\,dy \tag{7.30}$$

$$= q_{i+\frac{1}{2},j+\frac{1}{2}} \frac{h}{2}(u_{i+1,j+1}^1 - u_{i,j+1}^1 + u_{i+1,j}^1 - u_{ij}^1),$$

where for any i and j, u_{ij}^1 is the value of the first velocity component at node (x_i, y_j). An analogous computation can be done for the second velocity component.

Summarizing, we get

$$\int_{\Omega} q_h \nabla \cdot \mathbf{u}_h\, d\Omega = \sum_{i,j=0}^{n-1} \int_{\mathcal{Q}_{ij}} q_h \nabla \cdot \mathbf{u}_h dxdy = \frac{h}{2} \sum_{i,j=0}^{n-1} q_{i+\frac{1}{2},j+\frac{1}{2}} \tag{7.31}$$

$$\left[u_{i+1,j+1}^1 - u_{i,j+1}^1 + u_{i+1,j}^1 - u_{ij}^1 + u_{i+1,j+1}^2 - u_{i+1,j}^2 + u_{i,j+1}^2 - u_{ij}^2\right].$$

After some manipulations (*summation by parts*), the same integral can be rewritten as

$$\int_{\Omega} q_h \nabla \cdot \mathbf{u}_h\, d\Omega =$$

$$\frac{h}{2} \sum_{i,j=1}^{n-1} u_{i,j}^1 \left[q_{i-\frac{1}{2},j-\frac{1}{2}} + q_{i-\frac{1}{2},j+\frac{1}{2}} - q_{i+\frac{1}{2},j-\frac{1}{2}} - q_{i+\frac{1}{2},j+\frac{1}{2}}\right] + \tag{7.32}$$

$$u_{i,j}^2 \left[q_{i-\frac{1}{2},j-\frac{1}{2}} - q_{i-\frac{1}{2},j+\frac{1}{2}} + q_{i+\frac{1}{2},j-\frac{1}{2}} - q_{i+\frac{1}{2},j+\frac{1}{2}}\right].$$

Let us consider now a discrete piecewise constant pressure q_h that alternatively assumes values $+1$ and -1 on any two adjacent squares. For instance, referring to Fig. 7.7, let us assume it takes value $+1$ in the gray elements and -1 in the white ones. One verifies by direct inspection of (7.32) that $(\nabla \mathbf{v}_h, q_h) = 0$ for any $\mathbf{v}_h$, therefore the *inf-sup* condition is not satisfied. This specific spurious pressure mode is also called *checkerboard mode*. We conclude that couple Q^1-Q^0 does not fulfill the *inf-sup* condition.   ◇

Exercise 7.1.5. Consider the steady Navier-Stokes problem

$$\begin{cases} (\mathbf{u} \cdot \nabla)\,\mathbf{u} - \nu \triangle \mathbf{u} + \nabla p = \mathbf{f} \text{ in } & \Omega \subset \mathbb{R}^2, \\ \nabla \cdot \mathbf{u} = 0 & \text{in } \Omega, \\ \mathbf{u} = \mathbf{g} & \text{on } \Gamma. \end{cases} \tag{7.33}$$

Show that the stability estimate found in Exercise 7.1.2 becomes meaningless for $\nu \to 0$. Verify this circumstance numerically with **Freefem++**. Consider the Poiseuille and lid-driven cavity problems introduced in Exercises 7.1.3 and 7.1.2 and perform Newton linearization at the differential level. On the resulting linearized differential problem, write a finite element formulation stabilized with the SUPG method.

Show that the formulation found not only stabilizes the solution in convection dominated regimes (i.e. small ν) but also introduces a quasi-compressibility term. For this reason it is stable irrespectively of the *inf-sup* condition.

Write the corresponding **Freefem++** code for both Poiseuille and lid-driven cavity problems and compute the solution with different values of ν. Suggested mesh size is $h = 0.05$.

Solution 7.1.5.

Mathematical analysis

If L is a characteristic length for the problem at hand, U a representative velocity value of the fluid, the non-dimensional *Reynolds number*

$$\mathrm{Re} = \frac{UL}{\nu}$$

weights the convective term against the diffusive one in Navier-Stokes problems, similarly to the Péclet number in advection-diffusion problems.

The stability estimate (7.14) becomes meaningless in at least two circumstances:

1. viscosity gets very small, so that constant α (which is proportional to ν) tends to 0;
2. boundary data $\mathbf{g}$ is very large.

Both cases correspond to high Reynolds number. Fluids with high Reynolds numbers in general feature instabilities that eventually give rise to turbulence. In this book we do not address turbulence models, however we want to introduce the extension of the stabilization methods introduced in Chapter 4 to the Navier-Stokes problems.

Numerical approximation

A possible indirect way for experiencing the effects of a decreasing viscosity is to check the number of iterations required by the Newton method introduced in Exercise 7.1.2 to converge. An increment of the convective term in comparison with the viscous one makes in fact the Newton method less contractive and the convergence region becomes smaller. More iterations will be required therefore by the method or even no convergence is attained (starting from the same initial guess) when the viscosity tends to 0.

In Table 7.1 we report the number of Newton iterations for Poiseuille and lid-driven cavity. Discretization is carried out with P^2 velocity elements and P^1 pressure elements. Stopping criterion is based on the test $||\mathbf{u}^{(k+1)} - \mathbf{u}^{(k)}||_{\mathbf{H}^1} \leq 10^{-5}$.

Table 7.1 Number of Newton iterations for P^2, P^1 elements with a decreasing viscosity. Dash indicates that convergence is not attained after 30 Newton iterations

Viscosity ν	Poiseuille	Lid-Driven Cavity
10^{-1}	2	4
10^{-2}	2	6
10^{-3}	2	-
10^{-4}	2	-
10^{-5}	3	-
10^{-6}	-	-

It is evident that when the viscosity decreases, convergence is slowed down or even prevented. This phenomenon is more evident for the lid-driven cavity problem, whose solution does not belong to the chosen finite element spaces.

In Fig. 7.8 we show the numerical solution for the lid-driven cavity problem with $\nu = 10^{-3}$ after 10 Newton iterations. Numerical instabilities are evident (Newton method actually is not converging).

As in Exercise 7.1.3, given an initial guess $\mathbf{u}^{(0)}$ Newton iterations require at each step to solve system

$$\begin{cases} -\nu \triangle \mathbf{u}^{(k+1)} + \left(\mathbf{u}^{(k+1)} \cdot \nabla\right) \mathbf{u}^{(k)} + \left(\mathbf{u}^{(k)} \cdot \nabla\right) \mathbf{u}^{(k+1)} + \nabla p^{(k+1)} = \\ \qquad \left(\mathbf{u}^{(k)} \cdot \nabla\right) \mathbf{u}^{(k)}, \\ \nabla \cdot \mathbf{u}^{(k)} = 0. \end{cases} \tag{7.34}$$

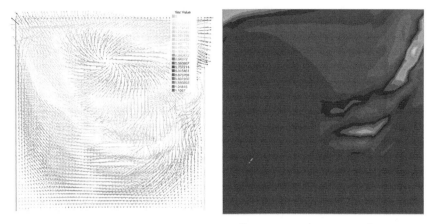

Fig. 7.8 Lid-driven cavity problem (velocity: P^2, pressure: P^1). $\nu = 10^{-3}$. 10 Newton iterations are performed, however numerical instabilities are evident

We introduce a lifting $\mathbf{G}$ of the boundary data $\mathbf{g}$ and set $\widetilde{\mathbf{u}} = \mathbf{u} - \mathbf{G}$. With the notation of Exercise 7.1.2, weak form of the problem reads: find $\widetilde{\mathbf{u}}^{(k+1)} \in V$, $p^{(k+1)} \in Q$ s.t. for any $\mathbf{v} \in V$ and $q \in Q$

$$a(\widetilde{\mathbf{u}}^{((k+1))}, \mathbf{v}) + b(\mathbf{v}, p^{(k+1)}) + c(\widetilde{\mathbf{u}}^{((k+1))}, \mathbf{u}^{(k)}, \mathbf{v}) +$$
$$c(\mathbf{u}^{(k)}, \widetilde{\mathbf{u}}^{((k+1))}, \mathbf{v}) + b(\widetilde{\mathbf{u}}^{((k+1))}, q) = \qquad (7.35)$$
$$(\mathbf{f}, \mathbf{v}) - c(\mathbf{u}^{(k)}, \mathbf{u}^{(k)}, \mathbf{v}) - a(\mathbf{G}, \mathbf{v}) - c(\mathbf{G}, \mathbf{u}^{(k)}, \mathbf{v}) - c(\mathbf{u}^{(k)}, \mathbf{G}, \mathbf{v}) - b(\mathbf{G}, q).$$

We have written the weak formulation in just one equation because this form is more prone to the application of a strongly consistent stabilization. We obtain again the two separate equations for momentum and mass conservation by selecting alternatively $\mathbf{v} = \mathbf{0}$ or $q = 0$. Formulation (7.35) is indeed obtained by summing the two contributions.

Standard Galerkin formulation replaces the functional spaces $V = \mathbf{H}_0^1(\Omega)$ and $Q = L_0^2(\Omega)$ with their finite element counterpart. For the sake of notation, in the sequel we omit the index h. It is understood that we are referring to discrete solutions (in subspaces V_h and Q_h for velocity and pressure respectively). Furthermore, we set $\boldsymbol{\beta} \equiv \mathbf{u}^{(k)}$, $\mathbf{w} \equiv \widetilde{\mathbf{u}}^{(k+1)}$. We remind that in the linearized problem $\mathbf{u}^{(k)}$ is given. So in the following $\boldsymbol{\beta}$ is considered as a given field. We do not assume it to be divergence-free, also to account for the considerations on the trilinear form made in Remark 7.1.

Set

$$\mathcal{F}(\mathbf{v}) \equiv (\mathbf{f}, \mathbf{v}) - c(\boldsymbol{\beta}, \boldsymbol{\beta}, \mathbf{v}) - a(\mathbf{G}, \mathbf{v}) - c(\mathbf{G}, \mathbf{G}, \mathbf{v}).$$

Equation (7.35) becomes

$$a(\mathbf{w}, \mathbf{v}) + c(\boldsymbol{\beta}, \mathbf{w}, \mathbf{v}) + c(\mathbf{w}, \boldsymbol{\beta}, \mathbf{v}) + b(\mathbf{v}, p) - b(\mathbf{w}, q) = \mathcal{F}(\mathbf{v}) - b(\mathbf{G}, q). \quad (7.36)$$

We have actually a linear advection-diffusion problem for $\mathbf{w}$. As it is well known (Chapter 4) we need numerical stabilization when the convective term is dominating. Let Ω_h be a mesh of domain Ω and K a generic element of $\mathcal{T}_h$. SUPG strongly consistent stabilization (see e.g. [Qua09], Chapter 10, [QV94], Chapters 8 and 9) requires to identify the skew-symmetric part of the operator to be stabilized. We refer to Exercise 4.1.1 for the definition.

In the present case, the operator we are considering is

$$\mathcal{L}(\mathbf{v}, q) \equiv \begin{bmatrix} -\nu \triangle \mathbf{v} + (\boldsymbol{\beta} \cdot \nabla) \mathbf{v} + (\mathbf{v} \cdot \nabla)\boldsymbol{\beta} + \nabla q \\ \nabla \cdot \mathbf{v} \end{bmatrix} \tag{7.37}$$

and we can write our linearized differential problem as

$$\mathcal{L}(\mathbf{w}, p) - \mathbf{b} = \mathbf{0}$$

where

$$\mathbf{b} = \begin{bmatrix} \mathbf{f} - (\boldsymbol{\beta} \cdot \nabla) \boldsymbol{\beta} - (\boldsymbol{\beta} \cdot \nabla) \mathbf{G} - (\mathbf{G} \cdot \nabla) \boldsymbol{\beta} - \nu \triangle \mathbf{G} \\ -\nabla \mathbf{G} \end{bmatrix}, \tag{7.38}$$

with homogeneous Dirichlet conditions on $\mathbf{w}$. Indeed, one may verify that (7.36) is the associated weak formulation. To identify the skew symmetric part $\mathcal{L}_{SS}$ of $\mathcal{L}$ we may exploit some results of Exercise 4.1.1 that may be readily extended to the vector case. Operator $-\nu\triangle$ is symmetric so it does not give any contribution to $\mathcal{L}_{SS}$. For the skew symmetric part of $\boldsymbol{\beta} \cdot \nabla \mathbf{w}$ we may apply the result of the cited exercise to each component $\sum_i \beta_i \frac{\partial w_j}{\partial x_i}$ to obtain $\mathcal{C}_{SS} = \boldsymbol{\beta} \cdot \nabla \mathbf{w} - \frac{1}{2}\nabla \cdot \boldsymbol{\beta}$. Indeed, we can verify that

$$_{V'} <\mathcal{C}_{ss}\mathbf{w}, \mathbf{v} >_V = \int_\Omega \left((\boldsymbol{\beta} \cdot \nabla) \mathbf{w} \cdot \mathbf{v} - \frac{1}{2} (\nabla \cdot \boldsymbol{\beta}) \mathbf{w} \cdot \mathbf{v} \right) d\Omega =$$

$$\int_\Omega \left(\frac{1}{2} (\nabla \cdot \boldsymbol{\beta}) \mathbf{v} \cdot \mathbf{w} - (\boldsymbol{\beta} \cdot \nabla) \mathbf{v} \cdot \mathbf{w} \right) d\Omega = -_V <\mathbf{w}, \mathcal{C}_{ss}\mathbf{v} >_{V'}$$

exploiting the fact that functions in V have null trace on the boundary.

Let us consider now operator $\mathcal{B}$ with arguments $[\mathbf{w}, p] \in V \times Q$ defined as

$$\mathcal{B}([\mathbf{w}, p]) \equiv \begin{bmatrix} \nabla p \\ \nabla \cdot \mathbf{w} \end{bmatrix}.$$

By the Green formula and exploiting again the null trace at the boundary, we have

$$\int_\Omega \mathcal{B}([\mathbf{w}, p]) [\mathbf{v}, q] \, d\Omega = (\mathbf{v}, \nabla p) + (\nabla \cdot \mathbf{w}, p) =$$
$$-(\nabla \cdot \mathbf{v}, p) - (\mathbf{w}, \nabla q) = -\int_\Omega \mathcal{B}([\mathbf{v}, q]) [\mathbf{w}, p] \, d\Omega.$$

Operator $\mathcal{B}$ is therefore skew-symmetric.

Finally, consider the operator $\mathcal{R} : \mathbb{R}^d \to \mathbb{R}^d$ defined as $\mathcal{R}\mathbf{w} \equiv (\mathbf{w} \cdot \nabla) \boldsymbol{\beta}$. To build its skew symmetric part we find its adjoint $\mathcal{R}^*$ first. Let's apply the definition

$$_{V'} < \mathcal{R}\mathbf{w}, \mathbf{v} >_V = \int_\Omega (\mathbf{w} \cdot \nabla) \boldsymbol{\beta} \cdot \mathbf{v} \, d\Omega =$$
$$\sum_{ij} \int_\Omega w_i \frac{\partial \beta_j}{\partial x_i} v_j \, d\Omega = \int_\Omega (\nabla^T \boldsymbol{\beta} \cdot \mathbf{v}) \cdot \mathbf{w} \, d\Omega =_V < \mathbf{w}, \mathcal{R}^* \mathbf{v} >_{V'},$$

where $\nabla^T \boldsymbol{\beta}$ is the tensor with component $(\nabla^T \boldsymbol{\beta})_{ij} = \dfrac{\partial \beta_j}{\partial x_i}$, i.e. the transpose of the gradient operator. Thus,

$$\mathcal{R}_{SS}\mathbf{v} = \frac{1}{2} (\mathcal{R}\mathbf{v} - \mathcal{R}^*\mathbf{v}) = \frac{1}{2} \left[(\mathbf{v} \cdot \nabla) \boldsymbol{\beta} - \nabla^T \boldsymbol{\beta} \cdot \mathbf{v} \right].$$

Summing all terms the skew-symmetric part of operator $\mathcal{L}$ is finally obtained:

$$\mathcal{L}_{SS}(\mathbf{v}, q) \equiv \begin{bmatrix} (\boldsymbol{\beta} \cdot \nabla) \mathbf{v} - \dfrac{1}{2} (\nabla \cdot \boldsymbol{\beta}) \mathbf{v} + \mathcal{R}_{SS}\mathbf{v} + \nabla q \\ \nabla \mathbf{v} \end{bmatrix}.$$

SUPG stabilization is obtained by summing to the left-hand side of (7.36) the term

$$\delta \sum_{K \in \mathcal{T}_h} \left(\mathcal{L}(\mathbf{w}, p) - \mathbf{b}, h_K^2 \mathcal{L}_{SS}(\mathbf{v}, q) \right)_K, \tag{7.39}$$

where $(\cdot, \cdot)_K$ indicates the $L^2(K)$ scalar product. Let us specify this term in the momentum equation. Set $q = 0$. On each element, SUPG stabilization reads

$$\delta h_K^2 \left(-\nu \triangle \mathbf{w} + (\boldsymbol{\beta} \cdot \nabla) \mathbf{w} + (\mathbf{w} \cdot \nabla)\boldsymbol{\beta} + \nabla p - \mathbf{b}_1, (\boldsymbol{\beta} \cdot \nabla) \mathbf{v} \right.$$
$$\left. - \frac{1}{2} (\nabla \cdot \boldsymbol{\beta}) \mathbf{v} + \mathcal{R}_{SS}\mathbf{v} \right)_K + \delta h_K^2 (\nabla \cdot \mathbf{w} + \nabla \cdot \mathbf{G}, \nabla \cdot \mathbf{v})_K. \tag{7.40}$$

In particular, term $\delta h_K^2 ((\boldsymbol{\beta} \cdot \nabla) \mathbf{w}, (\boldsymbol{\beta} \cdot \nabla) \mathbf{v})_K$ is the so-called *streamline upwind* term.

Stabilization term on the mass equation is obtained by setting in (7.39) $\mathbf{v} = \mathbf{0}$. We get

$$\delta h_k^2 (-\nu \triangle \mathbf{w} + (\boldsymbol{\beta} \cdot \nabla) \mathbf{w} + \mathcal{R}\mathbf{w} + \nabla p - \mathbf{b}_1, \nabla q)_K. \tag{7.41}$$

Weak form of the stabilized mass conservation equation reads therefore

$$b(\mathbf{w}, q) - \delta \sum_{K \in \mathcal{T}_h} h_K^2 \left(-\nu \triangle \mathbf{w} + (\boldsymbol{\beta} \cdot \nabla)\mathbf{w} + \mathcal{R}\mathbf{w} + \nabla p - \mathbf{b}_1, \nabla q\right)_K$$

$$+ \boxed{h_K^2 \left(\nabla p, \nabla q\right)_K} = -b(\mathbf{G}, q).$$

Boxed term is actually stabilizing with respect to the *inf-sup* condition being the element-wise discretization of a quasi-compressibility term of the type seen in (7.3) and discussed in Exercise 7.1.3. This shows that stabilization methods for dominating convection can allow to circumvent the LBB condition.

Finally, notice that at convergence of the Newton iterations the final solution is strongly consistent with the original Navier-Stokes problem. As a matter of fact, when $\mathbf{w} = \mathbf{u}^{(k+1)} \to \mathbf{u}_h$ and e $p^{(k+1)} \to p_h$ we have also $\boldsymbol{\beta} = \mathbf{u}^{(k)} \to \mathbf{u}_h$ and the converged solution solves problem

$$a(\mathbf{u}_h + \mathbf{G}, \mathbf{v}) + b(\mathbf{v}, p_h) + c(\mathbf{u}_h + \mathbf{G}, \mathbf{u}_h + \mathbf{G}, \mathbf{v})$$

$$+ b(\mathbf{u}_h + \mathbf{G}, q) = (\mathbf{f}, \mathbf{v}) -$$

$$\delta \sum_{K \in \mathcal{T}} ((-\nu \triangle(\mathbf{u}_h + \mathbf{G}) + ((\mathbf{u}_h + \mathbf{G}) \cdot \nabla)(\mathbf{u}_h + \mathbf{G}) + \nabla p_h - \mathbf{f}, \mathcal{L}_{SS}(\mathbf{v}, q))_K$$

$$+ (\nabla \cdot (\mathbf{u}_h + \mathbf{G}), \mathcal{L}_{SS}(\mathbf{v}, q))_K). \tag{7.42}$$

It may be verified that this system of equation is satisfied by the exact solution, so the method is strongly consistent. Indeed, the additional term introduced by stabilization is proportional to the residual of the original problem, that vanishes in correspondence of the exact solution.

The algebraic form of the stabilized problem is reported for the sake of completeness. At each Newton iteration the following system is solved

$$\begin{bmatrix} \widehat{K} & \widehat{D}^T \\ \widetilde{D} & -\delta S \end{bmatrix} \begin{bmatrix} \mathbf{W} \\ \mathbf{P} \end{bmatrix} = \begin{bmatrix} \mathbf{F}_1 \\ \mathbf{F}_2 \end{bmatrix}. \tag{7.43}$$

$\widehat{K}$ is the algebraic counterpart of the linearized and stabilized Navier-Stokes problem, $\widetilde{D}$ includes effects of stabilization in the mass equation on the divergence term, $\widehat{D}^T$ includes the effects on the pressure gradient. They are computed in the standard way, we do not give the expression for the sake of brevity.

Notice that the matrix acting on the pressure in the momentum equation and the one operating on the velocity in the mass equation are no longer one the transpose of the other. Finally, observe that matrix S is symmetric and positive definite. It is possible to verify that the matrix of system (7.43) is non-singular for any choice of velocity and pressure subspaces.

Finally, we stress that for piecewise linear finite elements, the Laplace term $\triangle \mathbf{w}$ inside the the stabilization vanishes.

Numerical results

In Program 37 we report the SUPG stabilized problem to be solved at each Newton iteration.

Program 37 - NavierStokesNewtonSUPG : Navier-Stokes problem: generic SUPG-Newton iteration

```
problem NewtonNavierStokesSUPG ([u1,u2,p],[v1,v2,q],solver=GMRES,
                                                    eps=1.e-12) =
 int2d(Th)(nu * ( dx(u1)*dx(v1) + dy(u1)*dy(v1)
      +       dx(u2)*dx(v2) + dy(u2)*dy(v2) ))
   + int2d(Th)(u1*dx(u1last)*v1 + u1*dx(u2last)*v2+ u2*dy(u1last)*v1
   + u2*dy(u2last)*v2
   + u1last*dx(u1)*v1 + u1last*dx(u2)*v2+ u2last*dy(u1)*v1
                                           + u2last*dy(u2)*v2)
   - int2d(Th)(u1last*dx(u1last)*v1+ u1last*dx(u2last)*v2
   + u2last*dy(u1last)*v1 + u2last*dy(u2last)*v2)
   - int2d(Th)(p*dx(v1) + p*dy(v2))
   - int2d(Th)(dx(u1)*q + dy(u2)*q)
// Momentum equation stabilization
   + int2d(Th)(delta*hTriangle^2*(-nu*(dxx(u1)+dyy(u1))+u1*dx(u1last)
   +u2*dy(u1last)+u1last*dx(u1)+u2last*dy(u1)+dx(p))*
     (u1last*dx(v1)+u2last*dy(v1)+0.5*(v2*dy(u1last)-v2*dx(u2last))))
   + int2d(Th)(delta*hTriangle^2*(-nu*(dxx(u2)+dyy(u2))+u1*dx(u2last)
   +u2*dy(u2last)+ u1last*dx(u2)+u2last*dy(u2)+dy(p))*
     (u1last*dx(v2)+u2last*dy(v2)+0.5*(v1*dx(u2last)-v1*dy(u1last))))
   - int2d(Th)(delta*hTriangle^2*(u1last*dx(u1last)+u2last*dy(u1last))*
     (u1last*dx(v1)+u2last*dy(v1)+0.5*(v2*dy(u1last)-v2*dx(u2last))))
   - int2d(Th)(delta*hTriangle^2*(u1last*dx(u2last)+u2last*dy(u2last))*
     (u1last*dx(v2)+u2last*dy(v2)+0.5*(v1*dx(u2last)-v1*dy(u1last))))
   + int2d(Th)(delta*hTriangle^2*(dx(u1)+dy(u2))*(dx(v1)+dy(v2)))
// Mass equation stabilization
   + int2d(Th)(delta*hTriangle^2*(nu*(dxx(u1)+dyy(u1))-u1*dx(u1last)
   -u2*dy(u1last)-u1last*dx(u1)-u2last*dy(u1)-dx(p))*dx(q))
   + int2d(Th)(delta*hTriangle^2*(nu*(dxx(u2)+dyy(u2))-u1*dx(u2last)
   -u2*dy(u2last)-u1last*dx(u2)-u2last*dy(u2)-dy(p))*dy(q))
   + int2d(Th)(delta*hTriangle^2*(u1last*dx(u1last)+
                                  u2last*dy(u1last))*dx(q))
   + int2d(Th)(delta*hTriangle^2*(u1last*dx(u2last)+
                                  u2last*dy(u2last))*dy(q))
   + on(1,2,4,u1=0,u2=0)
   + on(3,u1=1.,u2=0);
```

In Table 7.2 the number of iterations required by the SUPG-Newton method for small viscosity values, when the non stabilized method does not converge ($h = 1/20$).

In the lid-driven cavity problem ($h = 1/40$), with $\nu = 10^{-3}$ P^2-P^1 finite element solution converges in 15 Newton iterations ($\delta = 100$), and a computation time of 1830 s. P^1-P^1 solution is attained in 14 Newton iterations ($\delta = 20$) and computation time of 260 s. Discretization with P^1-P^1 features

Table 7.2 Number of SUPG-Newton iterations ($h = 1/20$) for Navier-Stokes Poiseuille problem with finite elements P^2-P^1 and P^1-P^1. Stabilization allows equal order interpolation

Viscosity ν	P^2-P^1	P^1-P^1
10^{-6}	8 ($\delta = 1$)	5 ($\delta = 1$)
10^{-7}	8 ($\delta = 1$)	5 ($\delta = 1$)
10^{-8}	8 ($\delta = 1$)	5 ($\delta = 1$)

smaller and better conditioned matrices, so that the solution is obtained with a lower computational effort (even if it is less accurate). In Figs. 7.9 and 7.10 we report velocity and pressure fields for the two cases.

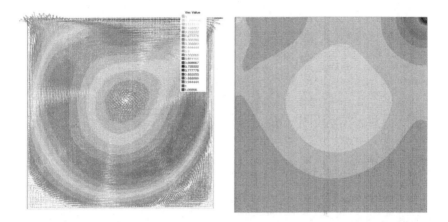

Fig. 7.9 P^2-P^1 solution of the lid-driven cavity problem with SUPG stabilization ($h = 1/40$, $\nu = 10^{-3}$). Convergence in 15 Newton iterations ($\delta = 100$)

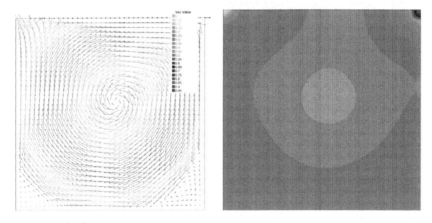

Fig. 7.10 P^1-P^1 solution of the lid-driven cavity problem with SUPG stabilization ($h = 1/40$, $\nu = 10^{-3}$). Convergence in 14 Newton iterations ($\delta = 20$)

In conclusion, stabilization allows equal order finite element approximation and reduces oscillations due to the convection. A good tuning of the parameter δ can be non trivial, as a trade-off between stability and accuracy. $\diamond$

Remark 7.4 Other ways for applying stabilization methods are based on a modification of the velocity finite element space (for instance P^1-*bubble finite elements*). These are available in `Freefem++`, see the manual of the code and the examples available at `www.freefem.org`.

7.2 Unsteady problems

Exercise 7.2.1. Consider the unsteady Stokes problem in the bounded 2D domain Ω

$$\begin{cases} \dfrac{\partial \mathbf{u}}{\partial t} - \nu\triangle\mathbf{u} + \nabla p = \mathbf{f} & \text{in } \Omega \times (0, T] \\ \nabla \cdot \mathbf{u} = 0 & \text{in } \Omega \times (0, T], \\ \mathbf{u} = \mathbf{g} \quad \text{on } \Gamma, \quad \mathbf{u} = \mathbf{u}_0 \quad \text{in } \Omega, \quad t = 0. \end{cases} \quad (7.44)$$

1. Write the weak formulation and an *a priori* estimate for the solution.
2. Discretize in time the problem with the Implicit Euler method. Show that the semi-discrete problem (continuous in space, discrete in time) is unconditionally stable.
3. Write the algebraic form of the problem after space discretization with *inf-sup* compatible finite elements, write the *pressure matrix method* for solving the associated linear system and formulate this method as an inexact LU factorization of the Stokes fully discretized system.
4. Write a `Freefem++` program for the pressure matrix method, suitably preconditioned. Test it on the lid-driven cavity problem (with $\nu = 0.01$).

Solution 7.2.1.

Mathematical analysis

We use the same notation of previous exercises. Weak formulation is obtained by multiplying the vector momentum equation by a function $\mathbf{v} \in V$ and the continuity equation by a scalar function $q \in Q$. Assume $\mathbf{f} \in L^2(0, T; \mathbf{L}^2(\Omega))$, and let $\mathbf{G}$ be a divergence free extension of the boundary data $\mathbf{g}$ (see Exercise 7.1.3). The weak formulation reads: find $\tilde{\mathbf{u}} \in L^2(0, T; V) \cap L^\infty(0, T; \mathbf{L}^2(\Omega))$ and $p \in L^2(0, T; Q)$ s.t. for any $t > 0$, for any $\mathbf{v} \in V$ and $q \in Q$ (where $V = H_0^1(\Omega)$ and $Q = L_0^2(\Omega)$), we have

$$
\begin{cases}
\left(\dfrac{\partial \tilde{\mathbf{u}}}{\partial t}, \mathbf{v}\right) + a(\tilde{\mathbf{u}}, \mathbf{v}) + b(\mathbf{v}, p) = (\mathbf{f}, \mathbf{v}) - \left(\dfrac{\partial \mathbf{G}}{\partial t}, \mathbf{v}\right) - a(\mathbf{G}, \mathbf{v}), \\
b(\tilde{\mathbf{u}}, q) = -b(\mathbf{G}, q) = 0,
\end{cases} \tag{7.45}
$$

with the initial condition $\mathbf{u}(\mathbf{x}, 0) = \mathbf{u}_0(\mathbf{x})$. As usual, $(\cdot, \cdot)$ is the $\mathbf{L}^2(\Omega)$ scalar product.

Take in (7.45) $\mathbf{v} = \tilde{\mathbf{u}}$ and $q = p$. Notice that

$$
\int_0^T \left(\frac{\partial \tilde{\mathbf{u}}}{\partial t}, \tilde{\mathbf{u}}\right) dt = \frac{1}{2} \int_0^T \frac{d}{dt} ||\tilde{\mathbf{u}}||_{\mathbf{L}^2}^2 dt = \frac{1}{2} ||\tilde{\mathbf{u}}(T)||_{\mathbf{L}^2}^2 - \frac{1}{2} ||\mathbf{u}_0||_{\mathbf{L}^2}^2.
$$

Since we have a Dirichlet problem, we can advocate the Poincaré inequality, so that for any $\mathbf{v} \in V$

$$
a(\mathbf{v}, \mathbf{v}) \geq \alpha ||\mathbf{v}||_V^2.
$$

After time integration of the two equations in (7.45) (still with $\mathbf{v} = \tilde{\mathbf{u}}$ and $q = p$) and subtracting the second of (7.45) from the first we get

$$
\frac{1}{2} ||\tilde{\mathbf{u}}(T)||_{\mathbf{L}^2}^2 + \alpha \int_0^T ||\tilde{\mathbf{u}}||_V^2 dt \leq \frac{1}{2} ||\mathbf{u}_0||_{\mathbf{L}^2}^2 + \int_0^T \left((\mathbf{f}, \tilde{\mathbf{u}}) - \left(\frac{\partial \mathbf{G}}{\partial t}, \tilde{\mathbf{u}}\right) - a(\mathbf{G}, \tilde{\mathbf{u}}) \right) dt.
$$

By the Young inequality and the continuity of $a(\cdot, \cdot)$ (i.e. $|a(\mathbf{u}, \mathbf{v})| \leq \gamma ||\mathbf{u}||_V ||\mathbf{v}||_V$) for any $t \in (0, T)$,

$$
(\mathbf{f}, \tilde{\mathbf{u}}) - \left(\frac{\partial \mathbf{G}}{\partial t}, \tilde{\mathbf{u}}\right) - a(\mathbf{G}, \tilde{\mathbf{u}}) \leq
$$

$$
\frac{1}{4\epsilon} \left(||\mathbf{f}||_{\mathbf{L}^2}^2 + ||\frac{\partial \mathbf{G}}{\partial t}||_{\mathbf{L}^2}^2 + \gamma ||\mathbf{G}||_V^2 \right) + \epsilon ||\tilde{\mathbf{u}}||_V^2
$$

with an arbitrary $\epsilon > 0$. For $\epsilon = \alpha/2$ we get

$$
||\tilde{\mathbf{u}}(T)||_{\mathbf{L}^2}^2 + \alpha \int_0^T ||\tilde{\mathbf{u}}||_V^2 dt \leq ||\mathbf{u}_0||_{\mathbf{L}^2}^2 + \frac{1}{\alpha} \int_0^T \left(||\mathbf{f}||_{\mathbf{L}^2}^2 + ||\frac{\partial \mathbf{G}}{\partial t}||_{\mathbf{L}^2}^2 + \gamma ||\mathbf{G}||_V^2 \right) dt
$$

showing that solution $\tilde{\mathbf{u}}$ is bounded in $L^\infty(0, T; \mathbf{L}^2(\Omega))$ and $L^2(0, T; V)$, so it belongs to the intersection of the two spaces. Since $\mathbf{u} = \tilde{\mathbf{u}} + \mathbf{G}$ under suitable hypothesis on $\mathbf{g}$ also the lifting belongs to the same spaces, and is bounded by the boundary data. So the stability estimate applies also to the "real" solution $\mathbf{u}$.

As discussed in Chapter 5, implicit Euler is a particular θ-method with $\theta = 1$. In view of time discretization, we subdivide the time interval $[0, T]$ with a time step Δt and collocate the problem in the nodes $t^k \equiv k\Delta t$ for $k = 1, 2, \ldots, N$ with $T = N\Delta t$. Time derivative is approximated by first order backward finite difference,

$$\frac{\partial \mathbf{u}}{\partial t}(t^{n+1}) \approx \frac{1}{\Delta t}\left(\mathbf{u}^{n+1} - \mathbf{u}^n\right).$$

For any $n \geq 0$, we have the following time discrete problem

$$\begin{cases} \dfrac{1}{\Delta t}\left(\mathbf{u}^{n+1} - \mathbf{u}^n\right) - \nu\triangle\mathbf{u}^{n+1} + \nabla p^{n+1} = \mathbf{f}^{n+1}, \\ \nabla \cdot \mathbf{u}^{n+1} = 0, \end{cases} \tag{7.46}$$

with the given initial condition $\mathbf{u}^0 = \mathbf{u}_0$. Weak form reads: find for any $n \geq 0$ $\tilde{\mathbf{u}}^{n+1} \in V$, $p^{n+1} \in Q$ s.t. for any $\mathbf{v} \in V$ and $q \in Q$ we have

$$\begin{cases} \dfrac{1}{\Delta t}\left(\tilde{\mathbf{u}}^{n+1}, \mathbf{v}\right) + a(\tilde{\mathbf{u}}^{n+1}, \mathbf{v}) + b(\mathbf{v}, p^{n+1}) = \\ \qquad \dfrac{1}{\Delta t}\left(\tilde{\mathbf{u}}^n, \mathbf{v}\right) + \left(\mathbf{f}^{n+1}, \mathbf{v}\right) - \left(\dfrac{\partial \mathbf{G}}{\partial t}\big|_{t^{n+1}}, \mathbf{v}\right) - a(\mathbf{G}^{n+1}, \mathbf{v}) \\ b(\tilde{\mathbf{u}}, q) = -b(\mathbf{G}^{n+1}, q) = 0. \end{cases}'$$

From now on, we set

$$\mathcal{F}^{n+1}(\mathbf{v}) \equiv \left(\mathbf{f}^{n+1}, \mathbf{v}\right) - \left(\frac{\partial \mathbf{G}}{\partial t}\big|_{t^{n+1}}, \mathbf{v}\right) - a(\mathbf{G}, \mathbf{v})$$

which is a linear and continuous functional in V.

A stability estimate is obtained by setting $\mathbf{v} = \tilde{\mathbf{u}}^{n+1}$ and $q = p^{n+1}$, and subtracting continuity equation from the momentum one,

$$\frac{1}{\Delta t}\left(\tilde{\mathbf{u}}^{n+1} - \tilde{\mathbf{u}}^n, \tilde{\mathbf{u}}^{n+1}\right) + a\left(\tilde{\mathbf{u}}^{n+1}, \tilde{\mathbf{u}}^{n+1}\right) = \mathcal{F}^{n+1}(\tilde{\mathbf{u}}^{n+1}). \tag{7.47}$$

By the coercivity of the bilinear form (which is a consequence of the Poincaré inequality) and the Young inequality to the right-hand side of (7.47), we get

$$\frac{1}{\Delta t}\left(\tilde{\mathbf{u}}^{n+1} - \tilde{\mathbf{u}}^n, \tilde{\mathbf{u}}^{n+1}\right) + \alpha||\tilde{\mathbf{u}}^{n+1}||_V^2 \leq \frac{1}{4\epsilon}||\mathcal{F}||_{V'}^2 + \epsilon||\tilde{\mathbf{u}}^{n+1}||_V^2 \tag{7.48}$$

with an arbitrary $\epsilon > 0$. We have the following identity (see Chapter 5)

$$2\left(\tilde{\mathbf{u}}^{n+1} - \tilde{\mathbf{u}}^n, \tilde{\mathbf{u}}^{n+1}\right) = ||\tilde{\mathbf{u}}^{n+1}||_{L^2}^2 + ||\tilde{\mathbf{u}}^{n+1} - \tilde{\mathbf{u}}^n||_{L^2}^2 - ||\tilde{\mathbf{u}}^n||_{L^2}^2.$$

Setting $\epsilon = \alpha/2$, we get

$$\frac{1}{\Delta t}\left(||\tilde{\mathbf{u}}^{n+1}||_{L^2}^2 + ||\tilde{\mathbf{u}}^{n+1} - \tilde{\mathbf{u}}^n||_{L^2}^2 + ||\tilde{\mathbf{u}}^n||_{L^2}^2\right) + \alpha||\tilde{\mathbf{u}}^{n+1}||_V^2 \le \frac{1}{\alpha}||\mathcal{F}||_{V'}^2. \quad (7.49)$$

Multiplying by Δt, summing the time index $n = 0, 1, \ldots, N-1$

$$\sum_{n=0}^{N-1}\left(||\tilde{\mathbf{u}}^{n+1}||_{L^2}^2 - ||\tilde{\mathbf{u}}^n||_{L^2}^2\right) = ||\tilde{\mathbf{u}}^N||_{L^2}^2 - ||\tilde{\mathbf{u}}^0||_{L^2}^2,$$

and exploiting the telescopic sum we finally get

$$||\tilde{\mathbf{u}}^N||_{L^2}^2 + \alpha\Delta t\sum_{n=0}^{N-1}||\tilde{\mathbf{u}}^{n+1}||_V^2 \le \frac{\Delta t}{\alpha}\sum_{n=0}^{N-1}||\mathcal{F}^{n+1}||_{V'}^2 + ||\mathbf{u}_0||_{L^2}^2 \quad (7.50)$$

holding for any Δt which is the unconditional stability estimate.

Finally, let V_h and Q_h be two finite dimensional subspaces of V and Q respectively. Set $\{\varphi_i\}$ ($i = 1, \ldots, d_\mathbf{u}$) and $\{\psi_j\}$ ($j = 1, \ldots, d_p$) the basis functions of the two subspaces. In particular, let us consider usual Lagrangian finite element spaces on a given mesh of Ω.

If we multiply momentum equation in (7.46) by φ_i and the continuity equation by ψ_j, integrating over Ω we get the system

$$\mathcal{A}\mathbf{y}^{n+1} = \mathbf{b}^{n+1}, \quad (7.51)$$

where $\mathbf{y}^k = [\mathbf{U}^k, \mathbf{P}^k]$ is the (velocity and pressure) vector of nodal values at time t^k. Matrix $\mathcal{A}$ is

$$\mathcal{A} = \begin{bmatrix} C & D^T \\ D & 0 \end{bmatrix}, \quad (7.52)$$

with $C = \dfrac{1}{\Delta t}M + \nu K$, M is the velocity mass matrix, $D_{ij} = \int_\Omega \psi_j \nabla \cdot \varphi_i \, d\Omega$ and K is the discretization of the velocity Laplace operator. Finally, we have

$$\mathbf{b}^{n+1} = \begin{bmatrix} \mathbf{b}_1^{n+1} \\ \mathbf{b}_2^{n+1} \end{bmatrix} \equiv \begin{bmatrix} \dfrac{1}{\Delta t}M\mathbf{U}^n + \mathbf{F}^{n+1} \\ -DG^{n+1} \end{bmatrix},$$

where $\mathbf{F}^k$ is the vector with components $F_i^k = \mathcal{F}^k(\varphi_i)$ and $\mathbf{G}^k$ is the vector of nodal values of the extension of the boundary data $\mathbf{g}$. Notice that C is a symmetric positive definite matrix, being the sum of two s.p.d. matrices, so it is non singular. For *inf-sup* compatible finite elements, system (7.51) is non singular. As in Exercise 7.1.3, we can indeed resort to a system for the pressure only

$$DC^{-1}D^T\mathbf{P}^{n+1} = DC^{-1}\mathbf{b}_1^{n+1} - \mathbf{b}_2^{n+1}. \quad (7.53)$$

Pressure matrix $DC^{-1}D^T$ is invertible (as shown in Exercise 7.1.3) under the *inf-sup* condition. Therefore, we first compute the unique pressure then the velocity, by solving the system

$$CU^{n+1} = b_1^{n+1} - D^T P^{n+1}. \tag{7.54}$$

System (7.51) is symmetric and not definite, having both positive and negative eigenvalues (as a consequence of the *constrained* minimization nature of the problem). This follows by the matrix identity (see [BGL05])

$$\begin{bmatrix} I & 0 \\ -DC^{-1} & I \end{bmatrix} \begin{bmatrix} C & D^T \\ D & 0 \end{bmatrix} \begin{bmatrix} I & -C^{-1}D^T \\ 0 & I \end{bmatrix} = \begin{bmatrix} C & 0 \\ 0 & -DC^{-1}D^T \end{bmatrix}$$

stating that $\mathcal{A}$ is similar to (so it has the same eigenvalues of) a block diagonal matrix. The latter has some positive eigenvalues (of C) and some negative ones (of $-DC^{-1}D^T$). Conjugate gradient method is therefore not recommended, since the matrix is not definite. GMRes or BiCGStab are more appropriate for solving system (7.51).

Remark 7.5 We may notice that our problem may be equally described by matrix

$$\mathcal{A}_1 \equiv \begin{bmatrix} C & D^T \\ -D & 0 \end{bmatrix}$$

by changing the sign of the right-hand side term b_2. This matrix is positive definite (since $q^T \mathcal{A}_1 q > 0$ for any non zero vector q) but not symmetric. So alternatively, we may use $\mathcal{A}_1$ to solve our problem, using an iterative method suited for positive (but not symmetric) matrices, like MINRES [Saa03]. Beware that even in this case one cannot use the (more efficient) conjugate gradient method, since the latter requires that the matrix be also symmetric.

Another possibility is to compute velocity and pressure separately, by solving at each time step the sequence of systems (7.53),(7.54). This method, called *pressure matrix method*, splits the entire problem into a sequence of sub-problems, in general better conditioned than the original one. In the current case of Stokes problem, matrix is s.p.d., so that systems (7.53),(7.54) can be solved with the Conjugate Gradient method. Notice that explicit computation of matrix $DC^{-1}D^T$ is not necessary, nor the computation of C^{-1}. These are actually expensive operations both in terms of CPU time and storage (due to the *fill in*: even if C is sparse its inverse is not necessarily so). Computation of $z = DC^{-1}D^T P$, required by any iterative method is carried out by the sequence of steps:

1. $y = D^T P$;
2. $Cw = y$;
3. $z = Dw$.

Step 2 is carried out either with an iterative method (like conjugate gradient, since the matrix is s.d.p) or with a direct method suited for sparse matrices (like the multifrontal solver implemented in the UMFPACK library[Dav04]). The latter, however, may become impractical for large problems. The pressure matrix method can be reinterpreted as a block LU factorization of $\mathcal{A}$,

$$\mathcal{A} = \begin{bmatrix} C & D^T \\ D & O \end{bmatrix} = \begin{bmatrix} L_{11} & O \\ L_{21} & L_{22} \end{bmatrix} \begin{bmatrix} I_{11} & U_{12} \\ O & I_{22} \end{bmatrix},$$

where I_{11} and I_{22} are the identity square matrices of size d_u and d_p, respectively. Matrices L_{11}, L_{21}, L_{22} and U_{12} are such that

$$L_{11} = C, \quad L_{11}U_{12} = D^T, \quad L_{21} = D, \quad L_{21}U_{12} + L_{22} = 0.$$

This yields the factorization

$$\mathcal{A} = \begin{bmatrix} C & O \\ D & -DC^{-1}D^T \end{bmatrix} \begin{bmatrix} I_{11} & C^{-1}D^T \\ O & I_{22} \end{bmatrix}. \tag{7.55}$$

Exploiting this splitting, solution to the Stokes system at each time step is obtained by the following stages, where, for the sake of simplicity, we have dropped the time index

$$\mathcal{A}\mathbf{y} = \mathbf{b} \Rightarrow \mathcal{L}\mathcal{U}\mathbf{y} = \mathbf{b} \Rightarrow \begin{cases} \mathcal{L}\mathbf{z} = \mathbf{b}, \\ \mathcal{U}\mathbf{y} = \mathbf{z}. \end{cases}$$

System $\mathcal{L}\mathbf{z} = \mathbf{b}$ amounts to solve

$$\begin{cases} C\mathbf{W} = \mathbf{b}_1, \\ DC^{-1}D^T\mathbf{P} = D\mathbf{W} - \mathbf{b}_2, \end{cases}$$

where $\mathbf{W}$ is the first block vector component of $\mathbf{z}$ and $\mathbf{P}$ is the second one. System $\mathcal{U}\mathbf{y} = \mathbf{z}$ reduces to

$$C\mathbf{U} = C\mathbf{W} - D^T\mathbf{P} = \mathbf{f}_1 - D^T\mathbf{P}.$$

Systems (7.53) and (7.54) correspond exactly to the three previous steps.

Pressure matrix method is still expensive since it requires to solve systems governed by C and $DC^{-1}D^T$ (the latter, as we have shown, still involves a linear system with matrix C). Moreover, matrix $DC^{-1}D^T$ is in general ill-conditioned and the condition number gets larger when the time step Δt decreases (see [QV94]). In the case of Stokes problem, computational costs are reduced whenever a Cholesky factorization of C is feasible (see [Dav08]), however for large problems the Cholesky factor may not fit the computer memory, so iterative methods (conjugate gradient in this case) are compulsory. Moreover, Cholesky factorization is not feasible for the Navier-Stokes case, when matrix C is no longer s.p.d. and changes at each time step (in addition, in this case we cannot use conjugate gradient, we need to resort, for instance, to GMRES).

To reduce computational cost when we use an iterative method we need to find a good preconditioner for $DC^{-1}D^T$. The ideal case is to find a preconditioner where the condition number of the preconditioned matrix is sensibly smaller than that of the original one, possibly with a bound of the condition number independent of the matrix size (in this case we say the the preconditioner is *optimal*). With an optimal preconditioner the number of iterations for a given tolerance are not only reduced but they are independent of the size of the problem. This is an open and active research field, see e.g. [QV94, BGL05, ESW05] and bibliography quoted therein. Here, we limit ourselves to some immediate observations.

Let us try to find a reasonable approximation of $DC^{-1}D^T$ to be used as preconditioner. Notice that $C^{-1} = \left(\frac{1}{\Delta t}M + \nu K\right)^{-1}$ so that a possible approximation reads $C^{-1} \approx \Delta t M^{-1} + \frac{1}{\nu}K$. Thus, an approximation to the pressure matrix is

$$DC^{-1}D^T \approx \Delta t D M^{-1} D^T + \frac{1}{\nu}DK^{-1}D^T. \tag{7.56}$$

This is still too difficult to be solved in an effective way. However, matrix $DK^{-1}D^T$ is the algebraic counterpart of operator

$$\frac{1}{\nu}\nabla \cdot (\Delta)^{-1}\nabla = \frac{1}{\nu}\nabla \cdot ((\nabla \cdot)\nabla)^{-1}\nabla$$

which is spectrally equivalent to the identity operator [QV94]. In a finite element setting this consideration leads to approximation

$$\frac{1}{\nu}DK^{-1}D^T \approx \frac{1}{\nu}M_p,$$

where we have used M_p to indicate the pressure mass matrix indicated with S in the previous exercises. For steady problems, a possible preconditioner of the pressure matrix is then $P_1 = \frac{1}{\nu}M_p$ (see [QV94], Chapter 9). For unsteady problems, however, this is not enough, since the condition number of the pressure matrix depends also on the time step. We note that the first term of (7.56),

$$P_2 = \Delta t D M^{-1} D^T,$$

is a sort of discretization of the Laplace operator $\Delta t \nabla \cdot (\nabla)$, and it is sometimes called *discrete Laplace operator*. If Δt small enough, thanks to the Nuemann expansion (1.20), the approximation $C^{-1} = (\Delta t^{-1}M + \nu K)^{-1} \approx \Delta t^{-1}M$ is justified, so P_2 is another possible preconditioner.

Another possible preconditioner of $DC^{-1}D^T$, following from (7.56), is therefore provided by the approximation

$$DC^{-1}D^T \approx P_2 + P_1 \equiv P.$$

In a preconditioned method we actually need the inverse of the preconditioner, we can then make the further step

$$P^{-1} = (P_2 + P_1)^{-1} \approx P_1^{-1} + P_2^{-1}.$$

This corresponds to the choice $P_3 = \left(P_1^{-1} + P_2^{-1}\right)^{-1}$. known as *Caouet-Chabard preconditioner*[5], which is one of the most popular preconditioner for both Stokes and Navier-Stokes problems (see [CC88]).

Numerical results

Code `StokesPressureMatrix`, available on line, implements in `Freefem++` the non-preconditioned pressure matrix method.

Programs `Pressure MassMP` e `DiscreteLaplacMP` implement the preconditioned formulations with either P_1 or P_2 separately.

In Table 7.3 we report the number of required iterations for the lid-driven cavity problem with $\nu = 0.01$ (Re= 100) for $h = 0.1$ and $h = 0.05$.

Table 7.3 Number of iterations required for the lid-driven cavity problem (Re= 100) for $h = 0.1$ and $h = 0.05$. Tolerance required for the convergence is 10^{-6}

Δt	$h = 0.1$			$h = 0.05$		
	Non-Precond	P_1	P_2	Non-Precond	P_1	P_2
0.1	$37 - 39$	$34 - 40$	$11 - 13$	$45 - 48$	$39 - 43$	$21 - 24$
0.01	$47 - 49$	$48 - 51$	$4 - 5$	$66 - 70$	$72 - 78$	$7 - 9$
0.001	$48 - 51$	$52 - 55$	$2 - 3$	$80 - 84$	$100 - 102$	$3 - 4$

Notice that P_1 is ineffective when the time step Δt decreases. However, we point out that *in the steady case*, for $h = 0.05$, the non-preconditioned case requires 40 iterations, while the one preconditioned with P_1 requires only 16 iterations. Preconditioner P_2 is effective also when the time step gets smaller. Notice however that the computational cost of the preconditioned iteration is by far larger for P_2 than P_1. It is therefore crucial to have a good implementation of the solver for P_2.

Finally, notice that estimate (7.50) holds also for the Navier-Stokes problem. Actually, for the given boundary conditions, trilinear form $c(\mathbf{w}, \mathbf{u}, \mathbf{v}) = \int_\Omega (\mathbf{w} \cdot \nabla) \mathbf{u} \cdot \mathbf{v} \, d\Omega$ is skew-symmetric, that means that $c(\mathbf{u}^{n+1}, \mathbf{u}^{n+1}, \mathbf{u}^{n+1}) = 0$. Note however, that Remark 7.1 still applies.   ◇

---

[5] Different possible implementations of this preconditioner are available. In `Freefem++` manual a different implementation is presented.

Exercise 7.2.2. Consider the discretized Stokes problem (7.51) introduced in Exercise 7.2.1, with matrix $\mathcal{A}$ given by (7.52). For solving this system we use a preconditioned Richardson iterative method, with preconditioner

$$\mathcal{Q} = \begin{bmatrix} C & 0 \\ D & \theta^{-1}M_p \end{bmatrix}, \qquad (7.57)$$

where M_p is the pressure mass matrix and θ is a real parameter. Calculate and analyze the iteration matrix. Implement the scheme and evaluate its performances on the lid-driven cavity problem with $\nu = 0.01$.

Solution 7.2.2.

Numerical approximation

Let us consider the discrete Stokes system (7.51) and solve it with the preconditioned Richardson scheme

$$\mathcal{Q}(\mathbf{y}_{k+1} - \mathbf{y}_k) = \mathbf{G} - \mathcal{A}\mathbf{y}_K, \qquad (7.58)$$

where $\mathcal{Q}$ is the preconditioner (7.57). A block-wise representation of (7.58) reads

$$C(\mathbf{U}_{k+1} - \mathbf{U}_k) = \mathbf{b}_1 - C\mathbf{U}_k - D^T\mathbf{P}_k \Rightarrow C\mathbf{U}_{k+1} = \mathbf{b}_1 - D^T\mathbf{P}_k \qquad (7.59)$$

and

$$D(\mathbf{U}_{k+1} - \mathbf{U}_k) + \theta^{-1}M_p(\mathbf{P}_{k+1} - \mathbf{P}_k) = \mathbf{b}_2 - D\mathbf{U}_k,$$

corresponding to

$$\theta^{-1}M_p(\mathbf{P}_{k+1} - \mathbf{P}_k) = \mathbf{b}_2 - D\mathbf{U}_{k+1}. \qquad (7.60)$$

Method (7.59), (7.60) is known as *Uzawa scheme* (see e.g. [QV94], Chapter 9). Notice that

$$\mathcal{Q}^{-1} = \begin{bmatrix} C^{-1} & 0 \\ -\theta M_p^{-1}DC^{-1} & \theta^{-1}M_p^{-1} \end{bmatrix},$$

so that the iteration matrix associated with the scheme reads

$$B = I - \mathcal{Q}^{-1}\mathcal{A} = \begin{bmatrix} 0 & -C^{-1}D^T \\ 0 & \theta M_p^{-1}DC^{-1}D^T \end{bmatrix}.$$

As it is well known from the theory of iterative methods, convergence is driven by the spectral radius of B. On the other hand, B is upper block triangular and its eigenvalues correspond to the ones of the diagonal blocks. Block $(1,1)$ is null, so it generates $d_\mathbf{u}$ null eigenvalues. Spectral radius of B corresponds therefore to the one of block $(2,2)$. Convergence depends on the

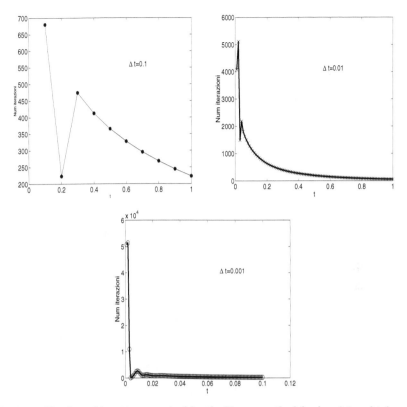

Fig. 7.11 Number of iterations required by the Uzawa method for $h = 0.1$ and tolerance for the convergence 10^{-6} for different values of the time step. Number of required iterations reduces remarkably when the solution approaches the steady state

effectiveness of $\theta^{-1}M_p$ as preconditioner of the pressure matrix $DC^{-1}D^T$. This issue has been faced in Exercise 7.2.1 for $\theta = \nu$, where we noticed that the scaled pressure mass matrix is a possible choice in the steady case. In the unsteady case, it is not effective, in particular for a decreasing time step Δt. However, from Exercise 7.2.1 we can improve the performances of the preconditioner thanks to the parameter θ. This parameter can be tuned for improving convergence properties. In the steady case, it is possible to prove that Uzawa method converges for $0 < \theta < 2\nu$ (see [QV94], Chapter 9).

Summarizing, Uzawa method may be ineffective for two reasons:

1. Richardson scheme is in general less effective than other iterative methods such as GMRES;
2. pressure mass matrix in the unsteady case is not a good preconditioner of $DC^{-1}D^T$.

Program `StokesUzawa`, available on line, implements the Uzawa method.
Consider the lid-driven cavity problem with $\nu = 0.01$ in the time interval $(0, 1]$, as in Exercise 7.2.1. In Fig. 7.11 we report the required iterations at each time step, with different values of the time step (P^2-P^1 elements). The number of required iterations is pretty high at the beginning and reduces when the solution gets closer to the steady state. The method is quite ineffective. $\diamond$

Exercise 7.2.3. Consider the homogeneous unsteady Navier-Stokes problem

$$\begin{cases} \dfrac{\partial \mathbf{u}}{\partial t} + (\mathbf{u} \cdot \nabla)\,\mathbf{u} - \nu \triangle \mathbf{u} + \nabla p = \mathbf{0} & \text{in} \quad Q_T \equiv \Omega \times (0, \infty) \\ \nabla \cdot \mathbf{u} = 0 \quad \text{in} \quad Q_T \end{cases} \tag{7.61}$$

($\Omega \equiv (0, 1) \times (0, 1)$) with the boundary conditions

$$\begin{cases} \mathbf{u} = \mathbf{0} & \text{on} \quad \Gamma_D, \\ p\mathbf{n} - \nu \nabla \mathbf{u} \cdot \mathbf{n} = -2\nu \mathbf{n} & \text{on} \quad \Gamma_{N_1}, \\ p\mathbf{n} - \nu \nabla \mathbf{u} \cdot \mathbf{n} = \mathbf{0} & \text{on} \, \Gamma_{N_2} \end{cases} \tag{7.62}$$

with the initial condition $\mathbf{u}_0 = \mathbf{0}$, where Γ_D is given by the sides $x_2 = 0$ and $x_2 = 1$ (with $0 \le x_1 \le 1$), Γ_{N_1} is the boundary $\{x_1 = 0, 0 \le x_2 \le 1\}$, Γ_{N_2} is the boundary $\{x_1 = 1, 0 \le x_2 \le 1\}$ and $\mathbf{n}$ is the outward normal unit vector to $\partial\Omega$ (see Fig. 7.12). Compute the steady analytic solution, assuming $u_2 = 0$. Then, modify on Γ_{N_1} the data as

$$p\mathbf{n} - \nu \nabla \mathbf{u} \cdot \mathbf{n} = 2\nu \sin(\omega t)\mathbf{n}$$

with $\omega \in \mathbb{R}$. Compute the unsteady solution, still assuming $u_2 = 0$ (Hint: expand u_1 according to its Fourier series).
Check the solutions with `Freefem++` for $\nu = 0.1$, $\omega = \pi$, P^2-P^1 finite elements and an implicit Euler time discretization. Draw the convergence curves for $h = 0.05$, $\Delta t = 0.1, 0.05, 0.025$.

Solution 7.2.3.

Steady solution can be computed as in Exercise 7.1.3. Since $u_2 = 0$, second component of the momentum equation reduces to $\dfrac{\partial p}{\partial x_2} = 0$, so that p is function only of x_1. Continuity equation yields $\partial u_1/\partial x_1 = 0$, so u_1 is function only of x_2. First momentum equation reduces to

$$\nu \frac{\partial^2 u_1}{\partial x_2^2} = \frac{\partial p}{\partial x_1}. \tag{7.63}$$

In fact, nonlinear term vanishes since $u_2 = 0$ and u_1 does not depend on x_1. left-hand side of (7.63) is a function of x_2, right-hand side is a function of x_1, so that equation holds only if the two sides separately are a constant c_1. We obtain

$$u_1 = \frac{c_1}{2\nu}x_2^2 + c_2 x_2 + c_3, \quad p = c_1 x_1 + c_4.$$

From the boundary conditions for u_1, we get $c_3 = 0$ and $c_2 = -\dfrac{c_1}{2\nu}$. Neumann boundary conditions apply on boundaries where we have outward unit vectors given by $[-1, 0]^T$ and $[1, 0]^T$ respectively, so

$$\nu \nabla \mathbf{u} \cdot \mathbf{n} = \pm \nu \frac{\partial u_1}{\partial x_1} = 0,$$

and the Neumann condition reduces to

$$p(0) = 2\nu, \quad p(1) = 0.$$

We have therefore $c_4 = 2\nu$ and $c_1 = -c_4 = -2\nu$. Solution reads therefore

$$u_1 = x_2 - x_2^2, \quad u_2 = 0, \quad p = 2\nu(1 - x_1),$$

which is again the Poiseuille solution.

In the case of sinusoidal-in-time data, we observe that time derivative occurs only in the first momentum equation, so the considerations on the second one and the continuity one are still valid. Pressure is function of x_1

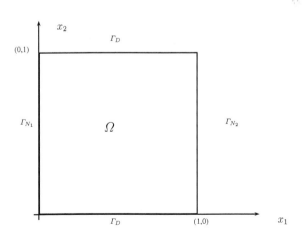

Fig. 7.12 Computational domain for Exercise 7.2.3

only, and u_1 only of x_2. The first momentum equation reads

$$\frac{\partial u_1}{\partial t} - \nu \frac{\partial^2 u_1}{\partial x_2^2} = -\frac{\partial p}{\partial x_1}.$$

Since pressure is a linear function of x_1, $\dfrac{\partial p}{\partial x_1}$ is constant in space and amounts to the difference between outlet and inlet pressure divided by the length of the domain, i.e.

$$\frac{\partial p}{\partial x_1} = -2\nu \sin \omega t.$$

First momentum equation reduces therefore to

$$\frac{\partial u_1}{\partial t} - \nu \frac{\partial^2 u_1}{\partial x_2^2} = 2\nu \sin \omega t.$$

Let us expand solution $u_1(x_2, t)$ (defined in a bounded interval of the independent variable x_2) as a Fourier series,

$$u_1(x_2, t) = \sum_{k=0}^{\infty} \gamma_k(t) \sin(k\pi x_2),$$

where we have only *sin* because of the boundary conditions. The first momentum equation gives

$$\sum_{k=0}^{\infty} \frac{\partial \gamma_k}{\partial t} \sin(k\pi x_2) + \nu \sum_{k=0}^{\infty} k^2 \pi^2 \gamma_k \sin(k\pi x_2) = 2\nu \sin(\omega t). \qquad (7.64)$$

Sinusoidal functions are *orthogonal* with respect to the L^2 scalar product,

$$\int_0^1 \sin(l\pi x_2) \sin(m\pi x_2) dx_2 = \begin{cases} 0 & \text{for} \quad l \neq m, \\ \dfrac{1}{2} & \text{for} \quad l = m, \end{cases}$$

for any integer l, m. Let us multiply (7.64) for $\sin(l\pi x_2)$ for any integer l and integrate over $0 \leq x_2 \leq 1$. By orthogonality and

$$\int_0^1 \sin(l\pi x_2) dx_2 = \begin{cases} \dfrac{2}{l\pi} & \text{for } l \quad \text{odd}, \\ 0 & \text{for } l \quad \text{even}, \end{cases}$$

we get a decoupled system of ordinary differential equations in the form

$$\gamma_l' + \nu l^2 \pi^2 \gamma_{2k+1} = \begin{cases} \dfrac{8\nu}{l\pi} \sin(\omega t) \text{ for } l \quad \text{odd}, \\ 0 \qquad\qquad \text{for } l \quad \text{even}, \end{cases} \qquad (7.65)$$

with the initial condition $\gamma_l(0) = 0$ (being $u_1(x_2, 0) = 0$). For even l we get the differential equations

$$\gamma' + b\gamma = 0$$

with $b = \nu l^2 \pi^2$. General solution $\gamma = Ce^{-bt}$ reduces to $\gamma = 0$ in view of the initial condition.

For odd index, equations (7.65) are in the form

$$\gamma' + b\gamma = A\sin(\omega t), \tag{7.66}$$

with $A = \dfrac{8\nu}{l\pi}$. As it is well known, the general solution to this problem is obtained by summing a particular solution to the general solution of the homogeneous problem. The right-hand side of (7.66), gives as particular solution $\gamma_{part} = \alpha\sin(\omega t) + \beta\cos(\omega t)$. Constant α and β make γ_{part} solution to (7.66). We find

$$(\alpha\omega + b\beta)\cos(\omega t) + (b\alpha - \beta\omega)\sin(\omega t) = A\sin(\omega t)$$

and consequently

$$\begin{cases} \alpha\omega + b\beta = 0, \\ b\alpha - \beta\omega = A, \end{cases}$$

yielding $\alpha = \dfrac{Ab}{b^2 + \omega^2}$, $\beta = -\dfrac{A\omega}{b^2 + \omega^2}$. Solution to (7.66) is therefore

$$\gamma = Ce^{-bt} + \frac{Ab}{b^2 + \omega^2}\sin(\omega t) - \frac{A\omega}{b^2 + \omega^2}\cos(\omega t).$$

From the initial condition $\gamma(0) = 0$ we find $C = \frac{A\omega}{b^2 + \omega^2}$. Therefore

$$\gamma = \frac{A}{b^2 + \omega^2}\left(\omega e^{-bt} - \omega\cos(\omega t) + \frac{Ab}{b^2 + \omega^2}\right)\sin(\omega t).$$

We finally obtain $u_1 = \displaystyle\sum_{k=0}^{\infty} \gamma_{2k+1}\sin((2k+1)\pi x_2)$ with

$$\gamma_{(2k+1)} = -8\nu\frac{\omega e^{(-\nu(2k+1)^2\pi^2)t} - \omega\cos(\omega t) + \nu(2k+1)^2\pi^2\sin(\omega t)}{(\nu^2(2k+1)^4\pi^4 + \omega^2)(2k+1)\pi}.$$

This is the unsteady counterpart of the Poiseuille solution for a constant-in-space and sinusoidal-in-time pressure drop and takes the name of 2D *Womersley solution*. In fact this problem has been originally solved in 1955 by Womersley for the 3D case of flow in a pipe, see [Wom55] .

Snapshots of the u_1 component of Womersley solution are reported in Fig. 7.13.

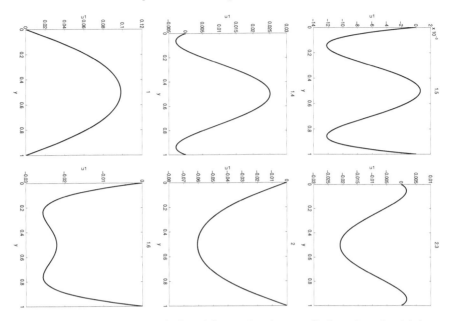

Fig. 7.13 Womersley solution (u_1) at different time instants (indicated on the right)

Numerical approximation

Let us discretize the problem in time. We assume a uniform time step Δt. With usual notation strong formulation of the problem discretized in time with the Implicit Euler method reads: for any $n \geq 0$ find $\mathbf{u}^{n+1}$ and p^{n+1} solving in Ω

$$
\begin{cases}
\dfrac{\mathbf{u}^{n+1} - \mathbf{u}^n}{\Delta t} + \left(\mathbf{u}^{n+1} \cdot \nabla\right) \mathbf{u}^{n+1} - \nu \triangle \mathbf{u}^{n+1} + \nabla p^{n+1} = \mathbf{0}, \\
\nabla \cdot \mathbf{u} = 0,
\end{cases}
\tag{7.67}
$$

with the boundary conditions (7.62) and the initial condition $\mathbf{u}_0 = \mathbf{0}$. Space discretization of the problem yields a nonlinear algebraic system. This can be linearized with Newton method, similar to the one in Exercise 7.1.2. This is an expensive approach, since we have Newton iterations nested in the time loop. An alternative solution is the semi-implicit method obtained by linearizing the nonlinear term with a time extrapolation of the occurrence of $\mathbf{u}^{n+1}$. For instance, since implicit Euler has a discretization error $O(\Delta t)$ and $\mathbf{u}^{n+1} = \mathbf{u}^n + O(\Delta t)$, the following discretization scheme has order 1, having a linear system to solve at each time step

$$
\begin{cases}
\dfrac{\mathbf{u}^{n+1} - \mathbf{u}^n}{\Delta t} + \boxed{(\mathbf{u}^n \cdot \nabla) \mathbf{u}^{n+1}} - \nu \triangle \mathbf{u}^{n+1} + \nabla p^{n+1} = \mathbf{0}, \\
\nabla \cdot \mathbf{u} = 0.
\end{cases}
\tag{7.68}
$$

After the usual space discretization, we finally get: for any $n \geq 0$ find $\mathbf{u}_h^{n+1} \in V_h$ and $p_h^{n+1} \in Q_h$ such that for any $\mathbf{v}_h \in V_h$ and $q_h \in Q_h$ we have

$$\frac{1}{\Delta t}\left(\mathbf{u}_h^{n+1}, \mathbf{v}_h\right) + a\left(\mathbf{u}_h^{n+1}, \mathbf{v}_h\right) + c\left(\mathbf{u}_h^n, \mathbf{u}_h^{n+1}, \mathbf{v}_h\right) + b(\mathbf{v}_h, p_h^{n+1}) =$$

$$\frac{1}{\Delta t}\left(\mathbf{u}_h^n, \mathbf{v}_h\right) + \int_{\Gamma_{N1}} s^{n+1}\mathbf{n} \cdot \mathbf{v}_h d\gamma,$$

$$b(\mathbf{u}_h^{n+1}, q_h) = 0,$$

being $\mathbf{u}_h^0$ an approximation in V_h of $\mathbf{u}_0$. In this formulation $s^{n+1} = 2\nu$ for the steady case and $s^{n+1} = 2\nu \sin(\omega t^{n+1})$ for the sinusoidal unsteady case. With the notation of Exercise 7.1.2, algebraic form of this problem reads

$$\begin{bmatrix} \frac{1}{\Delta t}M + K + C(\mathbf{U}^n) & D^T \\ D & 0 \end{bmatrix} \begin{bmatrix} \mathbf{U}^{n+1} \\ \mathbf{P}^{n+1} \end{bmatrix} = \begin{bmatrix} \mathbf{F}^{n+1} \\ 0 \end{bmatrix}.$$

Numerical results

In snapshot 38 we report the `Freefem++` implementation of the semi-implicit method for the Navier-Stokes problem.

Program 38 - SemiImplicitNavierStokes : Navier-Stokes: semi-implicit Euler time step

```
problem SINavierStokes ([u1,u2,p],[v1,v2,q],solver=GMRES,eps=1.e-10) =
    int2d(Th)(dti*u1*v1 + dti*u2*v2 + nu*( dx(u1)*dx(v1) + dy(u1)*dy(v1)
    dx(u2)*dx(v2) + dy(u2)*dy(v2) ))
  + int2d(Th)(u1last*dx(u1)*v1 + u1last*dx(u2)*v2+
    u2last*dy(u1)*v1 + u2last*dy(u2)*v2)
  + int2d(Th)(0.5*(dx(u1last)+dy(u2last))*(u1*v1))
  + int2d(Th)(0.5*(dx(u1last)+dy(u2last))*(u2*v2))
  - int2d(Th)(p*dx(v1) + p*dy(v2)) - int2d(Th)(dx(u1)*q + dy(u2)*q)
  - int2d(Th)(dti*u1last*v1 + dti*u2last*v2)
  - int1d(Th,4)((2*nu*sin(omega*t)*v1)) + on(1,3,u1=0.,u2=0.);
for (i=1;i<=nmax;i++)
{t=dt*i; SINavierStokes;
 u1last=u1; u2last=u2;}
```

With this code for the Womersley problem, we have the $L^\infty(0, T, \mathbf{L}^2(\Omega))$ velocity errors reported in Fig. 7.14 (logarithmic scale) for $\Delta t = 0.4, 0.2, 0.1$ and 0.05. Expected linear convergence is outlined by the dashed reference line.   ◊

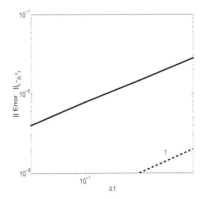

Fig. 7.14 $L^\infty(0, T, \mathbf{L}^2(\Omega))$ velocity errors for the Womersley solution

Exercise 7.2.4. Consider Stokes problem (7.44) with homogeneous boundary conditions ($\mathbf{g} = \mathbf{0}$). Write the time-discrete implicit Euler method and the Chorin-Temam projection method steps. Moreover:

1. Write the equation really solved by the Chorin-Temam velocity field, pointing out the associated error.
2. Show that the Chorin-Temam method is unconditionally stable.
3. Apply this method to the problem where Ω is a circular domain with radius 1, time interval is $(0, 10]$, initial conditions are null for both velocity and pressure and the forcing term in the momentum equation is the vector $\mathbf{f} = [-x_2, x_1]^T$; check the time stability of the method and compute the velocity divergence for different values of h.

Solution 7.2.4.

Numerical approximation

The implicit Euler scheme as shown in Exercise 7.2.1 gives:

$$\begin{cases} \dfrac{1}{\Delta t}\left(\mathbf{u}^{n+1} - \mathbf{u}^n\right) - \nu\triangle\mathbf{u}^{n+1} + \nabla p^{n+1} = \mathbf{f}^{n+1}, \\ \nabla \cdot \mathbf{u}^{n+1} = 0. \end{cases}$$

As we have seen, this is an unconditionally stable scheme.

Chorin-Temam scheme (see e.g. [Qua09], Chapter 10, [Qua93], Chapter 7, [QV94], Chapter 10) moves from the so called *Helmholtz decomposition principle* or *Ladyzhenskaja Theorem* (for the proof see [Qua93]), stating that a vector $\mathbf{w} \in \mathbf{L}^2(\Omega)$ can be split as $\mathbf{w} = \mathbf{u} + \nabla\varphi$ where $\mathbf{u} \in \mathbf{H}^1(\Omega)$ with $\nabla \cdot \mathbf{u} = 0$ and φ regular enough.

Chorin-Temam method exploits this principle for splitting computation of velocity and pressure, with an additional error (*splitting error*) and however a remarkable computational cost reduction. The steps of the methods are:

1. Given a velocity field $\mathbf{u}^n$, compute $\mathbf{w}$ by solving

$$\frac{1}{\Delta t}(\mathbf{w} - \mathbf{u}^n) - \nu\triangle\mathbf{w} = \mathbf{f}^{n+1}, \qquad (7.69)$$

   with boundary conditions $\mathbf{w} = \mathbf{0}$ on $\partial\Omega$.
2. Given $\mathbf{w}$, solve in Ω

$$-\Delta t\triangle p^{n+1} = -\nabla \cdot \mathbf{w}, \qquad (7.70)$$

   with the condition on the boundary

$$\nabla p^{n+1} \cdot \mathbf{n} = \mathbf{0}. \qquad (7.71)$$

3. Velocity field $\mathbf{u}^{n+1}$ is obtained by a direct application of the Helmholtz decomposition principle, i.e.

$$\mathbf{u}^{n+1} = \mathbf{w} - \Delta t\nabla p^{n+1}. \qquad (7.72)$$

Since the pressure in Chorin-Temam method solves a Poisson problem, we need to assume at least $H^1(\Omega)$ regularity.

Let us investigate the error introduced by the splitting (after discretization in space we will have of course discretization errors).Continuity equation is exactly satisfied in the time-discrete/space-continuous problem by construction. In fact, $\mathbf{u}^{n+1}$ is divergence free as a consequence of the Helmholtz decomposition principle.

Splitting error affects therefore momentum equations and boundary conditions. The latter involve both velocity (Helmholtz principle allows the specification of the normal component solely) and pressure (Neumann conditions (7.71) are not required by the original problem). Summing member-wise (7.69) and (7.72) we find the equation actually solved by $\mathbf{u}^{n+1}$,

$$\frac{1}{\Delta t}\left(\mathbf{u}^{n+1} - \mathbf{u}^n\right) - \nu\triangle\mathbf{w} + \nabla p^{n+1} = \mathbf{f}^{n+1},$$

where it is evident that Laplace operator is applied to $\mathbf{w}$ and not to $\mathbf{u}^{n+1}$. Using (7.72) to eliminate $\mathbf{w}$, we can formally derive the modified momentum equation

$$\frac{1}{\Delta t}\left(\mathbf{u}^{n+1} - \mathbf{u}^n\right) - \nu\triangle\mathbf{u}^{n+1} + \nabla p^{n+1} + \boxed{\nu\Delta t\triangle\nabla p^{n+1}} = \mathbf{f}^{n+1}. \qquad (7.73)$$

Boxed term is the error introduced by the splitting in the Chorin-Temam method. This equation together with the incompressibility condition $\nabla \cdot \mathbf{u}^{n+1} = 0$ is the differential problem actually solved when one uses Chorin-Temam method, with the associated boundary conditions. From (7.73) we deduce that the Chorin-Temam method is *consistent* at the differential level, since the splitting error vanishes when the time step tends to 0.

For proving the stability of the method, let us multiply (7.69) by $\mathbf{w}$ and integrate over Ω. Integrating by parts the Laplace term, exploiting the identity $((\mathbf{w} - \mathbf{u}^n), \mathbf{w}) = \frac{1}{2}||\mathbf{w}||_{L^2}^2 + ||\mathbf{w} - \mathbf{u}^n||_{L^2}^2 - \frac{1}{2}||\mathbf{u}^n||_{L^2}^2$ and the Young inequality, we obtain the estimate

$$||\mathbf{w}||_{L^2}^2 + \Delta t||\nabla\mathbf{w}||_{L^2}^2 = \frac{\Delta t}{2\nu}||\mathbf{f}^{n+1}||_{L^2}^2 + ||\mathbf{u}^n||_{L^2}^2. \tag{7.74}$$

Multiply now (7.72) by $\mathbf{u}^{n+1}$. Since

$$-\Delta t \int_\Omega \nabla p^{n+1}\mathbf{u}^{n+1}\,d\Omega = \Delta t \int_\Omega p^{n+1}\nabla \cdot \mathbf{u}^{n+1}\,d\Omega = 0,$$

we get

$$||\mathbf{u}^{n+1}||_{L^2} \le ||\mathbf{w}||_{L^2}. \tag{7.75}$$

From inequalities (7.74) and (7.75) we finally get

$$||\mathbf{u}^{n+1}||_{L^2}^2 - ||\mathbf{u}^n||_{L^2}^2 \le \frac{\Delta t}{2\nu}||\mathbf{f}||_{L^2}^2. \tag{7.76}$$

Summing for $n = 0, 1, \ldots N - 1$ and exploiting the telescopic property of the sum, we finally obtain

$$||\mathbf{u}^N||_{L^2}^2 \le \frac{\Delta t}{2\nu}\sum_{n=0}^{N-1}||\mathbf{f}^{n+1}||_{L^2}^2 + ||\mathbf{u}^0||_{L^2}^2$$

$$\le \frac{T}{2\nu}\max_{n=0,\ldots,N-1}||\mathbf{f}^{n+1}||_{L^2}^2 + ||\mathbf{u}^0||_{L^2}^2 \tag{7.77}$$

which is the required unconditional stability estimate. It bounds the final solution (and in fact the solution at any time step because of (7.75)) independently of Δt.

<hr>

Numerical results

An implementation of the Chorin Temam method is given in Program 39.

<hr>

Program 39 - ChorinTemam : Stokes problem: Chorin-Temam method

```
problem CT1 ([w1,w2],[v1,v2],solver=GMRES,eps=1.e-10) =
    int2d(Th)(dti*w1*v1 + dti*w2*v2+nu*(dx(w1)*dx(v1)+dy(w1)*dy(v1)
  + dx(w2)*dx(v2) + dy(w2)*dy(v2) ))
  - int2d(Th)(dti*u1last*v1 + dti*u2last*v2)
  - int2d(Th)(f1*v1 + f2*v2)   + on(1,w1=0.,w2=0.);

problem CT2 (p,q,solver=GMRES,eps=1.e-10) =
    int2d(Th)(dt*(dx(p)*dx(q)+dy(p)*dy(q)))
```

```
    + int2d(Th)(dx(w1)*q + dy(w2)*q);

problem CT3 ([u1,u2],[v1,v2],solver=Cholesky,init=1) =
    int2d(Th)(u1*v1 + u2*v2)- int2d(Th)(w1*v1 + w2*v2)
  + int1d(Th,1)(dt*p*(v1*N.x+v2*N.y))
  - int2d(Th)(dt*p*dx(v1) + dt*p*dy(v2));


for (i=1;i<=nmax;i++)
{
  t=dt*i; CT1; CT2; CT3;
  u1last=u1; u2last=u2;
}
```

In order to check the unconditional stability of the Chorin-Temam method,
we solve the problem with a large time step. We collocate 200 nodes on the
external circumference (corresponding approximately to a step $h \approx 0.03$) and
select the time steps $\Delta t = 0.1, 0.2, 1, 5$. Numerical solutions do not blow up,
featuring different discretization errors and however similar solutions. For
instance, in Fig. 7.15 we report isolines of velocity moduli for $\Delta t = 0.1$ (left)
and $\Delta t = 5$ (right).   ◇

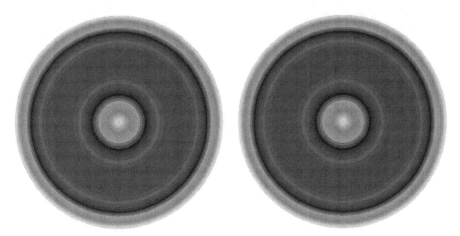

Fig. 7.15 Stokes problem with the Chorin-Temam method: velocity isolines at time $T =$
10 on a mesh with step $h = 0.03$ with time step $\Delta t = 0.1$ (left) and $\Delta t = 5$ (right).
Maximum velocity is 0.456958 on the left, 0.45027 on the right. Despite the large time step
on the right simulation, solution is acceptable, to confirm numerically the unconditional
stability

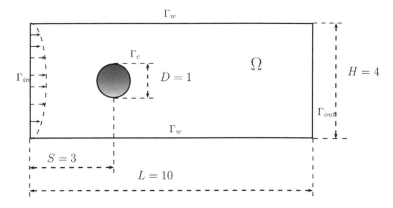

Fig. 7.16 Computational domain for Exercise 7.2.5

Exercise 7.2.5. Consider the domain Ω represented in Fig. 7.16. The cylinder in the domain is fixed. We assume that an incompressible fluid (described by the Navier-Stokes equations) is entering the channel through section Γ_{in} of the boundary with a parabolic profile. On the downstream section Γ_{out} we assume homogeneous Neumann conditions to hold. On the walls of the channel Γ_w as well as on the cylinder surface Γ_c we assume homogeneous Dirichlet conditions. Write a code for solving the problem with the *incremental Chorin-Temam* method. Assume in particular that the viscosity of the fluid is $\nu = 0.005$, the maximum inflow velocity is $U_m = 1.5$ and simulate the flow over 100 time units. As initial conditions, take the solution of the steady Stokes problem in the same domain. Comment on the result.

Solution 7.2.5.

Numerical approximation

As we have seen in the previous exercise, Chorin-Temam method relies upon the Helmholtz decomposition principle. In the present case, we have two differences in comparison with Exercise 7.2.4.

1. On Γ_{out} we have homogeneous Neumann conditions. The original Chorin-Temam method was conceived for Dirichlet problems. Velocity Dirichlet conditions give rise to Neumann condition for the pressure field computed in the second step of the method. When Neumann conditions are prescribed for the velocity, the corresponding conditions for the pressure problem are not trivially derived by the Helmholtz decomposition principle. We need to resort to approximate conditions. A possible empirical choice is to assume

Dirichlet conditions for the pressure on the portions of the boundary where Neumann conditions are prescribed.

2. In the incremental version of the Chorin-Temam method, the pressure at time step $n + 1$ is split into two contributions, $p^{n+1} = \delta p + \sigma(p, n)$, where $\sigma(p, n)$ is an extrapolation of the pressure based on the available solution at the previous time steps $n, n - 1, \ldots$ A typical choice is $\sigma(p, n) = p^n$. The "actual" unknown is δp. The advantage of this approach is that the error induced by empirical conditions is confined to the pressure increment rather than affecting the pressure field directly.

The incremental Chorin-Temam scheme for the time-discrete problem at hand is therefore given at time t^{n+1} by the following steps. Let Δt be the (constant) time step.

1. *Intermediate velocity* $\mathbf{w}$ *computation*:

$$\frac{\mathbf{w} - \mathbf{u}^n}{\Delta t} - \nu \Delta \mathbf{w} + \mathbf{u}^n \cdot \nabla \mathbf{w} = \nabla p^n \tag{7.78}$$

$$\mathbf{w} = \mathbf{0} \quad \text{on } \Gamma_{in,w,c}, \quad \nu \nabla \mathbf{w} \cdot \mathbf{n} = p^n \quad \text{on } \Gamma_{out}.$$

2. *Pressure increment* δp *computation*:

$$\Delta t \Delta \delta p = \nabla \cdot \mathbf{w} \tag{7.79}$$

$$\nabla \delta p \cdot \mathbf{n} = 0 \quad \text{on } \Gamma_{in,w,c}, \quad \delta p = 0 \quad \text{on } \Gamma_{out}.$$

3. *End-of-step computation of* $\mathbf{u}^{n+1}$ *and* p^{n+1}:

$$\mathbf{u}^{n+1} = \mathbf{w} - \Delta t \delta p \tag{7.80}$$

$$p^{n+1} = p^n + \delta p.$$

For the space discretization, we select a pair of finite dimensional spaces (V_h, Q_h) that fulfills the inf-sup condition. In particular, here we resort to the so called (P^1bubble-P^1) elements. This means that each velocity component v_i is assumed to be in a space of piecewise linear functions augmented by a cubic function vanishing on the boundary of each element (this justifies the term "bubble"), i.e. $v_i(x, y) = p_{1,i}(x, y) + \alpha_i \beta(x, y)$ where $p_{1,i}$ is a linear function on $\mathcal{K}$, α is a constant and $\beta(x, y)$ vanishes on $\partial \mathcal{K}$. Numerical pressure field is piecewise linear. This pair is proven to be inf-sup compatible.

We report the implementation of the method in Program 40. In the mesh generation instruction

```
mesh Th=buildmesh(a(50)+b(20)+c(50)+d(20)+cyl(-50));
```

notice the negative sign in `cyl(-50)`, prescribing that the domain to be meshed is external to the cylinder.

As done in the previous exercise, the projection step for the end-of-step velocity computation is carried out referring to a weak formulation.

> **Program 40 - Incremental ChorinTemam** : Navier-Stokes problem: Incremental Chorin-Temam method

```
border a(t=0.,10.)x=t;y=0.;label=1;
border b(t=0.,4.)x=10.;y=t;label=2;
border c(t=10.,0.)x=t;y=4.;label=1;
border d(t=4.,0.)x=0.;y=t;label=4;
border cyl(t=0.,2*pi)x=3+0.5*cos(t);y=2.+0.5*sin(t);label=1;
mesh Th=buildmesh(a(50)+b(20)+c(50)+d(20)+cyl(-50));
fespace Vh2(Th,P1b), Qh(Th,P1);
Vh2 u2,v2,u2last,w2,u1,v1,u1last,w1;
Qh dp=0,p,q;
real H=4, nu=0.005, Umax=1.5, toll=1.e-6, Tfin=100.0,dt=0.1,dti=1./dt,t=0.;
int i,nmax=Tfin*dti;
func uin = (H-y)*y*4*Umax/H^2;

problem CT1 ([w1,w2],[v1,v2],solver=GMRES,eps=1.e-6) =
    int2d(Th)(dti*w1*v1 + dti*w2*v2
    + nu * ( dx(w1)*dx(v1) + dy(w1)*dy(v1)+  dx(w2)*dx(v2) + dy(w2)*dy(v2) ))
  + int2d(Th)(u1last*dx(w1)*v1+u1last*dx(w2)*v2)
  + int2d(Th)(u2last*dy(w1)*v1+u2last*dy(w2)*v2)
  - int2d(Th)(dti*u1last*v1 + dti*u2last*v2)
  - int2d(Th)(p*dx(v1) + p*dy(v2))
  + on(1,w1=0.,w2=0.) + on(4,w1=uin,w2=0.);

problem CT2 (dp,q,solver=GMRES,eps=1.e-6) =
    int2d(Th)((dx(dp)*dx(q)+dy(dp)*dy(q)))
  + int2d(Th)(dti*dx(w1)*q + dti*dy(w2)*q)
  + on(2,dp=0.);

problem CT3 ([u1,u2],[v1,v2],solver=Cholesky,init=1) =
    int2d(Th)(u1*v1 + u2*v2) - int2d(Th)(w1*v1 + w2*v2)
  + int1d(Th,1)(dt*dp*(v1*N.x+v2*N.y))
  + int1d(Th,2)(dt*dp*(v1*N.x+v2*N.y))
  + int1d(Th,4)(dt*dp*(v1*N.x+v2*N.y))
  - int2d(Th)(dt*dp*dx(v1) + dt*dp*dy(v2));

// Initialization (Initial conditions)
solve Stokes ([u1,u2,p],[v1,v2,q],solver=GMRES,eps=1.e-6) =
  int2d(Th)(nu*( dx(u1)*dx(v1)+dy(u1)*dy(v1)+dx(u2)*dx(v2)+dy(u2)*dy(v2) )
    - p*dx(v1) - p*dy(v2) - dx(u1)*q - dy(u2)*q)
  + on(1,u1=0.,u2=0.)   + on(4,u1=uin,u2=0.);

u1last=u1;u2last=u2;dp=0.;

for (i=1;i<=nmax;i++)
{
 t=dt*i;
CT1;
CT2;
CT3;
p=p+dp; dp=0.;
  u1last=u1;u2last=u2;
}
```

Numerical results

The flow past a circular cylinder is a fascinating test problem. A excellent description of the phenomena involved can be found in [Van82]. For the conditions prescribed in our problem, the flow after the cylinder is not asymptotically tending to a steady state (as it happens at low Reynolds number, when the boundary conditions are independent of time) but presents regular oscillations of the wake. This is clearly evident from the Figs. 7.17, 7.18, 7.19 and 7.20 illustrating the velocity fields at different instants of the simulation. At larger Reynolds number we have transition to turbulence and specific numerical methods (which normally require a much finer mesh) need to be adopted for capturing the flow dynamics accurately.

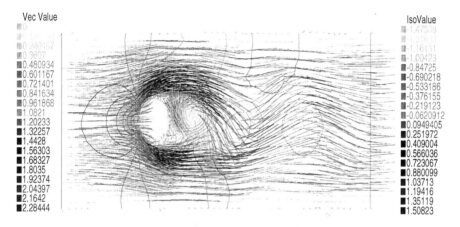

Fig. 7.17 Flow past a cylinder simulation: velocity and pressure isolines at $t = 85$ time units

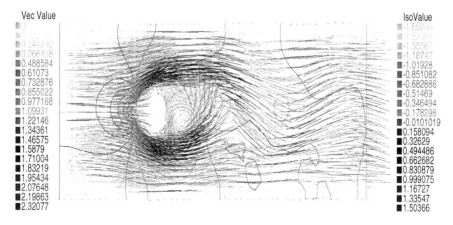

Fig. 7.18 Flow past a cylinder simulation: velocity and pressure isolines at $t = 90$ time units

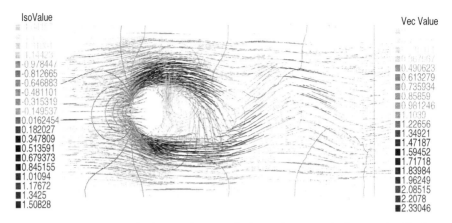

Fig. 7.19 Flow past a cylinder simulation: velocity and pressure isolines at $t = 92$ time units

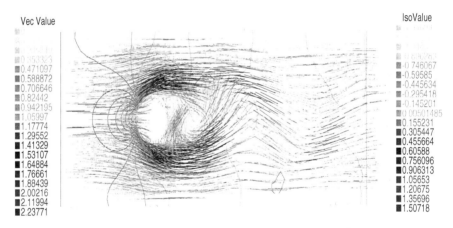

Fig. 7.20 Flow past a cylinder simulation: velocity and pressure isolines at $t = 98$ time units

For illustration purposes, we report in Fig. 7.21 the streamlines in a 3D simulation of the flow past a cylinder carried out with LifeV. The test case features Re = 500 and the streamlines are displayed at $t = 3.7$ and $t = 4.6$, respectively. This test has been proposed as a benchmark for numerical solvers in Computational Fluid Dynamics [STD+96]. The C++ code for this simulation is available at the website of LifeV.   ◇

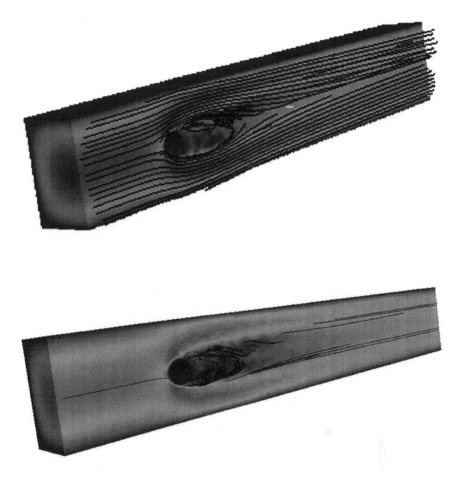

Fig. 7.21 Simulations of flow past a circular cylinder in 3D with the LifeV code. Streamlines at $t = 3.7$ time units (top) and $t = 4.6$ time units

Part IV
Appendices

A

The treatment of sparse matrices

This appendix is intended to readers interested in the numerical implementation of the methods discussed in the book. It aims to review how matrices originating from the discretization of a differential problem by the finite element method (or finite volumes/differences) can be handled by a computer program. Indeed, in real life computations the discretization of PDEs by any of the cited methods leads to the solution of large systems of linear equations whose matrix is sparse. And the efficient storage of sparse matrices requires to adopt special techniques.

We will also recall some practical strategies for dealing with Dirichlet-type conditions in finite element codes.

As far as numerical techniques for solving linear systems are concerned, the reader may refer to the vast literature on the matter, for instance [GV96, QSS00, Saa03, Sha08].

A matrix is *sparse* if it "contains a large number of zeros". Better said, if the number of non-null entries (also called non-zeros) is $O(n)$. This means that the average number of non-zero entries in each row is bounded independently from n. Indeed, what is important is that the location of the zero elements is known a-priori, so we can avoid reserving storage for them. A non-sparse matrix is also said *full*: obviously here the number of non-zero elements is $O(n^2)$.

A.1 Storing techniques for sparse matrices

The distribution of non-zero elements of a sparse matrix may be described by the *sparsity pattern*, defined as the set $\{(i,j) : A_{ij} \neq 0\}$. Alternatively, one may consider the matrix graph, where nodes i and j are connected by an edge if and only if $A_{ij} \neq 0$.

A representation of the pattern can be obtained through the MATLAB command spy (see Fig. A.1 for an example). The sparsity of a finite element matrix is a direct consequence of the small-support property of the finite

Formaggia L., Saleri F., Veneziani A.: Solving Numerical PDEs: Problems, Applications, Exercises. DOI 10.1007/978-88-470-2412-0_A, © Springer-Verlag Italia 2012

element basis functions. Thus, the sparsity pattern depends on the topology of the adopted computational grid, on the kind of finite element chosen and on the indexing of the nodes. It is completely known before the actual construction of the matrix. Therefore, the matrix can be stored efficiently by excluding the terms that are *a-priori* zero.

The use of adequate storage techniques for sparse matrices is fundamental, especially when dealing with large-scale problems typical of industrial applications. Let us make an example. Suppose we want to solve the Navier-Stokes equations on a two-dimensional grid formed by 10.000 vertexes with finite elements P^2-P^1(and this is a rather small problem!). By using the results of Exercise 2.2.4 and the relations of (2.10) we deduce that the number of degrees of freedom is around 10^5 for the pressure and 4×10^5 for each component of the velocity. The associated matrix will then be 90000×90000. If we had to store all 8.1×10^9 coefficients, using the usual double precision (8 bytes to represent each floating point number), around 60 *Gigabytes* would be necessary! This is too much even for a very large computer. Modern operative systems are able to employ areas larger than the available RAM by using the technique of virtual memory (also known as "*paging*"), which saves part of the data on a mass storage device (typically the hard disc). However, this does not solve our problem because paging is extremely inefficient.

In case of a three-dimensional problem the situation becomes even worse, since the number of degrees of freedom grows very rapidly as the grid gets finer, and nowadays it is customary to deal with millions of degrees of freedom.

Therefore to store sparse matrices efficiently we need data formats that are more compact than the classical table (*array*). The adoption of sparse formats, though, may affect the speed of certain operations. Indeed with these formats we cannot access or search for a particular element (or group of elements) directly, as happens with **array**, where the choice of two indexes i and j allows to determine directly where in the memory the wanted coefficient A_{ij} is located[1].

On the other hand, even if the operation of accessing an entry of a matrix in sparse format (like a matrix-vector multiplication) turns out to be less efficient, by adopting a sparse format we will nevertheless access only non-zero elements, thus eschewing futile operations. That is why, in general, the sparse format is preferable in terms of computing time as well, as long as the matrix is sufficiently sparse (and this is usually the case for finite element, finite volume and finite difference descretizations).

---

[1] The efficiency in accessing and browsing an *array* actually depends on the way the matrix is organized in the computer memory and on the operating system's ability to use the processor's *cache* memory proficiently. To go into details is beyond the scope of the present book, but the interested reader may refer to [HVZ01], for example.

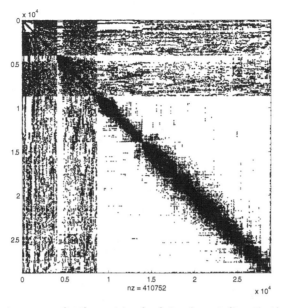

Fig. A.1 Sparsity pattern for the matrix of a finite-element discretization of a vectorial problem in three dimensions. The number of elements is around 9×10^8, of which only about 4.2×10^4 non-zero. The pattern was obtained with MATLAB's **spy** command

We can distinguish different kinds of operations on a matrix, the most important ones being:

1. *accessing a generic element*: this is sometimes called "random access";
2. *accessing the elements of a whole row*: important when multiplying a matrix by a vector;
3. *accessing the elements of a whole column*, or equivalently, of a row in the transpose matrix. This is relevant for operations such as symmetrizing the matrix after imposing Dirichlet conditions, as we will see in Section A.2;
4. *adding a new element to the matrix pattern*: this is no major issue if one builds the pattern at the beginning, and does not change it during the computations. It becomes critical if the pattern is not known beforehand or it can change throughout the computations. This happens for instance with grid adaptation techniques.

It is important to characterize formats for sparse matrices by the computational cost of these operations and by how the latter depends on the matrix size. That different formats exist for sparse matrices is due, historical reasons aside, precisely to the fact that there is no format that is simultaneously optimal for all above operations, and be at the same time efficient in terms of storage capacity.

In the sequel we will review the most common formats, including those used by MATLAB, FREEFEM and some important linear algebra libraries,

like SPARSEKIT [Saa90], PETSC [BBG+01], UMFPACK [DD97, Dav04] or
AztecOO [HBH+05, Her04]. For completeness, we also mention a document
describing the HARWELL-BOEING format [DGL92]. This is not so much a
format for storage on a computer, but rather one meant for writing and
reading sparse matrices on files. The reader interested in software and tools for
operating on large matrices and related examples and bibliography may refer
to the *Matrix Market* web site (`http://math.nist.gov/MatrixMarket`).

We have to remark that square matrices generated by finite-element codes
have certain fixed features "by construction":

1. even if the matrix is not symmetric, its sparsity pattern is. This is because
   an element A_{ij} is in the pattern if the intersection of the support of the
   basis functions associated to nodes i and j has a non-zero measure. And
   this is obviously a symmetric property;
2. diagonal elements are in most of the cases non-zero, so we can assume
   they belong to the pattern.

In fact, in a finite element matrix (i, j) belongs to the pattern if nodes i and j
share a common element. Note that by using this definition it is possible that
we allocate storage for elements which may eventually be zero: we exclude
only elements which are *a-priori* zero. We also have to point out that not
always the matrices of concern are square: think of the matrices D and D^T
of Chapter 7. Some of the formats we will describe are suitable only for the
square case, and cannot be employed in general.

As reference example we will consider the matrix that might have arisen
using linear finite elements on the grid of Fig. A.2, left. The pattern of this
matrix is shown on the right. In particular, the matrix A in "full" format

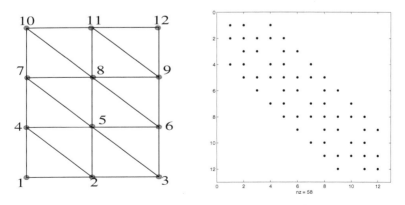

Fig. A.2 Example of grid with linear finite elements and pattern of the associated matrix.
Note that the pattern depends on the numeration chosen for the notes

(*array*) could be

$$
A = \begin{bmatrix}
101. & 102. & 0. & 103. & 0. & 0. & 0. & 0. & 0 & 0. & 0. & 0. \\
104. & 105. & 106. & 107. & 108. & 0. & 0. & 0. & 0. & 0. & 0. & 0. \\
0. & 109. & 110. & 0. & 111. & 112. & 0. & 0. & 0. & 0. & 0. & 0. \\
113. & 114. & 0. & 115. & 116. & 0. & 117. & 0. & 0. & 0. & 0. & 0. \\
0. & 118. & 119. & 120. & 121. & 122. & 123. & 124. & 0. & 0. & 0. & 0. \\
0. & 0. & 125. & 0. & 126. & 127. & 0. & 128. & 129. & 0. & 0. & 0. \\
0. & 0. & 0. & 130. & 131. & 0. & 132. & 133. & 0. & 134. & 0. & 0. \\
0. & 0. & 0. & 0. & 135. & 136. & 137. & 138. & 139. & 140. & 141. & 0. \\
0. & 0. & 0. & 0. & 0. & 142. & 0 & 143. & 144. & 0 & 145. & 146. \\
0. & 0. & 0. & 0. & 0. & 0. & 147. & 148. & 0. & 149. & 150. & 0. \\
0. & 0. & 0. & 0. & 0. & 0. & 0. & 151. & 152. & 153. & 154. & 155. \\
0. & 0. & 0. & 0. & 0. & 0. & 0. & 0. & 156. & 0 & 157. & 158.
\end{bmatrix}, \quad (A.1)
$$

where the values of the matrix elements are not relevant for this discussion, and indeed they have been just made up to allow to identify them easily.

In the sequel n will always be the matrix' size, nz the number of non-zero entries. Moreover, we will adopt the convention of indexing entries of matrices and vectors (arrays) starting[2] from 1. To estimate how much memory the matrix occupies we have assumed an integer occupies 4 bytes, and a real number (floating point representation) 8 bytes (double precision)[3]. Hence storing the matrix of Fig. A.1, which has $n = 12$ and $nz = 58$, would require $12 \times 12 \times 8 = 1152$ bytes if stored as an *array*. At last, A_{ij} will denote the entry of the matrix A on row i and column j.

A.1.1 The COO format

The format by *coordinates*, (*COO*rdinate format) is conceptually the simplest, even though it is poorly efficient in terms of both memory space and access to a generic element.

This format uses three arrays which we denote I, J and A. The first two describe the pattern: precisely, in the generic kth place of I and J we store the row and column indexes of the coefficient whose value is stored at the same position in A. Hence I, J and A all have as many elements as the number of non-zero elements nz.

---

[2] Some programming languages (e.g., C and C++) number arrays from 0, so to use this convention it suffices to subtract 1 from our indexes.

[3] On a 64 bit architecture also the integers may use up 8 bytes.

In this way the space occupied is $(4 + 4 + 8) \times nz$ bytes. For the matrix A in (A.1), a possible coding in COO format reads

$$I = [1, 1, 1, 2, 2, 2, 2, 2, 3, 3, 3, 3, 4, 4, 4, 4, 4, 5, 5, 5, 5, 5, 5, 5, 6, 6, 6, 6, 6,$$
$$7, 7, 7, 7, 7, 8, 8, 8, 8, 8, 8, 8, 9, 9, 9, 9, 9, 10, 10, 10, 10, 11, 11, 11, 11,$$
$$11, 12, 12, 12]$$

$$J = [1, 2, 4, 1, 2, 3, 4, 5, 2, 3, 5, 6, 1, 2, 4, 5, 7, 2, 3, 4, 5, 6, 7, 8, 3, 5, 6, 8, 9, 4,$$
$$5, 7, 8, 10, 5, 6, 7, 8, 9, 10, 11, 6, 8, 9, 11, 12, 7, 8, 10, 11, 8, 9, 10, 11, 12,$$
$$9, 11, 12]$$
$$\text{(A.2)}$$

$$A = [101., 102., 103., 104., 105., 106., 107., 108., 109., 110., 111., 112., 113.,$$
$$114., 115., 116., 117., 118., 119., 120., 121., 122., 123., 124., 125., 126.,$$
$$127., 128., 129., 130., 131., 132., 133., 134., 135., 136., 137., 138., 139.,$$
$$140., 141., 142., 143., 144., 145., 146., 147., 148., 149., 150., 151., 152.,$$
$$153., 154., 155., 156., 157., 158.] ,$$

requiring 928 bytes. Clearly, the three arrays can contain the same elements in different order. This format does not guarantee rapid access to an element, nor to rows or columns. Finding the generic element of the matrix from the row and column indexes normally requires a number of operations proportional to nz. In fact, it is necessary to go through all elements of I and J until one hits those indexes, using expensive comparison operations. There is a way, though at a higher storing price, to use specific techniques to store the indexes in special search data structure, and reduce the cost to $\mathcal{O}(\log_2(nz))$.

The operation of multiplying a matrix and a vector can be done directly, by running through the elements of the three arrays. We show a possible code for the product $\mathbf{y} = \mathbf{A}\mathbf{x}$ using the MATLAB syntax[4]

```
y=zeros(nz,1);
for k=1:nz
 i=I(k); j=J(k);
 y(i)=y(i) + A(k)*x(j);
end
```

The additional cost of this operation, compared to the analogue for a full matrix, depends essentially on *indirect addressing*: accessing y(i) requires first of all to access I(k). Furthermore, the access and update of the arrays x and y does not proceed by consecutive elements, a fact that would greatly reduce the possibility of optimizing the use of the processor's *cache*. Recall, however, that now we operate only on non-zero elements, and that, in general, $nz << n^2$.

An advantage of this format is that it is easy to add a new element to the matrix. In fact, it is enough to add a new entry to the arrays I, J and A. That

---

[4] We use MATLAB syntax for simplicity, yet normally these operations are coded in a compiled language, like C of FORTRAN, for efficiency reasons.

is why COO is often used when the pattern is not known a priori. Obviously, to do so, it is necessary to handle memory allocation in a suitable dynamical way.

A generalization of the COO format uses an *associative array* or a *hash table* to construct the map $(i, j) \rightarrow A_{ij}$. In the C++ language, for instance, one may adopt the `map` container of the standard library for this purpose [SS03]. In some linear algebra packages, like `Eigen` (`xxx.eigen.org`) or `Aztecoo` for instance, it is possible to build a sparse matrix dynamically, and in this case a "COO-type" format is used internally. When one knows that the pattern will not change anymore, the matrix can be "finalized" with a conversion to a more efficient, yet more static, format.

A.1.2 The *skyline* format

The format called *skyline* was among the first used to store matrices arising from the method of finite elements. The idea, schematically depicted in Fig. A.3, left, is to store the blue area formed, on each row, by the elements between the first and last non-zero coefficient. It is clear that this forces to store some null entries, in general. This extra cost will be small if the matrix has non-zero entries clustered around the diagonal. Indeed, algorithm have been developed, the most known one being probably the Cuthill-McKee algorithm, to cluster non zero elements by permuting the rows and columns of the matrix, see [Saa03] for details,

We will explain how this format applies to symmetric matrices, and then generalize it.

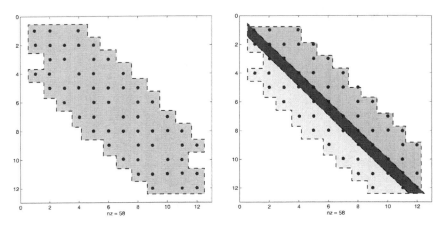

Fig. A.3 Skyline of a matrix (right). On the left, the decomposition in lower triangular, diagonal and upper triangular parts

Skyline for symmetric matrices. If a matrix is symmetric we can store
only its lower triangular part (diagonal included). Or we can store the diagonal on an auxiliary array and treat the off-diagonal entries separately. The
latter choice has the advantage of allowing the direct access to the diagonal
elements. If we opt for this solution, the skyline format is given by three
arrays, D, I and AL. In D we store diagonal entries, in AL all skyline elements
in succession and row-wise (except the diagonal), i.e. the light-coloured area.
This can clearly include null coefficients. The kth component of the array
I tells (technically, "*points to*") where the $(k + 1)$th row of AL begins: all
elements of AL from position I(k) to I(k+1)-1 are the off-diagonal elements
belonging to row $k + 1$, in increasing column order. In this way the first row
is not stored, since it only has the diagonal element, I(k) points to the first
non-zero element on the $(k+1)$th row, I(k+1)-1 points to the element $A_{k+1,k}$,
and the difference I(k+1)- I(k) tells how many off-diagonal elements on row
$k + 1$ belong to the skyline. A quick computation allows to verify that the
first non-zero element on row $k > 1$ is the one on column k-I(k) - I(k-1).

Supposing, for example, we wish to store the symmetric matrix constructed
from the lower triangular part in A as of (A.1), corresponding to the Matlab
instructions tril(A)+tril(A,-1)'. Then

$$D = [101., 105., 110., 115., 121., 127., 132., 138., 144., 149., 154., 158.]$$

$$I = [1, 2, 3, 6, 9, 12, 15, 18, 21, 24, 27, 30]$$

$$AL = [104., 109., 113., 114., 0., 118., 119., 120., 125., 0., 126., 130., 131., 0.,$$
$$135., 136., 137., 142., 0., 143., 147., 148., 0., 151., 152., 153., 156., 0., 157.].$$

Note that in the nth place of the array I we have left a pointer at the
beginning of an hypothetical second row. In this way I(n) -1 is the total
number of elements in the skyline. Moreover, we can compute the number of
skyline elements using I(n) $-$ I(n-1), for the last row as well. The product
$\mathbf{y} = \mathbf{A}\mathbf{x}$ is computed as follows (MATLAB syntax),

```
y=D.*x;
for k=2:n
 nex = I(k)-I(k-1);
 ik  = I(k-1):I(k)-1;
 jcol= k-nex:k-1;
 y(k)   = y(k)+dot(AL(ik),x(jcol));
 y(jcol)= y(jcol)+AL(ik)*x(k);
end
```

Observe the need to operate symmetrically on rows and columns to exploit
the fact that only the lower triangular part was stored in AL.

As said, the memory needed to store the matrix in this format, depends on
how effectively the skyline reproduces the actual pattern. In the case under
scrutiny the array AL contains 29 real numbers, to which we add the fixed

length n of the arrays D and I, in our case 12. The first has real numbers, the second integers, so storing our matrix requires 376 bytes. A direct comparison with the *COO* format is not possible as in the previous section's example the matrix was non-symmetric. One can exploit the possible symmetry also with *COO* by storing only the lower triangular part (the multiplication algorithm between matrix and vector changes accordingly). In this case, with *COO* we would store 35 coefficients and use 560 bytes. So *skyline* apparently looks more convenient: but if the coefficients are not well clustered around the diagonal the memory space used by *skyline* would increase quickly as n increases.

Skyline for general matrices. As with non-symmetric matrices in the general case, a reasonable way to proceed is to split A into the diagonal D, the strictly lower triangular part E and strictly upper triangular part F . Using the Matlab syntax, these matrices would be defined as D=diag(diag(A)); E=tril(A,-1); F=triu(A,1). As the pattern of A is symmetric, the *skyline* of E coincides with that of F^T, hence we will store E and F^T (and D) with the previous technique. In this way there is no need to duplicate the array I, this being the same for both triangular parts. Therefore we can use two arrays of length n, still denoted D and I, and two real-valued arrays of length equal to the skyline dimension, called E and FT (containing E and F^T respectively). In the example, this would necessitate of 608 bytes.

$$D = [101., 105., 110., 115., 121., 127., 132., 138., 144., 149., 154., 158.],$$

$$I = [1, 2, 3, 6, 9, 12, 15, 18, 21, 24, 27, 30],$$

$$E = [104., 109., 113., 114., 0., 118., 119., 120., 125., 0., 126., 130., 131., 0., \\ 135., 136., 137., 142., 0., 143., 147., 148., 0., 151., 152., 153., 156., 0., 157.],$$

$$FT = [102., 106., 0., 107., 103., 116., 111., 108., 122., 0., 112., 0., 123., 117., \\ 133., 128., 124., 139., 0., 129., 0., 140., 134., 150., 145., 141., 155., 0., 146.].$$

The product matrix-vector $\mathbf{y} = A\mathbf{x}$ now reads

```
y=D.*x;
for k=2:n
  nex  = I(k)-I(k-1);
  ik   = I(k-1):I(k)-1;
  jcol = k-nex:k-1;
  y(k)    = y(k)+dot(E(ik),x(jcol));
  y(jcol)= y(jcol)+FT(ik)*x(k);
end
```

We should observe that in this format the access to diagonal entries is direct, and the cost of extracting a row is independent of the matrix' size.

Indeed, the fact that the data relative to a row are stored consecutively in the memory allows the system to optimize the processor's *cache* memory when multiplying a matrix by a vector. In the example above `icol` and `ik` contain all indexes corresponding to the columns of row `k`, so the scalar product `dot(E(ik),x(jcol))` and the multiplication vector-constant `FT(ik)*x(k)` can be optimized[5].

The extraction of column is, vice versa, an expensive operation that requires many comparisons, and whose cost grows linearly in n.

Being able to access diagonal entries directly has certain advantages. For instance we will see that methods to impose essential boundary condition based on penalization (Section A.2.2) only need the access to diagonal elements.

A.1.3 The CSR format

The problem with the skyline format is that the memory used depends on the numeration of elements and is in general impossible to avoid the unnecessary storage of zero elements. Renumeration algorithm may be very inefficients for large scale problems.

For these reasons other formats have been developed that render memory space independent of the numeration of degrees of freedom. The format `CSR` (*Compressed Sparse Row*) is one of them, and can be seen as a compressed version of `COO` that renders it more efficient, but also as an improved *skyline*, that stores non-zero elements only. The format uses three arrays:

1. The real-valued array `A` of length nz, containing the non-zero entries of the matrix, ordered row-wise: in the example it coincides with the array `A` written in (A.2).
2. The integer-valued array `J` of length nz, whose entry $J(k)$ indicates the column of the element `A(k)`. In our case it coincides with the `J` in (A.2).
3. The array `I` of length n containing "pointers" to the rows. Essentially, `I(k)` gives the position where the kth row in `A` and `J` begins, as shown in Fig. A.4.

In many practical applications, the array `I` is of length $n + 1$ so that the number of non-zero entries on row k is always `I(k+1)-1-I(k)`. To make this hold the last element `I(n+1)` will contain $nz + 1$ and in this way we also have that nz=`I(n+1)-I(1)`. The format CSR stores the matrix using $4 \times (nz + n + 1) + 8 \times nz$ bytes. For the example we have

$$I = [1, 4, 9, 13, 18, 25, 30, 35, 42, 47, 51, 56, 59] \tag{A.3}$$

while `J` and `A` are the same as in (A.2). Notice again that this particular fact is not general. With COO the order of the indexes in `I`, `J` can be arbitrary.

---

[5] These operations are contained in the library BLAS (Basic Linear Algebra Subroutines), which furnishes highly optimized functions for some elementary matrix operations.

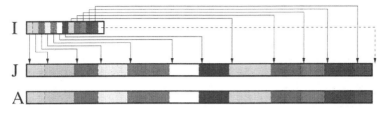

Fig. A.4 The format CSR. The picture refers to the numerical example discussed in the text. In CSR the elements of I point to J and A, respectively telling where the column indexes and the values of a given row begin

The memory space required for the example is of 748 bytes. However, the gain in storage requirement with this format becomes more evident with large sparse matrices.

This format suits square and rectangular matrices alike, and allows a quick extraction the i-th row: it is sufficient to consider the elements of A lying between I(i) and I(i+1)-1. Less immediate is column extraction, which requires localizing on each row the values of J corresponding to the wanted column. If we adopt no particular ordering, the operation has a cost proportionate to nz. If, instead, column indexes of each row in J are ordered, e.g. in increasing order as in our example, with a binary-search algorithm the extraction cost for a column lowers; more precisely, it becomes proportional to $n \log_2(m)$, where m is the mean number of elements on each row. Analogously, the access to a generic element has normally a cost proportional to m, yet if we order columns it reduces to $\log_2 m$.

A further variant is to store in the first element of the part of J corresponding to a given row the index of the diagonal element. In this way A(I(k)) provides the coefficient A_{kk} directly.

The matrix-vector product $\mathbf{y} = \mathbf{A}\mathbf{x}$ is given by

```
y=zeros(n,1);
% y=A(I(1:n)).*x if the  diagonal is stored first
for k=1:n
 ik=I(k):I(k+1)-1;
 % ik=I(k)+1:I(k+1)-1; if the  diagonal is stored
 %                     first
 jcol =J(ik); y(k)=y(k)+dot(A(ik),x(jcol));
end
```

A.1.4 The CSC format

Evidently, there is a corresponding format *CSC* (Compressed Sparse Column) that stores matrices by ordering them column-wise, so it is easy to extract a column as opposed to rows. Here the roles of vectors I and J is exchanged compared with the CSR format. It is the format preferred, for instance, by the UMFPACK library.

When performing matrix-times-vector operations with a sparse matrix ordered by columns it is preferable to compute the result as a linear combination of the columns of the matrix. Indeed, if c_i indicates the i-th column of matrix A we have that $Ax = \sum_i x_i c_i$. Therefore, the matrix-vector product $y = Ax$ on a CSC matrix may be computed as

```
y=zeros(n,1);
for k=1:n
  xcoeff=x(k);
  jk=I(k):I(k+1)-1;
  ik=J(jk);
  y(ik)=y(ik) + xcoeff * A(jk)';
end
```

A.1.5 The MSR format

The format MSR (*Modified Sparse Row*) is a special version of *CSR* for square matrices whose diagonal elements are always contained in the pattern (as it happens in general for matrices generated by finite elements). Diagonal entries can be stored in one single array, since their indexes are implicitly known from their position in the array. As for the *symmetric skyline*, only off-diagonal elements are stored in a special fashion, i.e. through a format akin to *CSR*.

In practice one uses two arrays, which we call V (Values) and B (Bind). In the first n entries of V we store the diagonal. The place $n+1$ in V is left with no significant value (the reason will become clear later). From place $n+2$ onwards off-diagonal elements are stored, row-wise. Hence V has length $nz + 1$. The array B has the same length as V: from $n+2$ to $nz+1$ are the column indexes of the elements stored in the corresponding places in V; the first $n+1$ point to where rows begin in subsequent positions. Therefore B(k), $1 \leq k \leq n$, contains the position *within the same array* B, and correspondingly in V, where the kth row begins to be stored (Fig. A.5, top). More exactly, column indexes of non-zero coefficients of row k will be stored between B(B(k)) and B(B(k+1))-1, while the corresponding values ranges between V(B(k)) and V(B(k+1))-1. The element B(n+1) plays the same role of I(n+1) in the format *CSR*: it points to a hypothetical row $n+1$. In this way nz=B(n+1)-1. The reason for sacrificing the element V(n+1) is now clear: one wants to set up an exact correspondence between the elements of V and B, starting from element $n+2$ till the last. The space needed is $12 \times (nz+1)$ bytes.

Concerning the example, the coding *MSR* reads as follows (the unused element in V is marked with $*$)

$$B = [14, 16, 20, 23, 27, 33, 37, 41, 47, 51, 54, 58, 60,$$
$$2, 4, 1, 3, 4, 5, 2, 5, 6, 1, 2, 5, 7, 2, 3, 4, 6, 7, 8, 3, 5, 8, 9, 4, 5, 8, 10,$$
$$5, 6, 7, 9, 10, 11, 6, 8, 11, 12, 7, 8, 11, 8, 9, 10, 12, 9, 11],$$

$$V = [101., 105., 110., 115., 121., 127., 132., 138., 144., 149., 154., 158., *,$$
$$102., 103., 104., 106., 107., 108., 109., 111., 112., 113., 114., 116., 117.,$$
$$118., 119., 120., 122., 123., 124., 125., 126., 128., 129., 130., 131., 133.,$$
$$134., 135., 136., 137., 139., 140., 141., 142., 143., 145., 146., 147., 148.,$$
$$150., 151., 152., 153., 155., 156., 157.]$$

which occupies 708 bytes.

The format *MSR* turns out to be very efficient in memory terms. It is one of the most "compact" formats for sparse matrices, reason for which it is used in several linear algebra libraries dealing with large problems. As already mentioned, the drawback is that it only applies to square matrices.

The product matrix-vector is coded as

```
y=V(1:n).*x;
for k=1:n
 ik=B(k):B(k+1)-1;
 jcol =B(ik);
 y(k)=y(k)+dot(A(ik),x(jcol));
end
```

As for computational efficiency, its features are similar to those of *CSR*: whereas accessing rows is easy, extracting a column is more expensive an operation, for it requires finding the column index in the array B. Here, too, the cost of an extraction can be reduced to being proportional to $n \log_2 m$ by ordering the columns corresponding to each row and adopting a binary search algorithm (m is still the mean number of columns per row).

We present in the sequel a non-standard variant (that actually works for *CSR* as well), based on adding a third array that allows to access columns in a time lapse that is independent of the sparse matrix size and without a search on indexes (and hence without conditional branches).

A non-standard modification of *MSR*. The modification presented here has been adopted by the serial version of the finite-element library LIFEV ([lif10]), and exploits the fact that matrices coming from the finite-element method have a symmetric pattern. This means that if we run through off-diagonal elements on row k and detect that the coefficient A_{kl} is the pattern (i.e. is non zero), the pattern will also contain A_{lk}, in row l. If the position of A_{lk} in B (and V) is stored in a "twin" array of the part of B from $n + 2$ to $nz + 1$, then we have a structure yielding the elements of a given column. Let us call this array CB (Column Bind): to extract the column in-

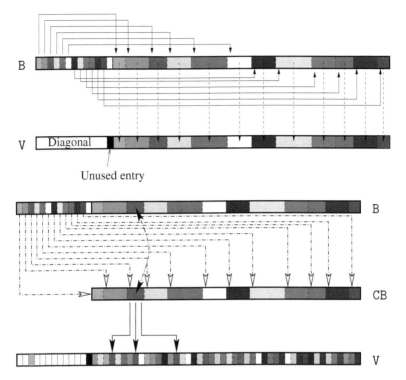

Fig. A.5 Above, the format MSR. The first components of B point (solid arrows) to column indexes contained in the second part of B, which in turn correspond to the array V (dashed arrows). Below: the modified *MSR* format: the array CB comes (dash-dotted arrows) from the first n elements of B. The elements of CB point to the elements of V belonging to a given column. For example, solid arrows denote in V the elements relative to the third column. The respective row indexes of these elements are located in the area corresponding to the second section of B (curved arrow). The elements of the third column are therefore the targets of the arrows, apart the one on the diagonal (which is highlighted in the first section of V)

dexed k it is enough to read the elements of CB between B(k)-(n+1) and B(k+1)-1-(n+1) (subtracting $n + 1$ *shifts* indexes, from those to which B points in V to those in CB). These elements point to the positions of B and V where one can find the corresponding row indexes and the matrix values, respectively.

Basically, the first positions of B point to CB, which in turn points to the off-diagonal entries in B and V (Fig. A.5 bottom). This double-pointing system, though burdensome, allows to access the column of a sparse matrix at a cost that is independent of nz and without demanding conditional branches, provided it is programmed properly.

The structure CB in our case is

$$CB = [16, 23, 14, 20, 24, 27, 17, 28, 33, 15, 18, 29, 37, 19, 21, 25, 34, 38, 41,$$
$$22, 30, 42, 47, 26, 31, 43, 51, 32, 35, 40, 48, 52, 54, 36, 44, 55, 58, 39,$$
$$45, 56, 46, 49, 53, 59, 50, 57].$$

The array B tells us that to extract the elements of the third column, say, we should find the pointers to the column elements in the positions between B(3)-(n+1)=20-13=7 and B(4)-1-(n+1)=22-13=9 in CB ($n + 1 = 13$ is the shift). At these places we read 17, 28, 33 corresponding to the positions in V of 106., 119., 125., that are precisely the off-diagonal elements of column number three.

Compared to MSR, the modified format requires storing $nz - n$ additional integers (the number of non-zero off-diagonal entries), so the memory is now $12 \times (nz + 1) + 4 \times (nz - n)$ bytes.

Again, the advantage of this sparse storage becomes important for larger sparse matrices, as the reader may verify easily.

A.2 Imposing essential boundary conditions

The demand for efficient storage of sparse matrices must come to terms with the need of accessing and manipulating the matrix itself. These operations are especially important to impose Dirichlet-type (essential) boundary conditions. In a finite-element code, in fact, the stiffness matrix is often generated ignoring essential boundary conditions, which are then introduced by modifying the algebraic system suitably. This happens because the assembling operation is characterized by several cycles in which it would not be efficient to introduce tests on the nature of a degree of freedom (whether boundary or not) and on the type of associated boundary condition.

Henceforth we shall denote by $\widetilde{A}$ and $\widetilde{b}$ the matrix and the source term *before* the essential boundary conditions are imposed.

A.2.1 Elimination of essential degrees of freedom

The way to impose Dirichlet conditions that is "closest to the theory" consists in eliminating freedom degrees corresponding to the nodes where the conditions should apply, since there the solution is known.

If k_D is the generic index of a Dirichlet node and g_{k_D} the (known) value of u_h at the node, eliminating the degree of freedom means that:

1. The columns of index k_D of $\widetilde{A}$ are erased, correcting the right-hand side. Better said, the rows of index $k_{nD} \neq k_D$ are "shortened" by eliminating all non-zero coefficients $A_{k_{nD}k_D}$, used to update the right-hand side to $\widetilde{b}_{k_{nD}} = \widetilde{b}_{k_{nD}} - A_{k_{nD}k_D}g_{k_D}$. Therefore the columns corresponding to the Dirichlet nodes are truly *eliminated* from $\widetilde{A}$.

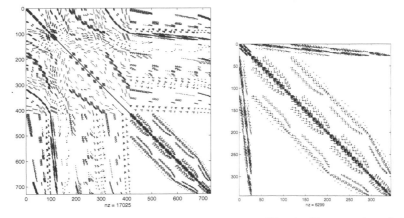

Fig. A.6 Elimination of Dirichlet freedom degrees on a 3D cube (from a real case). The matrix before the boundary conditions (left), and afterwards (right)

2. The rows of index k_D in the matrix and right-hand side are erased from the system, eventually producing a square matrix A and a source **b** with size equal to the problem's effective number of freedom degrees.

Operation 1 coincides actually with lifting the boundary datum, cf. the discussion of Chapter 3.

The only advantage of this procedure is that we eventually have to solve a system with just the "true" unknowns of our problem. However is has many practical downsides. First, the complexity of the implementation, since in general the numbering of the Dirichlet nodes is arbitrary. Furthermore, it alters the pattern, which could be inconvenient in case we wanted to share it between several matrices to save memory space. This happens if the problem has several unknowns, each with its own boundary condition. Eventually, considering that normally one requires to have the solution *at all nodes* with the original numbering for post-processing, one must store the array that allows to recover it.

A.2.2 Penalization technique

If one rates techniques on how involved their programming is, at the opposite end of the scale to the above is the so-called node-based *penalization*. The basic idea is to add a term **hv** to the diagonal elements in $\widetilde{A}$ matching the rows of index k_D relative to Dirichlet degrees, and correspondingly add the term $\mathbf{hv}g_{k_D}$ to the source element k_D.

In this way the equation relative to row k_D becomes

$$\sum_{j=1}^{N} \widetilde{a}_{k_D j} u_j + \mathbf{hv}u_{k_D} = \widetilde{b}_{k_D} + \mathbf{hv}g_{k_D}.$$

If the coefficient **hv** is sufficiently large (**hv** stands for *high value*), the effect of the perturbation is to make the equation an approximation of $\mathbf{hv}u_{k_D} = \mathbf{hv}g_{k_D}$, whose solution is, naturally, $u_{k_D} = g_{k_D}$. Indeed, if a_{K_d} denotes the maximum absolute value of matrix elements in row k_D and $a_{K_d}/\mathbf{hv} < \mathbf{eps}$, **eps** being the machine epsilon number[6], the computer will in fact "see" $\mathbf{hv}u_{k_D} = \mathbf{hv}g_{k_D}$. In this way we impose the wanted condition without changing the problem's dimension nor the matrix pattern.

This approach (adopted by the software FREEFEM++) has simplicity as its asset: the only requirement is the access to diagonal elements. Its weak point rests on the fact that to have an accurate approximation of the boundary datum the value **hv** must be much large enough (**FreeFem** sets to 10^{30} by default). This, generally speaking, will degrade the matrix condition number, since it introduces eigenvalues of order **hv**. However, the new introduced eigenvalues are all clustered around this value, so the situation is better than what one may think at first sight.

Other (more complex) penalization techniques operates on the variational problem. Notably, we mention the Nitsche's method, which provides a penalty formulation which is consistent and maintains optimal convergence rate also for high order elements. More details on the Nitsche's method may be found in [Ste95].

A.2.3 "Diagonalization" technique

A third option, which neither alters the pattern nor necessarily introduces ill-conditioning for the system, is to consider the Dirichlet condition as an equation of the form $\alpha u_{k_D} = \alpha g_{k_D}$ to replace row k_D of the original system. Here, $\alpha \neq 0$ is a suitable coefficient, often taken equal to 1 or to the average of the absolute values of the elements of row k_D (to avoid degrading the condition number). This substitution is performed by setting to zero the row's off-diagonal elements except for the diagonal one, which is set to α, without modifying the sparsity pattern. Accordingly, the corresponding element in the right hand side is set to αg_{k_D}.

The operation requires access to the sole rows, so it is efficient in formats like *COO*, *CSR* or *MSR* (Fig. A.7 left). This approach, that we termed *diagonalization* (not to be confused with the usual meaning in linear algebra), is without doubt a good compromise between easy programming and control of the conditioning of the problem.

Its major fault is to destroy the symmetry of the matrix (if the original matrix was). If one wishes to keep that symmetry (for instance, in order to use a Cholesky decomposition), then it is necessary to modify the columns as well, and consequently the source term. A possible strategy to address this issue is explained in the next section. When using a Krylov-based iterative method

---

[6] **eps** is the largest floating point number so that $1 + \mathbf{eps} = 1$ in floating point arithmetic.

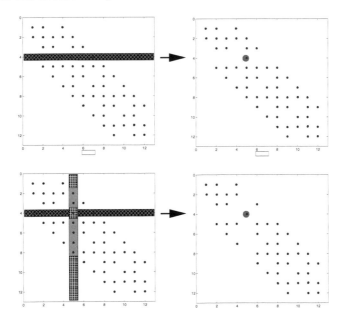

Fig. A.7 Treatment of essential conditions (e.g. on row 4 of the example matrix) using diagonalization: basic version (above) and symmetric one (below). For the symmetric version the column vector (bar the diagonal element) is used to update the source

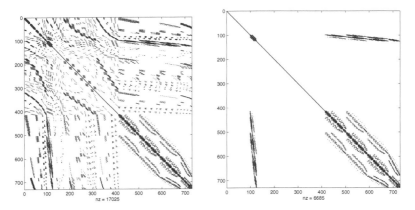

Fig. A.8 The effects of symmetric diagonalization on a 3D grid (from a real case): on the left, before the boundary conditions are imposed; on the right, afterwards

of solution, it is worth noticing that this loss of symmetry does not affect the performances of the solver, as it has been proved in [EG04], Chapter 8.

Symmetric diagonalization. Once we have "diagonalized" in the sense of the previous Section, we can think of modifying the column k_D in a way similar to what seen in Section A.2.1 (Fig. A.7 right): in practice we set to zero the elements $\widetilde{a}_{k_{nD}k_D}$ for all $k_{nD} \neq k_D$, and update the corresponding source term by adding to it $-A_{k_{nD}k_D}g_{k_D}$. The main difference with Section A.2.1 consists in annihilating off-diagonal coefficients in the columns corresponding to Dirichlet nodes, instead of erasing the whole columns of the system.

The technique just described (and adopted by LifeV) attains a most favorable balancing between easy programming and mathematical stability of the algebraic system. The con is that it needs an efficient access to columns. Therefore it can be convenient to use formats that warrant efficient access to columns as well, like the modified *MSR*.

A.2.4 Essential conditions in a vectorial problem

In presence of vectorial problems it can happen to impose essential conditions not on a single component in the vector of unknowns, but rather on a linear combination. Consider for example a Navier-Stokes problem, and suppose we want to impose a velocity condition like $\mathbf{u}^T\mathbf{n} = g$ on the boundary Γ_D, where $\mathbf{n}$ is the normal vector. The operation involves a linear combination of the components of $\mathbf{u}$: in 2D, $u_x n_x + u_y n_y = g$. If the normal is parallel to a coordinate axis we fall back into the case of a single component prescribing the condition. An example is $\mathbf{n} = [1,0]^T$, forcing an essential condition on the first component: $u_x = g$. In general, though, $\mathbf{n}$ is arbitrary.

Let us then see how we can set up the problem by focusing on one boundary node. If the condition applies to more nodes (as normally happens), the procedure described below should be carried out at each node.

Suppose $\mathbf{U} \in \mathbb{R}^{N_h}$ contains all the problem's unknowns (hence, in particular, all the components of the vector), including those relative to the degrees of freedom at which we will impose the boundary condition. We shall only consider conditions that can be written in the form

$$\mathbf{N}^T\mathbf{U} = g, \tag{A.4}$$

where $\mathbf{N} \in \mathbb{R}^{N_h}$. Returning to the example, if we imposed $\mathbf{u}_i^T\mathbf{n} = g$ then $\mathbf{N}$ would have all components null except those corresponding, in $\mathbf{U}$, to the position of the velocity components $u_{i,x}$ and $u_{i,y}$ at node i, where they would equal n_x and n_y, respectively.

For simplicity (and without loss of generality) we shall assume $\mathbf{N}$ to be a unit vector, $\mathbf{N}^T\mathbf{N} = 1$. We will make use of

$$\mathbf{Z} = \mathbf{N}\mathbf{N}^T \in \mathbb{R}^{N_h \times N_h},$$

of components $Z_{ij} = N_i N_j$; this matrix enjoys the following properties[7]:

1. it is symmetric;
2. it has *rank one*, ie the set of $\mathbf{v} \in \mathbb{R}^{N_h}$ such that $\mathbf{Z}\mathbf{v} \neq \mathbf{0}$ is a vector space of dimension one. In fact, $\mathbf{Z}\mathbf{v}$ is the orthogonal projection of $\mathbf{v}$ along $\mathbf{N}$ because, by definition, $\mathbf{Z}\mathbf{v} = (\mathbf{N}^T \mathbf{v})\mathbf{N}$. Therefore $\mathbf{Z}\mathbf{v} = \mathbf{0}$ for all $\mathbf{v}$ orthogonal to $\mathbf{N}$, hence such that $\mathbf{N}^T \mathbf{v} = 0$.

Now take the matrix $\widetilde{\mathbf{A}}$ and the right-hand side $\widetilde{\mathbf{b}}$ of our problem before the boundary condition (A.4) is imposed. To do so we can use Lagrange multipliers. The method consists in adding a further unknown λ and solve

$$\begin{cases} \widetilde{\mathbf{A}}\mathbf{U} + \lambda\mathbf{N} = \widetilde{\mathbf{b}}, \\ \mathbf{N}^T\mathbf{U} = g. \end{cases} \tag{A.5}$$

This problem has an extra unknown, but a series of algebraic manipulations will eliminate λ and reduce it to a system in $\mathbf{U}$ only, of the form $\mathbf{A}\mathbf{U} = \mathbf{b}$ where

$$A = \widetilde{A} - Z\widetilde{A} + Z\widetilde{A}Z, \quad b = \widetilde{b} - Z\widetilde{b} + gZ\widetilde{A}N. \tag{A.6}$$

In addition, one can prove the system can be further simplified to become

$$\left[\widetilde{A} - Z\widetilde{A} + \alpha Z\right]\mathbf{U} = \widetilde{b} - Z\widetilde{b} + g\alpha\mathbf{N}, \tag{A.7}$$

where the parameter α can be chosen so to not ill-condition the system (although, in practice, one chooses $\alpha = 1$ often).

Looking at how matrix and source in (A.7) are built, the operation $\widetilde{A} - Z\widetilde{A}$ is nothing more that a generalized version of the annihilation seen in Section A.2.3. The addition of αZ generalizes the introduction of the diagonal term α, which in the case of a (non-trivial) linear combination of unknowns entails a change in the original pattern.

Likewise, the operations on the source term correspond to replacing the original component of the right-hand side along $\mathbf{N}$ with $g\alpha\mathbf{N}$. Actually, one can check easily that if the components of $\mathbf{N}$ are zero except for $N_i = 1$, the procedure corresponds *exactly* to set to zero the whole ith row except the diagonal term, and modify the source as we saw in Section A.2.3.

This routine has two shortcomings. The first is unavoidable and is due to the distinct patterns of A and $\widetilde{A}$. Usually, the boundary condition (A.4) constrains some components of the solution that are instead uncoupled in the "unconstrained" system. We have seen that the majority of formats for sparse grids are not efficient when the pattern is altered (save for COO, which is however less efficient than other formats). One possibility to steer clear of the issue is to take this kind of boundary conditions into account already in the preliminary phase, when the pattern is identified.

---

[7] In concrete cases neither the matrix Z, nor $\mathbf{N}$, need be constructed explicitly, but they are useful algebraic tools that allow to describe the procedure in a concise and accurate way.

The second fault is related to the fact that A, as of (A.6), just like the matrix of (A.7), is not symmetric in general, even if $\widetilde{A}$ is. The operations considered so far, in fact, correspond to acting only on rows. So, if we want to keep the matrix symmetric it could be necessary to generalize the "symmetric diagonalization", a topic we will not discuss for lack of space.

One final, practical remark is that if one adopts iterative methods to solve the linear system it is unnecessary to construct A explicitly: one can use, instead, the definition in terms of $\widetilde{A}$ and $\mathbf{N}$ directly. What one needs is to implement efficiently the products $\mathbf{N}^T\mathbf{v}$ and $\mathbf{Z}\mathbf{v}$, for any given $\mathbf{v}$.

B

Who's who

Many inequalities, properties and spaces disseminating the text come with name tags referring to mathematicians from several different backgrounds. It can be interesting to put these figures in a historical perspective to get a glimpse, if only superficial, on the advancements made by the mathematical sciences during the last centuries. That is why we have collected at the end of the book the biographical data of some of the most cited people.

Stefan Banach, 1892–1945. Born in Krakow, Banach studied engineering. He obtained an assistant position at the University of Lvov in 1920, after defending the dissertation "On Operations on Abstract Sets and their Application to Integral Equations", which is considered the starting point of functional analysis. In 1924 he became Professor of Mathematics in virtue of his contributions to measure theory. In the subsequent decade Banach made tremendous advances in the theories of integration, measure and vector spaces. He introduced and characterized the notion of complete normed linear space, nowadays known as Banach space in his honor. He died on August 31st 1945 in Lvov, Ukraine.

Augustin Louis Cauchy, 1789–1857. Born a few days after the outbreak of the French Revolution, Cauchy was encouraged to pursue mathematical studies by Lagrange, who was a friend of the family. After becoming a civil engineer he participated in the project of the fleet with which Napoleon wanted to invade England, and held positions in several institutions like the Collège de France and the École Polytechnique. From 1813 he engaged full-time in mathematical research, becoming one of the pioneers of mathematical analysis. Among his enormous contributions to the subject, gathered in 789

Formaggia L., Saleri F., Veneziani A.: Solving Numerical PDEs: Problems, Applications, Exercises. DOI 10.1007/978-88-470-2412-0_B, © Springer-Verlag Italia 2012

articles, we just mention those relative to the definitions of limit and integral, the theory of complex variables and the convergence of infinite series.

Richard Courant, 1888–1972. Born in Lublin, Germany, of a Jewish family, Courant obtained the doctorate in 1910 at Göttingen, immediately becoming Hilbert's assistant. WWI caused a break in Courant's scientific activities. In 1922 he published a first book on functional analysis based on the lectures of Hurwitz, who had died in 1919. In 1924 Courant published with Hilbert an important text on mathematical physics. In 1925 he started working on a second book, while the Mathematical Institute he had founded in Göttingen a few years earlier was taking the first steps. Hitler's rise to power changed Courant's plans, and forced him to abandon Germany for the U.S. where, starting from 1935, he held a permanent position in New York, and where he later founded an Institute of Mathematics on the model of Göttingen's one. The method of finite elements is one of Courant's main contributions, and was already present in embryo in the 1922 volume and in a note dating 1924. The name 'finite elements' is not due to Courant, but appeared first in 1960.

John Crank, 1916–. Born in Hindley, UK, Crank studied at the University of Manchester from 1934 to 1938. After graduation, he worked as a mathematical physicist at the Courtaulds Fundamental Research Laboratory from 1945 to 1957, and then as Professor of Mathematics at Brunel University from 1958 to 1981. His main contribution is to be found in the numerical solution of heat conduction problems; these studies, in collaboration with Phyllis Nicolson, eventually led to the method known as the Crank-Nicolson method.

Johann Peter Gustav Lejeune Dirichlet, 1805–1859. Born in Düren (now Germany, at the time under the Napoleonic empire), Dirichlet completed his undergraduate education in Paris, showing an early inclination towards mathematics. His first published article, related to the celebrated theorem of Fermat, gave him immediate fame. In 1825 he decided to return to Germany, where he received a doctorate ad honorem from the University of Cologne and the Habitation to teach at the University of Breslau. However, he did not settle down and instead embarked on a long tour that brought him to Berlin first, and then Italy where he spent some time. His contributions to mathematics are impressive. In particular, we mention analytical number theory and the theory of Fourier series, which took off with Dirichlet's work.

One article on the Laplace problem relative to the stability of the solar system led him to the problem that nowadays bears his name.

Leonhard Euler, 1707–1783. Born in Basel, Switzerland, Euler began studying theology in 1723. Despite being always a fervent Lutheran, he was lukewarm towards the clerical life and was quickly prompted to study mathematics upon suggestion of the mathematician Johann Bernoulli, who was friend of Leonhard's father. This decision led to the making of one of the most prolific mathematicians of all time. It is a hard task to keep record of all his contributions; we owe him, for example, the notation $f(x)$ for functions (1734) or the use of the letter e for natural logarithms. His work spans from modern analytic geometry (Euler was the first to consider the sine and the cosine as functions, and not just chords as Ptolemy did) to differential calculus, from continuum mechanics to gravitational problems. In particular, he is considered the founder of analytic mechanics following the 1765 treatise *Theory of the Motions of Rigid Bodies.* His academic career started in 1727 at St.Petersburg's Academy of Sciences, where he became Professor of Physics in 1730 and where he stayed until his death. Euler's scientific output is so colossal that the Academy continued to publish his work for fifty years after he had died.

Alessandro Faedo, 1913–2000. Born in Chiampo (Vicenza) in 1913, Faedo graduated in mathematics in Pisa, where he obtained the Chair of Mathematical Analysis at the Scuola Normale Superiore. Faedo is known especially for the detailed study of Galerkin's method, also known as method of Faedo-Galerkin. A great intuition of his was to understand, already in the '50s, the profound impact that computers would have in research and real life. For this reason he projected and promoted the creation of the Centro Studi Calcolatrici Elettroniche, built the first all-Italian computer, and later went on to found the CNUCE with IBM, to answer the growing demands of the newborn Computer Science, as it was known at the time. He was among the promoters of the first university degree in computer science in Italy, at the University of Pisa.

Maurice Fréchet, 1878–1973. Born in 1878 in Maligny, France, Fréchet was a student of Hadamard and wrote in 1906 an important dissertation in which he introduced the notion of metric space (although the name is due to Hausdorff) and formulated the abstract theory of compactness. He was appointed Professor of Mechanics at Poitiers (1910-1919), then became Professor of Analysis at Strasbourg (1920-1927), after which he moved to the University of Paris. Fréchet made crucial advancements in the fields of statistics, probability and analysis, even though his main contributions are to be found in topology and the theory of abstract spaces.

Kurt Otto Friedrichs, 1901–1982. Friedrichs was born in 1901 in Kiel, Germany. He became Courant's assistant in Göttingen, then Professor at Braunschweig in 1932. His main field of interest was that of PDEs in mathematical physics and fluid dynamics especially. He used the method of finite differences to prove the existence of solutions. When forced to flee Germany in 1937, he emigrated to the U.S., meeting up with Courant who had escaped there earlier.

Boris Grigorievich Galerkin, 1871–1945. Galerkin was born in Polotsok, Belarus, from a very poor family. Amid great difficulties he was able to pursue higher studies. He attended St.Petersburg's Polytechnic, working first as a private tutor then as a draftsman. After graduating in 1899, he worked as engineer for several firms where he oversaw the construction of many industrial complexes across Europe. In 1914 he switched to academia. A year later he published the first paper on what is now universally recognized as *Galerkin's method.* In 1920 he became Head of the Department of Structural Mechanics at St.Petersburg's Polytechnic, and in the meantime Professor of Elasticity for the Institute of (Tele)communication Engineering and Structural Mechanics at the University of St.Petersburg. Galerkin had a primary role together with Steklov and Bernstein, among others, in the relaunch, in 1921, of the Mathematical Society of St.Petersburg, whose activities had been interrupted during the October Revolution. He is very famous still today for his studies of thin plates, the subject of a 1937 monograph. From 1940 till his death Galerkin was Head of the Institute of Mechanics at the Soviet Academy of Sciences.

George Green, 1793–1841. Born in Sneinton, UK, Green is responsible for the mathematical systematization of the theory of solid elastic bodies. His main achievement is the treatise *On the Application of Mathematical Analysis to the Theories of Electricity and Magnetism* (1828), contain-

ing the so-called Green's theorem (a special case of Gauss' theorem in the plane), discovered simultaneously with Ostrogradskiĭ in Russia. Green was the first to recognize the importance of the potential function in an article dating 1828. He introduced the function that bears his name as a way to solve boundary-value problems. He also worked on the propagation of light-and sound waves.

Thomas Hakon Grönwall, 1877–1932. Born in Dylta, Sweden, after becoming a civil engineer Grönwall worked in Germany from 1902 to 1903, then emigrated to the U.S. where he worked for several firms. From 1913 he started doing mathematics at Princeton University, and obtained striking results at the crossroads of pure and applied mathematics. From 1925 he was member of the Physics Department at Columbia University in New York. His con-

tributions are in classical analysis (Fourier series, Gibbs phenomena, Laplace and Legendre series), integro-differential equations, analytical number theory, mathematical physics, atomic physics and chemistry. His name is especially remembered in relationship to the well-known inequality (Grönwall's lemma) that he formulated in 1919.

David Hilbert, 1862–1943. Born in Königsberg, Prussia, Hilbert was a doctoral student of Minkowski at the University of his birth place. There he became Professor in 1893, only to obtain the Chair of Mathematics at Göttingen in 1895. He is considered one of the paramount figures of the whole history of mathematics. In his opus *Foundations of Geometry* (1899) he was the first to lay out a rigorous collection of geometrical axioms, proving that his system-

atization was self-consistent. His main achievements concern number theory, mathematical logic, differential equations and the three-body problem. His intervention at the Paris International Congress of 1900 is very famous: there he stated 23 problems that the mathematicians of the XXI century should consider. Those questions have been known, ever since, as Hilbert's problems and some still remain unsolved today.

Peter D. Lax, 1926–. Peter David Lax is one of the greatest living mathematicians, both in pure and applied mathematics. He gave important contributions to integrable systems, fluid dynamics and shock waves, conservation laws of hyperbolic type, scientific calculus and numerical analysis. He spent much of his professional life at the Mathematics Department of the Courant Institute of Mathematical Sciences at New York University. He is member of the U.S. National Academy of Sciences. He won the National Medal of Science in 1986, the Wolf Prize in 1987 and the prestigious Abel Prize in 2005.

Arthur N. Milgram, 1912–1961. Arthur N. Milgram received his PhD from the University of Pennsylvania, where he worked under the supervision of John Kline on the "Decomposition and dimension of Closed Sets in R^n". Beyond the Lax-Milgram Lemma (published in the Annals of Mathematical Studies published by Princeton University Press in 1954 - see the picture to side) gave contributions in combinatorics, differential geometry, topology and Galois theory. He worked at Syracuse University in the 1940s and 1950s, then moved to the University of Minnesota at Minneapolis, where co-founded the group working on partial differential equations.

Henri Lèon Lebesgue. Born on June 28th, 1875 in Beauvais, and passed away on July 28th, 1941 in Paris, Lebesgue formulated measure theory in 1901, and later generalized the theory of Riemann integrals. Apart from this, his main contributions concern the fields of topology, Fourier analysis and the solution of other several problems relevant in the applications.

Hans Lewy, 1904–1988. Born in Breslau, Germany, Lewy received the doctorate in Göttingen under Richard Courant in 1926, where he worked for the ensuing six years. During that time he obtained together with Courant and Friedric many of mathematically-relevant results on the numerical stability of certain classes of differential equations. Subsequently he published a series of fundamental papers on the calculus of variations and PDEs, thus completely solving the initial value problem for non-linear hyperbolic equations in two independent variables. Forced to flee Germany in 1930, he emigrated to the U.S. where he worked at Brown University, and then at Berkeley until 1972.

Claude-Louis Navier, 1785–1836. Born in Dijon, Navier lost his father at an early age, and was raised by his maternal uncle Emiland Gauthey, one of the foremost French civil engineers. The young Claude-Louis was thus pushed to enroll in the École Polytechnique, where he followed the lectures of Fourier, whom he later befriended. In 1804 Navier entered the École des Ponts et Chaussées, where he graduated with honors in two years. A few years later he succeeded to his uncle, who had meanwhile passed away, in the Corps des Ponts et Chaussés. In 1819 he began teaching applied mechanics at the École des Ponts et Chaussées. Subsequently he became Professor at the École Polytechnique, the position Cauchy had held. Proof of his fame as an expert in building roads and bridges is, for example, his pioneering theory of suspension bridges, which had been constructed–until then–on the basis of empirical knowledge. His name is, however, related to the equations governing viscous fluids, which he presented in 1822. Among the many awards he was conferred, the most prestigious was becoming, in 1824, a member of the Academy of Sciences of Paris In 1831 he also became Knight of the French *Legion of Honor.*

Phyllis Nicolson, 1917–1968. Born in Macclesfield, UK, Nicolson received a Ph.D. at Manchester University. She became lecturer at Cambridge's Girton College in 1946, and after her husband's death in a train crash, she was appointed to fill his lectureship in Physics at Leeds University. She is known for her collaboration with John Crank on the solution of the heat equation.

Henri Poincaré, 1854–1912. Born in Nancy, after a childhood characterized by muscle problems and diphtheria, Henri entered the Lycée of Nancy in 1862, where he studied for 11 years and soon become among the best pupils in every subject taught there. In 1873 he began the École Polytechnique. After graduating in 1875 he continued studying at the École des Mines, after which he spent some time as mineral engineer in Vesoul. He started the doctoral school in Mathematics under the supervision of Charles Hermite. Immediately after defending his thesis he started teaching mathematical analysis at the University of Caen. Two years later he was offered a position at the Science Faculty of Paris (1881). In 1886 he obtained the Chair of Mathematical Physics and Probability at the Sorbonne. Subsequently he was also appointed Chair at the École Polytechnique, where the lectured a different course every year, including optics, fluid dynamics, astronomy, probability. He stayed at the École Polytechnique until he died, aged 58.

George Gabriel Stokes, 1819–1903. Born in Skreen, Ireland, Stokes was educated in Dublin, then in Bristol where he studied mathematics. Then he entered Cambridge's Pembroke College, and his teacher William Hopkins directed him to the study of hydrodynamics. During 1842–1845 Stokes published articles on the internal friction of fluids. At that time it was difficult, in England, to find out about the research of overseas mathematicians. Despite Stokes had realized that some of his ideas were contained in the work of other people (esp. Navier), he nonetheless thought the originality of his approach deserved publication. In 1849 he was offered the Lucasian Chair in Mathematics at Cambridge University. In 1851 he became Fellow of the Royal Society, and Secretary in 1853. During those years Stokes worked on many subjects, including hydrodynamics (motion of a pendulum in a fluid), fluorescence, and the theory of Fraunhhofer lines in the solar spectrum. From 1857 he devoted himself to experimentally–more than theoretically–flavored investigations.

Sergei Lvovich Sobolev, 1908–1989. Born in St.Petersburg, Sobolev is one of the leading figures of modern mathematical analysis. His studies on the spaces that have inherited his name, introduced in 1930, gave immediate birth to a novel branch of functional analysis. We owe him the notion of generalized functions (distributions), the present-day variational formulation of elliptic problems, the study of numerical quadratures (integration) in several dimensions, and also a host of norm inequalities in function spaces which turned out to be crucial for subsequent developments. At the young age of 31 he became an effective member of the USSR Academy of Sciences. He worked on the solution of hard problems in mathematical physics important in the applications, as well. His most celebrated publication is *Applications of functional analysis in mathematical physics* from 1962.

References

[AC97] Avgoustiniatos E. and Colton C. (1997) Effect of external oxygen mass transfer resistances on viability of immunoisolated tissue. *Ann. NY Acad. Sci.* 831: 145–167.

[AF03] Adams R. A. and Fournier J. J. F. (2003) *Sobolev Spaces.* Pure and Applied Mathematics (Amsterdam) 140. Elsevier/Academic Press, Amsterdam, second edition.

[All25] Allievi L. (1925) *Theory of water-hammer*, vol. 1. Typography R. Garroni, Rome.

[BBG+01] Balay S., Buschelman K., Gropp W., Kaushik D., Knepley M., McInnes L. C., Smith B., and Zhang H. (2001) PETSc Web page. http://www.mcs.anl.gov/petsc.

[BD70] Boyce W. and DiPrima R. (1970) *Introduction to Ordinary Differential Equations.* John Wiley, New York.

[Bea88] Bear J. (1988) *Dynamics of Fluid in Porous Media.* Courier Dover Publication.

[BGL05] Benzi M., Golub G., and Liesen J. (2005) Numerical solution of saddle point systems. *Acta Numerica* 14: 1–137.

[BP84] Brezzi F. and Pitkaranta J. (1984) On the stabilization of finite element approximations of the stokes problem. *Efficient solutions of elliptic systems, Notes on Numerical Fluid Mechanics* 10: 11–19.

[Bre11] Brezis H. (2011) *Functional Analysis, Sobolev Spaces and Partial Differential Equations.* Universitext. Springer, New York.

[BS02] Brenner S. C. and Scott L. R. (2002) *The Mathematical Theory of Finite Element Methods.* Texts in Applied Mathematics 15. Springer-Verlag, New York, second edition.

[CC88] Cahouet J. and Chabard J. (1988) Some fast 3D finite element solvers for the generalized Stokes problem. *International Journal for Numerical Methods in Fluids* 8(8): 869–895.

[CHQZ88] Canuto C., Hussaini M., Quarteroni A., and Zang T. (1988) *Spectral Methods in Fluid Dynamics*. Springer Series in Computational Physics. Springer-Verlag, New York.

[Cia78] Ciarlet P. (1978) *The Finite Element Method for Elliptic Problems*. Studies in Mathematics and its Applications 4. North-Holland Publishing Co., Amsterdam.

[Dav04] Davis T. A. (June 2004) Algorithm 832: Umfpack v4.3 – an unsymmetric-pattern multifrontal method. *ACM Trans. Math. Softw.* 30: 196–199.

[Dav08] Davis T. A. (2008) User's guide for suitesparseQR, a multifrontal multithreaded sparse QR factorization package. *ACM Trans. Math. Software.*

[DD97] Davis T. and Duff I. (1997) An unsymmetric-pattern multifrontal method for sparse LU factorization. *SIAM J. Matrix Analysis and Applications* 19(1): 140–158.

[DFM02] Deville M., Fischer P., and Mund E. (2002) *High-order methods for incompressible fluid flow*. Cambridge Monographs on Applied and Computational Mathematics 9. Cambridge University Press.

[DGL92] Duff I., Grimes R., and Lewis J. (October 1992) Users guide for the Harwell-Boeing sparse matrix collection. Technical Report TR/PA/92/86, CERFACS.

[EG04] Ern A. and Guermond J.-L. (2004) *Theory and Practice of Finite Elements*. Applied Mathematical Sciences 159. Springer-Verlag, New York.

[ESW05] Elman H. C., Sylvester D. J., and Wathen A. J. (2005) *Finite Elements and Fast Iterative Solvers with Applications in Incompressible Fluid Dynamics*. Numerical Mathematics and Scientific Computation 8. Oxford University Press, Oxford.

[Eva10] Evans L. (2010) *Partial Differential Equations*. Graduate Studies in Mathematics 19. American Mathematical Society, Providence, RI, second edition.

[Far93] Farlow S. (1993) *Partial Differential Equations for Scientists and Engineers*. Courier Dover.

[FG00] Frey P. and George P.-L. (2000) *Mesh Generation. Application to finite elements*. Hermes Science Publishing, Oxford.

[FP99] Ferziger J. and Perić M. (1999) *Computational Methods for Fluid Dynamics*. Springer-Verlag, Berlin, revised edition.

[FQV09] Formaggia L., Quarteroni A., and Veneziani A. (eds) (2009) *Cardiovascular Mathematics. Modeling and simulation of the circulatory system*. Modeling, Simulation and Applications 1. Springer, Milan.

[GR86] Girault V. and Raviart P. (1986) *Finite element methods for Navier-Stokes equations: Theory and algorithms*. Springer Series in Computational Mathematics 5. Springer-Verlag, Berlin New York.

[GV96]   Golub G. and Van Loan C. (1996) *Matrix computations*. Johns Hopkins Studies in the Mathematical Sciences 3. Johns Hopkins University Press.

[HBH⁺05] Heroux M.A., Bartlett R.A., Howle V.E., Hoekstra R.J., Hu J.J., Kolda T.G., Lehoucq R.B., Long K.R., Pawlowski R.P., Phipps E.T., Salinger A.G., Thornquist H.K., Tuminaro R.S., Willenbring J.M., Williams A., and Stanley K.S. (2005) An overview of the Trilinos project. *ACM Trans. Math. Softw.* 31(3): 397–423.

[HCB05]   Hughes T., Cottrell J., and Bazilevs Y. (2005) Isogeometric analysis: CAD, finite elements, NURBS, exact geometry and mesh refinement. *Computer Methods in Applied Mechanics and Engineering* 194(39–41): 4135–4195.

[Her04]   Heroux M. (July 2004) AztecOO user guide. Sandia Laboratory Report N. SAND2004-3796. http://software.sandia.gov/trilinos/.

[Hig02]   Higham N. (2002) *Accuracy and Stability of Numerical Algorithms*. Society for Industrial and Applied Mathematics (SIAM), Philadelphia, PA, second edition.

[HVZ01]   Hamacher V., Vranesic Z., and Zaky S. (2001) *Computer Organization*. McGraw-Hill, Inc. New York.

[Joh87]   Johnson C. (1987) *Numerical Solution of Partial Differential Equations by the Finite Element Method*. Cambridge University Press, Cambridge.

[Kel11]   Keller J.P. (2011) The spread of rabies in raccoons: numerical simulations of a spatial diffusion model. Honor thesis, Department of Mathematics and Computer Science, Emory University, Atlanta (USA).

[KGV11]   Keller J., Giorda L.G., and Veneziani A. (2011) Numerical simulation of space continuous SEI models for raccoon rabies diffusion in a realistic landscape. In preparation.

[Lad63]   Ladyzhenskaya O.A. (1963) *The Mathematical Theory of Viscous Incompressible Flow*. Gordon and Breach Science Publishers, New York.

[Leo09]   Leoni G. (2009) *A first course in Sobolev spaces*. Graduate Studies in Mathematics 105. American Mathematical Society.

[LeV90]   LeVeque R. (1990) *Numerical Methods for Conservation Laws*. Lectures in Mathematics ETH Zürich. Birkhäuser Verlag, Basel.

[LeV02]   LeVeque R. (2002) *Finite Volume Methods for Hyperbolic Problems*. Cambridge Texts in Applied Mathematics. Cambridge University Press, Cambridge.

[lif10]   (2010) Lifev user on-line manual. http://www.lifev.org.

[LM54]   Lax P. and Milgram A. (1954) *Parabolic Equations*. Annals of Mathematics Studies 33.

[Pro97]   Prohl A. (1997) *Projection and Quasi-Compressibility Methods for Solving the Incompressible Navier-Stokes Equations*. Advances in Numerical Mathematics. B. G. Teubner, Stuttgart.

[PV09]   Pietro D. D. and Veneziani A. (2009) Expression templates implementation of continuous and discontinuous Galerkin methods. *Computing and visualization in science* 12(8): 421–436.

[QSS00]   Quarteroni A., Sacco R., and Saleri F. (2000) *Numerical Mathematics*. Texts in Applied Mathematics 37. Springer-Verlag, New York.

[Qua93]   Quartapelle L. (1993) *Numerical solution of the incompressible Navier-Stokes equations*. International Series of Numerical Mathematics 113. Birkhäuser Verlag, Basel.

[Qua09]   Quarteroni A. (2009) *Numerical Models for Differential Problems*. Springer, Milan.

[QV94]   Quarteroni A. and Valli A. (1994) *Numerical Approximation of Partial Differential Equations*. Springer Series in Computational Mathematics 23. Springer-Verlag, Berlin.

[QV99]   Quarteroni A. and Valli A. (1999) *Domain Decomposition Methods for Partial Differential Equations*. Numerical Mathematics and Scientific Computation. The Clarendon Press Oxford University Press, New York. Oxford Science Publications.

[RR04]   Renardy M. and Rogers R. (2004) *An Introduction to Partial Differential Equations*. Texts in Applied Mathematics 13. Springer-Verlag, New York, second edition.

[Saa90]   Saad Y. (1990) SPARSKIT: A basic tool kit for sparse matrix computations. Technical Report RIACS-90-20, Research Institute for Advanced Computer Science, NASA Ames Research Center, Moffett Field, CA. http://www-users.cs.umn.edu/~saad/software/SPARSKIT/sparskit.html.

[Saa92]   Saad Y. (1992) *Numerical Methods for Large Eigenvalue Problems*. Algorithms and Architectures for Advanced Scientific Computing. Manchester University Press, Manchester.

[Saa03]   Saad Y. (2003) *Iterative Methods for Sparse Linear Systems*. Society for Industrial and Applied Mathematics, Philadelphia, PA, second edition.

[Sal08]   Salsa S. (2008) *Partial Differential Equations in Action*. Universitext. Springer, Milan.

[Sch]   Schöberl J.Netgen website http://www.hpfem.jku.at/netgen/.

[Sel84]   Selberherr S. (1984) *Analysis and Simulation of Semiconductor Devices*. Springer-Verlag, Wien New York.

[Sha08]   Shapira Y. (2008) *Matrix-based multigrid*. Numerical Methods and Algorithms 2. Springer, New York, second edition. Theory and applications.

[Slo73]   Slotboom J. (1973) Computer-aided two dimensional analysis of bipolar transistor. *IEEE Trans. Electron Devices* ED-20: 669–673.

[SS03]   Schildt H. and Schildt H. (2003) *C/C++ Programmer's reference*. McGraw-Hill/Osborne.

[STD$^+$96] Schaefer M., Turek S., Durst F., Krause E., and Rannacher R. (1996) Benchmark computations of laminar flow around a cylinder. *Notes on numerical fluid mechanics* 52: 547–566.

[Ste95]   Stemberg R. (1995) On some techniques for approximating boundary conditions in the finite element method. *Journal of Computational and Applied Mathematics* 63: 139–148.

[Sto48]   Stommel H. (1948) The westward intensification of wind-driven ocean currents. *Trans. Amer. Geophys. Union* 29(202).

[Str03]   Strang G. (2003) *Introduction to Linear Algebra*. Wellesley Cambridge Press.

[Str04]   Strikwerda J. (2004) *Finite Difference Schemes and Partial Differential Equations*. Society for Industrial and Applied Mathematics (SIAM), Philadelphia, PA, second edition.

[SV09]   Salsa S. and Verzini G. (2009) *Equazioni a derivate parziali*. Springer, Milan.

[Tem84]   Temam R. (1984) *Navier-Stokes Equations*. Studies in Mathematics and its Applications 2. North-Holland Publishing Co., Amsterdam, third edition.

[Tem95]   Temam R. (1995) *Navier-Stokes Equations and Nonlinear Functional Analysis*. CBMS-NSF Regional Conference Series in Applied Mathematics 66. Society for Industrial and Applied Mathematics (SIAM), Philadelphia, PA, second edition.

[Tur99]   Turek S. (1999) *Efficient Solvers for Incompressible Flow Problems*. Lecture Notes in Computational Science and Engineering 6. Springer-Verlag, Berlin.

[TW05]   Toselli A. and Widlund O. (2005) *Domain Decomposition Methods – Algorithms and Theory*. Springer Series in Computational Mathematics 34. Springer-Verlag, Berlin.

[Van82]   Van Dyke M. (1982) *An Album of Fluid Motion*. Parabolic Press, Stanford, CA.

[Wom55]   Womersley J. (1955) Method for the calculation of velocity, rate of flow and viscous drag in arteries when the pressure gradient is known. *The journal of physiology* 127(3): 553.

[ZZ92]   Zienkiewicz O. and Zhu J. (1992) The superconvergent patch recovery and a posteriori error estimates. Part 1: The recovery technique. *International Journal for Numerical Methods in Engineering* 33(7): 1331–1364.

Subject Index

Unitext – La Matematica per il 3+2

As of 2004, the books published in the series have been given a volume number. Titles in grey indicate editions out of print.
As of 2011, the series also publishes books in English.

A. Bernasconi, B. Codenotti
Introduzione alla complessità computazionale
1998, X+260 pp, ISBN 88-470-0020-3

A. Bernasconi, B. Codenotti, G. Resta
Metodi matematici in complessità computazionale
1999, X+364 pp, ISBN 88-470-0060-2

E. Salinelli, F. Tomarelli
Modelli dinamici discreti
2002, XII+354 pp, ISBN 88-470-0187-0

S. Bosch
Algebra
2003, VIII+380 pp, ISBN 88-470-0221-4

S. Graffi, M. Degli Esposti
Fisica matematica discreta
2003, X+248 pp, ISBN 88-470-0212-5

S. Margarita, E. Salinelli
MultiMath - Matematica Multimediale per l'Università
2004, XX+270 pp, ISBN 88-470-0228-1

A. Quarteroni, R. Sacco, F.Saleri
Matematica numerica (2a Ed.)
2000, XIV+448 pp, ISBN 88-470-0077-7
2002, 2004 ristampa riveduta e corretta
(1a edizione 1998, ISBN 88-470-0010-6)

13. A. Quarteroni, F. Saleri
    Introduzione al Calcolo Scientifico (2a Ed.)
    2004, X+262 pp, ISBN 88-470-0256-7
    (1a edizione 2002, ISBN 88-470-0149-8)

14. S. Salsa
    Equazioni a derivate parziali - Metodi, modelli e applicazioni
    2004, XII+426 pp, ISBN 88-470-0259-1

15. G. Riccardi
    Calcolo differenziale ed integrale
    2004, XII+314 pp, ISBN 88-470-0285-0

16. M. Impedovo
    Matematica generale con il calcolatore
    2005, X+526 pp, ISBN 88-470-0258-3

17. L. Formaggia, F. Saleri, A. Veneziani
    Applicazioni ed esercizi di modellistica numerica
    per problemi differenziali
    2005, VIII+396 pp, ISBN 88-470-0257-5

18. S. Salsa, G. Verzini
    Equazioni a derivate parziali – Complementi ed esercizi
    2005, VIII+406 pp, ISBN 88-470-0260-5
    2007, ristampa con modifiche

19. C. Canuto, A. Tabacco
    Analisi Matematica I (2a Ed.)
    2005, XII+448 pp, ISBN 88-470-0337-7
    (1a edizione, 2003, XII+376 pp, ISBN 88-470-0220-6)

20. F. Biagini, M. Campanino
    Elementi di Probabilità e Statistica
    2006, XII+236 pp, ISBN 88-470-0330-X

21. S. Leonesi, C. Toffalori
    Numeri e Crittografia
    2006, VIII+178 pp, ISBN 88-470-0331-8

22. A. Quarteroni, F. Saleri
    Introduzione al Calcolo Scientifico (3a Ed.)
    2006, X+306 pp, ISBN 88-470-0480-2

23. S. Leonesi, C. Toffalori
    Un invito all'Algebra
    2006, XVII+432 pp, ISBN 88-470-0313-X

24. W.M. Baldoni, C. Ciliberto, G.M. Piacentini Cattaneo
    Aritmetica, Crittografia e Codici
    2006, XVI+518 pp, ISBN 88-470-0455-1

25. A. Quarteroni
    Modellistica numerica per problemi differenziali (3a Ed.)
    2006, XIV+452 pp, ISBN 88-470-0493-4
    (1a edizione 2000, ISBN 88-470-0108-0)
    (2a edizione 2003, ISBN 88-470-0203-6)

26. M. Abate, F. Tovena
    Curve e superfici
    2006, XIV+394 pp, ISBN 88-470-0535-3

27. L. Giuzzi
    Codici correttori
    2006, XVI+402 pp, ISBN 88-470-0539-6

28. L. Robbiano
    Algebra lineare
    2007, XVI+210 pp, ISBN 88-470-0446-2

29. E. Rosazza Gianin, C. Sgarra
    Esercizi di finanza matematica
    2007, X+184 pp, ISBN 978-88-470-0610-2

30. A. Machì
Gruppi - Una introduzione a idee e metodi della Teoria dei Gruppi
2007, XII+350 pp, ISBN 978-88-470-0622-5
2010, ristampa con modifiche

31. Y. Biollay, A. Chaabouni, J. Stubbe
Matematica si parte!
A cura di A. Quarteroni
2007, XII+196 pp, ISBN 978-88-470-0675-1

32. M. Manetti
Topologia
2008, XII+298 pp, ISBN 978-88-470-0756-7

33. A. Pascucci
Calcolo stocastico per la finanza
2008, XVI+518 pp, ISBN 978-88-470-0600-3

34. A. Quarteroni, R. Sacco, F. Saleri
Matematica numerica (3a Ed.)
2008, XVI+510 pp, ISBN 978-88-470-0782-6

35. P. Cannarsa, T. D'Aprile
Introduzione alla teoria della misura e all'analisi funzionale
2008, XII+268 pp, ISBN 978-88-470-0701-7

36. A. Quarteroni, F. Saleri
Calcolo scientifico (4a Ed.)
2008, XIV+358 pp, ISBN 978-88-470-0837-3

37. C. Canuto, A. Tabacco
Analisi Matematica I (3a Ed.)
2008, XIV+452 pp, ISBN 978-88-470-0871-3

38. S. Gabelli
Teoria delle Equazioni e Teoria di Galois
2008, XVI+410 pp, ISBN 978-88-470-0618-8

39. A. Quarteroni
Modellistica numerica per problemi differenziali (4a Ed.)
2008, XVI+560 pp, ISBN 978-88-470-0841-0

40. C. Canuto, A. Tabacco
Analisi Matematica II
2008, XVI+536 pp, ISBN 978-88-470-0873-1
2010, ristampa con modifiche

41. E. Salinelli, F. Tomarelli
Modelli Dinamici Discreti (2a Ed.)
2009, XIV+382 pp, ISBN 978-88-470-1075-8

42. S. Salsa, F.M.G. Vegni, A. Zaretti, P. Zunino
Invito alle equazioni a derivate parziali
2009, XIV+440 pp, ISBN 978-88-470-1179-3

43. S. Dulli, S. Furini, E. Peron
Data mining
2009, XIV+178 pp, ISBN 978-88-470-1162-5

44. A. Pascucci, W.J. Runggaldier
Finanza Matematica
2009, X+264 pp, ISBN 978-88-470-1441-1

45. S. Salsa
Equazioni a derivate parziali – Metodi, modelli e applicazioni (2a Ed.)
2010, XVI+614 pp, ISBN 978-88-470-1645-3

46. C. D'Angelo, A. Quarteroni
Matematica Numerica – Esercizi, Laboratori e Progetti
2010, VIII+374 pp, ISBN 978-88-470-1639-2

47. V. Moretti
Teoria Spettrale e Meccanica Quantistica – Operatori in spazi di Hilbert
2010, XVI+704 pp, ISBN 978-88-470-1610-1

48. C. Parenti, A. Parmeggiani
Algebra lineare ed equazioni differenziali ordinarie
2010, VIII+208 pp, ISBN 978-88-470-1787-0

49. B. Korte, J. Vygen
Ottimizzazione Combinatoria. Teoria e Algoritmi
2010, XVI+662 pp, ISBN 978-88-470-1522-7

50. D. Mundici
Logica: Metodo Breve
2011, XII+126 pp, ISBN 978-88-470-1883-9

51. E. Fortuna, R. Frigerio, R. Pardini
Geometria proiettiva. Problemi risolti e richiami di teoria
2011, VIII+274 pp, ISBN 978-88-470-1746-7

52. C. Presilla
Elementi di Analisi Complessa. Funzioni di una variabile
2011, XII+324 pp, ISBN 978-88-470-1829-7

53. L. Grippo, M. Sciandrone
Metodi di ottimizzazione non vincolata
2011, XIV+614 pp, ISBN 978-88-470-1793-1

54. M. Abate, F. Tovena
Geometria Differenziale
2011, XIV+466 pp, ISBN 978-88-470-1919-5

55. M. Abate, F. Tovena
Curves and Surfaces
2011, XIV+390 pp, ISBN 978-88-470-1940-9

56. A. Ambrosetti
Appunti sulle equazioni differenziali ordinarie
2011, X+114 pp, ISBN 978-88-470-2393-2

57. L. Formaggia, F. Saleri, A. Veneziani
Solving Numerical PDEs: Problems, Applications, Exercises
2011, X+434 pp, ISBN 978-88-470-2411-3

The online version of the books published in this series is available at SpringerLink.
For further information, please visit the following link:
http://www.springer.com/series/5418